Climate Change and Resilient Food Systems

Vinaya Kumar Hebsale Mallappa •
Mahantesh Shirur
Editors

Climate Change and Resilient Food Systems

Issues, Challenges, and Way Forward

Springer

Editors
Vinaya Kumar Hebsale Mallappa
Department of Agricultural Extension & Communication
Anand Agricultural University
Anand, Gujarat, India

Mahantesh Shirur
Center for Agriculture Extension Policy and International Centre of Excellence
National Institute of Agricultural Extension Management (MANAGE)
Hyderabad, Telangana, India

ISBN 978-981-33-4540-9 ISBN 978-981-33-4538-6 (eBook)
https://doi.org/10.1007/978-981-33-4538-6

This Springer imprint is published by the registered company Springer Nature Singapore Pte Ltd.
The registered company address is: 152 Beach Road, #21-01/04 Gateway East, Singapore 189721, Singapore

We dedicate this book to the 'farmer' who is most affected by changing climate yet working tirelessly amidst it to feed the world

Preface

The knowledge of both climate change and food system can allow us to ensure the resilient food system in the society. This book is primarily the outcome of our shared concern and the subsequent discussions about the ill effects of climate change since our days as PhD research scholars. At times, we felt overwhelmed by the enormity of the challenges posed by global climate change issues. Yet, we never stopped thinking about the environment, the sustainability and the food security issues in teaching, research and extension activities in which we were involved in our respective capacities.

We were aware of the information and huge data available on climate change, food and nutrition security in different regions of the globe. However, our inquiries revolved around the issue of making the food production, processing and consumption more resilient to the challenges of changing climate. Several researchers, through their research and accumulated experiences, must have developed their own insights from their long work experience. We thought that collating and making available all such experiences and insights from cross-sectional disciplines to the young readers, research scholars and policy formulators will be a worthy idea. Therefore, we included the topics starting from soil and water conservation, production aspects, processing, to value chain along with the role of allied sectors, animal husbandry, fisheries and forestry, etc.

The chapters included in this book bring into focus all the diverse views about the ill effect of climate change and alternatives in solving the impending world food crisis. It also shows the plethora of opportunities in the agriculture sector to adapt and its contribution in mitigating climate change. We also included different dimensions of climate change and resilient issues. We gratefully acknowledge all the authors' contributions. All the authors were quick enough to grasp the idea behind the book and carefully stride with the right focus on the topic and closely related issues. We welcome all the critique and suggestions while taking more of such work in the coming days.

Anand, Gujarat, India — Vinaya Kumar Hebsale Mallappa
Hyderabad, Telangana, India — Mahantesh Shirur

About the Book

The changing climate impacts everyone in the globe in more than one way. There are growing concerns of adverse impacts of climatic variability on food supplies prominently from the early 1990s, due to doubling of the number of extreme weather-related disasters. Consequently, crop shortfalls, natural disasters, famines and food supply emergencies throughout the world have underscored the importance of the study on the climate-food interactions encircling the complex mechanisms involved. The book throws light on different views of climate change, the effect of changing climate on food production and distribution system and the contribution of agriculture in adapting and mitigating to climate change. In the wake of climate change, the book also discusses several holistic strategies for resilient food and nutritional security. All the chapters explore and present an array of technologies and practical policies, which promise to reduce the adverse impact of changing climate on food production. They need to be heard without further delay in achieving the food and nutritional security to meet the targets identified under sustainable development goals.

Contents

About the Editors

Vinaya Kumar Hebsale Mallappa is an Assistant Professor at the Department of Agricultural Extension and Communication of Anand Agricultural University, Anand, India. He received his doctorate in Agricultural Extension from University of Agricultural Sciences, Bangalore, with ICSSR Doctoral Fellowship and secured four university gold medals. His recent publications are "Predictive factors to avoid farming as a livelihood" (Journal of Economic Structures, 2019), "Impact of rainfall variability and trend on rice yield in Coastal Karnataka" (Journal of Agrometeorology, 2017), and "Assessing decision-making and economic performance of farmers to manage climate-induced crisis in Coastal Karnataka" (Climatic Change, 2017). His research interests include the climate-induced crisis management, surface and groundwater management, food and nutritional security, and sustainable development.

Mahantesh Shirur is working as Dy. Director (Agricultural Extension) at the National Institute of Agricultural Extension Management (MANAGE), Hyderabad. He is looking after the Center for Agriculture Extension Policy, PPP in Extension and International Center for Excellence in Agricultural Extension at MANAGE. Prior to joining MANAGE, Dr. Mahantesh Shirur was working as scientist (Agricultural Extension) at ICAR—Directorate of Mushroom Research, Solan and ICAR—Indian Institute of Soil and Water Conservation, RC, Ballari. He has expertise on transfer of technology, impact assessment, on-farm trials, and ICT application in agriculture. His contribution in the development of digital content and video documentaries on mushroom cultivation is popular as learning and teaching aids. He worked on developing and standardizing the training modules on mushroom cultivation technology at ICAR-DMR, Solan, India. He has authored 30 research papers, 2 reports, 1 book, 7 popular articles, 4 book chapters, and many conference proceedings. Dr. Mahantesh Shirur was the Program Director of the prestigious Feed The Future India Training (FTF ITT) and Indian Technical Economic Cooperation (ITEC) International and national training programs of MANAGE during 2018–2020.

Abbreviations

AAS	Agromet advisory services
ACP	Agricultural crop production
ACZ	Agroclimatic zone
ADA	American Dietetic Association
AgriDSS	Agricultural decision support system
AI	Artificial intelligence
AICMIP	All India Coordinated Mushroom Improvement Project
AICRPAM	All India Coordinated Research Project on Agrometeorology
AMF	Arbuscular mycorrhizal fungi
AMFU	Agro-meteorological field unit
ATMA	Agricultural Technology Management Agency
AWC	Anganwadi Centres
BCM	Bromo-chloromethane
BDO	Block Development Officer
BEE	Bureau of Energy Efficiency
BES	2-Bromo-ethanesulfonate
BFAR	Bureau of Fisheries and Aquatic Resources
BNF	Biological nitrogen fixation
BOD	Bio-oxygen demand
BRM	Beneficial rhizospheric microorganism
CA	Conservation agriculture
CAI	Computer-assisted instruction
CBOs	Community-based organizations
CC	Climate change
CCAP	Climate Change Action Plan
CDR	Complex-diverse-risk-prone
CD-ROM	Compact disc read-only memory
CGIAR	Consultative Group on International Agricultural Research
CH_4	Methane
COVID-19	Coronavirus disease 2019
CPR	Common property resources
CRIDA	Central Research Institute for Dryland Agriculture
CS	Carbon sequestration

CSA	Climate-smart agriculture
CSM	Crop simulation model
CV	Climate variability
CWWG	Crop Weather Watch Group
DEFRA	Department for Environment, Food and Rural Affairs
DMR	Directorate of Mushroom Research
DRDA	District Rural Development Agency
DSS	Decision support system
ESS	Ecosystem service
EWE	Extreme weather events
FAO	Food and Agricultural Organization
FARMS	Farm activity record management system
FF	Fossil fuel
FFS	Farmer field schools
FGD	Focus group discussion
FIF	Field Information Facilitator
FILE	Farmer income-led extension
FLW	Food loss and wastages
FMD	Foot and mouth disease
FNS	Farmer Nutrition Schools
FNS	Food and nutrition security
FPOs	Farmer Producer Organisations
FWB	Farmers' Weather Bulletin
GDP	Gross domestic product
GECAFS	Global Environmental Change and Food Systems
GEO	Global Environmental Organization
GHG	Greenhouse gas
GIS	Geographic information system
GISS	Goddard Institute for Space Studies
GWP	Global warming potential
HAIC	Haryana Agro-Industrial Corporation
HDPE	High-density polyethylene
HLPE	High-Level Panel of Experts
HS	Heat stress
IAA	Indole acetic acid
IARI	Indian Agricultural Research Institute
ICAR	Indian Council of Agricultural Research
ICDS	Integrated Child Development Services
ICRISAT	International Crops Research Institute for the Semi-arid Tropics
ICT	Information and communication technologies
IFLED	Innovative farmer-led extension delivery
IFPRI	International Food Policy Research Institute
IFS	Integrated farming system
IGFRI	ICAR-Indian Grassland and Fodder Research Institute

IIHR	Indian Institute of Horticultural Research
IITA	International Institute of Tropical Agriculture
IMD	India Meteorological Department
IoT	Internet of things
IPCC	Intergovernmental Panel on Climate Change
ITK	Indigenous technology knowledge
JFM	Joint forest management
KINEROS	Kinematic runoff and erosion
KSB	Potassium-solubilizing bacteria
KSHMTA	Knowledge Systems and Homestead Agriculture Management in Tribal Areas
KSM	Potassium-solubilizing microorganism
KVK	Krishi Vigyan Kendra
LCA	Life cycle assessment
LCT	Life cycle thinking
LGP	Length of growing period
LiDAR	Light detection and ranging
LISEM	Limburg soil erosion
LPA	Long period average
MAAS	Micro-level Agromet Advisory Services
MDG	Millennium Development Goals
MGNREGA	Mahatma Gandhi National Rural Employment Guarantee Act
MH	Mycorrhiza helper
MIRCEN	Microbial Resources Centre Network
MLP-FFA	Multilayer Perceptron Algorithm Integrated with Firefly Optimizer Algorithm
MNRE	Ministry of New and Renewable Energy
MSSRF	MS Swaminathan Research Foundation
NAPCC	National Action Plan on Climate Change
NARI	Nutri-Sensitive Agricultural Resources and Innovations
NARS	National Agricultural Research System
NASA	National Aeronautics and Space Administration
NCAER	National Council of Applied Economic Research
NCAP	National Centre for Agricultural Economics and Policy Research
NCFC	National Crop Forecasting Centre
NCMRWF	National Centre for Medium-Range Weather Forecast
NDC	Nationally determined contribution
NGCR	National Centre for Genetic Resources
NGOs	Nongovernment organizations
NICRA	National Innovations in Climate Resilient Agriculture
NMSA	National Mission for Sustainable Agriculture
NUE	Nutrient use efficiency
OF	Organic farming

PA	Precision agriculture
PAG- ASA	Philippine Atmospheric Geophysical and Astronomical Services Administration
PGPF	Plant growth-promoting fungi
PGPR	Plant growth-promoting rhizobacteria
PLW	Postharvest losses and wastage
PSM	Phosphate-solubilizing microorganism
PVC	Polyvinyl chloride
RCTs	Resource-conserving technologies
RE	Renewable energy
RFID	Radio-frequency identification
RFS	Resilient food system
RICMS	Rice-integrated crop management systems
RO	Reactive oxygen
RS & GIS	Remote sensing and geographic information system
RTF	Ready to fruit
RUR	Rossum's Universal Robots
RUSLE	Revised Universal Soil Loss Equation
SAARC	South Asian Association for Regional Cooperation
SAD	Sustainable agricultural development
SAMS	Smart agricultural machinery systems
SAPCC	State Action Plans on Climate Change
SDGs	Sustainable Development Goals
SDSN	Sustainable Development Solutions Network
SF	Smart farming
SHGs	Self-help groups
SI	Sustainable intensification
SLR	Sea level rise
SMC	State Agromet Centre
SMEs	Small and medium enterprises
SMS	Short messaging service
SOC	Soil organic carbon
SOM	Soil organic matter
SRI	System of Rice Intensification
SSA	Sub-Saharan Africa
SWAT	Soil and Water Assessment Tool
TBI	Tree-based intercropping
THI	Temperature humidity index
TPA	Tonnes per annum
UCD	User-centred design
UHT	Ultrahigh temperature
UNDP	United Nations Development Programme
UNESCO	United Nations Educational, Scientific and Cultural Organization
UNFCCC	United Nations Framework Convention on Climate Change

USEPA	US Environmental Protection Agency
USGCRP	US Global Change Research Program
USLE	Universal Soil Loss Equation
VATICA	Value Addition and Technology Incubation Centres in Agriculture
VFA	Volatile fatty acids
VRI	Vavilov Research Institute
WEPP	Water Erosion Prediction Project
WHO	World Health Organization
WMO	World Meteorological Organization
WRI	World Resources Institute
WTO	World Trade Organization
WUE	Water use efficiency
ZEPT	Zero energy polytunnel

Building a Resilient Food System: Challenges and a Way Forward

1

Vinaya Kumar Hebsale Mallappa and S. C. Babu

Abstract

In a growing crisis, various internal and external factors threaten food and nutritional security in the developing world. These include ample processes of global climate change, urbanization, the rapid growth of population, unexpected shocks, natural disasters, economic, political crisis, etc. In this context, we are discussing various factors responsible for food and nutritional insecurity in developing countries. To tackle the existing turmoil, we have developed a theoretical framework for the resilient food and nutritional system and cogitate how this possibly will be implemented through stakeholder participation to ensure the resilience of the food system for an individual in society and nation as a whole. Resilience is conceptualized as the capacity of an individual to recover quickly from the shocks and intricacy of whole food systems. Also, from a sustainable perspective, it is viewed as personal, social, psychological, political, economic, technological, and communication factors operating at various scales. It presents the prospect of eliminating weaknesses, deal with future shocks, and build a resilient food system through institutional, technological, capacity building, and policy interventions.

Keywords

Crisis · Food system · Resilience · Sustainable development · Shocks

V. K. Hebsale Mallappa (✉)
Department of Agricultural Extension and Communication, Anand Agricultural University, Anand, Gujarat, India
e-mail: vinay@aau.in

S. C. Babu
IFPRI, Washington, DC, USA
e-mail: s.babu@cgiar.org

V. K. Hebsale Mallappa, M. Shirur (eds.), *Climate Change and Resilient Food Systems*, https://doi.org/10.1007/978-981-33-4538-6_1

1.1 Introduction

The linkages between agricultural advance, industrial growth, and sociopolitical stability have been long recognized. While there has been undoubtedly significant progress during this century in increasing food production, the development has been uneven concerning regions, crops, and categories of farmers. Fluctuations in production have also been too wide to provide the necessary margin of safety. The need for a portion of global food and nutritional security system, which will help to insulate the human race from the threat of hunger and famine, has been realized and stressed in many national and international seminars, conferences, and other food and climate change congress forums. Sharp rhetoric at such gatherings does not, however, shorten the steps needed to build such a system. Experience teaches us that global cooperation can alone provide the necessary motivation for enduring to build a resilient food system (RFS). Fortunately, we now find several common denominators emerging, which could provide the basis for global collective action. These include the Global Environmental Organization (GEO), Intergovernmental Panel on Climate Change (IPCC), World Trade Organization (WTO), etc.

The different nations are blessed with different natural endowments. Countries, which were once regarded as hopeless deserts, have now thriving economies because of their fossil fuel reserves. Similarly, countries in the tropics and subtropics, which were formerly considered to be agriculturally backward and even classified based on the triage principle into relative grades of hopelessness, are now known to have the potential for becoming significant "green power" countries because they endowed sunlight and water. Hence, our spaceship earth is so structured that different components of it have complementary strengths.

The necessary life support systems in the world constitute a way of a common heritage (Swaminathan 1981). The tropics and subtropics form the major reservoirs of genetic variability in crops and farm animals. The temperate countries, in contrast, have developed methodologies for collecting and maintaining such variability (e.g., gene banks). Global atmospheric effects like increased temperature and enhanced CO_2 concentration may adversely affect affluent countries. The considerations of survival, nations, whether poor or rich, will have to develop a system of organized cooperation based on principles of symbiosis and synergy. Therefore, we should continuously seek and identify areas, which will bring nations together irrespective of their political ideologies into a working partnership just to save our earth from annihilation. How can we work toward achieving sustainable development objective? Initially, each nation should begin a well-planned and sustained effort in building a resilient food system. The solution to the food and nutritional insecurity problem must create holistic development programs at the community level. Some of the essential components of a resilient food system are described below:

1.1.1 Ecological Security

Steps for accomplishing ecological security would comprise measures for guarding the underlying assets of farming and minimizing the burden on natural resources (Walker et al. 1969; Pimm 1984). The government alone, however, cannot promote ecological security. It has to be a joint sector activity involving the people and private and public agencies. Hence there is a need to create local-level community management eco-development associations. These associations should actively work to build societal awareness on judicious use of a scarce natural resource (land and water) and waste management. It will be more effective if they involve schools and colleges in organizing community-level programs (Walker et al. 2002). Such associations should also operate collectively on waste exchanges and collect and recycle all organic wastes. The economic benefits of eco-development and waste recycling could provide the motivation necessary for attracting public attention and participation in sustainable development.

1.1.2 Technological Security

Growth with stability should be the aim of agricultural development programs. These would call for the breeding of high yield, cum-high stability varieties, and similar measures in animal husbandry and fisheries. The research system should be capable of promoting appropriate early warning-cum-timely action programs. Pest survey and surveillance systems, as well as soil, water, plant, and animal healthcare programs, will have to be developed. Here again, the total involvement of the local farming community will be necessary (bottom-up approach). Since, in the case of pest epidemics, political and administrative boundaries may not always provide the basis for appropriate early warning and control systems, it will be necessary for all the countries concerned with the problem to get together and form a regional control grid. The FAO-sponsored program to improve national and local locust management is an excellent example of the value of such cooperation.

1.1.3 PostHarvest Technology and Building Grain Reserves

A mismatch amid production and postharvest technologies frequently results in losses to both the producer and the consumer. Indian farmers incur around rupees 93,000 crore per year in the postharvest losses. If they cut postharvest losses, few core people can be fed for a year. Therefore, all operations after harvest, such as drying, storage, processing, transport, and marketing, will have to receive integrated attention. Depending upon possibilities, every country should maintain a reasonable grain reserve. India, for example, derived immense benefit during the unprecedented drought year of 1979 from the grain reserve of about 20 million tonnes built up by the government. Presently (January 2020), 237.15 lakh tonnes of rice and 327.96 lakh tonnes of wheat are stored. As per the buffer stock norm, only 76 lakh tonnes of rice and 138 lakh tonnes of wheat are required. A decentralized strategy of grain storage would also help under conditions of a free market economy to prevent panic purchase when conditions for crop growth are unfavorable and distress sale by poor farmers with no holding capacity when the harvests are good. In fact, a decentralized plan for storing grain as well as water at appropriate locations all over the country should be an essential element of the food security system of nations whose agricultural fortunes are closely linked to rainfall distribution. Besides, the unforeseen situations, such as the pandemic caused by COVID-19 in 2020, can be efficiently managed with decentralized grain reserves.

1.1.4 Social Security

Including India, many developing countries witness the paradox of grain surplus and widespread hunger. Even when there is food in the market, the lack of purchasing power leads to undernutrition and malnutrition. Therefore, in countries where agricultural production keeps ahead of population growth, the food and nutrition problem could be better stated in terms of person days of employment rather than in metric tonnes of food grains. The right to work should hence become an integral part of the plan for food security (e.g., Mahatma Gandhi National Rural Employment Guarantee Act 2005). This is why the integration of employment generation as an explicit aim in land and water use plans assumes relevance. Social security measures should include programs like food for work for abled persons, food for nutrition for young children, pregnant and nursing mothers, old and infirm persons, and rural development programs designed to provide the minimum needs in the field of drinking water, education, health, and environmental sanitation. Besides, there has to be a detailed manpower planning and employment generation strategy in rural areas based on a careful analysis of the possibilities for:

1. Agriculture, horticulture, sericulture, forestry, animal husbandry, and fisheries: land- and water-based occupations.
2. Agro-processing enterprises and village industries: non-land occupations.
3. Provision of relevant services.

In developed countries, the service sector tends to provide excellent employment opportunities. In fact, as agriculture advances, more and more persons working on a daily wage status tend to be employed in the secondary and tertiary sectors of the country's economy (Swaminathan 1981). The standard of living has depended on the capability of farming to release the workforce to other more industrial pursuits.

Social security measures are as much needed for farmers and fishermen as for consumers. Suitable devices to insulate farmers against losses due to factors beyond their control will have to be developed. Old-age pension and insurance measures to prevent marginal and small farmers and agricultural labor suffering when they become old and infirm are necessary (Vinaya Kumar et al. 2019).

1.1.5 Nutrition Education

Even when people have the requisite purchasing power, there may still occur several forms of nutritional diseases arising from specific causes such as vitamin A-induced blindness, iron deficiency, anemia, goiter, etc. Such nutritional disorders attributable to well-identified causes can be easily eliminated within a specific time frame through concerted efforts in nutrition education and intervention. Drinking water supply and mycotoxins in food need particular attention. Developments in preventive and curative medicine and improvements in environmental sanitation and safe drinking water supply will lead to continuous improvement in the average lifespan of a human being. Family planning programs will hence have to become an integral part of national food and nutritional security systems.

1.1.6 Population Stabilization

Depending upon the situation prevailing in each country concerning the population-natural resources equation, an appropriate population policy will have to be developed and implemented. In countries like India, planning for economic development will be a futile exercise without the widespread adoption of the small family norm. Ecological security cannot be achieved, either, without arresting the rapid growth of human populations (Walker et al. 2002). It would be advisable in all countries where the food-population equation is not favorable, to have at the community-based committee on food and nutritional security to provide the necessary political and policy guidance to ensure the involvement of administrative, academic, and local communities into a well-coordinated and cooperative action program.

However, the poverty trap in which a majority of developing countries find themselves in is an example of a feedback loop. The poverty of the land is reflected in the poverty of the mind. Opportunities for accelerated agricultural production at home are neglected. Instead, much time is spent in seeking food aid from the rich nations or in spending valuable foreign exchange in buying from the rich countries commodities, which can be easily grown by helping small farmers to derive benefit from the vast untapped production reservoir existing in most tropical and subtropical farming systems. Developing countries are doubly hit in this process. Fast-growing population, the small size of farms, high input prices coupled with an absence of measures to insulate farmers from risks, absence of appropriate land and livestock reform, poor land and water management, exploitative marketing systems, and a host of other problems have led to a situation where many low-income countries find themselves unable to feed their populations unaided. Since the majority of small and marginal farmers generate a little surplus, they cannot generate capital for being invested in enhancing the productivity of land and livestock. High interest rates and urban-centered public policies enhance the unattractive nature of investments in land and water-based occupations. This is unfortunate since 98% of the global food supply is land-dependent, the oceans providing only the remaining 2% (FAO 2019). The dilemma of finding adequate financial resources for elevating and stabilizing agricultural productivity in the midst of an unfavorable environment for investment in rural professions will have to be resolved in many developing countries if agriculture is to move forward (Vinaya Kumar and Shivamurthy 2018).

The farmer works under severe handicaps. His is a very risky profession, and he will not know what the fruits of his labor will be until the grains have been safely harvested and stored. In many developing nations, the moneylender is also the merchant, and the gap between what the farmer gets for his produce and what the consumer pays is very wide. As is said most often, the farmer is the only man in our economy who buys everything at retail, sells everything at wholesale, and pays the freight both ways.

The finite nature of earth's essential resources underlines the economic and ecological interdependence of nations. The key to lasting human happiness lies in strengthening such symbiotic bonds. Building a strong national food security system is a fundamental prerequisite for developing symbiotic links among nations (Swaminathan 1981; Walker and Salt 2006). The global food and nutritional security system will then follow. To understand this, we should know the factors which determine food availability.

1.2 Factors Are Affecting Food Availability

The nutritional disease is the relative deficiency of one or more specific nutrients. The various factors, which determine the availability of nutrients for human consumption, are of major significance. Insufficient food supplies are even now an important cause of malnutrition in many of the developing countries, especially in seasons and years of scarcity. Moreover, per capita, food production in most of these countries is actually decreasing as population growth outstrips agricultural improvement.

1.2.1 Physical Factor

The pertinent factors in the physical environment influencing food production obviously include the nature and fertility of the soil; topography and type of climate; the frequency of natural disasters such as floods, droughts, and storms; and access to lakes, rivers, and the sea.

1.2.2 Biological Factor

Among limitations to food production imposed by the biological environment are the species of plants that can be grown under the physical conditions of the region and the types of animals the land can support. Closely related considerations are the plant and animal diseases, the predators in the environment, and the extent to which they can be controlled.

The biological environment will also affect a man's capacity to produce food. Regions in which parasitic diseases such as malaria, filariasis, trypanosomiasis, schistosomiasis, and hookworm are highly endemic may have fertile soil, good rainfall, and high potential productivity but remain underdeveloped until public health measures improve the biological environment of the man himself.

It has already been pointed out that social measures change the biological environment so that infections and predators will less affect a man and his domestic plants and animals. The physical environment can also be altered by individual and community effort, as is seen in the irrigation of deserts and the terracing of hills to obtain level, water-retaining land. Fertilizers, soil conditioners, and soil-building crops such as legumes can improve the fertility of the soil.

Differences in technical knowledge and its application explain to a great extent, the impressive discrepancies in agricultural production among areas of similar soil and climate (Vinaya Kumar et al. 2017b). Similarly, where fish resources are available, the efficiency of catching, storing, and processing fish–all aspects of the social environment–will largely determine the availability and cost of fish and marine products. Eventually, much food will also be produced microbiologically and synthetically, independently of better climate or agricultural land.

1.2.3 Political Factors

Politics is a feature of the social environment, and governmental actions significantly affect food production and availability, sometimes determining whether or not malnutrition will occur. The governmental policy may stimulate or discourage food production in a number of ways, including subsidies, restrictions on the use of land, selective taxes, price controls, and even collectivization. In the United States, price supports the encouragement of producer cooperatives, and the research and extension activities of land-grant colleges have significantly contributed to increased agricultural production.

It is hardly necessary to point out the extent to which civil disturbances, revolutions, and wars and their aftermath have contributed to malnutrition and even to famine through interference with food production. Throughout history, civilian populations have suffered varying degrees of malnutrition in times of war. Today political unrest and guerilla warfare are seriously damaging agriculture in many countries. Yet, food shortages themselves lead to such unrest.

1.2.4 Economic Factor

Economic aspects of the environment are major determinants of the kinds of foods produced and the efficiency of the process. Farmers, in general, tend to produce whatever promises the greatest return for their efforts. In India and other developing countries, however, farmers frequently are unaware of many desirable crops they could grow; they are unfamiliar with the methods for cultivating these crops or lack access to essentials of modern agricultural production (scale neutral technologies) such as improved seeds and breeds, fertilizers, pesticides, machinery, and the credit to acquire them. Ignorance and poverty are, therefore, major links in the chain of multiple causes leading to widespread food shortages in the developing countries.

Many other economic factors influencing the availability of food could be cited even where local food supplies are insufficient; much agricultural land is used for production largely for export of cash crops such as coffee, rubber, and cotton. For many of these crops, synthetic substitutes are already being produced. The land would simply not be available today to produce plant sources of the amounts of dyes, rubber, and fibers, which we now obtain synthetically. This trend can be expected to continue.

Crops such as cassava or manioc, which fulfill the primary need for calories though they lack sufficient protein, have replaced food crops of higher protein value (balanced diet) in some areas because of the ease and cheapness with which they can be grown and because they can be left in the ground until sold or consumed. Cassava may prevent overt hunger, but it is responsible for much protein deficiency in young children.

1.2.5 Food Conservation Factor

Food production is, of course, a primary determinant of food availability; the quality of food conservation may determine whether or not the quantities produced suffice for human needs. In most developing countries, the application of modern principles of food storage, processing, and packaging could make as significant an immediate contribution to food supplies as comparable efforts to increase primary food production (ISUCAED 1967).

Even in the United States, where pesticides are used extensively, standing crop losses from insects alone are estimated to represent nearly 5 billion dollars annually. FAO (2019) reported that each year globally, around 20–40% per annum of crop production is lost to pests. Annually, it cost approximately $220 billion in a global economy. It is not surprising, then, that in developing countries where there is little or no insect or rodent control, either in the field or during storage, more than half of the food produced is lost before it reaches the consumer.

In India, the recurrent food shortages, even in years of normal rainfall, could be converted to exportable food surpluses through the control of preventable storage waste. In some of the godowns or storage warehouses, a third of the grain is consumed by rats while it is being held for high-priced sale during the dry season. Overall preventable waste of locally produced grain in India and countries with similar storage practices is probably in the range of 10–15%.

The seasonal scarcity and marked seasonal fluctuations in the cost of food in developing regions are additional factors in the occurrence of nutritional disease. These are related, in turn, to climatic cycles, lack of control of insects and rodents, poor conservation and storage, and profiteering by intermediaries.

When no satisfactory means of preservation are available for perishable foods in developing countries, these foods, when they are in season, may be so cheap as to return little profit to the producer. They are then unobtainable for the rest of the year. Several tropical fruits and vegetables are in this category. Furthermore, if farmers cannot store their grain crops properly, they must sell their harvest promptly to mediators and may even need to purchase part of it back later at a higher cost. If storage losses are high and administrative restraints are few, cereal grains, which are the main food staple, maybe several times higher in price in the season of scarcity than at the time of harvest.

Efficient storage procedures are a significant contributory factor to the grain surpluses of the United States. Methods of handling and storing food, thus, share a place with food production in assuring adequate food supplies.

1.2.6 Food Distribution Factor

Distribution may be equally important as production and conservation in determining the availability and cost of food to population groups. In developing countries, the phenomenon of surpluses in one region and scarcity in another is very common. For example, fresh fish may be available near the seacoast and fruit abundant in tropical lowlands, but because of poor preservation and distribution, these foods may be scarce and costly in the highlands of the same country.

Food distribution is hindered by such geographical characteristics as mountain ranges, deserts, and jungles and is facilitated by accessibility to harbors, inland waterways, railroads, highways, roads, and even trails. Indeed, one cause of malnutrition is the inefficient distribution systems in developing countries; improving them can prevent much malnutrition.

1.3 Means to Increase the Availability of Nutrients

A cultural characteristic of man is his ability to apply science and technology to free himself from the limitations that nature imposes on animals and primitive man. The production of synthetic nutrients is a dramatic example. Nutrients need no longer be obtained only from food in its natural state.

It is now possible virtually to eliminate certain highly prevalent nutritional diseases from entire population groups by adding synthetic nutrients in which they are deficient to staple foods. For example, the addition of thiamine to rice could prevent beriberi, which is still prevalent in some Far Eastern countries. Niacin added to cornmeal is useful in avoiding pellagra in Yugoslavia, and its addition to wheat flour was a factor in the disappearance of pellagra from the Southern United States. The addition of iodine to salt has been repeatedly demonstrated to be effective in preventing endemic goiter.

Enrichment of wheat and wheat flour with the limiting essential amino acid lysine, or of corn flour with both lysine and tryptophan, can nearly double the otherwise reduced utilization of their proteins. Methionine is already added to soybean-based animal rations in ordinary commercial practice and can benefit legume-based foods for human consumption as well.

Not only the amino acids but also all of the essential vitamins and minerals are now available in chemically pure form, and even synthetic sources of energy are feasible. A pure organic compound, 1,3-butanediol, which is produced on a large scale for the plastics industry from a petroleum fraction, has been demonstrated to be an excellent substitute for carbohydrate or fat in the rations of experimental animals. Another compound, 2,4-dimethyl heptanoic acid, has been used successfully in preliminary experiments, and many more such compounds can be devised. It is only a matter of time until palatable and interesting synthetic foods containing all essential nutrients are formulated. Eventually, they will undoubtedly play a role in meeting the nutritional requirements of man.

Attention should also be given to the entire range of unconventional sources of food for human consumption. In addition to the utilization of oilseed protein and marine resources, there should be consideration of the feasibility of single-cell protein from yeasts and bacteria grown on energy substrates from petroleum or natural gas without the need for agricultural land. Long before synthetic foods are economically feasible, single-cell protein should be competitive with fish protein concentrate and dried milk solids.

1.4 Factors Are Affecting the Consumption of Food

By now, it should be clear that neither the production nor availability of food per se or the precise nutrient requirements of the individual can in themselves explain the occurrence of nutritional disease. The environmental influences on the actual consumption of available food are often the most significant. Physical and biological characteristics of the environment are only of minor importance in this regard compared with social factors. Even where food is available, or could be made so, irrational dietary habits arising from ignorance and food prejudices are often a cause of malnutrition. The classic example is the development of malnutrition in young children as a result of poor dietary practices during and after weaning. Many cases of kwashiorkor are due, not to a primary lack of essential protein-rich foods in the household but to prejudice against giving these foods to infants and toddlers. In some cultures, certain foods are considered too strong for the young child. For example, in Peru, foods are classified as hot, cold, heavy, and light. Hot and heavy foods, usually those rich in protein, are not considered appropriate for young children and sick persons. In one form or another, the classification of foods into groups, which do not combine well or are believed to have special properties is prevalent. In India, for instance, the two main classes of food are hot and cold, and some are considered unsuitable for the feeding of young children.

Personal Factors Personal beliefs and taboos surround many foods of animal origin. Eggs are believed by various cultures to produce worms, to be too "strong" or too "hot" for young children, to cause sterility, or even to embody spirits. Religious taboos, such as Muslim strictures against eating pork, and the Hindu belief in the sacredness of the cow, may complicate efforts to improve dietary habits. The belief that cow's milk causes diarrhea in young infants and, therefore, should be avoided presumably stems from the almost universal contamination and adulteration of milk in primitive societies. It is hard to establish confidence in milk in many developing countries since it is still sometimes a source of infectious diarrhea.

A tragic and common cause of malnutrition in young children is the practice of feeding them highly diluted milk. When an impoverished mother is told by medical or public health advisors to give her malnourished child milk, she either cannot afford adequate quantities or believes that adding a small amount of milk powder or evaporated milk to a glass of water is wholly sufficient. The child, of course, goes rapidly downhill.

The protein needs of infants and young children can be adequately met from a variety of conventional and unconventional vegetable and animal protein sources. The biggest obstacle of low cost, protein-rich mixtures, incorporating oilseed meals, fish protein concentrates, or food yeast is not the development and production of such combinations but the difficulty of persuading parents to buy them and feed them to their children.

Food like corn and beans have great traditional value in some cultures but do not provide sufficient proteins even in their most favorable combinations for the growth and development of young children. Due to the low protein concentration and quality of such diets, the young child cannot eat the large quantity of these foods necessary to provide the amount of protein required.

Nowadays, new foods of poorer nutritive value than those presently consumed acquire prestige although they contribute little or nothing to the nourishment of the individual. In some instances, they even lead to malnutrition. Substituting highly milled wheat flour for whole-grain cereal and machine-milled polished rice for hand-pounded rice has been responsible for much malnutrition. More recently, the purchase of carbonated beverages instead of foods that would improve the diet has sometimes had tragic consequences.

Alcoholism is important in the development of acute malnutrition in some individuals. Not only does it waste money better spent on food, but it substitutes the "empty" calories of alcohol for food calories. By "empty" calories are meant those sources of body energy which are devoid, or nearly so, of essential protein, essential fatty acids, minerals, and vitamins.

Economic Factors It is very important. The relatively high cost of food, particularly of the so-called protective foods such as milk, meat, and eggs, is a major factor in the appearance of malnutrition in any population. Too often, people do not have sufficient income to purchase foods of good quality in quantities necessary to nourish them properly. Measures that will either raise income high enough to cover the cost of quality foods or lower the prices of such foods to the purchasing power of larger sectors of the population are often required to prevent malnutrition.

Political Factors It affects not only the production and distribution of food but also its ultimate consumption. In developing countries, subsidies and price controls can make it easier for people to buy food. Systems of food rationing influence the consumption of foods by the individual, and enrichment programs can ensure that certain essential nutrients are consumed without changes in food habits. Obviously, national policies relating to the distribution of food surpluses and the controls arising out of the war, civil disturbances, and natural disasters have a direct impact on individual food consumption.

1.5 Means to Combating Malnutrition

Since the occurrence of malnutrition in any human population depends upon multiple causes, it follows that there are multiple approaches to its prevention. In each situation, the first essential is to identify the factors involved and look for the approaches which are most feasible to ameliorate a given nutritional inadequacy.

The problem is not only to determine the nutrient deficiencies in a population but also to identify the other factors involved. The incidence of kwashiorkor, for example, may be reduced by making available dried skim milk or a protein-rich vegetable mixture, by preventing the infections responsible for precipitating the syndrome, and by educating parents to share family food with the preschool child. Agricultural, economic, educational, political, medical, and public health measures each help to prevent malnutrition since all of these factors are likely to be involved in its causation. It is not necessary, however, to alter all of these factors. Changing even one may be sufficient to prevent malnutrition in some situations.

1.6 Challenges and Way Forward Are the Issues for Resilient Food System

The primary question posed for those of us who speculate on world food production potential is how critical is the crisis? We look ahead today from a different base point from that of only 2–3 years ago when we were preoccupied with crop surpluses and less alert to prospects of inadequate worldwide reserves of food grains.

Judgments today reflect serious concern not only about long-term food needs but also the near term of massive starvation in some African and Asian countries unless there is a vast setup in the use of essential agricultural inputs, including conservation agriculture, protected cultivation, and integrated approaches for sustainable development. Therefore, it should be concerned not only with enhanced efforts to meet today's requirement but also with those requirements ahead.

1.6.1 Challenges for Resilient Food System

Some of the facts we have already discussed. Here we will discuss the institutional environment, which is lacking in most of the developing countries. More than 570 million small farmers throughout the world have been living in a socio-economic environment of severe poverty for centuries. The lack of a institutional framework becoming an obstacle to rural progress, to agricultural production and, the emancipation of the peasant. We can't overcome rural poverty unless we deal with it through an understanding of the specific institutional pattern responsible for formulating effective plans and policies.

Rural poverty has many aspects (access to basic facilities and infrastructure), each of which impedes progress. The institutional obstacles appear in widely different combination and with different degree of severity in various countries and even in various regions in the same country. Hence, the strategy for overcoming these obstacles must be adopted to local conditions. There is no universal panacea for accelerating agriculture progress.

In many rural areas, farmers work in an institutional environment that places two rather formidable obstacles in the way toward progress: (1) a very weak bargaining position in the market and (2) a lack of opportunities for securing economies of scale on their small individual farms. These obstacles call for some kind of organized group action of cooperation.

1.6.2 Cooperatives in the Developing World

In India, many cooperatives, which have failed, were set up with great hopes. Why did they fail? Stories abound of incompetent managers, corruption, and clerks disappearing with the till; of farmers continuing to patronize their old merchants; and of governments coddling farmers with subsidies and accepting inferior quality products or getting management snarled up with red tape. Many of these stories are true.

But, farmers have much to learn in group action dealing with modern marketing and production problems, and they can learn only by doing–by making mistakes and learning from them. What alternatives are there to overcome the institutional obstacles cooperatives are intended to meet? Persuading merchants to pay farmers the highest price they can afford rather than the lowest price they have to pay? No farmer would believe in the efficacy of such persuasion. The only other realistic alternative might be direct government purchase and marketing of farm products (ISUCAED 1967; Gunderson et al. 1997).

In fact, this alternative is quite tempting for many a government since it provides a ready source of revenue and many government jobs. But is it an alternative more conducive to rural development and more in the interest of farmers than cooperatives struggling along with inefficiency and other imperfections? Maybe, but these imperfect cooperatives do offer farmers opportunities for developing in themselves the behavioral qualities of self-discipline and social responsibility required for group action in modern society. Membership participation in a marketing cooperative involves a farmer is a much more concrete and practical terms than participation in educational or social or political organizations. It will take time to learn, but learn he will, if often the hard way.

Indeed, well-managed cooperatives can bring small farmers practically all the benefits of economies of scale which large-scale plantations and collective farms enjoy and still retain the decentralization of initiative and managerial responsibility among many farmers which is so vital for agricultural progress in view of the nature of the production process and its spatial dispersion. This has been demonstrated impressively by the milk cooperative, Anand Milk Union Limited (Amul), India.

1.6.3 Farmers' Cooperatives to Improve Bargaining Power

The basic purpose of marketing cooperatives is to strengthen the bargaining power of the farmer in the market (ISUCAED 1967). Without a cooperative, he often has no other choice than selling to the local dealer in the village, who may also be his landlord and his money lender. He must usually sell all his disposable market surplus at harvest time when prices are lowest, and a good part of it is used to repay accumulated debts. Under these conditions, mainly if he is illiterate and cannot check the merchants' accounting, he cannot help but accept any price offered to him. Yet, if he is to adopt modern techniques and produce more, he needs prices sufficiently attractive to give him incentives to do so and money for buying the modern inputs.

A cooperative gives him an alternative market outlet and sometimes also storage space for part of his crops for future sale at higher prices for which he can receive credit at low-interest rates. Since the manager of the cooperative is responsible for the farmer members, the farmer feels he can trust the cooperative that his sale transactions are carried out in his interest. This, at least, is the way a cooperative ought to function. And if it does, it effectively strengthens the farmers' bargaining power in the market and stimulates progress.

1.6.4 Farmers' Cooperatives to Improve Farm Income

Cooperatives not only strengthen the farmers' bargaining power in the market; they also bring him benefits of economies of scale. By handling the combined output of many small farmers, crops can be graded for quality, and higher prices can be paid for the right-quality products. The pooling of produce as well as purchase orders from many farmers permits the use of large-scale storage and often also processing and packing facilities, which bring higher prices to farmers for things they sell and permit lower-priced bulk purchases of many things farmers buy. These improve cost-price relationships and again provide a potent stimulus for the modernization of farming. Cooperatives can also bring economies of scale to small farmers in production, in the use of specific critical modern inputs, such as agricultural machinery and specialized equipment and technical advice combined with production credit–which is a most potent combination of inputs for introducing innovations into farming techniques.

1.6.5 A Way Forward

In most agricultural areas in the developing countries where small farmers prevail, where population pressure is heavy, and where local trade is in control of a small elite of landlords, plantation owners, merchants, and lawyers, the setting up of marketing cooperatives are necessary despite all the difficulties. This requires strong government support with a trained workforce and finances for organizing and managing cooperatives and educating farmers in membership participation and responsibilities for at least a period of 5 years. It might well be that neither cooperatives nor a government purchase organization could be made to function properly until the area had undergone a thorough land reform in which the intimidating power of landlords and creditors over farmers had been broken. These obstacles are taken care of in the recent land reforms in India.

Where these obstacles are not as severe, other alternatives for strengthening the farmers' bargaining power might be found, such as price supports enforceable on private trade, publicly controlled storage and grading facilities, current price information readily accessible to farmers, and various tax and credit measures to encourage merchants to improve the efficiency of their facilities and operations.

Very often, these measures are not in conflict with cooperatives, and the interests of farmers and agricultural progress might be best served by fostering farmers' cooperatives side by side with improvements in the functioning of private trade (ISUCAED 1967). There are usually some farm products and many consumer goods that are more efficiently handled by private trade than by cooperatives. Experience in many countries shows that cooperatives and private trade in competition with each other not only strengthen the farmers' bargaining power but improve the efficiency of both types of marketing enterprises.

Here again, no general principle concerning the role of private trade in rural areas can be established because so much depends on the local power structure, on the bargaining position of the small farmers and farm laborers, and on the degree and kind of control exercised by merchants and their attitudes toward modernization of the marketing system. In many developing countries, marketing cooperatives are handling an increasing share of the farmers' staple food and major export crops, often including processing and packaging. In contrast, farmers continue to sell their livestock and fruit and vegetable products to private merchants and dealers. Similarly, modern inputs such as fertilizers, pesticides, and machinery will continue to be distributed mainly through private trade as long as merchants succeed in meeting the quickly rising demand efficiently and at prices which the great majority of small farmers can afford to pay. To this end, governments can help local merchants by such measures as extension education in marketing, information services on prices, new sources of supplies, new transport and storage facilities, and new types of products–especially modern inputs in farming at the specific time at which farmers need them–but also new consumer goods which might be of special interest to farmers. Perhaps the most important single factor which will determine the role of local merchants in fostering agricultural progress is their ability to win the confidence of the small farmers and to make them feel that merchants respect them as citizens and want to serve them as customers, rather than dominate their fate and exploit their weak position. This change in attitudes of local merchants may be fast in some areas and slow in others; to win the farmers' confidence, it must manifest itself in the merchant's behavior toward small farmers, which might well include supporting the farmers' interest in schools, land reforms, and even cooperatives.

Two aspects are very important for agricultural progress which cannot be served by improving the functioning of the private trade but for which cooperatives are specifically best suited: (1) bring economies of scale to small farmers and (2) offering farmers concrete opportunities for developing the self-discipline, social responsibility, and cooperative attitude which are requisite to participation in any group organization and public affairs.

1.7 Building a Resilient Food System at the Community Level

A dynamic agricultural system capable of responding to the needs of today as well as tomorrow can be built only under conditions where the development strategy is in harmony with sociocultural and socio-economic conditions, as well as the agro-ecological features prevailing in different parts of developing countries (Swaminathan 1981; Walker et al. 2002). A great agricultural asset of most of the developing countries in the tropics and subtropics is the existence of a sizeable untapped production reservoir even at current levels of technology. In India, the arable land is the second largest in the world. It accounts for about 159.7 million hectares. The second largest producer of cereals (rice and wheat) globally, but in terms of productivity, it is quite low than the global average. How can this potential be tapped in a manner that does not do damage to the long-term production potential of land and water? For the achievement of sustained agricultural advance, there is a need to pay concurrent and integrated attention to the following three major areas of action:

1. Development of an economically viable and ecologically sound technological package that can help to bring about continuous improvement in the productivity of the terrestrial and aquatic farming system.
2. Development and introduction of a package of services which can help all farmers and fishermen to drive economic advantage from new technology irrespective of the size of their holdings, their innate input mobilizing potential, and risk-taking capacity.
3. Introduction of a package of public-private policies which can help to stimulate and sustain both production and consumption.

Agriculture starts moving forward only when appropriate packages of technology, services, and public-private policies are introduced in a symbiotic manner. The major ingredients which will have to go into the making of these packages:

Dynamic agricultural research, education, and extension system is a must for promoting a vibrant RFS. Too often, developing countries have launched ambitious agricultural production programs without appropriate research and training infrastructure. Invariably in such projects, success tends to be short-lived. Cynical comments usually follow such disappointing exercises in agrarian development. Agriculture is basically a location-specific phenomenon. Therefore, even if suitable concepts and materials can become available through international research efforts, it will not be possible to provide the kind of advice and assistance the farmers need without the support of a strong national agricultural research system. Experience in yield improvement in countries like Japan characterized by a small size of average landholdings indicates that productivity improvement is a slow process requiring the building up of the necessary infrastructure both at the production and postharvest phases (Swaminathan 1981).

In India, dramatic progress has been witnessed during the Green Revolution in the improvement of wheat and rice production in parts of the country like Punjab. When high-yielding and management-responsive varieties of wheat carrying the Norin dwarfing genes became available in the mid-1960s, wheat production in Punjab rose dramatically. In India, as a whole, wheat production went up from about 12 million tonnes in 1964–1965 to about 104 million tonnes in the year 2018–2019. However, the production advance has been dramatic only in areas where certain preconditions for technology to take root existed. For example, in Punjab, these major prerequisites, viz., owner cultivation, land consolidation, and rural communication, already existed when high-yielding varieties of wheat and rice became available. In other words, the substrate requirements for the adoption and diffusion of technology must be fulfilled if sustained progress is to take place.

For research to be able to find meaningful solutions to farmers' problems, it is evident that the scientific program should be organized both on an agro-ecological region and problem-solving basis (Walker and Salt 2006). Research institutions should have both short-term and long-term goals. Whenever there is a need for an immediate step up in production, the research organization should be able to adapt and adopt suitable existing technology in operation designed to purchase time. However, simultaneously it should be done, which will help to maximize production and minimize risks. The technology development should also bear in mind that a poor farmer bases his decisions not on yield per hectare but on stable income per hectare. So profit-maximizing technology characterized by the stability of income and low risk appeals to him more than just production-maximizing technology. This would then call for close integration of socio-economic research with technology development. At the same time, policymakers should understand that there will hardly be any differences between ivory tower research and applied research of great value if arrangements are not made for the transfer of technology. For example, the farmer will not be interested in a radio or television program on a high-yielding variety of a crop plant if seeds of that variety are not available to him. In other words, agricultural extension has to be so designed that the transfer of knowledge and the transfer of the inputs essential to apply that knowledge in the field are synchronized in time. This calls for a highly orchestrated effort on the part of scientists, extension workers, and agricultural administrators in charge of input supply. Also, to farmers, seeing is believing. Hence, demonstrations in farmers' fields of new technology, both on a factor basis (i.e., fertilizer application, pest control, salinity reclamation, etc.) and the basis of a system (i.e., crop-livestock, agriculture-aquaculture, agroforestry, etc.), are potent instruments of extension. Care must, however, be taken to organize the demonstrations in poor farmers' fields since the success of demonstrations in rich farmers' fields will tend to be attributed to affluence rather than to technology (Swaminathan 1981).

Agricultural science is now in a fascinating state of development. Several old practices based primarily on empirical observations and intuitive understanding of problems are now giving way to more science-based production systems. For example, in the field of plant breeding, individual plant performance used to be the method of choice in selection programs in the past. Today population performance is the major index of selection. Many other criteria, such as the following, have been introduced:

1. High productivity per unit of time, water, energy, and air space.
2. High photosynthetic ability.
3. Low photorespiration (where relevant).
4. Photoperiod- and thermo-insensitivity (where relevant).
5. High response to nutrients and other inputs of cultural energy.
6. Multiple resistance to pests.
7. Better nutritive and storage quality.
8. Crop canopies that can retain and fix maximum CO_2.
9. Suitability for incorporation in multiple and intercropping systems.
10. Suitability for improved postharvest technology.

As the pressure of population on land increases, it will be necessary to design farming systems capable of optimizing the returns not only from soil and water but also from air space. In other words, three-dimensional crop canopies involving a vertical dimension will have to be designed. More research will have to be done on root systems in relation to soil profile characteristics. Cooperative combinations of intercropping and mixed farming will have to be developed. The principles to be adopted in three-dimensional crop planning are the same as already introduced in composite fish culture involving the optimum use of a cubic volume of water. Some of these possibilities are not open in temperate countries covered with snow during several months of the year. Hence such research is unlikely to be done in those countries. It is for the scientists in the tropics and subtropics to look at the potential offered by their own environment and help to derive full benefit from it. National research systems should investigate the factors which cause instability in production. These factors can be broadly grouped into three categories:

1. Weather aberrations.
2. Pest epidemics.
3. Public-private policies.

It is possible now to mitigate to some extent the undulations in production caused by weather conditions. For undertaking relevant research programs in this field, a strong agro-meteorological research base will be necessary. The most likely weather patterns in an area will have to be worked out on the basis of an analysis of the available information from the past. In chronically drought-prone areas, the only enduring solution is the introduction of irrigation wherever possible. Even where irrigation is not available, some yield improvement can be brought about through appropriate land use and water harvesting and conservation techniques. So far, there has been very little work on the development of risk-minimizing technology. Such technology will involve:

1. Development of alternative cropping strategies (location-specific strategies) to suit different weather probabilities.
2. Standardization of crop life-saving techniques.
3. Development of compensatory production programs in off seasons and in areas with assured irrigation.

It will be necessary to build appropriate seed and fertilizer reserves to put these techniques into practice.

For example, if the available moisture is not sufficient to take a cereal crop, it may be possible to take an early maturing grain legume crop. However, cereal-legume intercropping is promising. This will, however, call for the building of a suitable seed reserve of the alternative crop. The investment made in such reserves will be more than compensated both by the increased production achieved under adverse conditions and by the morale-uplifting impact of such technology on farmers and population alike.

Agricultural scientists should pay specific attention to maximizing the return from the most limiting production factor in each agro-ecological area. Where land and not water is limiting, the strategy should aim at improving productivity per unit of land. Conversely, where water and not the land is limiting, the aim should be to maximize production from a liter of water. Similarly, the research system should look into every aspect of the production-consumption chain so that products that are produced at considerable effort are not damaged, either quantitatively or qualitatively, before they reach the consumer. An action-reaction analysis will also be necessary at every phase of technology development. Irrigation without an adequate understanding of soil profile characteristics can lead to problems of salinization. Water reservoirs can also become the breeding grounds of vectors of important human and animal diseases. If a nation has a capable and responsive research system, then all these problems can be faced and solved successfully. Soil and plant health problems of different farming systems need local research and solutions.

In light of climate change and scarce land and water resources, it is essential to build resilient food and nutritional system to bring food security to the regional and national with global integration.

Resilience was initially termed in the ecological perspective (Holling 1973) and has recently been anticipated as a way of exploring the complex dynamic systems, including socio-economic ones (Levin et al. 1998). The resilience approach has lately been adopted into food and nutritional security literature. The main aim of the RFS is to measure individual capability to sustain the adverse effects of future shocks, as a relevant factor of vulnerability study. The first three United Nations Sustainable Development Goals (UN-SDGs[1]) are no poverty, zero hunger, and good health and well-being (UN 2015). This is a clear indication of the renewed effort to achieve food and nutrition security and develop sustainable agricultural practices at all levels of the organization. In other words, the achievement of these SDGs requires strategies that are not only multipronged and sustainable (Tinarwo et al. 2018) but also resilient. Food system resilience is a significant factor that has become a priority in recent decades in light of the food and financial crises of 2007–2008 and 2011 (Pinstrup-Andersen 2014). This trend requires development initiatives to move from providing support for relief efforts to building the means for a RFS that can withstand shocks and stresses. Therefore, achieving food system resilience is imperative in the process of attaining food and nutrition security.

Food and nutritional system approaches are progressively seen as a way to improve its outcomes and resilience, to deal with challenging priorities, and to address the multifaceted association that exists within (internal) and between (external) components of food and nutritional systems. We have developed a conceptual framework that explains building resilience through food systems. The definition of food and nutritional insecurity has a direct impact on the approach used to measure it, and the theoretical model developed in this paper considers resilience to be an internal and external factor defined according to four building blocks: significant factors, interceding factors, action as strategies, access, and community empowerment for individual or family to respond and adapt to future shocks.

Most of the research in food and nutritional security in the wake of climate change has stressed on refining analysis approaches to understanding and forecast of adverse effects and their scale. In specific, the present food and nutritional insecurity predicting a capacity concentrate on establishing capabilities to withstand future crises and shocks, based either on selected ecological, political, technological, production, and consumer indicators that are considered causes of shocks or on individual vulnerability patterns that reflect a behavioral pattern. These methods provide alerts on promising food and nutritional insecurity risks rising from value changes in selected socio-economic factors (Buchanan and Davies 1995). The most frequently used definition of vulnerability is that of the IPCC (2001), which states

[1]https://www.un.org/sustainabledevelopment/sustainable-development-goals/ (for a brief information about UN SDGs refer Annexure I)

Table 1.1 Food system resilience queries

Question regarding resilience	Answer
Of what?	Food system activities to deliver outcomes
To what?	Food system stresses and shocks
For whom?	Food system actors
Over what time period?	Short-/long-term interruptions

Source: Authors compilation based on Ingram (2018)

that "vulnerability is the degree to which a system is vulnerable to and incompetent to cope with uncertainty shocks."

In order to holistically achieve food and nutritional security, food systems must accurately provide the means for food availability, food access, and food utilization (Ingram 2018). These outcomes are reliant on various factors, comprising resilience. Food system resilience indicates the ability of "people, communities, countries, and global institutions [to] prevent, anticipate, prepare for, cope with, and recover from shocks and not only bounce back to where they were before the shocks occurred but become even better off" (IFPRI 2014).

There are a variety of ways to approach resilience through a food systems perspective. Table 1.1 illustrates the four critical queries of food system resilience based on Ingrams' (2018) definition of the concept.

The first question regarding food system resilience refers to the various activities that contribute to sufficient food utilization, access, and availability. Such activities include storing, producing, processing and packaging, retailing and wholesaling, consuming, and disposing and reusing (Ingram 2018). Resilience enhancement occurs at several levels and tends to target these activities.

The second question addresses the central importance of building resilience within food systems: stresses and shocks. Stresses are pressures or tensions exerted on a system and include demography, social and cultural norms, and climate. Meanwhile, shocks are sudden, surprising events affecting a system, and include trade, food scares, and extreme weather (Ingram 2018). This key distinction between stresses and shocks is an important consideration because depending on its categorization as stress or a shock, every event requires a different resilience approach. Together, stresses and shocks illustrate the importance of incorporating resilience practices into the various activities within a food system.

The third question discusses the importance of identifying the several stakeholders involved in food systems. These actors include the input industry, farmers and fisherman, traders, processors, and consumers (Ingram 2018). It also addresses other parties, such as nonprofit organizations, political actors, and other sectors. Collaboration and communication among these stakeholders are vital in fostering rigidity in the network of actors in the food system. Consequently, such efforts will contribute to building resilience.

Lastly, the fourth question takes into account the time frame during which resilience plays a key role in influencing food system processes. Short-term interruptions, which tend to be shocking, include fishing and agricultural activities, critical ingredient shortfall, and changing consumer shopping patterns due to food scares. Long-term interruptions, which tend to be stresses, include natural resource degradation, energy price changes, and changes in dietary preferences (Ingram 2018).

Resilience research directed to date in economically advanced countries has primarily focused on handling crisis, unexpected changes, uncertainty, and disruptions, such as issues caused by the climate change and nutritional crisis (Benton et al. 2012; Wood et al. 2014; Hodbod and Eakin 2015; Vinaya Kumar and Shivamurthy 2018). It headed in the development of sustainable approaches in a food system to support lacking capabilities, such as socio-economic-ecological dimensions that allow actors to recognize, understand, and face future shocks. This helps to build sustainable food and nutritional security among individual in society (Pingali et al. 2005; Wood et al. 2014; Babu and Blom 2014a). However, in India, the predominant challenge concerns the supply and distribution system, and postharvest loss account for about more than 40% while enabling inclusive socio-economic welfare of the society (Ericksen 2008b). Hands-on preparation of resilience food systems that draws on the cooperative effort of stakeholders (public-private partnership) recognizing that empathizing the functioning of socio-economic-ecosystem factors is not sufficient for explaining the operation of the resilient food and nutritional system as a whole (Babu and Blom 2014b; Tendall 2015). An integrated view incorporates the analysis and design of supporting institutions and policies as well. But, to our knowledge, there is a shortage of field-based studies that comprise food system actors and critical interventions in the understanding and identification of their contribution in making a RFS on a global and regional level. The conceptual framework gathered the views of actors and factors (direct and indirect linkage) on food system resilience from all levels. Resilience deals a holistic way of viewing the evolution of cultural, social, political, and economic systems; it offers a means of assessing and employing the sustainability of such regimes toward a systems capacity to respond to stresses and shocks inventively and constructively.

1.8 A Conceptual Framework for Resilience of Food and Nutritional System

According to FAO (2014) report on the 805 million smallholders, approximately 791 million live in developing countries. Nearly 2 billion live on less than $2 per day and spend 50–80% of their income on food, which makes them vulnerable to food price volatility (World Bank 2014). A population predicted to be 9.6 billion people by 2050 will require double today's food demand (UN 2014). Approximately 32% of global food production is wasted, which, when converted into calories, equates to a comprehensive food loss and waste amounting to about 25% of all food produced (Lipinski et al. 2013). About 80% of food consumed in the developing countries is produced by the nearly 500 million marginal and small farmers (FAO 2014). Women contribute around 60–70% of the labor force for food production but gain little or no financial and land ownership accountability. Agriculture accounts for nearly 40% and 48% of employment globally and in India, respectively. The one who depends directly on agriculture has to respond to long-term climate change and the direct impact of natural disasters, including droughts and floods.

Conventional agricultural practices, reduced production, decreased productivity, and increased postharvest losses, coupled with a lack of infrastructural development in a rural area, poor food supply chain system, inequitable land tenure systems, and income disparities among members of the society, have contributed to around 805 m people still being food insecure. The increasing incidents of natural disasters, economic shocks, pressure on natural resources (land and water), and explosive food prices are causing additional stress on food and nutritional security among individuals in a society. Hence there is a need to develop a desirable approach assisting rural development programs to respond to the diverse needs of rural resource–poor in sustainably achieving food and nutritional security (Ericksen 2008a).

Agriculture has been able to adjust to recent changes in climate and vulnerability (Vinaya Kumar et al. 2017c), though improved innovations are needed to ensure the rate of adaptation of sustainable agricultural and the associated factor system can keep pace with changing climate over the years. The adverse effects of climate change on the food system will be influenced by the agrarian's adaptive responses to local climate and weather stressors. Adaptation can ensue from the individual (farmer) scale to a global level. Various determinants influence the adaptive responses at the societal and individual level (McCarthy et al. 2001; Vinaya Kumar et al. 2017c).

The importance of employing social-economic-ecological resilience concept in food and nutritional systems is twofold: first, to define direct and indirect factors that help to achieve a state in which food and nutritional security for individual and at the regional and national level is promising and, second, to deliver insights into how to sustain the food system in this necessary regime. However, the resilience of food and nutritional security is distinctive from the broader conceptualizations of resilience in social-economic-ecological systems because of the inherently normative nature of food systems: humans need food to survive, and, thus, system sustainability is typically a prime policy objective for food-nutritional system management. However, society also desires food systems that can strengthen sustainably, i.e., availability, access, utilization, and stability, and also concerning sustainable management while responding flexibly to shocks and crises. Current failure in meeting food and nutritional security objectives can be understood as the lack of constructive interventions to consider the full and differential dimensions of food system functions such as socio-economic, cultural, ecological, political, psychological, market, and technological factors at suitable scales (Table 1.2). In this paper, we are focusing on functional and sustainable diversity as two critical attributes of resilient food and nutritional systems. Achieving resilient food and the nutritional system will necessitate policy, technology, and institutional and capacity-building interventions for enhanced response diversity, building dynamic multiple avenues to fulfill resilient food-nutritional system objectives.

Table 1.2 Resilience factors are beneficial to assess food and nutritional systems resilience. These factors are drawn from an extensive review of the literature and other line disciplines. Therefore, the need to validate both qualitatively and quantitatively with interdisciplinary collaborative work (case study and behavioral analysis to improve the capacity of an individual)

The resilience of the food system		
Factors	Internal	External
1. Social/cultural	• Education level • Religion • Caste system • Changes in lifestyle and trend • Conflict • Leadership	• Population demographics • Distribution of wealth • Laws affecting social values • Religious issues • Ethical issues • Social rules
2. Economical	• Infrastructure • Input cost • Leadership value • Innovation opportunities • Waste reduction • Saving • Asset • Land and livestock	• Monetary policies • Inflation • Unemployment • Consumer preferences • Development of new market
3. Environmental	• Soil fertility • Erosion • Forest resources • Drainage facilities • Salinization	• Precipitation • Temperature • Increased CO_2 • Relative humidity • Wind velocity • Flood • Drought
4. Food production	• Cropping system • Cropping intensity • Crop productivity • Crop competition • Limited access to improved varieties	• Uncertainty • Pest and disease outbreak • Input availability • Change in the growing season
5. Processing and marketing	• Market planning • Grading • Intermediaries • Branding • Market segmentation • Value chain	• Pricing policies • Government control • Distribution channel • Market competition • Trade policies • Consumer demand • Transport
6. Consumer	• Income • Occupation • Lifestyle • Personality • Attitude • Motivation • Perception	• Culture • Family • Reference group • Food safety • Social class
7. Political/legal	• Stability of government • Government expenditure levels • Government leadership • Corruption level • Services • Infrastructure	• Tax policies • Bureaucracy issues • Law and regulation • Employment and operational law

Multiple interacting factors influence the food system of a given household. A food system hence comprises all the phases from food production to consumption, through intermediating factors. The conceptual definition is general and broad since it entails multifold dimensions and scales. Figure 1.1 illustrates this theoretical model for food system resilience. The multidimensionality of the food system always contains at least three components: the resource base that safeguards the food supply, the intermediating base that secures the strategies, and the socio-economic-ecological part that depends on this resource base that provides the food demand (Fig. 1.1). Nevertheless, the dynamics of food systems, different factors have to analyze through internal and external approaches that redirect the resilience of the food system at an individual and societal level.

Direct and interceding factors influence the process of transitioning from an exposed food system to a RFS. The food system factors refer to variables that directly consider the food system and include changes to the consumer, processing, markets, food production, environment, technological advancements, social and cultural beliefs, and political and legal components. Interceding factors are those that have a role in affecting the direct food system factors. In other words, interceding factors are the underlying concerns that differentiate various countries' food systems. Such factors include attitude, perception, topography, policies, religious issues, and law and regulation. Eventually, these two types of factors come together in shaping actions that aim to achieve food and nutrition security through resilience building of food systems. These actions include, but are not limited to, health strategies, livelihood strategies, economic strategies, community participation, and integrated management. These actions take place within the three subsystems of a food system–policy, institutional, and production (Babu and Blom 2014b). As a result, such actions bring about intermediate outcomes that signal a nearing achievement toward a RFS.

The efforts are made to develop a workable conceptual framework to assess the factors responsible for the RFS recognize its strengths and weaknesses. Also, to identify a critical interventions to establish a dynamic resilient food system at a community level. System resilience is anticipated to be improved by system's direct and indirect factors such as social, economic, financial, environmental, production, processing, technological, market, consumer, and political consideration. The interventions such as policy, institutional, technical, and capacity building will assess governance capacity, public-private-partnership, exposure to the crisis, self-capacity, individual learning capacity, as well as the existence of a suitable institutional framework with equitable rights and technological advantages (Fig. 1.2).

Various factors have been proposed to assess these resilience qualities in multiple systems and frameworks: livelihoods, income, household food security, natural resource management, etc. On the other hand, to what degree are these factors valid and adaptable for the resilience food system? To date, there is a lack of understanding of how such factors affect food system resilience, whether the association between the factors and resilience is necessarily linear and always positive, what levels of these factors are desirable? How many different factors interact (direct and indirect relationship) to strengthen or reduce the resilience? Building resilience

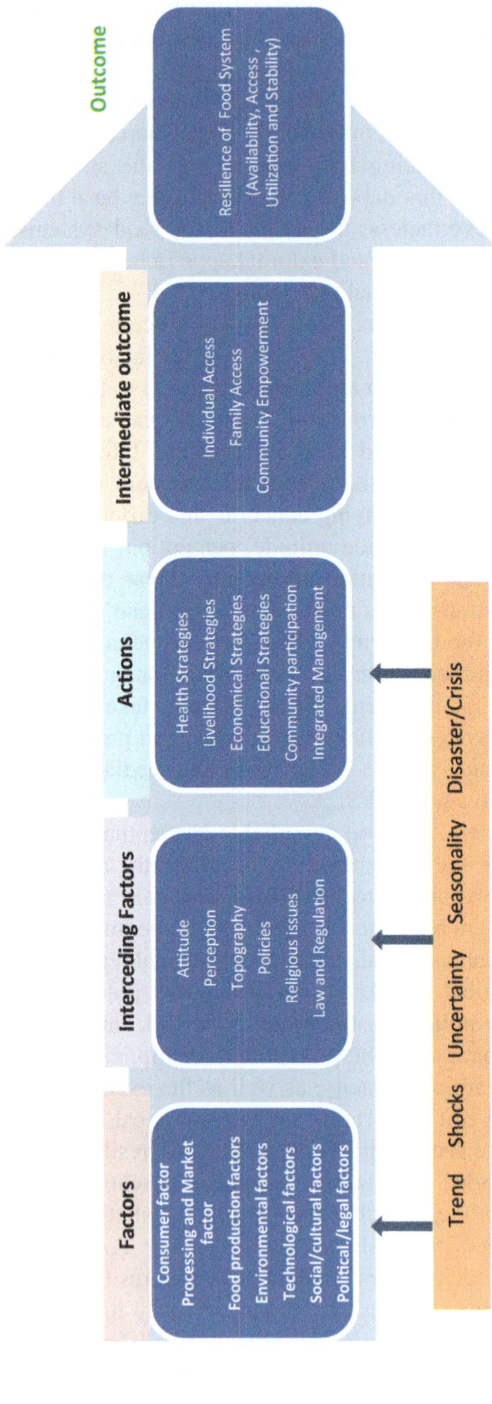

Fig. 1.1 Resilience of food system: a conceptual framework (Source: authors' compilation)

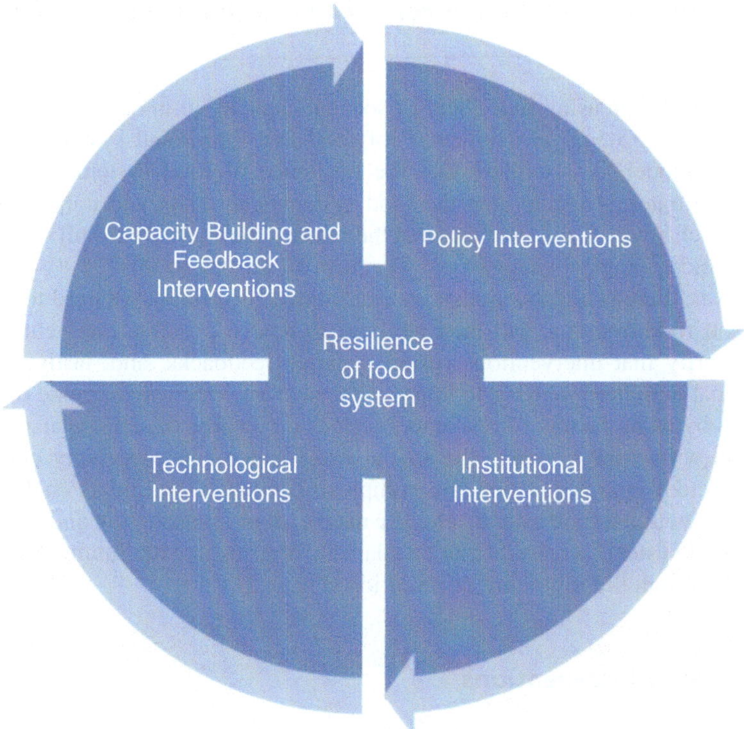

Fig. 1.2 Interventions to a build resilient food system (source: authors compilation)

in food systems necessities the identification, understanding, validation and assessment of food system resilience attributes and factors (direct and indirect). To strengthen our conceptual knowledge of resilience food systems, much more empirical data, both qualitative and quantitative, are needed (FSIN 2014). Collecting these first-hand data, though, is no simple academic task but must be conducted within an interdisciplinary framework with key actor involvement (collective approach), to ensure that the food system is conclusively attributed: essential institutions, structures, linkage, mechanisms, drivers, outcomes, and key actors are to be considered. These factors can be compared with measures of food system performance in the study area over the timeframe, before and after disaster or shocks, to determine whether the food system is behaving resiliently and where resilience can be improved in the food system. The cycle for assessing field-based case studies using factors, attributes, and indicators and enhancing the resilience of those case studies provides a concrete example of the RFS action conceptualized in Fig. 1.1. In the durable, these studies must enable a more functional formulation of what contributes to the RFS.

1.9 Interventions as a Lever for the Resilience of the Food System

The method of increasing individual or societal capacity to understand and predict shocks, by "improving the resilience" of food and nutritional systems. With the conceptual model, it is evident that the resilience of food systems means both internal and external factors that the food systems can adapt and transform themselves in such a way that no matter what the future shocks, they can still sustain with access, avoid ecological changes, etc. But how do we impart this?

However, there is a pathway that we can follow. First of all, we need to map and understand food systems, and then we have to design possible interventions. We need to verify that interventions curtail negative feedbacks, and, lastly, we need policies that help food systems to improve themselves. One way to guide the whole process is by using factors of resilience, such as the degree of self-organization and regulation within the systems, the spatial and temporal heterogeneity among the interconnected systems, the dynamic, complexity, redundancy, and diversification of food chains within a food system, etc. We need a lot of investigation and elucidation of how food systems work and can be adapted for a sustainable future. This can be achieved through the following intervention (Fig. 1.2).

1.9.1 Policy Interventions

Policy interventions enhance food system resilience through policies that build resilience capacity through community empowerment, social safety nets, and adequate warning analytical tools (Babu and Blom 2014a; Tinarwo et al. 2018). Such interventions must be open, empirical-based, transparent, participatory, and democratic (Babu and Blom 2014b). To ensure resilient food and nutritional system at the local and regional level requires policymakers to go beyond the question of accessibility of food at the national and global scale, which is whether there is adequate food at the local level to feed all individual. Some cogitations are required, including the following: physical access, economic access, social access, quality and safety access, and physiological access at the societal and individual level. The most important determinants of food and nutrition security are enough food supply produced by or accessible to households; access to income and markets to get food, also on land and inputs for food production; and therefore the adequacy of sanitation, health, and feeding practices, and the support of social claims to food. National policies must be adopted that withstand these critical priorities in political processes so that they're not moved aside in times of crisis or maybe of steady growth.

Policymakers need to identify determinants to improve the resilience of the food system. For a sustainable change, it may be vital to address both sides of the food supply as well as individual or society access to food. During policy interventions, it is essential to take into account the various determinants of the resilience of food interact (food production, processing, transport, retail outlets, and consumers). It needs to identify points of intervention for improving the community food supply.

The policy interventions to improve individual access to food depend on their resources and capacity to acquire and use food. Access to food varies on individual, family, and societal factors, many of which are associated with social and economic determinants of health, employment, income, education, residence, and social inclusion.

To increase individual access to food, it is often necessary to address the financial and capacity that are required to access and prepare food to enhance income, financial assistance, transport, and storage, preparation, etc.. The sustainable (sufficient, reliable, safe, nutritious, acceptable food to everyone in the society) policy interventions needed to improve access to food would also adopt a community development approach, to ensure that the target population will be able to mobilize itself in the future shocks and to continue to obtain or generate the resources required for themselves.

1.9.2 Institutional Intervention

Institutional interventions refer to structural changes within the relationships between various stakeholders. For example, such an intervention may include the establishment and management of multi-stakeholder partnerships. In India, agricultural policies are unsuccessful because they do not form the permitting environments in which marginal and small producers and their groups operate. Policymakers need to improve their understanding of the determinants that can enable or inhibit marginal and small producer cooperative performance. To plan capable food resilience policies, decision-makers need to be aware of the crucial function of rural businesses in accomplishing the resilience of the food and nutritional system. To improve the accessibility, quality, variety, and price of food, policymakers and stakeholders need to engage with the institutional linkage of the food system. Institutional interventions aiming to improve aspects of the food supply that support the food resilience of individuals, households, and communities usually depend on collaboration and partnership between sectors of government, private and nongovernment (public-private partnership).

The institutional obstacles appear in widely different combinations and with different degrees of severity in various countries and even in various regions in the same country. Hence the strategy for overcoming these obstacles must be adapted to local conditions. There is no universal panacea for accelerating the resilience of the food system. In recent years the progressive forces pressing for rural education, cooperatives, and land reforms have grown more reliably and faster in central government circles and in cities than in rural villages. The examination of the institutional obstacles goes a long way to explain why it is that so many governments have passed well-intentioned progressive legislation but were not able to implement it effectively on the local farm and village level. The prospects for a faster rate of introducing technical innovations are becoming better every year and that the experiment stations, extension services, farmers' cooperatives, farmers' associations, and land reforms must play a decisive role in accelerating this process.

For this challenging task, on-farm action projects through trails and demonstrations on cultivators' fields, reenforced by pilot schemes for increasing the number of cooperating farmers and for making modern inputs and credit facilities available, offer the best hope for success in modernizing farming techniques. Also, there is a need to enhance smallholder bargaining power, build their skill and competencies, allow them to access information and technologies, and besides to engage in policymaking and partnerships on an equal footing with government and the private sector. In this way, they can significantly improve their livelihoods and food and nutritional security directly through increased food production and income stability and indirectly through empowerment and political action. Such interventions are essential, and the presences of three different relations—namely, bonding, bridging, and linking—are vital for an efficient and sustainable institutional building process to the resilience of food systems at the local and national levels. And this success is necessary for winning the race between food and population.

1.9.3 Technological Intervention

Technological interventions contribute to food system resilience through initiatives that enhance research and innovation mechanisms (Babu and Blom 2014a). Application of technology is one of the vital means of increasing productivity, primarily in subsistence farming, where factors of production and resources are highly scarce and vulnerable to a complex set of factors. The technology adoption decision is strongly influenced by the availability of education, credit, services, income, farm size, etc. Within this frame condition, producer choice depends on their needs; benefit accruing and cost incurred on it would be the principal motivating factors for the acceptance or rejection of a particular technology (Karki 2004).

Technological interventions seek to minimize the possible impact of a crisis event that may occur (Vinaya Kumar and Shivamurthy 2018). Strategies that develop different sources of income for a family can allow it to respond better to future shocks, through technological interventions like drought resistant, pest, and disease-resistant varieties, multi-cropping, mixed farming, integrated farming approach, natural resource management, precision farming, storage, value chain, etc. Strategies for after the shocks need to be managed by relief, emergency response, and safety net approaches to increase households' access to food. Not only does this assistance have a direct impact on family well-being, but it also helps to preserve its assets, and thus it is potential for long-term food resilience.

1.9.4 Capacity-Building Intervention

Lastly, capacity-building interventions are crucial for resilience building within food systems. Capacity-building programs for existing institutions can assess the human resources of a food system and provide capacity-development approaches for building resilience (Babu and Blom 2014a). Examples of these programs include training workshops, partnerships, technical assistance, and improved monitoring and evaluation (Babu and Blom 2014b). Capacity building is a key concept in ensuring the sustainability of resilience within a food system. Capacity building is an area, which still needs to progress on many fronts, despite the various governmental and private sector efforts in the past. Attempts to rectify the absurdities in this regard have not made very satisfactory results. Our reaction is based more on conjecture than awareness, knowledge, and research. There is more than sufficient material available with policymakers in crisis management. However, the translation of disaster management theory into action has always been uneven and superficial. Identifying the need to modernize efforts on capacity building (knowledge, skill, attitude, and aspiration) in shock management to build and strengthen the capacity of individuals or households of shocks empathizing, prediction, preparedness, mitigation, adaptation, and recovery at the grassroots level (McCarthy et al. 2001; Vinaya Kumar et al. 2017a).

Capacity building is essential to crisis management. The resilience in shock mitigation and recovery cannot be successful without building adequate capabilities at the individual level. In a broad perspective, resilience focuses mainly on a capacity-building (understanding capacity, know and avoid shocks, adaptive capacity; alternative options in the face of crisis and transformative capacity; alternative livelihood options through empowering themselves) approach that enhances the competency and problem-solving capabilities of a community so that they would be able to apply their attained knowledge and skills to solve food crisis and improve their living standard. Sustainable interventions to improve access to food would also adopt a capacity-building approach to safeguard that the household can mobilize itself in the future and to continue to obtain the resources required.

1.10 Conclusion and Policy Implication

Looking ahead, the scenario in global agriculture is not a pleasant one. The population of the world and in India is predicted to be 9.7 billion and 1.67 billion, respectively, by 2050. Population stabilization is the number one national task. Sunlight and green plants provide the primary tools both for feeding the population and for deriving benefit from the vast human resource. Since the unpleasant nature of the agricultural scenario is mainly man-made, the process is reversible, provided we do not remain silent onlookers of this dismal landscape. We have the technology and the capability to shape a new landscape where development proceeds without the destruction of ecological assets and where hunger becomes a problem of the past. Human achievement in science and development is not a museum of finished constructions. We can add and alter if there is the requisite blend of brain and will.

Our policies are not appropriately equipped to deal with three sets of variables: (1) organizational aspects regarding how to get resources where they are most needed and how to improve the production process and food accessibility; (2) motivational, incentive, and socio-economic aspects regarding how to make people be willing and able to modernize and expand diversified production; and (3) welfare aspects regarding how to guide income distribution into a desirable pattern in which poverty is minimized and increasing the individual purchasing power. It is inappropriate to leave the organizational aspects solely to the assumption of the competitive market and the motivational aspects solely to the assumption of the profit motive and to neglect welfare aspects almost entirely, as detailed facts about them must be gathered and analyzed. They must be incorporated into our studies as key variables or normative parameters rather than ceteris paribus conditions or vague axiomatic assumptions.

The differences in frequency and severity of nutritional disease with time, place, and individual can be appropriately understood only through analysis of the epidemiological factors involved. Given the complexity of the factors so regularly responsible for nutritional insecurity and malnutrition in developing countries, an ecological approach to the prevention and control of malnutrition is one which has the best chance of success. It requires determining not only nutrient deficiencies but also the physical, social, economic, and biological aspects of the environment which influence nutrient requirements; affects the production, conservation, and distribution of food; and modifies the actual consumption of food by individuals and populations. This is one of the holistic approaches that will guide the development of programs through which the causative factors most susceptible to change in a given society can be altered.

Finally, it should be strongly emphasized that the relative availability of food for human consumption depends on population size and growth. Whether this is considered a biological or social influence on food availability, a significant determinant of the future per capita availability of food will be the rate of population increase.

References

Babu SC, Blom S (2014a) Building capacity for resilient food systems. International Food Policy Research Institute (IFPRI), Washington, DC

Babu SC, Blom S (2014b) Capacity development for resilient food systems: issues, approaches, and knowledge gaps. Building resilience for food and nutrition security, 2020 conference. Paper 6. International Food Policy Research Institute, Washington, DC. http://ebrary.ifpri.org/cdm/ref/collection/p15738coll2/id/128151

Benton T, Gallani B, Jones C, Lewis K, Tiffin R (2012) Severe weather and UK food chain resilience. Food research partnership: resilience of the UK food system subgroup. Global Food Security. UK Government Office for Science, London. http://www.foodsecurity.ac.uk/assets/pdfs/frpsevere-weather-uk-food-chain-resilience.pdf

Buchanan-Smith M, Davies S (1995) Famine early warning and response: the missing links? ITDG, London

Ericksen PJ (2008a) Conceptualizing food systems for global environmental change research. Glob Environ Chang 18:234–245

Ericksen PJ (2008b) What is the vulnerability of a food system to global change? Ecol Soc 13(2):14

FAO (2014) State of world food and agriculture. FAO, Rome. http://www.fao.org/publications/sofa/2014

FAO (2019) New standards to curb the global spread of plant pests and diseases. FAO, Rome. http://www.fao.org/news/story/en/item/1187738/icode/

Food Security Information Network (FSIN) (2014) A Common analytical model for resilience measurement causal framework and methodological options. Resilience measurement technical working group, Technical series no. 2
Gunderson LH, Holling CS, Peterson G, Pritchard L (1997) Resilience in ecosystems, institutions and societies. Discussion paper number 92, Beijer International Institute for Ecological Economics, Stockholm
Hodbod J, Eakin H (2015) Adapting a social-ecological resilience framework for food systems. J Environ Stud Sci 5:474–484. https://doi.org/10.1007/s13412-015-0280-6
Holling CS (1973) Resilience and stability of ecological systems. Annu Rev Ecol Syst 4:1–23
Ingram J (2018) Food system resilience. IFPRI brownbag seminar. Washington DC: International Food Policy Research Institute (IFPRI)
International Food Policy Research Institute (IFPRI) (2014) Building resilience for food and nutrition security 2020. International Food Policy Research Institute (IFPRI), Washington, DC
Iowa State University center for Agricultural and Economic Development (ISUCAED) (1967) Alternatives for balancing world food production and needs. The Iowa State University Press, Iowa
IPCC (2001) Climate change 2001: impacts, adaptation, and vulnerability, summary for policymakers. IPCC, Geneva
Karki LB (2004) Impact of project intervention on rural households in Nepal: assessment of socio-economic and environmental implications. Giessen, Weikersheim
Levin SA et al (1998) Resilience in natural and socio-economic systems. Environ Dev Econ 3 (2):222–235
Lipinski B et al. (2013) Reducing food loss & waste. http://bit.ly/1kmWB9L
McCarthy JJ, Canziani OF, Leary NA, Dokken DJ, White KS (eds) (2001) Climate change 2001: impacts, adaptation and vulnerability. Cambridge University Press, Cambridge
Pimm SL (1984) The complexity and stability of ecosystems. Nature 307:321–326
Pingali P, Alinovi L, Sutton J (2005) Food security in complex emergencies: enhancing food system resilience. Disasters 29:S5–S24. https://doi.org/10.1111/j.0361-3666.2005.00282.x
Pinstrup-Andersen P (2014) Political economy of food crisis. Oxford University Press, Oxford
Swaminathan MS (1981) Building a national food security system. Indian Environmental Society, New Delhi
Tendall et al (2015) Food system resilience: defining the concept. Glob Food Sec 6:17–23. https://doi.org/10.1016/j.gfs.2015.08.001
Tinarwo J, Babu SC, Iyappan K (2018) Improving food system resilience through better governance: lessons from multi-stakeholder partnerships in Zimbabwe. Internation Food Policy Research Institute (IFPRI), Washington, DC
UN (2014) Department of Economic and Social Affairs. http://bit.ly/1FnSDH0
UN (2015) Sustainable development goals. Sustainable development knowledge platform. https://sustainabledevelopment.un.org/sdgs
Vinaya Kumar HM, Shivamurthy M (2018) Adaptation to mitigate climate-induced crisis by farmers in coastal zone of Karnataka, India. Curr J Appl Sci Technol 30(5):1–13. https://doi.org/10.9734/CJAST/2018/41134
Vinaya Kumar HM, Shivamurthy M, Lunagaria MM (2017a) Rainfall trend analysis and adaptation strategies to manage climate-induced crisis in coastal zone of Karnataka, India. J Sci Res Rep 13 (5):1–11. https://doi.org/10.9734/JSRR/2017/32709
Vinaya Kumar HM, Shivamurthy M, Lunagaria MM (2017b) Impact of rainfall variability and trend on rice yield in coastal Karnataka. J Agrometeorol 19(3):286–287
Vinaya Kumar HM, Shivamurthy M, Govinda Gowda V, Biradar GS (2017c) Assessing decision-making and economic performance of farmers to manage climate-induced crisis in coastal Karnataka (India). Clim Change 142(1):143–153. https://doi.org/10.1007/s10584-017-1928-x
Vinaya Kumar HM, Chauhan NB, Patel DD, Patel JB (2019) Predictive factors to avoid farming as a livelihood. Econ Struct 8(10):1–19. https://doi.org/10.1186/s40008-019-0141-7

Walker B, Salt D (2006) Resilience thinking: sustaining ecosystems and people in a changing world. Island, Washington
Walker BH, Holling CS, Carpenter SR, Kinzig AP (1969) Resilience, adaptability and transformability in social–ecological systems. Ecol Soc 9(2):5
Walker B et al (2002) Resilience management in social-ecological systems: a working hypothesis for a participatory approach. Conserv Ecol 6(1):14. https://doi.org/10.5751/es-00356-060114
Wood SA, Jina AS, Jain M, Kristjanson P, DeFries RS (2014) Smallholder farmer cropping decisions related to climate variability across multiple regions. Glob Environ Chang 25:163–172. https://doi.org/10.1016/j.gloenvcha.2013.12.011
World Bank (2014). http://www.worldbank.org/en/topic/poverty/overview

From an Empty-Plate Lunch to Silk-Stocking Dinner: Some Futuristic Approaches in Agriculture

2

Reshma Gills and J. P. Sharma

Abstract

Vicissitudes are an inevitable part of the developmental process. Agriculture, for eras, which has been known for its contribution to the livelihood of the human population over the world, is not an exception to this fact. The anthropogenic development happened in all other sectors of the global economy like population growth, industrialization, urbanization, changed preference in consumption pattern, climate changes, etc. created and accelerated many folded hurdles in the agriculture. These hurdles are conflicting production and consumption system, post-harvest losses, food wastage, food crisis, triple burden of malnutrition, etc. which need immediate action to maintain its sustainability. This chapter highlights the need and scope of futuristic approaches, including climate-smart agriculture, nutritional farming, nontraditional and urban farming, diversified agriculture, robotics in agriculture, ICT-led agriculture, post-production, post-harvest management, logistics and value chain development, surplus management, waste to wealth in agriculture, etc. The chapter also includes the policy options and its imperatives for farmer income-led agriculture, nutritional-sensitive agriculture and farming to reach the unfed population, etc. to conquer the emerging challenges in agriculture.

R. Gills (✉)
Agricultural Extension, Socio-Economic Evaluation and Technology Transfer Division, ICAR-CMFRI, Kochi, Kerala, India
e-mail: Reshma.Gills@icar.gov.in

J. P. Sharma
Sher-e-Kashmir University of Agricultural Sciences and Technology of Jammu (SKUAST-Jammu), Jammu, Jammu and Kashmir, India
e-mail: vc@skuast.org

V. K. Hebsale Mallappa, M. Shirur (eds.), *Climate Change and Resilient Food Systems*, https://doi.org/10.1007/978-981-33-4538-6_2

Keywords

Development · Population · Urbanization · Climate change · Malnutrition

2.1 Agriculture as a Development Process

The ethnicity of domestication of plants and herbs among humankind is the origin of most important science and art, which is known as agriculture. Farming practices and agriculture have a pivotal changeover in human history, which occurred several times independently around the world. From the time of the ancient era till date, agriculture is believed to have been developed at manifold epochs in multiple areas with different phases. Agriculture has created all the social and economic development existing in the world through its magical power to feed and satisfy human needs. Even the trade and the market economy through which the world has its existence and developed nations have its dominance over other countries now are the mere product of the agriculture (Rostow 1960). The majestic role of the agriculture sector in creating and buffering employment and livelihood options is well figured (more than 65% of the working adults) by different developmental organizations. Hence, the development happens in the agriculture sector has been termed as one of the most powerful gears (assumed with two- to fourfold more effective than any other options) to battle against the dangerous levels of poverty experienced in different developing and underdeveloped nations and to create a prosperous food bowl to cater with an anticipated populace of 9.7 billion persons by 2050 (World Bank 2019a).

2.2 Challenges in Agriculture: An Overview

In the present development period, this has been called as Anthropocene Age (Slaughter 2012) due to the increased human influence in all the spears during the last 100 years, which created many hurdles in agriculture and the production system. What is notable is that the modus operandi through which mankind was trying to beat the challenges of production enhancement created an unmindful dangerous situation to the existed balanced production system and environment before, along with the irreversible cost to the contemporary and upcoming generation (Solbrig and Solbrig 1994). The first adverse example of this was the reduction of the enormous biological diversity of nature at genetic and species level due to the selected and preferred cultivation along with recurrent chapters of introgression (Turcotte et al. 2017; Walters et al. 2018; Oliver et al. 2019). Numerous other conflicts and challenges are there in present-day agriculture.

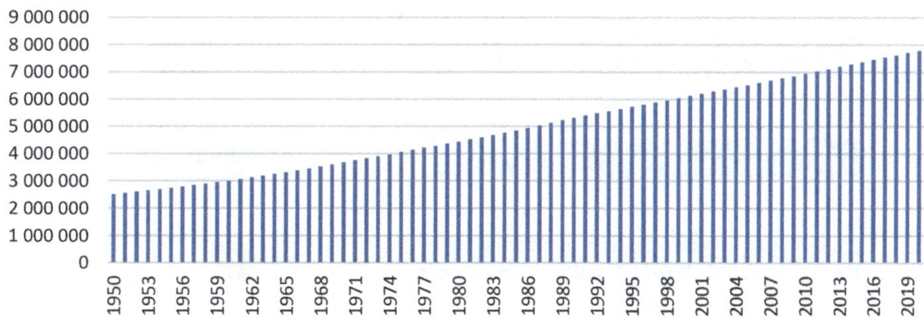

Fig. 2.1 World population (source: UN 2019a)

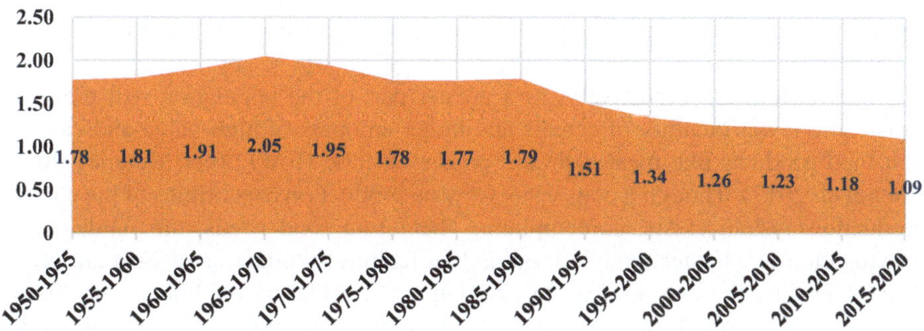

Fig. 2.2 Average annual rates of population change (%) (source: UN 2019b)

2.2.1 Population-Induced Growing Global Food Demand

According to the Malthusian theory (1798), the dynamicity of the population growth is exponential, but food production increases in a slow arithmetic ratio. As quantified by UN (2019a), the total global population is 7.7 billion as of 2019 (Fig. 2.1) and is projected to be 9.7 billion in 2050 and 10.9 billion in 2100, respectively.

As indicated in Fig. 2.2, the population growth is happening at a decreasing rate (peak point was attained in 1968 at a growth rate of 2.1%/annum) and could be reached at a steady state by the end of 2100. Still, by that time, the population pressure may be too huge to withstand the food production and carrying capacity of the planet (FAO 2017b).

Along with the population growth, increased population density is also negatively affecting the buffering and production capacities of the agricultural land (Gavin 2015). In a mise-en-scenes with modest economic and developmental growth rate, the increased population pressure is likely to push up global food demand by 70% between 2005 and 2050 as per the present projections (UN 2019c).

In accordance with the ongoing and perceived global dietary transition (Fig. 2.3), annual production of cereal (including rice, wheat, maize and minor millets) and meat will be vital to upsurge to about 3 BT and 470 MT from 2.1 billion and

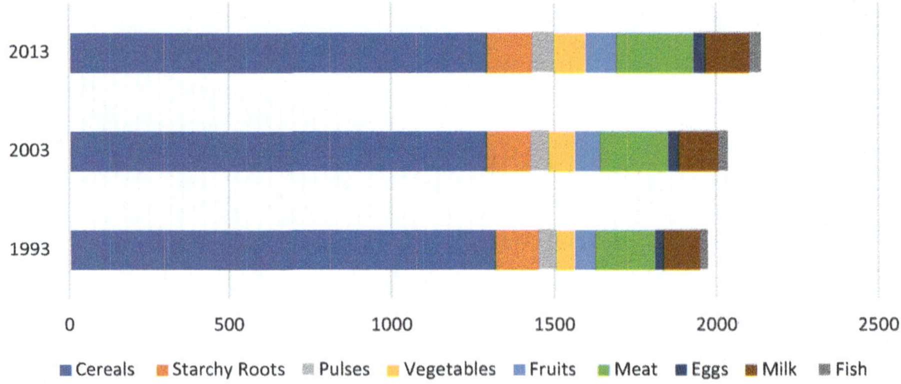

Fig. 2.3 Trends in global food consumption (kcal/person/day) (data source: FAOSTAT 2019)

200 million tonnes, respectively, as a greater part of the population will prefer to consume a fewer quantity of cereals and larger amounts of high-value and calorie-rich food products like meat, fish and processed food (FAO 2009; Kearney 2010; Roos et al. 2017). Though a consensus view on production capabilities of the current production system exists, contemplating the future food needs, the fundamental interrogation is whether today's food and agricultural production system are proficient in feeding the ever-increasing population. Considering the hitches associated with population growth-driven food demand, an intensification of productivity of the existing food production system at the extensive margin is essential (McNicoll 1984; Bongaarts 2011). It will further add burdens on already-constrained natural resources (land, water, environment, etc.) and drive to more land and water degradation, deforestation and climatic aberrations.

2.2.2 Urbanization-Led Drift in Food Demand and Increased Production System Struggle

Urbanization is a contemporary phenomenon happening mainly in developing nations. While observing the population dynamics, a major fraction of the world's population (more than half) is already residing in urban and near urban areas (Satterthwaite et al. 2010; Seto et al. 2012). Change in the urban-rural population composition is well evident with the data that, during 1900, there were 6.7 rural denizens to each urban denizen, and the present moment the rural-urban population ratio is decreasing at an increasing rate (Fig. 2.4). According to the projection, by the end of 2025, it will be in the ratio of 3:2 (Satterthwaite et al. 2010).

Urbanization has many direct and indirect effects on the agricultural production system like the population federalization-induced conversion of farmland to urban uses and more production pressure on the agricultural system (Berry 1978; Faulkner 2004; Deng et al. 2006; McDonald 2008; Fita et al. 2013; Pandey and Seto 2015; Smit 2016; Abu et al. 2019) for the diversified products due to alteration in dietary

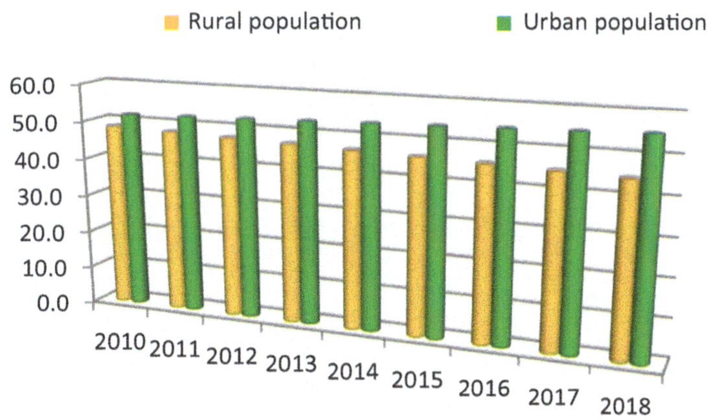

Fig. 2.4 Urban-rural population composition (data source, FAOSTAT 2019)

needs and demands of the urban people (John et al. 2019). Urban and rural food demand varies on many levels. Urban food consumption is higher (up to 70%) as compared with the rural areas (FAO 2017b) with more proportion of protein-rich animal-based (sourced) foods and value-added processed foods (Vorley and Lancon 2016). Urbanization has changed the consumption basket of the urban areas from conventional to high-priced (Beaulac et al. 2009; Jessica et al. 2019; Larson et al. 2009) and convenience-based (Kiran and Savneet 2018; Abu et al. 2019). The increased anthropogenic intervention to realize the changing food demand is further adding to the environmental pollutions and population-induced climatic changes (Xiaowei et al. 2015; Yan et al. 2016; Patra et al. 2018).

2.2.3 Climate Change and Reduced Crop Yields

A devastating scientific unanimity is observed that the Earth's environment and related conditional factors (climate) are changing with a dangerous impetus as an upshot of unintended anthropogenic activity on the planet other than the natural pace of its change. This may lead to some adverse effects like an escalation in the observed concentration of greenhouse gases (GHGs), i.e. CH_4, CO_2 and N_2O, warming (Table 2.1) athwart over the globe along with further gushing ill effects in the form of escalation in seawater level due to melting of ice, a shift in rainfall pattern, increased occurrence of weather-induced calamities, etc. (Jayaraman 2011; Shakeel et al. 2009). More than 25% of greenhouse gas emissions recorded is estimated as a spillover of the irresponsible practices coupled with land-use change in agriculture and forestry. Thus, climate change is identified and now treated as the most apocalyptic global ecological and environmental challenge with repercussions for natural bio-networks, agriculture and health, which humanity is facing (Parikh and Parikh 2002; Senapati et al. 2013). In most of the underdeveloped and developing countries, agriculture is the main livelihood option for more than half of the

Table 2.1 Year-wise world temperature change (data source, FAOSTAT 2019)

Year	2011	2012	2013	2014	2015	2016	2017	2018
Temperature change (°C)	0.871	1.007	0.948	1.026	1.333	1.572	1.366	1.191
Emissions (CO_2eq) gigagrams	677681.2	682826.1	684284.5	693386.7	690341.1	684634.9	693871.8	704100.2

population, and the changes in climate involve multifaceted interactions which are altering livelihoods of people with adverse impacts due to low penetration of climate risk management practices, poor coping mechanisms, high dependency on monsoons and smallholding nature among them. An increase in temperature would increase crop respiration rate, reduce crop duration, alter the pattern of pest attack and nature of weed intensification, accelerate mineralization process in soil and diminish fertilization use efficiency. All these could noticeably affect crop yield for the long run (Reddy and Pachepsky 2000; Pathak et al. 2003; Popova and Kercheva 2005).

Other than this, intense weather conditions at the stage of sowing and harvesting are the main menace for agriculture. In agriculture, the economic impact of climate change is myriad. It has a direct impact on the price of the commodity and its supply, demand induced and the trade cycle, the profitability of the farm and comparative advantages/disadvantages of the farm produce in terms of time and space. Indirectly it points to the challenging happenstances of food security and the nutritional stability of the population (Islam and Wong 2017). The scale and territorial scattering of such climate-persuaded weather vagaries may affect the existing production systems' ability to broaden the food production range, which is required to nourish the mushrooming population projected for the middle of the next century (Shakeel et al. 2009). According to the World Bank (2017a), 80 million people per day are nearly affected by drought-induced food shortage. Smallholder and sustenance sharecroppers, herder and fisherfolk in many of the developing countries may not have the resilience to deal with climate change meritoriously, due to abridged adaptive capability and higher threats of climate defencelessness (Tubiello 2012; Campbell et al. 2016; OXFAM 2018; Kunmin 2018).

2.2.4 Triple Burden of Malnutrition

The triple burden of malnutrition (Pinstrup 2007; Gomez et al. 2013; Gillespie and Van den Bold 2017) entailing the co-existing conditions of undernutrition (stunted and underweight), overweight and obesity and micronutrient deficiencies is a significant peril in the present global food consumption patterns. Recent data showed that one in every three persons suffers from one or other form of malnutrition. Almost 1 billion persons munch too little calories, not less than 3 billion don't have adequate nutrients and above 2.5 billion eat too copious as required (IFPRI 2018). Nearly 821 million people (approximately 1 in 9) are undernourished in the world (Table 2.2). Along with almost 3 million childhood deaths reported all over the world are correlated with the imperfections in the nutrition, especially malnutrition (NCD 2018; UNICEF 2019; FAOSTAT 2019). More than 260 million women of reproductive age are affected by iron-amenable anaemia (WHO 2018), which will ultimately lead to an unhealthy future generation. Though food is considered as the common entry point for nutrition and agriculture, there is often some obvious significant disconnect between the two. It is mainly due to the fact that food availability does not ensure abundant or adequate nutrition always (World Bank

Table 2.2 Number of people undernourished and severely food insecure (million)

Year		2010	2011	2012	2013	2014	2015	2016	2017	2018
Undernourished	World	822.3	814.4	807.6	800.1	788.8	785.4	796.5	811.7	821.6
	Africa	199.8	201.7	204.6	207.1	212.1	217.9	234.6	248.6	256.1
	Northern America and Europe	–	–	–	–	–	–	–	–	–
	Latin America and the Caribbean	40.7	39.5	38.8	38.4	38	39.1	40.4	41.7	42.5
	Asia	572.1	563.5	554.7	545	529.2	518.7	512.3	512.4	513.9
Severely food insecure	World	–	–	–	–	585	568.2	600.4	657.6	704.3
	Africa	–	–	–	–	210.7	226.7	268.2	287.5	277
	Northern America and Europe	–	–	–	–	16.1	16.3	13.4	13.6	10.6
	Latin America and the Caribbean	–	–	–	–	–	–	–	–	–
	Asia	–	–	–	–	305.9	280	264.8	288.5	353.6

Source: FAOSTAT (2019)

2015; Hawkes et al. 2017). Triple burden of the malnutrition relics is a daunting challenge and a mulish barrier in almost all the poverty eradication efforts.

2.2.5 Alarming Level of Post-Harvest Losses

Post-harvest losses including food loss and wastages (FLW) are a great concern to food security (including the availability, accessibility and affordability) and environmental and economic sustainability faced by the global food production system (Shafiee-Jood and Cai 2016; Shee et al. 2019; Kuyu et al. 2019) (Fig. 2.5). Though the accurate estimate of the FLW in quantifiable terms is difficult to get (Landry et al. 2018; Aamir et al. 2018), the meta-analysis and multiregional studies done by different international agencies under the aegis of FAO observed that more than 30% (about 1.3 billion tonnes per year) of the total food produced globally is being wasted (Gills et al. 2015; FAO 2017d; Banjaw 2017).

The food and agriculture produce waste quantification shows that wastage is maximum (35.75%) at the food after cooking or in the plate (Parfitt et al. 2010; Hodges et al. 2011; FAO 2011; Kummu et al. 2012; Gills et al. 2015; Dusoruth et al. 2018; Aamir et al. 2018; Landry et al. 2018). Food wastage has a straightforward impact on food supply chain difficulties with reduced availability and accessibility to food of a major portion of the population, low income to the farmers due to decreased selling price and increased purchase price, enhanced greenhouse emission accounted for wasting degradation and increased agricultural activities to produce more food to feed the ravenous stomach. Data from a populous and developing country like India shows a dangerous level of post-harvest losses (Lisa and Adel 2015). It is almost 30% in cereals, 40–50% in horticulture crops including fruits and

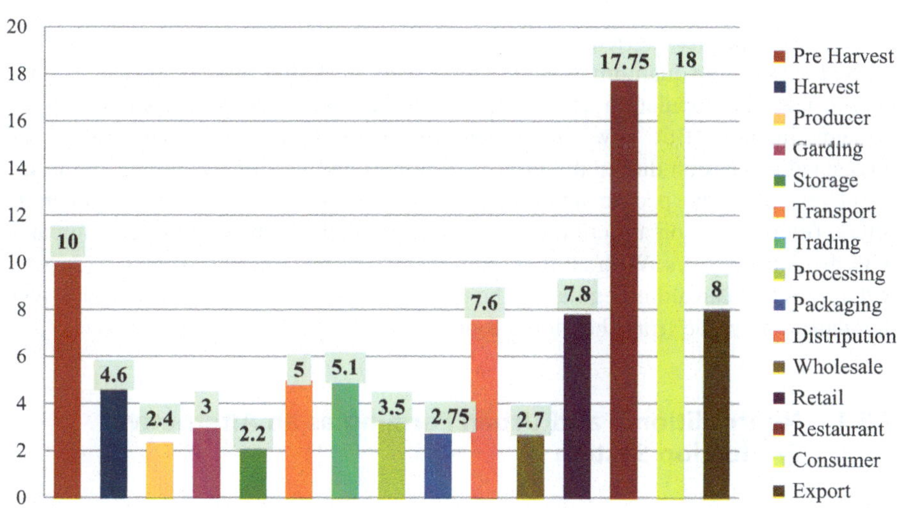

Fig. 2.5 % Post-harvest losses globally (in median value) (data source, FAO 2019b)

vegetables, 20% in oil crops, dairy and meat sector and almost 35% in fish (FAO 2015). A study in the SAARC region showed that horticultural crops are showing greater vulnerability to the post-harvest wastage. Estimated losses are as exemplified as 20, 29, 38, 46 and 52% in mandarin, banana, mango, tomato and cauliflower, respectively, in the SAARC region (FAO 2017g).

2.2.6 Low Income and Low Social Status Offer to Farmers

Even though agriculture has all the pride to accept the nomination of the mother of civilization and social development, many imperfections are still there in agriculture sector. Due to the overstress given by the officials of the national and international leaders to the production strategies rather on the farmers' affairs, majority of the developing and under developed nations negligibly recognized or rather ill-treated the farmers who have contributed and sacrificed their life and youth in the sun and rain to create food to feed and to create a market for the economy of their own country among other nations. According to the UN (2019c), two-thirds of impoverished employed workers worldwide are agricultural workers. Spending on food and consumable showed a contrasting trend as compared to the income level of the people in developed and developing nations. In a developed country, families spent 15–25% of their income on food, whereas it is almost 50–80% in developing nations (Bourne 2014). Malnutrition and associated mortality showed an exponentially increased relation to the high food cost and insufficient quantity of food.

2.3 Pragmatic Solutions and Strategies for Overcoming the Challenges

In the past centuries, agriculture has been described as a 'Godly profession' and blessed with a large quantum of natural resources, workforce and less pressure on the land to feed the population as the population was too small as compared with the present situation. But now the agriculture sector is the most vulnerable and threatened profession due to the limited resources and lack of workforce due to the disinterest from the people in farming due to the less remuneration it generally offers. But pressure on agriculture to feed the population is many folds as compared with the ancient years. With all these reasons, the farmers need to switch to smarter techniques that can aid in regulating the proper use of land, water and energy to feed the planet and elude from the global food crisis.

2.3.1 Nontraditional and Urban Farming as an Alternative Production System

As indicated earlier, over the last three to four decades, the world has witnessed continuing and increasing phenomenon of urbanization. Population explosion, combined with a massive migration, particularly of rural populations to urban centres and expansions of the existing cities, is the main reason for this. Since the settlement of the roving old generation of the human race into communities, conurbations have been indissolubly coupled with the crops that sustain them. But the urbanization led to severe poverty in some of the area, especially where the jobless migrants are there. Along with the economic down slopes, increased demand for the food for feeding all the urban population with the shrinking rate of agriculture land and less involvement of people in production activities in turn resulted in malnutrition and starvation among the city people. Here the prominence of urban and peri-urban agriculture stands for.

Urban and peri-urban agriculture can be defined as the mounting of crops or plants and rearing of fish or animals in a domesticated environment within and around metropolises and townships. It includes all the activities like production, processing and marketing of the agricultural produce using reclaimable and reusable natural resources, urban trash and waste. In fact, one can say glorious farming is what gave rise to the first cities in human history. Though the term of urban agriculture is more pronounced in recent years, it has an extended date back history in the process of human civilization. Some of the examples are 'Eridu' the first city to ever exist located in Mesopotamia built over 7000 years ago, on the extremely fertile shoreline of Tigris and Euphrates rivers (Peeter 2015); 'Machu Picchu', the ancient Incan city built around 1450–1460 well-known for hundreds of man-made terraced and irrigated crop fields surrounding the mountain city (Alconini 2005; Wright 2016); 'Chinampa agricultural system', the so-called 'floating gardens' of Aztecs which were found on the shallow lake beds in the Valley of Mexico constructed during 1519 (Altieri and Koohafkan 2004; Antonio 2017; Roland 2019); etc.

Backyard gardens or kitchen gardens (cultivation of crops on the property of a home), allotment gardens (vegetable garden of medium size in the municipal area), vertical farming (producing food in vertically piled stratums, vertically inclined planes, other plant-growing substrata integrated with other structures in a combination of a controlled or modified environment for agriculture), fertigation, aeroponic and aquaponic (Despommier 2013; Ritu and Janakiram 2016), tactical gardens (a keyhole garden) rooftop gardens (farming on roofs) window farming, community garden (a piece of land (single or combination of multiple small pieces) planted communally by a group of people in a region), educational gardens (urban agriculture developed by any educational institution) micro-gardens (intensive cultivation practice of a wide range of crop) and aquafarming (cultivation of fisheries in wastewater), etc. are innovative and contemporary approaches added to the urban and peri-urban farming with many fold benefits (Hendrickson and Porth 2012; Specht et al. 2014). It gives food for the city dwellers, reduced pressure on the rural land and farmers, reduction in pollution, reduced poverty in urban areas, and moreover tension relief and health benefits for the rural people who engaged in farming activities.

Another nontraditional method of crop cultivation, which gained the attention due to its wide range of benefits on income, environment and heritage in the near past, is organic farming (IFOAM 2009; EC 2012). Organic farming is a sustainable and eco-friendly farming practice because it restores the nutrient and carbon to the soil resulting in higher nutrient density in the soil as well as higher crop yield (Srivastava et al. 2018). Furthermore, organic agriculture can be more purely defined in comparison to sustainable agricultural practices and takes its allusion theme in ecological fortification and conservation (Gills 2012). For example, organic farming is fundamentally not contingent upon peripheral contributions as external inputs. This curtails the usage of different resources in the farming practices followed in the crop production system and limits the nutrient freight in the soil, which in turn clues to the reduced water pollution level due to dripping effect of over-fertilization and resulted eutrophication due to increased concentration of nitro-phosphorous elements.

2.3.2 Internet of Things for the Better Agriculture

From time immemorial, the agricultural world is a field for scientific development in the form of machines and tools. Recent day's agriculture is witnessing the robotic or artificial intelligent era of precision farm operations. The word robotics was made known to the public through a play titled Rossum's Universal Robots (RUR) by Czech writer Karel Capek in 1921 (STANFORD 1998). Throughout history, the robot was pragmatic in mimicking the behaviour of humans and habitually manages tasks in a similar manner. Intelligence that allows the robot to behave in a humanistic way is a programme or system of programmes. To satisfy the food demand in the challenging situation of increasing climatic aberrations, reducing soil productivity and adverse environmental impacts, the food production system must turn to a smart farming system by embrace with advanced technologies, especially the Internet of Things (IoT) with automated and connected devices. It is the way of the temporal and spatial application of capital-intensive and hi-tech modern Information and Communication Technologies (ICTs) (big data analysis, artificial intelligence (AI), machine learning, etc.) in on- and off-field activities. Precision agriculture, weather monitoring, information system, IT-based post-harvest produce handling, livestock monitoring, etc. are able to detect minute changes in humidity, light, moisture and temperature with the help of high precision sensitive gears called sensors. In accordance with the adaptive and experimental nature of every scientific discipline, agriculture science has made many strides in the engulfment of the advancement of ICT in its various fields, and the rate of change in the adoption is upward sloping too. Almost 84% of IoT deployments happening in the world are addressing the SDGs[1] in some or other way (UN 2019d) with the following distribution: 25% to manufacturing, innovation, and infrastructure (SDG 9), 19% to smart metropolises and publics (SDG 11), 19% to affordable and clean energy (SDG 7), 7% to good health and wellbeing (SDG 3) and 5% to responsible production and consumption (SDG 12). Data on the IoT penetration shows that IoT is very less utilized in the

[1]Refer Appendix I.

production system, which would have a direct impact on poverty alleviation. Hence harnessing the power of machine learning and artificial intelligence is essential to make a smart way of the food production system. According to Andrew (2016), IoT device connexions predicted in the agriculture world will reach 225 million by 2024 as compared with 75 million in 2020 (projected) and 30 million in 2015 with a compound annual adoption growth rate of 20%. IoT applications are expected to produce a mediocre of 4.1 million data plugs per day from an average farm in 2050, as compared with 190,000 data points in 2014, which can able to create a smart intelligence bank to make management decisions to enhance the productivity of agriculture system further.

IoT in agriculture has many fold benefits like efficiency excellence by virtue of the informed decisions; cultivable area expansion to skyscrapers' walls, abandoned containers and every unutilized point due to its artificial intelligent mechanism to conserve and utilize the resources efficiently; and agility in food supply with added components of cleanliness and assured quality (FAO 2017a; World Bank 2017b). Agricultural androids (robots) mechanize sluggish, tedious and murky tasks of farmers, countenancing them to focus more on recuperating inclusive production in terms of yields and profits (RIA 2017). Some of the AI-based applications in agriculture are smart harvesting, precision farming, mechanized grading, smart irrigation, quality-controlled processing, variable rate technology, smart greenhouses, etc. The custom of AI has made the agriculture shrewder than ever with a touch of intelligence, which is as good as a human in many cases. Smart agriculture can be used to augment food productivity. It can act as an agent to address the issue in the food supply chain and make the farms virtually more connected and more intelligent in decision-making. From drones to sovereign tractors to robotic artilleries, the technology is being positioned in the creation of more interactive and innovative applications. The following section will give an insight into the robotic technology (IoT) applications in agriculture.

2.3.2.1 Planting and Seeding

Life of all the plants begins as seeds in a field. The outmoded and most common means for seeds sowing are to fling them by broadcasting by hand or by use of mechanical broadcaster connected to a tractor. This throws many kernels around the field while the tractor pushes at a sturdy pace. This is not an efficient method of implanting as it causes more seed rate through some sort of the wastage of seeds. Seeding and planting through autonomous AI robots with geo-mapping (soil mapping) are the most efficient way of planting. In this method, a map is made which displays the soil properties (texture, quality, moisture, density, etc.) at every point in the particular field. The tractor with mechanical seeding accessory places these seeds or transplants the seedlings at accurate locations and in calculated depths so that each has the best chance of budding, e.g. farm guru, developed by Tech Mahindra. It is a solar-powered portable unit, to assist farmers from the time of sowing, through the cropping season and then to post-harvest phase.

2.3.2.2 Weed Control and Intercultural Operations

Herbicide and pesticide spraying in the field to control the weeds and crop pests generally creates wastage and severe environmental pollutions also. If AI-induced robots are doing this job, then it became a much more efficient method. Micro-spraying robots with AI can distinguish weeds from the crops and then spray a beleaguered droplet of herbicide with calculated active ingredient onto them. Some of the weeding robots don't even want to use chemicals to destroy it, but they could detect plants as it is pushed by a tractor and then mechanically digs the spaces between plants to deracinate the weeds. Some others are using lasers to kill the weeds. Likewise, AI-assisted irrigation robots can effectively and efficiently irrigate the target plants through autonomously navigating between rows of crops and dispensing water unswervingly at the base of each vegetal and also have the benefit as they are able to admittance to the areas where other machines cannot.

2.3.2.3 Harvesting of the Produce

Garnering and picking are some of the most venerated areas in agriculture where robotic applications are very crucial due to the exactness and promptness that robots can accomplish. It will help to mend the yield losses and trim down squander from crops being leftward in the field. The artificial intelligence integrated into it helps the machine to detect the fruit, senses its maturity and mellowness, moves to clutch it and gently detaches it (only the ripe fruit) from the tree or plant without any damage. The robot uses a combination of machine vision and gesticulation forecasting algorithms to distinguish and locate the ripe fruits which to be harvested. These machines can detect injured, smashed, ruined, diseased and unripe fruits and have the inbuilt intelligence to pack the harvested fruits in different boxes according to size.

2.3.2.4 Primarily Surveillance

Monitoring and reporting of the field and climatic conditions are very important for the crop cultivation practices to make adjustment in the planting and other operations. Sensors and geo-mapping technologies are letting farmers to get a much advanced and sophisticated level of information about their crops and the microclimate of the field. Drones, the robotic machine used for this purpose, has hardware and analysis software to do the field surveillance and analysis. The farmer who is using the drone can move the appliance to the field and commence the observation activities with the help of software attached to tablet, computer or smartphone and view the collected and composed crop-related data in real time as a tangible entity. The airborne shadowing engines can detect stunted crops, ciphers of pest or weed injury, aridness and many other variables that are part of the exertion of agricultural activities general. For recording the accurate climatic conditions, ground-based robots are being used as they are talented to get nearer to the crops. These data points, which an artificially intelligent machine can make, will enrich the farmer with decision-relevant information, which in turn enhances the predictability of the production models.

2.3.2.5 Processing of Farm Produces

One of the most time-consuming processes in the processing is the sorting of fresh produce before processing. This will help to decide which product can be made from the farm produce, e.g. from potato products like French fries, potato chips and hash browns. Sorting of the products is generally done on the basis of size, colour and texture. Another important thing in the food processing industry is the cleaning of the utensils. AI-induced robots use ultrasonic sensing and ocular fluorescence imaging to measure food scum and microbial wreckage in a portion of paraphernalia and then optimize the cleaning procedure. This will give most hygiene equipments for the artificial processing intelligence which can be used to decide about which new products one should create through an unimaginable number of ways of blending and mixing of an endless variety of options of flavours, spices and ingredients that exist.

2.3.3 Price Forecast in Agriculture

Artificial intelligence can integrate into the price-predicting model, which forecasts the upcoming commodity influx and their corresponding prices in a specified area. This model makes use of strategies and remote sensing data from geo-satellite zones to imagine the future price of the commodity. The historical data pertaining to soil character information, fertility status of different crops, etc. are generally used in the forecasting models to predict particular crop price data, e.g. the Microsoft Price Forecasting tool.

2.3.4 Diversification and Commercialization of Agriculture

Since agriculture sectors are struggling to create a sustainable economic impact on most of the underdeveloped and developing nation, it is necessary to reduce distress in farming, and focus should be more on diversification and commercialization of the agricultural products which can yield high value in low volume (Singh et al. 2006; Rahman 2009; Meena et al. 2018; Deogharia 2018). Planned diversification in the crop husbandry and crop-based product development is an innovative technique to augment the revenue of the farmers in a lucrative way and a concave to give extensive options and crop portfolio in the production which has the potential to help in alleviating poverty, conserving the environment and creating employment (Hayami and Keijiro 1995; Haque 1996; Joshi et al. 2004; Weinberger and Lumpkin 2007; Maggio et al. 2018). As the tendency of rice consumption per person in a decreasing line with the rapid urbanization and income growth (Ito 1989; Huang and Cristina 1993), diversification is essential to meeting the changing dietary demand (Delgado and Siamwalla 1999; Joshi et al. 2003; FAO 2004). According to Chand (1996), diversification in agriculture can be indicated as intentional changes in crop mix, activity mix and enterprise mix at farm level, especially at household level, by diverting the resources from single crop to a mixture of different crops or livestock.

Flower cultivation, honey bee cultivation, fisheries, mushroom cultivation, protected cultivation of exotic vegetables, nonfarm activities, etc. are some of the remunerative undertakings in this area (Barrett et al. 2001; Basantaray and Nancharaiah 2017). Farmers' income can be improved with increased productivity, decreased cost of production and increased risk-bearing ability against climate change-induced yield losses. Consequently, there is a need to boost up diversified agriculture along with auxiliary agricultural undertakings (Smith et al. 2008; UNFCCC 2009; Khanam et al. 2018). Crop rotation is another crop diversification action in which the suitable arrangement of consecutive crops in a pattern is followed to facilitate the different plants to pull nutrients in different magnitudes from various echelons (Quiroz and Valdes 1995; Rahman 2009). Thus the crop diversification in crop rotation model can augment the production and reduce the cultivation cost as in the case of legume-cereal rotation. It is also a proven solution of the most widespread farm level rejoinders to climate erraticism and alteration (UNFCCC 2009; Smith et al. 2008).

2.3.5 Climate-Smart Agriculture: Adaptation and Mitigation

Food and nutritional crisis are intensified by the extreme changes observed in the climatic conditions (World Bank 2019b). It is estimated that climate change will add up to 122 million more to the population in extreme poverty group by 2030 (FAO 2016). To tackle the catena of food security challenges and climate-related encounters, an integrative approach is needed (Ericksen 2008; Ingram 2011; CCAFS and UNFAO 2014; Sapkota et al. 2015; Olorunfemi et al. 2019). Climate-smart agriculture (CSA) is the new terminology in advanced agriculture options that has an unequivocal focus on addressing climate change-related agricultural challenges (FAO 2010; World Bank 2019b).

'Adaptation' and 'mitigation' are two vital terms that are central in the management of climate changes in socio-agricultural-related scenarios (IPCC 2001; Tol 2005; Laukkonen et al. 2009; CIFOR 2011; FAO 2012a). Climate adaptation denotes to the capability of a system to amend to variability and extremeness observed in the climatic conditions which can cause moderate to potential mutilation to all components of a production system and the ability to identify the options to cope with the cost and consequences produced by the human-induced climatic changes and to take advantage of opportunities (Pan and Zheng 2010). In contrast, climate mitigation is any deed made to enduringly eradicate or diminish the long-term risk and perils of climate change to human life, living components and any property of the production system (Tol 2005; Lu 2013; Duguma et al. 2014).

In recent days, the adaptation strategies were recognized as important as mitigation plans by the numerous international negotiations done on the topic of climate change (Campillo et al. 2017; OECD 2018; UNFCCC 2019). As on today's estimate, the extent of climate change committed by humankind is at a certain level, especially in terms of increased temperature, and even the jam-packed level of possible mitigation efforts and strategies will not prevent the anticipated upsurge in temperature till 2100. Therefore, instead of the mitigation efforts, the adaptation measures to deal with adverse climate variability and changes need greater attention during the planning and implementation stages of climate and environment policy formulation, research interventions and institutional development (OECD 2018; Vogel and Meyer 2018). Adaptation is progressions done by the societies; thereby, they make themselves well competent to cope with an ambiguous and uncertain future (Jo-Ellen et al. 2005; IPCC 2007). Climate change adaptation strategies help in choosing the right measures and options to reduce the undesirable effects of climate change through making appropriate alterations and changes in the production system. It can also be defined as the act of choosing or exploiting the positive effect or changes that happened in nature by dint of climate change. With an eye toward developing and target appropriate adaptation measures, it is very much imperative to recognize regions that are comparatively more affected by the adverse effect of climate change. Generally, the nature of subsistence farming and social-cultural barriers like a low level of formal education will limit the adaptive capacity of farmers. Hence, for the climate change adaptation process, there is a need to develop and implement simple, economically viable and culturally acceptable strategies and technologies. Furthermore, the research extension and governance system should effectively transfer the climate change adaptation knowledge to the end-users. There must be intentional efforts from the government side to make new technologies, institutional, social and economic resources accessible by the farmers through integrating them with the existing resources base of farmers. To address the impact of climate change, the research and development wing of the nation should be armed enough. Many alternatives resilient to climate change are needed to be developed and transferred to the farmers filed for wider adaptation. These climate-smart technologies are crop cultivars tolerant to abiotic stresses like heat and salinity changes and resistant to flood (Chakraborty et al. 2014), frost and drought, improving soil and water management practices, altering crop management practices (mixed cropping, crop diversification, relay cropping, multiple cropping, etc.), adopting innovative farm management techniques such as resource-conserving technologies (RCTs) (Altieri and Koohafkan 2008; Lin 2011; Pathak et al. 2012; Adhikari et al. 2018) improving pest and disease management practices followed, improved and accurate weather forecasting and efficient governance (Godfray and Garnett 2014), crop insurance, hitching the indigenous technical knowledge of farmers with the scientific know-how, etc.

2.3.5.1 Climate-Smart Technologies

The first major challenge to adapt to the climate-persuaded changes is to maintain the yield stability at the present level. The development of improved crop varieties with advanced and improved yield potential and resistance or tolerance to multiple stresses (drought, flood, salinity) is the key to it. Climate change is making not only an adverse effect on the abiotic element but also many abiotic stresses like shortening of the growing season; changes in the production environment and microclimate, too, are the part of it. It is very crucial to develop tolerance to multiple abiotic stresses for the food crops through multiple breeding programmes (Pathak et al. 2012). Farmers must be provided with cultivars of the broad genetic base to fight against the climate change-related biotic and abiotic challenges and to minimize the risks of climatic aberrations (Fita et al. 2015). To make a better production system, farmers should use (access to) a basket of technological options. Examples of some climate-smart verities of different crops popular in Indian subcontinental conditions are (1) rice, Pusa Basmati 1509, Pusa Sugandh 2, Pusa Basmati 1121, etc., and (2) wheat, HD 3086 (Pusa Gautami), HD 3090 (Pusa Amulya), H D 2985 (Pusa Basant), etc. (Hema et al. 2014).

2.3.5.2 The Resource-Conserving Technologies (RCTs)

It involves practices or cultivation techniques that augment resource use or input use efficiency and can deliver immediate, distinguishable and palpable economic benefits to the farmers like reduced production cost due to enhanced savings in irrigation, fuel, nutrients and labour requirements and improved level of yield by well-timed crops establishment, etc. (Gupta and Seth 2007; Pathak et al. 2012; Soumya et al. 2018; Patle et al. 2019). Water-saving mechanisms (UNFCCC 2006) like narrow/broad bed planting of crops with crop residues can enhance farmers' income through reduced cost of cultivation and enhanced production (Altieri and Nicholls 2013). It can also provide resistance against lodging of crops due to rains and hailstorm, which are rousing in an unusual trend in recent years. Zero tillage, another resource conservation technique experimented in rice-wheat cropping system, permits farmers to plant wheat immediately after the harvest of rice; hence the crop could escape from the terminal heat stress (Pathak et al. 2011, 2012). Direct seeded rice (DSR), organic farming and conservation agriculture, which reduce the water wastage and soil evaporation through cover cropping and mulching with the crop residues, are some of the other extensively accepted RCTs (Pathak et al. 2012; Srinivasa et al. 2016). Management of rice straw burning (a serious issue that is contributing to the GHG emission) can be done through various on-site soil composting practices to make organic fertilizer with the use of effective micro-organisms. Similarly, the practice of the use of by-products and leftovers (e.g. rice straw and hull) as fuel for cooking, inputs for the paper industry, etc. are some of the greener techniques which can indirectly reduce the adverse effect of climate change.

2.3.5.3 Carrot and Stick Approach in Knowledge Integration

Farmers are ever experimenting, and their knowledge about the change in the climate is very crucial to include in the scientific knowledge realm. They can observe the change in the climate by the disappearance of some varieties of plants and animals. They are correlating the long-term changes with their knowledge domain and finding and validating some solutions from nature itself to adapt and mitigate it. Since the climate change effect on different cropping systems is different, it is better to adopt location-specific technologies and farmers' ITKs (MoSTE 2015; FAO 2019a). Upscaling of farmers' own climate-smart technologies in partnership with state departments and research institutions is also an important step. Recognizing custodian farmers for preserving the climate-smart varieties and genome of different crops suitable for the climate change scenario is a significant step forward to motivate the farmers to adopt the mitigation and adaptation practices and increase their zeal in climate-smart farming. Like awards, the penalty for the environment-unfriendly activities should be realized and firmly watched. Like the carbon exchange and trade between the low carbon emission countries and high carbon emission, to lessen the ill upshot of climate change community-level assessment of GHGs generation and its monitoring should be done.

2.3.6 Nutrition Farming as a Hunger Fighter

The main issue of malnutrition faced by the hunger population over the globe today is the access to safe and nutrient-rich food than the availability of food (Das et al. 2014; FAO 2017f; Vijaya et al. 2017; UNICEF 2019). The nutrition-sensitive agriculture focused on the following significant ideologies like increasing availability and accessibility of the food, encouraging sustainable and diversified production, enhancing the nutrition content and making the food more nutritious (FAO 2014; Nagarajan et al. 2014). Nutri-farms or nutri-gardens attained considerable attention in the near past as a local solution for malnutrition predominant in the majority of the developing and undeveloped countries. Nutri-garden provides some essential nutri-food and additional income by incorporating the nutrition goals in critical entry points (crop diversification and reduced cost of cultivation) (Jaenicke and Virchow 2013). Nutri-gardens can be created with the integration of high-nutrient-value vegetable and fruit crops and biofortified crop varieties in cultivation (Bouis et al. 2013; Bouis and Saltzman 2017), along with good agricultural practices for productivity enhancement and soil quality improvement (Ruel et al. 2018). Biofortification in different food products showed a positive impact on nutritional security aspects (Low et al. 2007; Bouis et al. 2013). Some of the dazzling examples are iron-biofortified pearl millet (Finkelstein et al. 2015; Yadava et al. 2017), zinc-biofortified wheat (Rosado et al. 2009; Singh and Velu 2017; Sazawal et al. 2018), vitamin A-biofortified maize (Fabiana et al. 2014; Menkir et al. 2018; Zuma et al. 2018), etc.

International organizations around the world with state support are making many approaches among the rural population in the participatory and target-oriented way to fight against all forms of malnutrition. For example, in India, many breeding programmes were started to increase the nutrient quality of the crop produce by National Agricultural Research System (NARS) in hand with the international crop organization. Some of the promising biofortified varieties developed by the NARS are CR Dhan 310 (protein-rich variety of rice) and DRR Dhan 45 (zinc-rich variety of rice); WB 02 and HPBW 01 (zinc- and iron-rich variety of wheat); Pusa Vivek QPM9 Improved (lysine- and tryptophan- and a provitamin A-rich hybrid of maize), Pusa HM4, Pusa HM8 Improved and Pusa HM9 Improved (lysine- and tryptophan-rich maize hybrid); Pusa Mustard 30 (low erucic acid variety of mustard) and Pusa Double Zero Mustard 31 (mustard variety with low erucic acid and low glucosinolate); Pusa Beta Kesari 1 (cauliflower variety with rich β-carotene content); and Bhu Sona (β-carotene-rich sweet potato) and Bhu Krishna (anthocyanin-rich sweet potato) (Yadava et al. 2017). Some of the innovative extension strategies developed by the Indian NARS like Knowledge Systems and Homestead Agriculture Management in Tribal Areas (KSHMTA), Nutri-Sensitive Agricultural Resources and Innovations (NARI) and Value Addition and Technology Incubation Centres in Agriculture (VATICA) are showing committed movements to stamp out the hunger, poverty and malnutrition (Randhir 2017; ICAR 2017; TAAS 2017; Paroda 2018) which can be adopted by other countries too.

2.3.7 Reaching Zero Hunger Through Zero Wastage and Surplus Management

Production perspectives that had attained commensurable techno-policy attention during the last decades merely cannot meet the nutrition demand, as it can address only the availability dimension. For assured nutrition, food should be accessible and affordable too. But the quantum of the post-harvest losses and wastage (PLW) makes it challenging to attain reasonably priced food by the majority of the poor people. The inducing factor for the post-harvest losses are varied by region, season and crops across the value chain (Hailu and Derbew 2015; Ndirangu et al. 2017; Macheka et al. 2018; Shee et al. 2019; Khader et al. 2019). But in general, studies conducted globally in different time and space horizons enlisted specific biotic and spatio-temporal independent abiotic and developmental factors. It includes playing up of the food production and supply system (physiological, physical, mechanical and hygienic conditions) and imperfections in the institutional and policy framework which are observed before and after the farm gate (Hodges et al. 2010; Arah et al. 2015; Chalak et al. 2016, 2018; FAO 2017e; Yahaya and Mardiyya 2019). Inefficient value chain management starting from the production to final consumption is a significant challenge that includes poor infrastructure for storage, marketing and processing which need immediate attention (Randela 2003; Godfray et al. 2010; Hengsdijk and de Boer 2017; Rahiel et al. 2018).

The lack of decision pertinent data (five Ws (the person (who), the time (when), form and context (what and which) and the place (where)), one H (how)) is a major reason for the post-harvest loss and food wastage issues, and domineering thing needs to be addressed by those paying attention in captivating battle to trim down post-harvest loss (Gills et al. 2016; Dumont et al. 2015). The first and foremost logical step in identifying a proper strategy for plummeting post-harvest losses and food wastage is the efficient analysis of each commodity production and handling system at the different entry points. Based on the analytical results, developing and developed countries need separate strategies to tackle the problem as the reasons and the stages of the losses may differ in both (Niewiara 2016; Ghosh et al. 2016). By refining farmers' custom of good agricultural practices, such as proper field management, careful handling of harvested crops, etc., significant reductions in food loss can be made. Infrastructure for appropriate storage of the harvested crops until it reached the final consumption points needs to be created in all the entry points in the value chain (Kumar and Kalita 2017). Cold storage, controlled and modified atmospheric storage facilities, etc. are proven post-harvest loss prevention technologies. However, derisory extension efforts and inadequate information flow from potential buyers to the producers limit the degree to which farmers are conscious of and actively implementing these innovative practices (Pearson et al. 2016). Training, demonstration and education about effective handling and storage practices, marketing strategies like collective marketing and target marketing, knowledge development and real-time communication in the supply chain, conscience building, etc. are particularly critical in diminishing the food wastage and post-harvest losses (APO 2006; Hodges et al. 2010; Bendito and Twomlow 2014; ACF 2014; Dumont et al. 2015; Calvo-Porral et al. 2016; FAO 2017c; NAAS 2019). On-farm marketing, farm-firm linkage and farmers' portals, etc. are needed to be emphasized for reducing the market imperfections and attaining more contribution for producers' share in consumer rupee (Reardon et al. 2009; Singla 2017). Tumbling both post-harvest losses and food waste level necessitates multiple stratagems, including behavioural modifications like changing consumption behaviour through increased consumer awareness and refining enticements among various supply chain participants (Kantor et al. 1997; Parfitt et al. 2010; Aschemann-Witzel et al. 2016; Calvo-Porral et al. 2016; Facchini et al. 2018; Aschemann-Witzel 2018; Boulet et al. 2019). The tactics for reducing waste and loss will undoubtedly be diverse in developed and developing regions because the core causes are different, but shortening waste and loss in all areas will yet be critical to plummeting hunger and nutritional insecurity and meeting future food demand.

Post-harvest losses and food wastages are reducing not only the nutritional food bowl very deprived but also creating a major concern of environmental pollution as most uneaten foodstuff ends up in the plonk. Waste disposal and its management is a serious issue which is diverting gigantic potions of national and international money reserve to make a clean environment. In agriculture, the waste created is generally in the biodegradable form. Though the best fit strategy is zero wastage, due to many reasons, especially delay in rectifications of identified inefficiencies, wastage in agriculture is still at its peak point. Now another strategy to reduce its negative impact on environment and livelihood sustainability is not to lose it as sheer waste but to view it as a laudable resource to rebuild the productivity of the land and profitability of agriculture through converting it into the wealthy product (waste to wealth). Green energy production, compost making, pelleting, integrated farming, etc. are some of the waste to wealth strategies recommended (Kathiravale and Muhd Yunus 2008; Obi et al. 2016; TERI 2017; Chunping et al. 2019).

2.3.8 Policy Options and Its Imperatives

Can we imagine a developed world without agriculture? The demand exerted on agriculture does not end at production; rather, it has an extended function in many directions. As the majority of the world nations primarily are with the agrarian economy, the sector must also contribute to economic prosperity, social security, nutritional safety and communal wellbeing of rural and urban areas. At the same moment, it has the responsibility to preserve and enrich the natural resources and biodiversity from the dark shadow of challenges like pollution, climate change, population growth, urbanization, etc. for the future generation too. For the sustained production and consumption system, a collective action from different nations, developmental organizations and community groups backed with policy support is needed. The foremost thing to make a revamped and prosperous agricultural sector in any country is to have a well-equipped extension system (Garforth and Lawrence 1997; FAO 2003; Budak and Yurdakul 2004; Baloch and Thapa 2017). Since the knowledge and information gaps identified as the major challenge to the productivity boosted, nutrition-oriented and post-harvest waste reduced production system, extensionists and the grass-root-level workers need to be provided with innovative teaching and communication aids to outreach the art and science of sustainable, climate-smart and nutrition farming practices to the farmers (Christoplos 2012; FAO 2012b; Fanzo et al. 2015). Technologies are the instruments to compact with the pressure of food demand induced by the growing population and changing food preference in a multiplier model (Beder 2000; Anadon et al. 2016; FAO 2003). But except for some, the majority of the technologies were not diffused properly among the farmers or end-users due to the lack of awareness and accessibility. Efforts are needed to be taken at the global level for the wider use of improved crop production, post-harvest management and marketing strategies with special emphasis on location-specific needs among the farmers.

For that, a welcoming approach integrated with lessons learned from past for collaboration in research, international science, technology development and innovation culture should be adopted by nations. Governments of different nations need to conduct a committed review about their food and nutritional security policies to identify the pragmatic obstacles and policy gaps, finding new initiatives for the international, inter-organization and interdepartmental collaborations. Agriculture science to prosper, research and extension system need to emphasis vigorous advocacy and promotional programmes directing to the youth which will help to retain them in the sector and also help them to make career options in it (Van Enst et al. 2014; Ravi et al. 2018; Zougmore et al. 2019; Nkiaka and Lovett 2019). One of the most helpful things policy can do is to form agenda of networking and build a platform for dialogue with relevant institutions providing multi-sectorial services that exist in most countries (Kavoossi 1991; Jackson et al. 2009). Similarly extension scientists at the regional level and field-level extension agent should make networks with other community-level organizations and agencies for reducing the information channel length and burden of the farmers and making farming as a profitable endeavour. The farmer who is the major contributor in the demand sensitive agriculture needs to be given the prime focus in any development policy directly affecting sustainable agriculture.

Along with the technological assistance, agricultural system of the nation should provide a wide range of services like social support, legal advice, financial consultation, spiritual guidance, job or career counselling, emotional counselling, vocational training, etc. to the farmers considering their social, cultural and personal needs. Strategies followed in agriculture need a paradigm shift from the production orientation to the framers' income orientation. It can be attained through policies supporting technological advancement for reduced cost of production and enhanced productivity, creating alternative livelihood options and off-season employment opportunities and promoting a transparent price discovery mechanism that provides assured and remunerative price for farm produce.

2.4 Conclusion

Global agriculture and production system is struggling to fill the empty food plates with food in many of the developing and underdeveloped nations. The dimension of conflict with man and production system is changing from production enhancement to nutrition, climate-smart, income and value chain oriented due to the population propelled demand drift. In this chapter, we attempted to give a global overview of the different challenges which are creating an alarming level of burden over the world agriculture, which has been characterized by fragmented land holding, degraded soil and water and fearsome level of reduction in the availability of the natural resources coupled with the anthropogenic adverse climatic condition. For providing a meal that is able to meet the essential requirements as per suggested standards to the hunger population, some target-oriented conceptual changes must be adopted in the coeval agricultural system. The suggested strategies like climate change adaptation and mitigation, nutritional-sensitive agriculture, post-harvest loss and wastage reduction through value chain creation, cold storage and innovations in the marketing system, using artificial and robotic intelligence in production and post-production arena, diversified and commercialized farming, etc., are discussed in a global perspective. Science cannot be able to make a real-time shift in the livelihood conditions as it hypothesized, if it is not backed up with a strong polices. The evidence of the failure of science, which has been not supported by the policy which addresses social capital development, can be figured out from many regions all over the globe (e.g. Bt. Brinjal in India). The nutritional and food demand of the growing population highlighted the urgent need of a shift in the agricultural policy orientation from cereal-centric to smart crop-centric worldwide. It is equally imperative to note that agriculture needs location-specific strategic actions than a blanket of suggestive measures, as each region and each production system has its own unique microclimate, resource abundance as well as a social system. The policy failures often associated with the inability of the governance system to realize the essentialities and needs of the social system in which the farmers are the key stakeholders of the agricultural system integrated. Hence it is urged to make a sustainable social system through capacity building, behavioural interventions, inter–/intradisciplinary linkages. These marketing innovations enable a favourable and conducive environment for the successful establishment of policy interventions.

2.5 Some Policy Suggestions

- The most important challenge, the future generation is going to face, is the climate change-induced, tangible and intangible natural calamities and menaces. Worldwide collective actions are needed to preserve the nature and natural resources in a sustainable way. Voluntary agreements and instrumental arrangements are needed for this. An increase in fuel efficiency by switching to more nature-friendly cooking arrangements, vehicles, etc. is one of the voluntary arrangements which individual or small state can practice. Farming can also be in a more climate-friendly manner by adopting the practice of carbon farming like cover cropping, zero tillage, crop diversification, etc. which may lead to reduced carbon emission through sequestering the carbon in the soil by the creation of soil organic matter. Different countries need to fix specific target coverage for carbon farming and encourage the farmers to follow it through regular capacity building, knowledge integration and providing financial assistance or insurance coverage to cope up with the profitability risk associated with this in the initial periods.
- Alarming rate in the current post-harvest losses and food wastage is another area, which needs policy attention. As mentioned in the chapter, the post-harvest losses in the agriculture sector occur in almost all the stages, starting from the field and farm to consumer level. Most of the developed nations have efficient post-harvest value chains for the proper handling of the agricultural produces. But the majority of the developing and underdeveloped nations are struggling to reduce the crop and produce losses at one end and to meet the food demand in the other end. In order to reduce the post-harvest losses, it is very essential to develop commodity-specific as well as community- or region-specific value chains and integrate it with the global value chains. While making strategic measures in the value chain development, it is important to give more emphasis on the reduced length of market chain with a lesser number of players in each, to get better price realization for both the producers and consumers. The underdeveloped countries are facing major food losses in the stage of on-farm storage and transportation to the sink. Lack of cold chain or cold storage facilities are the main trigger in it. The government should give more emphasis on the development of infrastructure for the cold storage facilities, accessible and affordable for the farmers to keep the produce fresh until it reaches the consumers or finds a better market. The government needs to facilitate the private investment or public-private partnership in the infrastructure development to share the cost and responsibility in the construction and maintenance. Another most important stage of the post-harvest food losses is after cooking or on plate. In a developed country like America, food lose on plate has been reported even as high as 30%. Along with creation of the habit of no food wastage at the individual level, state needs to promote the community-owned kitchens or take away outlets to distribute the huge quantum of the food which is consumable but may be wasted from marriage functions as well as get together to needy people. It will help to feed the hungry mouth instead of feeding the bins. Similarly, the government can provide some public accessible refrigerator facilities near the cities in which the person (living in flats or

apartments) who has the excess cooked food can keep it instead of wasting and the needy person can take and consume it.

- Malnutrition and triple burden associated with the insufficient intake of the nutrient food are other dangerous situations, which need the government and policy support to tackle with as this may destroy the human wealth of the nation. If the malnutrition of the city dwellers is mainly due to the changed food habits and more dependency on the junk food; the villagers, daily wagers, and poor people have a different story to tell: it can be narrated as the lack of food, lack of access to nutrients, etc. lead them to a malnourished status. While developing any policies, the government should give proper attention to balance these two scenarios. Promotion of nutri-smart villages or Nutri-gardens, which are self-sufficient to produce nutrient-rich grains, fruits and vegetables and also able to provide value-added products to the city dwellers, is one solution for it. By this, the farmers can meet the local nutrient demand of their villages and also get a good market for the farm products, which are the ingredients for the health drinks, biscuits or nutrient powders. The government should fix the basic guidelines and standards for the nutri-smart villages and also should provide inputs and conceptual orientation for efficiently developing the same while considering the topography, nutrient status and market demand. Since the nutrient-rich coarse cereals like maize, pearl millet, *Sorghum*, finger millet, etc. can be cultivated in nutrient-poor soil without any compromise in yield and nutrient quality, the government should promote the farming, intake and marketing of such nutrient-rich coarse grains, which are otherwise known as poor man's cereals, in a wider way. The government should take a strategic policy decision in the area of biofortification with due consideration on the ethical values and the societal norms. More resource investment needs to be given for biotechnological researches, which aims to develop the nutri-food through the differential coding of the plant genes. Policies must be made in more targeted approaches to identify and map locally available nutrient-rich food commodities (nutrient food map) and food consumption pattern (food intake map) in different areas to make correlation and further exploration of the possibilities to provide a balanced diet through a diversified food basket approach.
- Titivated and productive extension system with capable grass-root-level workers is an essential component of any development programme. Though the extension system's history shows the evolution of target-specific, commodity-specific, orientation-specific and pragmatic variations in the world extension methods, the principle norm of all those models is to reach the farmers in a most efficient and effective way to increase the production. The extension system, which is the bridge between the scientific rationality, government policies and the farm reality, needs to be strengthened both in terms of quality and quantity. The suggested FILE framework must have a convergence of different pillars of profitable and sustainable agriculture like different institutions, which support agricultural activities, infrastructure, technologies and, finally, incentive structures. Monetization of farmers' produce must be done through different strategic policies; those regulate the marketing through encouragement of collective (Farmer Producer

Organizations (FPOs) or Community-Based Organization (CBOs)) and direct market structure (online agri. Market platforms) creation. Policies must aim to reorient and strengthen the service function offer to the farmers and recognize agriculture as an enterprise through diversification and value addition. The extension system, which is the bridge between the scientific rationality, government policies and the farm reality, needs to be strengthened both in quality and quantity. The farmer to extension person ratio should be shortened by recruiting the qualified extension staffs or by giving intensive training to the para-extension workers who could be proclaimed as extension workers after this training. Extension workers need to be equipped with innovative and user-friendly ICT tools, which help them to deliver the knowledge, technologies, governing policies, etc. in the farmer's door steps in a cost- and time-effective way. It will also help them to amass the feedback on different policy options as firsthand information and to communicate it with the governing body for further refinement.

- Most importantly, there must be a common platform to share, discuss and refurbish scientific ideas for the development and maintenance of the sustainable production system. Cross-linkages need to be created in farmer to farmer, farmers to extension persons, extension person to extension person, farmer to extension person to governing body, etc. through development of knowledge value chains. The current agricultural production system needs strong social capital along with the technologies to cope up with the present challenges and to meet future demands. There must be solid efforts to evaluate and upscale the innovative ideas and traditional knowledge of the farmers in different areas like crisis management, coping mechanisms, technological advancement, nutritional strategies, etc. The manpower and wisdom of the innovative farmers can be integrated with the existing extension delivery system as Innovative Farmer-Led Extension Delivery (IFLED) model in which innovative farmers are the key extension persons at the village level in the technology transfer and refinement paradigm. While considering the farmers as an extension person or change agent, there must be strong policies to evaluate and upgrade the innovative farmers' services and innovation without diluting the conceptual framework and the farmers' rights associated with it.

References

Aamir M, Ahmad H, Javaid Q, Hasan SM (2018) Waste not, want not: a case study on food waste in restaurants of Lahore, Pakistan. J Food Prod Mark 24(5):591–610. https://doi.org/10.1080/10454446.2018.1472695

Abu HA, Cavinato MER, Lindemer A, Lagerkvist CJ (2019) Urban sprawl, food security and agricultural systems in developing countries: a systematic review of the literature. Cities 94:129–142. https://doi.org/10.1016/j.cities.2019.06.001

ACF (2014) Post-harvest losses and strategies to reduce them. Technical paper on post-harvest losses

Adhikari S, Baral H, Nitschke C (2018) Adaptation to climate change in Panchase Mountain ecological regions of Nepal. Environments 5(3):42. https://doi.org/10.3390/environments5030042

Alconini S (2005) Machu Picchu: unveiling the mystery of the Incas Hispanic. Am Hist Rev 85 (3):500–502. https://doi.org/10.1215/00182168-85-3-500

Altieri MA, Koohafkan P (2004) Globally important ingenious agricultural heritage systems (GIAHS): extent significance and implications for development. www.fao.org/docrep/015/ap021e/ap021e.pdf. Accessed 14 Oct 2019

Altieri MA, Koohafkan P (2008) Enduring farms: climate change, smallholders and traditional farming communities. Third World Network (TWN), Penang. http://www.fao.org/docs/eims/upload/288618/Enduring_Farms.pdf

Altieri MA, Nicholls CI (2013) The adaptation and mitigation potential of traditional agriculture in a changing climate. Clim Change 140(1):33–45. https://doi.org/10.1007/s10584-013-0909-y

Anadon LD, Chan G, Harley AG, Matus K, Moon S, Murthy SL, Clark WC (2016) Making technological innovation work for sustainable development: table S1. Proc Natl Acad Sci USA 113(35):9682–9690. https://doi.org/10.1073/pnas.1525004113

Andrew M (2016) Why IoT, big data & smart farming are the future of agriculture. https://www.businessinsider.com/internet-of-things-smart-agriculture-2016-10?IR=T. Accessed 17 Nov 2019

Antonio V (2017) Chinampa: an ancient agricultural system. CAFNRM/Agriculture Club Newsletter 1. https://hilo.hawaii.edu/nihopeku/2018/02/02/chinampa-an-ancient-agricultural-system/

APO (2006) Postharvest management of fruit and vegetables in the Asia-Pacific region. https://www.apo-tokyo.org/00e-books/AG-18_PostHarvest/AG-18_PostHarvest.pdf. Accessed 22 Nov 2019

Arah IK, Amaglo H, Kumah EK, Ofori H (2015) Preharvest and postharvest factors affecting the quality and shelf life of harvested tomatoes: a mini review. Int J Agron 2015(6):1–6. https://doi.org/10.1155/2015/478041

Aschemann-Witzel J (2018) Helping you to waste less? Consumer acceptance of food marketing offers targeted to food-related lifestyle segments of consumers. J Food Prod Mark 24 (5):522–538. https://doi.org/10.1080/10454446.2018.1472693

Aschemann-Witzel J, De Hooge I, Normann A (2016) Consumer-related food waste: role of food marketing and retailers and potential for action. J Int Food Agribus Mark 28(3):271–285. https://doi.org/10.1080/08974438.2015.1110549

Baloch MA, Thapa GB (2017) Review of the agricultural extension modes and services with the focus to Balochistan, Pakistan. J Saudi Soc Agric Sci 18(2):188–194. https://doi.org/10.1016/j.jssas.2017.05.001

Banjaw TD (2017) Review of post-harvest loss of horticulture crops in Ethiopia, its causes and mitigation strategies. J Plant Sci Agric Res 2(1):1–4. https://www.imedpub.com/plant-sciences-and-agricultural-research

Barrett C, Reardon T, Webb P (2001) Nonfarm income diversification and household livelihood strategies in rural Africa: concepts, dynamics, and policy implications. Food Policy 26 (4):315–331. https://doi.org/10.1016/s0306-9192(01)00014-8

Basantaray AK, Nancharaiah G (2017) Relationship between crop diversification and farm. Agric Econ Res Rev 30:45–58. https://doi.org/10.5958/0974-0279.2017.00021.0

Beaulac J, Kristjansson E, Cummins S (2009) A systematic review of food deserts, 1966–2007. Prev Chronic Dis 6(3):A105

Beder S (2000) The role of technology in sustainable development. In: Herkert JR (ed) Social, ethical, and policy implications of engineering, selected readings. IEEE, New York, pp 230–235

Bendito A, Twomlow S (2014) Promoting climate smart approaches to post-harvest challenges in Rwanda. Int J Agr Sustain 13(3):222–239. https://doi.org/10.1080/14735903.2014.959329

Berry D (1978) Effects of urbanization on agricultural activities. Growth Chang 9(3):2–8. https://doi.org/10.1111/j.1468-2257.1978.tb01024.x

Bongaarts J (2011) Rising food prices raise spectre of Malthus, people and planet. Population Media Center (PMC), South Burlington. https://www.populationmedia.org/2011/05/12/rising-food-prices-raise-spectre-of-malthus/. Accessed 18 Oct 2019

Bouis HE, Saltzman A (2017) Improving nutrition through biofortification: a review of evidence from HarvestPlus, 2003 through 2016. Glob Food Sec 12:49–58. https://doi.org/10.1016/j.gfs.2017.01.009

Bouis H, Low J, McEwan M, Tanumihardjo S (2013) Biofortification: evidence and lessons learned linking agriculture and nutrition. FAO, Rome. http://www.fao.org/3/a-as660e.pdf. Accessed 17 Oct 2019

Boulet M, Wright B, Williams C, Rickinson M (2019) Return to sender: a behavioural approach to reducing food waste in schools. Australas J Environ Manag 26:1–19. https://doi.org/10.1080/14486563.2019.1672587

Bourne MC (2014) Food security: postharvest losses. In: Neal KVA (ed) Encyclopedia of agriculture and food systems, 2nd edn. Academic Press, Amsterdam, pp 338–351

Budak DB, Yurdakul O (2004) Sustainable agricultural development through extension education. Asian J Plant Sci 3:215–218. https://doi.org/10.3923/ajps.2004.215.218

Calvo-Porral C, Medin AF, Losada-Lopez C (2016) Can marketing help in tackling food waste?: proposals in developed countries. J Food Prod Mark 23(1):42–60. https://doi.org/10.1080/10454446.2017.1244792

Campbell BM, Vermeulen SJ, Aggarwal PK, Corner-Dolloff C, Girvetz E, Loboguerrero AM, Ramirez-Villegas J, Rosenstock T, Sebastian L, Thornton P, Wollenberg E (2016) Reducing risks to food security from climate change. Glob Food Sec 11:34–43. https://doi.org/10.1016/j.gfs.2016.06.002

Campillo G, Mullan M, Vallejo L (2017) Climate change adaptation and financial protection: synthesis of key findings from Colombia and Senegal. OECD environment working papers, no. 120, OECD Publishing, Paris, https://doi.org/10.1787/0b3dc22a-en. Accessed 19 Nov 2019

CCAFS and UNFAO (2014) Questions & answers: knowledge on climate-smart agriculture. United Nations Food and Agriculture Organisation (UNFAO), Rome. https://hdl.handle.net/10568/42432

Chakraborty D, Watts CW, Powlson DS, Macdonald AJ, Ashton RW, White RP, Whalley WR (2014) Triaxial testing to determine the effect of soil type and organic carbon content on soil consolidation and shear deformation characteristics. Soil Sci Soc Am J 78(4):1192–1200. https://doi.org/10.2136/sssaj2014.01.0007

Chalak A, Daher CA, Chaaban J, Abiad MG (2016) The global economic and regulatory determinants of household generation of food waste: a cross-country analysis. Waste Manag 48:418–422. https://doi.org/10.1016/j.wasman.2015.11.040

Chalak A, Daher CA, Abiad MG (2018) Generation of food waste in the hospitality and food retail and wholesale sectors: lessons from developed economies. Food Sec 10(5):1279–1290. https://doi.org/10.1007/s12571-018-0841-0

Chand R (1996) Ecological and economic impact of horticultural development in the Himalayas: evidence from Himachal Pradesh. Econ Pol Wkly 31(26):A93–A99

Christoplos I (2012) Climate advice and extension practice. Geografisk Tidsskrift-Danish Journal of Geography 112(2):183–193. https://doi.org/10.1080/00167223.2012.741882

Chunping X, Nasrollahzadeh M, Maurizio S, Issaabadi Z, Luque R (2019) Waste-to-wealth: biowaste valorization into valuable bio(nano)materials. Chem Soc Rev 48:4791–4822. https://doi.org/10.1039/c8cs00543e

CIFOR (2011) Climate change and forests in the Congo Basin: synergies between adaptation and mitigation. https://www.cifor.org/fileadmin/fileupload/cobam/ENGLISHDefinitions%26ConceptualFramework.pdf. Accessed 18 Nov 2019

Das PK, Bhavani RV, Swaminathan MS (2014) A farming system model to leverage agriculture for nutritional outcomes. Agric Res 3(3):193–203. https://doi.org/10.1007/s40003-014-0119-5

Delgado C, Siamwalla A (1999) Rural economy and farm income diversification in developing countries. In: Peters GH, Von Braun J (eds) Food security, diversification and resource management: refocusing the role of agriculture. Ashgate Publishing Company, Brookfield, pp 126–143

Deng X, Huang J, Rozelle S, Uchida E (2006) Cultivated land conversion and potential agricultural productivity in China. Land Use Policy 23(4):372–384. https://doi.org/10.1016/j.landusepol.2005.07.003

Deogharia PC (2018) Diversification of agriculture: a review. J Econ Soc Develop 15(1):46–59

Despommier D (2013) Farming up the city: the rise of urban vertical farms. Trends Biotechnol 31 (7):388–389. https://doi.org/10.1016/j.tibtech.2013.03.008

Duguma LA, Wambugu SW, Minang PA, van Noordwijk M (2014) A systematic analysis of enabling conditions for synergy between climate change mitigation and adaptation measures in developing countries. Environ Sci Policy 42:138–148. https://doi.org/10.1016/j.envsci.2014.06.003

Dumont MJ, Orsat V, Raghavan V (2015) Reducing postharvest losses. In: Madramootoo C (ed) Emerging technologies for promoting food security: overcoming the world food crisis. Woodhead Publishing, Cambridge, pp 135–154

Dusoruth V, Peterson HH, Schmitt J (2018) Estimating a local food waste baseline. J Food Prod Mark 24(5):654–680. https://doi.org/10.1080/10454446.2018.1472698

EC (2012) Organic agriculture information note. https://ec.europa.eu/europeaid/sites/devco/files/study-organic-agriculture-201206_en_5.pdf. Accessed 2 Nov 2019

Ericksen PJ (2008) What is the vulnerability of a food system to global environmental change? Ecol Soc 13(2):14. https://doi.org/10.5751/es-02475-130214

Fabiana FM, Palmer AC, Finkelstein JL, Haas JD, Murray-Kolb LE, Wenger MJ, Birol E, Boy E, Pena-Rosas JP (2014) Are biofortified staple food crops improving vitamin A and iron status in women and children? New evidence from efficacy trials. Adv Nutr 5(5):568–570. https://doi.org/10.3945/an.114.006627

Facchini E, Iacovidou E, Gronow J, Voulvoulis N (2018) Food flows in the United Kingdom: the potential of surplus food redistribution to reduce waste. J Air Waste Manage Assoc 68 (9):887–899. https://doi.org/10.1080/10962247.2017.1405854

Fanzo J, Marshall Q, Dobermann D, Wong J, Merchan RI, Jaber MI, Souza A, Verjee N, Davis K (2015) Integration of nutrition into extension and advisory services. Food Nutr Bull 36 (2):120–137. https://doi.org/10.1177/0379572115586783

FAO (2003) Agricultural extension, rural development and the food security challenge. http://www.fao.org/3/a-y5061e.pdf. Accessed 5 Dec 2019

FAO (2004) Agricultural diversification: opportunities and constraints. Rome, Italy In: FAO rice conference 04/CRS.15 http://www.fao.org/rice2004/en/pdf/warda.pdf. Accessed 8 Dec 2019

FAO (2009) How to feed the world in 2050. http://www.fao.org/fileadmin/templates/wsfs/docs/expert_paper/How_to_Feed_the_World_in_2050.pdf. Accessed 18 Oct 2019

FAO (2010) Climate-smart agriculture: policies, practices and financing for food security, adaptation and mitigation. FAO, Rome. http://www.fao.org/3/i1881e/i1881e00.pdf. Accessed 22 Nov 2019

FAO (2011) Global food losses and food waste extent, causes and prevention. FAO, Rome. http://wwwfaoorg/docrep/014/mb060e/mb060e00pdf. Accessed 25 Oct 2019

FAO (2012a) Climate change adaptation and mitigation challenges and opportunities in the food sector. FAO, Rome. http://www.fao.org/3/i2855e/i2855e.pdf. Accessed 17 Nov 2019

FAO (2012b) Incorporating climate change considerations into agricultural investment programmes: a guidance document. FAO, Rome. http://www.fao.org/3/i2778e/i2778e.pdf. Accessed 7 Nov 2019

FAO (2014) Nutrition-sensitive agriculture. Second international conference on nutrition 19-–21 November 2014. http://www.fao.org/3/a-as601e.pdf. Accessed 12 Oct 2019

FAO (2015) Save food: global initiative on food loss and waste reduction. http://www.fao.org/save-food/resources/keyfindings/en/

FAO (2016). The state of food and agriculture: climate change, agriculture and food security. http://www.fao.org/3/a-i6030e.pdf. Accessed 17 Nov 2019

FAO (2017a) The future of food and agriculture—trends and challenges. FAO, Rome. http://www.fao.org/3/a-i6583e.pdf . Accessed 21 Nov 2019

FAO (2017b) The possibilities of Internet of Things (IoT) for Agriculture. http://www.fao.org/e-agriculture/news/possibilities-internet-things-iot-agriculture. Accessed 17 Nov 2019

FAO (2017c) Policy measures for managing quality and reducing post-harvest losses in fresh produce supply chains in South Asian countries. http://www.fao.org/3/a-i7954e.pdf . Accessed 27 Nov 2019

FAO (2017d) Global initiative on food loss and waste. http://www.fao.org/3/a-i7657e.pdf . Accessed 27 Nov 2019

FAO (2017e) Food loss and waste in the food supply chain. http://www.fao.org/3/bt300e/BT300E.pdf. Accessed 18 Nov 2019

FAO (2017f) Sustainable farming systems for food and nutrition security. http://www.fao.org/3/I8603EN/i8603en.pdf. Accessed 18 Oct 2019

FAO (2017g) The state of food and agriculture: leveraging food systems for inclusive rural transformation. FAO, Rome. http://www.fao.org/3/a-i7658e.pdf. Accessed 23 Nov 2019

FAO (2019a) Handbook on climate information for farming communities—What farmers need and what is available. FAO, Rome, p 184. http://www.fao.org/3/ca4059en/ca4059en.pdf?eloutlink=imf2fao

FAO (2019b) Food loss and waste database. http://www.fao.org/food-loss-and-food-waste/flw-data/en/. Accessed 20 October 2019

FAOSTAT (2019) Food and agriculture data. http://www.fao.org/faostat/en/#home

Faulkner S (2004) Urbanization impacts on the structure and function of forested wetlands. Urban Ecosyst 7:89–106. https://doi.org/10.1023/B:UECO.0000036269.56249.66

Finkelstein JL, Mehta S, Udipi SA, Ghugre PS, Luna SV, Wenger MJ, Michael J, Murray-Kolb LE, Przybyszewski EM, Haas JD (2015) A randomized trial of iron-biofortified pearl millet in school children in India. J Nutr 145(7):1576–1581. https://doi.org/10.3945/jn.114.208009

Fita A, Rodríguez-Burruezo A, Boscaiu M, Francis ZN, Romanus DD, Raphael KK (2013) Urbanisation and its impact on agricultural lands in growing cities in developing countries: a case study of tamale in Ghana. Modr Soc Sci J 2(2):256–287

Fita A, Rodríguez-Burruezo A, Boscaiu M, Prohens J, Vicente O (2015) Breeding and domesticating crops adapted to drought and salinity: a new paradigm for increasing food production. Front Plant Sci 6(978):1–14. https://doi.org/10.3389/fpls.2015.00978

Garforth C, Lawrence A (1997) Supporting sustainable agriculture through extension in Asia. https://www.odi.org/sites/odi.org.uk/files/odi-assets/publications-opinion-files/2927.pdf. Accessed 3 Dec 2019

Gavin W (2015) Global challenges facing the agriculture sector. Open Access Government. https://www.openaccessgovernment.org/global-challenges-facing-agriculture-sector/13064/. Accessed 23 Nov 2019

Ghosh PR, Fawcett D, Sharma SB, Poinern GEJ (2016) Progress towards sustainable utilisation and management of food wastes in the global economy. Int J Food Sci 2016:1–22. https://doi.org/10.1155/2016/3563478

Gillespie S, van den Bold M (2017) Agriculture, food systems, and nutrition: meeting the challenge. Global Chall 1(3):1600002. https://doi.org/10.1002/gch2.201600002

Gills R (2012) An analytical study of organic cardamom growers in Idukki district of Kerala. Indian Agricultural Research Institute, New Delhi. http://krishikosh.egranth.ac.in/handle/1/75751

Gills R, Sharma JP, Bhardwaj T (2015) Achieving zero hunger through zero wastage: an overview of present scenario and future reflections. Indian J Agric Sci 85(9):1127–1133

Gills R, Sharma JP, Burman RR, Sharma RR, Kar A (2016) Comparative analysis of vegetable production, value-addition and marketing in National Capital Region. Indian J Hort 73(1):87–93. https://doi.org/10.5958/0974-0112.2016.00024.4

Godfray HCJ, Garnett T (2014) Food security and sustainable intensification. Philos Trans R Soc B 369(1639):20120273. https://doi.org/10.1098/rstb.2012.0273

Godfray HCJ, Beddington JR, Crute IR, Haddad L, Lawrence D, Muir JF, Pretty J, Robinson S, Thomas SM, Toulmin C (2010) Food security: the challenge of feeding 9 billion people. Science 327(5967):812–818. https://doi.org/10.1126/science.1185383

Gomez MI, Barret CB, Terri TR, Pinstrup-Andersen P, Meerman J, Croppenstedt A, Carisma B, Thompson B (2013) Post-green revolution food systems and the triple burden of malnutrition. Food Policy 42:129–138. https://doi.org/10.1016/j.foodpol.2013.06.009

Gupta R, Seth A (2007) A review of resource conserving technologies for sustainable management of the rice–wheat cropping systems of the Indo-Gangetic plains (IGP). Crop Prot 26 (3):436–447. https://doi.org/10.1016/j.cropro.2006.04.030

Hailu G, Derbew B (2015) Extent, causes and reduction strategies of postharvest losses of fresh fruits and vegetables—a review. J Biol Agri Healthcare 5:49–65

Haque T (1996) Diversification of small farms in India: problems and prospects. In: Haque T (ed) Small farm diversification: problems and prospects. National Centre for Agricultural Economics and Policy Research, New Delhi

Hawkes C, Demaio AR, Branca F (2017) Double-duty actions for ending malnutrition within a decade. Lancet Glob Health 5(8):745–746. https://doi.org/10.1016/S2214-109X(17)30204-8

Hayami Y, Keijiro O (1995) Beyond the green revolution: agricultural development strategy into the new century. In: Jock RA (ed) Agricultural technology: policy issues for the international community. CAB International, World Bank, Wallingford, Oxon

Hema B, Gills R, Sharma JP (2014) Climate resilient technologies for sustainable agriculture. In: Shukla JP (ed) Technologies for sustainable rural development: having potential of socio – economic upliftment (TSRD–2014). Allied Publishers, New Delhi, pp 256–264

Hendrickson MK, Porth M (2012) Urban agriculture—best practices and possibilities. http://extension.missouri.edu/foodsystems/documents/urbanagreport_072012.pdf. Accessed 16 Oct 2019

Hengsdijk H, de Boer WJ (2017) Post-harvest management and post-harvest losses of cereals in Ethiopia. Food Sec 9(5):945–958. https://doi.org/10.1007/s12571-017-0714-y

Hodges RJ, Bernard M, Knipschild H, Rembold F (2010). African postharvest losses information system—a network for the estimation of cereal weight losses. In Carvalho MO, Fields PG, Adler CS et al. (Ed). Proceedings of the 10th international working conference on stored product protection held on 27 June–2 July 2010, Estoril, Portugal (Julius-Kühn-Archiv, 425, pp 956–962). Julius Kühn-Institut, Berlin

Hodges RJ, Buzby JC, Bennett B (2011) Post-harvest losses and waste in developed and less developed countries: opportunities to improve resource use. J Agric Sci 149(S1):37–45. https://doi.org/10.1017/S0021859610000936

Huang J, Cristina CD (1993) Demand for cereal grains in Asia: the effect of urbanization. Agric Econ 8(2):107–124. https://doi.org/10.1016/0169-5150(92)90025-t

ICAR (2017) New initiatives-AE. https://www.icar.org.in/content/new-initiatives-ae. Accessed 15 Oct 2019

IFOAM (2009) High sequestration, low emission, food secure farming. Organic agriculture—a guide to climate change & food security. https://www.ifoam-eu.org/sites/default/files/page/files/ifoam_ifoameu_policy_climate_food_security_dossier_2009.pdf. Accessed 2 Nov 2019

IFPRI (2018) To address the triple burden of malnutrition, focus on food systems and demand. http://www.ifpri.org/blog/address-triple-burden-malnutrition-focus-food-systems-and-demand. Accessed 15 Oct 2019

Ingram J (2011) A food systems approach to researching food security and its interactions with global environmental change. Food Security 3(4):417–431. https://doi.org/10.1007/s12571-011-0149-9

IPCC (2001) Climate change 2001. Synthesis report. Cambridge University Press. https://www.ess.uci.edu/researchgrp/prather/files/2001IPCC_SyR-Watson.pdf. Accessed 17 Nov 2019

IPCC (2007) Climate change 2007: impacts, adaptation and vulnerability. Fourth assessment report. Intergovernmental Panel on Climate Change Secretariat. Geneva, Switzerland. https://www.ipcc.ch/site/assets/uploads/2018/03/ar4_wg2_full_report.pdf. Accessed 20 Nov 2019

Islam MS, Wong AT (2017) Climate change and food in/security: a critical nexus. Environ 4 (2):1–15. https://doi.org/10.3390/environments4020038
Ito S, Peterson W, Grant W (1989) Rice in Asia: is it becoming an inferior good? Am J Agric Econ 71(1):32–42. https://doi.org/10.2307/1241772
Jackson RJ, Minjares R, Naumoff KS, Shrimali BP, Martin LK (2009) Agriculture policy is health policy. J Hunger Environ Nutr 4(3–4):393–408. https://doi.org/10.1080/19320240903321367
Jaenicke H, Virchow D (2013) Entry points into a nutrition-sensitive agriculture. Food Secur 5 (5):679–692. https://doi.org/10.1007/s12571-013-0293-5
Jayaraman T (2011) Climate change and agriculture: a review article with special reference to India. Revi Agrian Studs 1(2):16–78
Jessica AD, Kate O, Amanda JC, Kaur H (2019) Household food consumption patterns and food security among low-income migrant urban farmers in Delhi, Jakarta, and Quito. Sustainability 11(5):1378. https://doi.org/10.3390/su11051378
Jo-Ellen P, Anne H, John D (2005) Climate change and adaptation. International Institute for Sustainable Development, Winnipeg. https://iisd.org/pdf/2005/climate_adaptation.pdf. Accessed 20 Nov 2019
John LCL, Andrew LA, Emmanuel CO (2019) An assessment of the impact of urbanization on agricultural land use in Juba city, central Equatoria State, Republic of South Sudan. J Appl Agric Econ Pol Analysis 2(1):22–30. https://doi.org/10.12691/jaaepa-2-1-4
Joshi PK, Gulati A, Birthal PS, Tewari L (2003) Agriculture diversification in South Asia: patterns, determinants, and policy implications. MSSD discussion paper no. 57. IFPRI, Washington, DC
Joshi PK, Gulati A, Birthal PS, Tewari L (2004) Agriculture diversification in South Asia: patterns, determinants and policy implications. Econ Pol Wkly 39(24):2457–2467. https://doi.org/10.2307/4415148
Kantor LS, Lipton K, Manchester A, Oliveira V (1997) Estimating and addressing America's food losses. Food Rev 20:2–12
Kathiravale S, Muhd Yunus MN (2008) Waste to wealth. Asia Eur J 6(2):359–371. https://doi.org/10.1007/s10308-008-0179-x
Kavoossi M (1991) Agriculture policy and development in Iran. J Int Food Agribus Mark 3 (3):19–41. https://doi.org/10.1300/j047v03n03_02
Kearney J (2010) Food consumption trends and drivers. Philos Trans R Soc B 365 (1554):2793–2807. https://doi.org/10.1098/rstb.2010.0149
Khader BFY, Yigezu YA, Duwayri MA, Niane AA, Shideed K (2019) Where in the value chain are we losing the most food? The case of wheat in Jordan. Food Sec 11(5):1009–1027. https://doi.org/10.1007/s12571-019-00962-7
Khanam R, Debarati B, Nayak AK (2018) Crop diversification: an important way-out for doubling farmers' income. Indian Farming 68(01):31–32
Kiran M, Savneet S (2018) Changing consumption basket in rural and urban areas- a journey from conventional food to convenience food. Pac Bus Rev Int 10(10):29–39
Kumar D, Kalita P (2017) Reducing postharvest losses during storage of grain crops to strengthen food security in developing countries. Foods 6(1):8. https://doi.org/10.3390/foods6010008
Kummu M, de Moel H, Porkka M, Siebert S, Varis O, Ward PJ (2012) Lost food, wasted resources: global food supply chain losses and their impacts on freshwater, cropland and fertilizer use. Sci Total Environ 438(1):477–489. https://doi.org/10.1016/j.scitotenv.2012.08.092
Kunmin K (2018) How do we prevent a food crisis in the midst of climate change? https://www.adborg/sites/default/files/publication/467376/adbi-pb2018-2pdf. Accessed 12 Oct 2019
Kuyu CG, Tola YB, Abdi GG (2019) Study on post-harvest quantitative and qualitative losses of potato tubers from two different road access districts of Jimma zone, South West Ethiopia. Heliyon 5(8):e02272. https://doi.org/10.1016/j.heliyon.2019.e02272
Landry C, Smith TA, Turner D (2018) Food waste and food retail density. J Food Prod Mark 24 (5):632–653. https://doi.org/10.1080/10454446.2018.1472697

Larson NI, Story MT, Nelsonb MC (2009) Neighbourhood environments: disparities in access to healthy foods in the U.S. Am J Prev Med 36(1):74–81. https://doi.org/10.1016/j.amepre.2008.09.025

Laukkonen J, Blanco PK, Lenhart J, Keiner M, Cavric B, Kinuthia-Njenga C (2009) Combining climate change adaptation and mitigation measures at the local level. Habitat Int 33(3):287–292. https://doi.org/10.1016/j.habitatint.2008.10.003

Lin BB (2011) Resilience in agriculture through crop diversification: adaptive management for environmental change. Bioscience 61(3):183–193. https://doi.org/10.1525/bio.2011.61.3.4

Lisa K, Adel AK (2015) Measuring postharvest losses of fresh fruits and vegetables in developing countries. http://postharvest.org/PEF_White_Paper_15-02_PHFVmeasurement.pdf

Low JW, Arimond M, Osman N, Cunguara B, Zano F, Tschirley D (2007) A food-based approach introducing orange fleshed sweet potato increased vitamin A intake and serum retinol concentrations in young children in rural Mozambique. J Nutr 137(5):1320–1327. https://doi.org/10.1093/jn/137.5.1320

Lu L (2013) Policies and technologies for the mitigation of climate change. Chin J Agri Res Reg Plan 1(34):71–75

Macheka L, Spelt EJH, Bakker EJ, van der Vorst JGAJ, Luning PA (2018) Identification of determinants of postharvest losses in Zimbabwean tomato supply chains as basis for dedicated interventions. Food Control 87:135–144. https://doi.org/10.1016/j.foodcont.2017.12.017

Maggio G, Sitko N, Ignaciuk A (2018) Cropping system diversification in eastern and southern Africa: identifying policy options to enhance productivity and build resilience. FAO Agricultural Development Economics Working Paper 18-05. Rome, FAO. www.fao.org/3/CA1562EN/CA1562EN.pdf. Accessed 18 Nov 2019

Malthus T (1798) An essay on the principle of population J. Johnson, in St. Paul's Church-Yard, London

McDonald RI (2008) Global urbanization: can ecologists identify a sustainable way forward? Front Ecol Environ 6(2):99–104

McNicoll G (1984) Consequences of rapid population growth: an overview and assessment. Popul Dev Rev 10:177–240. https://doi.org/10.2307/1973081

Meena BP, Shirale AO, Biswas AK, Lakaria BL, Jha P, Gurav PP, Wanjari RH, Patra AK (2018) Diversified agriculture for higher productivity and profitability–a review. Agric Rev 39 (2):104–112. https://doi.org/10.18805/ag.R-1679

Menkir A, Palacios-Rojas N, Alamu O, Paes A, Cristina M, Thanda D, Maziya-Dixon B, Mengesha W, Ndhlela T, Guimarães O, Evaris P (2018) Vitamin A-biofortified maize: exploiting native genetic variation for nutrient enrichment. Science Brief: Biofortification No. 2 CIMMYT, IITA, EMBRAPA, HarvestPlus, and Crop Trust. Bonn, Germany https://doi.org/10.22004/ag.econ.283983. Accessed 2 Nov 2019

MoSTE (2015) Indigenous and local knowledge and practices for climate resilience in Nepal, mainstreaming climate change risk management in development. Ministry of Science, Technology and Environment (MoSTE), Kathmandu

NAAS (2019) Saving the harvest: reducing the food loss and waste. Policy brief no 5. FAO National Academy of Agricultural Sciences, New Delhi, p 10. http://naasindia.org/documents/Saving%20the%20Harvest.pdf

Nagarajan S, Bhavani RV, Swaminathan MS (2014) Operationalizing the concept of farming system for nutrition through the promotion of nutrition-sensitive agriculture. Curr Sci 107 (6):959–964

NCD (2018) The triple burden of global malnutrition. https://www.usdairy.com/news-articles/the-triple-burden-of-global-malnutrition. Accessed 14 Oct 2019

Ndirangu SN, Kanali C, Mutwiwa U, Kituu G, Kamwere M, Mungatu J (2017) Determinants of postharvest losses among high moisture content vegetables traders in Kenya. J Post Harvest Technol 05(2):37–46

Niewiara M (2016) Postharvest loss: global collaboration needed to solve a global problem. i-ACES 2(1):29–36

Nkiaka E, Lovett JC (2019) Strengthening the science-policy interface for climate adaptation: stakeholder perceptions in Cameroon. Reg Environ Chang 19:1047–1057. https://doi.org/10.1007/s10113-018-1441-4

Obi FO, Ugwuishiwu BO, Nwakaire JN (2016) Agricultural waste concept, generation, utilization and management. Niger J Technol 35(4):957–964

OECD (2018) Innovative approaches to building resilient coastal infrastructure. OECD Publishing, Paris. https://doi.org/10.1787/9c8a13a0-en. Accessed 20 Nov 2019

Oliver S, William VN, Logan K, Emma M, Alan C, Pamela R, Chris S, Roselyn W, Siva S, Guy B, David J, Dorian QF, Robin GA (2019) A domestication history of dynamic adaptation and genomic deterioration in sorghum. Nat Plants 5:369–379. https://doi.org/10.1101/336503

Olorunfemi TO, Olorunfemi OD, Oladele OI (2019) Borich needs model analysis of extension agents' competence on climate smart agricultural initiatives in south West Nigeria. J Agr Educ Ext 26:1. https://doi.org/10.1080/1389224X.2019.1693406

OXFAM International (2018) 5 natural disasters that beg for climate action. https://www.oxfam.org/en/campaigns/grow/5-natural-disasters-beg-climate-action. Accessed 18 Nov 2019

Pan J, Zheng Y (2010) Analytical framework and policy implications on adapting to climate change. China population. Resour Environ 20(10):1–5. https://doi.org/10.3969/j.issn.1002-2104.2010.10.001

Pandey B, Seto KC (2015) Urbanization and agricultural land loss in India: comparing satellite estimates with census data. J Environ Manage 148:53–66. https://doi.org/10.1016/j.jenvman.2014.05.014

Parfitt J, Barthel M, Macnaughton S (2010) Food waste within food supply chains: quantification and potential for change to 2050. Philos Trans R Soc B 2365(1554):3065–3081. https://doi.org/10.1098/rstb.2010.0126

Parikh JK, Parikh K (2002) Climate change: India's perceptions, positions, policies and possibilities. Climate Change and Development OECD, New Delhi

Paroda RS (2018) Reorienting Indian agriculture: challenges and opportunities. CABI, Boston

Pathak H, Aggarwal PK, Singh SD, Das S, Singh Y, Singh B, Kamra SK, Mishra B, Sasthri ASRAS, Aggarwal HP, Das DK, Gupta RK (2003) Trends of climatic potential and on-farm yields of rice and wheat in the Indo-Gangetic Plains. Field Crop Res 80(3):223–234. https://doi.org/10.1016/s0378-4290(02)00194-6

Pathak H, Saharawat YS, Gathala M, Ladha JK (2011) Impact of resource-conserving technologies on productivity and greenhouse gas emissions in the rice-wheat system. Greenhouse Gas Sci Technol 1:261–277. https://doi.org/10.1002/ghg.27

Pathak H, Aggarwal PK, Singh SD (2012) Climate change impact, adaptation and mitigation in agriculture: methodology for assessment and application. Indian Agricultural Research Institute, New Delhi. http://www.nicra.iari.res.in/Data/Climate%20Change%20Impact,%20Adaptation%20and%20Mitigation%20Print.pdf

Patle GT, Kumar M, Khanna M (2019) Climate-smart water technologies for sustainable agriculture: a review. J Water Clim Change. https://doi.org/10.2166/wcc.2019.257

Patra S, Sahoo S, Mishra P, Mahapatra SC (2018) Impacts of urbanization on land use/cover changes and its probable implications on local climate and groundwater level. J Urban Manag 7(2):70–84. https://doi.org/10.1016/j.jum.2018.04.006

Pearson D, Mirosa M, Andrews L, Keer G (2016) Reframing communications that encourage individuals to reduce food waste. Commun Res Prac 3(2):137–154. https://doi.org/10.1080/22041451.2016.1209274

Peeter E (2015) Was Eridu the first city in Sumerian mythology? https://www.researchgate.net/publication/298043082_Was_Eridu_The_First_City_in_Sumerian_Mythology. Accessed 14 Oct 2019

Pinstrup AP (2007) Agricultural research and policy for better health and nutrition in developing countries: a food systems approach. Agric Econ 37(S1):187–198. https://doi.org/10.1111/j.1574-0862.2007.00244.x

Popova Z, Kercheva M (2005) CERES model application for increasing preparedness to climate variability in agricultural planning-risk analyses. Phys Chem Earth 30(1–3):17–124. https://doi.org/10.1016/j.pce.2004.08.025

Quiroz JA, Valdes A (1995) Agricultural diversification and policy reform. Food Policy 20 (3):245–255. https://doi.org/10.1016/0306-9192(95)00016-8

Rahiel HA, Zenebe AK, Leake GW, Gebremedhin BW (2018) Assessment of production potential and post-harvest losses of fruits and vegetables in northern region of Ethiopia. Agric & Food Secur 7:29. https://doi.org/10.1186/s40066-018-0181-5

Rahman S (2009) Whether crop diversification is a desired strategy for agricultural growth in Bangladesh? Food Policy 34(4):340–349. https://doi.org/10.1016/j.foodpol.2009.02.004

Randela R (2003) The incidence of post-harvest problems among small farmers surveyed in three regions of the Limpopo province. Agrekon 42(2):163–180. https://doi.org/10.1080/03031853.2003.9523618

Randhir S (2017) Role of innovation platforms & participatory research in agricultural development in India. https://orgprints.org/32404/36/singh-2017-tipi-preconfrence.pdf. Accessed 15 Oct 2019

Ravi N, Nithya VG, Muttanna (2018) Agricultural extension through nutrition sensitive lens: Potential of fruits and vegetables as a source of nutrition. Res J Agric Sci 9(5):945–950

Reardon T, Barrett CB, Berdegue JA, Swinnen JFM (2009) Agrifood industry transformation and small farmers in developing countries. World Dev 37(11):1717–1727

Reddy VR, Pachepsky YA (2000) Predicting crop yields under climate change conditions from monthly GCM weather projections. Environ Model Software 15(1):79–86. https://doi.org/10.1016/s1364-8152(99)00011-0

RIA (2017) Robotics in agriculture: types and applications. https://www.robotics.org/blog-article.cfm/Robotics-in-Agriculture-Types-and Applications/74. Accessed 21 November 2019

Ritu J, Janakiram T (2016) Vertical gardening: a new concept of modern era. In: Patel NL, Chawla SL, Ahlawat TR (eds) Commercial horticulture edition, Ist edn. New India Publishing Agency, New Delhi

Roland E (2019) Chinampas: an urban farming model of the aztecs and a potential solution for modern megalopolis. Hort Technol 30:13–19. https://doi.org/10.21273/HORTTECH04310-19

Roos E, Bajzelj B, Smith P, Patel M, Little D, Garnett T (2017) Greedy or needy? Land use and climate impacts of food in 2050 under different livestock futures. Glob Environ Chang 47:1–12. https://doi.org/10.1016/j.gloenvcha.2017.09.001

Rosado JL, Hambidge KM, Miller LV, Garcia OP, Westcott J, Gonzalez K, Conde J, Hotz C, Pfeiffer W, Ortiz-Monasterio I, Krebs NF (2009) The quantity of zinc absorbed from wheat in adult women is enhanced by biofortification. J Nutr 139(10):1920–1925. https://doi.org/10.3945/jn.109.107755

Rostow WW (1960) The stages of economic growth: a non-communist manifesto. Cambridge University Press, Cambridge, pp 4–16

Ruel MT, Quisumbing AR, Balagamwala M (2018) Nutrition-sensitive agriculture: what have we learned so far? Glob Food Secur 17:128–153. https://doi.org/10.1016/j.gfs.2018.01.002

Sapkota TB, Jat ML, Aryal JP, Jat RK, Khatri-Chhetri A (2015) Climate change adaptation, greenhouse gas mitigation and economic profitability of conservation agriculture: some examples from cereal systems of indo-Gangetic Plains. J Integr Agric 14(8):1524–1533. https://doi.org/10.1016/s2095-3119(15)61093-0

Satterthwaite D, McGranahan G, Tacoli C (2010) Urbanization and its implications for food and farming. Philos Trans R Soc B 365(1554):2809–2820. https://doi.org/10.1098/rstb.2010.0136

Sazawal S, Dhingra U, Dhingra P, Dutta A, Deb S, Kumar J, Devi P, Prakash A (2018) Efficacy of high zinc biofortified wheat in improvement of micronutrient status, and prevention of morbidity among preschool children and women - a double masked, randomized, controlled trial. J Nutr 17:86. https://doi.org/10.1186/s12937-018-0391-5

Senapati MR, Bhagirathi B, Sruti RM (2013) Impact of climate change on Indian agriculture & its mitigating priorities. Am J Environ Prot 1(4):109–111. https://doi.org/10.12691/env-1-4-6

Seto KC, Reenberg A, Boone CG, Fragkiasd M, Haasee D, Langankef T, Marcotulliog P, Munroeh DK, Olahi B, Simonk D (2012) Urban land teleconnections and sustainability. Proc Natl Acad Sci U S A 109(20):7687–7692. https://doi.org/10.1073/pnas.1117622109

Shafiee-Jood M, Cai X (2016) Reducing food loss and waste to enhance food security and environmental sustainability. Environ Sci Technol 50(16):8432–8443. https://doi.org/10.1021/acs.est.6b01993

Shakeel AK, Sanjeev Kumar MZ, Hussain KN (2009) Climate change, climate variability and Indian agriculture: impacts vulnerability and adaptation strategies. In: Singh SN (ed) Climate change and crops, environmental science and engineering. Springer, Berlin Heidelberg. https://doi.org/10.1007/978-3-540-88246-62

Shee A, Mayanja S, Simba E, Stathers T, Bechoff A, Bennett B (2019) Determinants of postharvest losses along smallholder producers maize and sweet potato value chains: an ordered probit analysis. Food Sec 11(5):1101–1120. https://doi.org/10.1007/s12571-019-00949-4

Singh R, Velu G (2017) Zinc-biofortified wheat: harnessing genetic diversity for improved nutritional quality. Science brief: biofortification no. 1 CIMMYT, HarvestPlus, and the global crop diversity trust. Bonn. https://www.harvestplus.org/sites/default/files/publications/Science%20Brief-Biofortification-1_Zinc%20Wheat_May2017.pdf. Accessed 2 Nov 2019

Singh NP, Ranjit K, Singh RP (2006) Diversification of Indian agriculture: composition, determinants and trade implications. Agric Econ Res Rev 19:23–36. https://doi.org/10.22004/ag.econ.57775

Singla N (2017) Innovations in agricultural marketing in India: a case study of supermarket in Punjab. In: Mani G, Joshi P, Ashok M (eds) Financing agriculture value chains in India. India studies in business and economics. Springer, Singapore

Slaughter RA (2012) Welcome to the anthropocene. Futures 44(2):119–126. https://doi.org/10.1016/j.futures.2011.09.004

Smit W (2016) Urban governance and urban food systems in Africa: examining the linkage. Cities 58:80–86. https://doi.org/10.1016/j.cities.2016.05.001

Smith P, Martino D, Cai Z, Gwary D, Janzen H, Kumar P, McCarl B, Ogle S, O'Mara F, Rice C, Scholes B, Sirotenko O, Howden M, McAllister T, Pan G, Romanenkov V, Schneider U, Towprayoon S, Wattenbach M, Smith J (2008) Greenhouse gas mitigation in agriculture. Philos Trans R Soc B 363(1492):789–813. https://doi.org/10.1098/rstb.2007.2184

Solbrig OT, Solbrig DJ (1994) So shall you reap: farming and crops in human. Am J Alternative Agr 10(01):18. https://doi.org/10.1017/s0889189300006056

Soumya S, Mangal DT, Aarti S, Kumar RM, Sreedevi B (2018) Resource conservation techniques for climate change mitigation and adaptation. Indian Farming 68(04):03–06

Specht K, Siebert R, Hartmann I, Freisinger UB, Sawicka M, Werner A, Thomaier S, Henckel D, Walk H, Dierich A (2014) Urban agriculture of the future: an overview of sustainability aspects of food production in and on buildings. Agric Hum Values 31(1):33–51. https://doi.org/10.1007/s10460-013-9448-4

Srinivasa R, Prabhaka CM, Maheswari M, Srinivasa RM, Sharma KL, Srinivas K, Prasad JVNS, Rama RCA, Vanaja M, Ramana DBV, Gopinat KA, Subba RAVM, Rejani R, Bhaskar S, Sikka AK, Alagusundaram K (2016) National innovations in climate resilient agriculture (NICRA), research highlights 2015–16. Central Research Institute for Dryland Agriculture, Hyderabad

Srivastava P, Singh R, Tripathi S, Singh H, Raghubanshi AS, Mishra PK (2018) A new insight into the warming potential of organically amended agro-ecosystems. Org Agr 8:275–284. https://doi.org/10.1007/s13165-017-0188-4

STANFORD (1998) Robotics: a brief history https://cs.stanford.edu/people/eroberts/courses/soco/projects/199899/robotics/history.html. Accessed 20 Nov 2019

TAAS (2017) Policy dialogue on incentives and strategies for scaling-out innovations for smallholder farmers: proceedings and recommendations. Trust for Advancement of Agricultural Sciences (TAAS), New Delhi, p 17. https://www.taas.in/documents/pub-activity-34.pdf. Accessed 5 Nov 2019

TERI (2017) Dousing the farm fires: turning crop waste into wealth for Punjab, Haryana and Delhi NCR. https://www.teriin.org/infographics/dousing-farm-fires-turning-crop-waste-wealth-punjab-haryana-and-delhi-ncr. Accessed 3 Nov 2019

Tol RSJ (2005) Adaptation and mitigation: trade-offs in substance and methods. Environ Sci Policy 8(6):572–578. https://doi.org/10.1016/j.envsci.2005.06.011

Tubiello F (2012) Climate change adaptation and mitigation: challenges and opportunities in the food sector. Natural resources management and environment department, FAO, Rome. Prepared for the high-level conference on world food security: the challenges of climate change and bioenergy, Rome, 3–5 June 2008

Turcotte MM, Araki H, Karp DS, Poveda K, Whitehead SR (2017) The eco evolutionary impacts of domestication and agricultural practices on wild species. Philos Trans R Soc B 372:20160033. https://doi.org/10.1098/rstb.2016.0033

UN (2019a) World population prospects 2019. https://population.un.org/wpp2019/Download/Standard/Population/. Accessed 21 Nov 2019

UN (2019b) World population prospects: average annual rate of population changes by region, subregion and country, 1950–2100 (percentage). https://population.un.org/wpp/. Accessed 21 Nov 2019

UN (2019c) The impact of rapid technological change on sustainable development. https://unctad.org/meetings/en/SessionalDocuments/ecn162019d2_en.pdf. Accessed 1 Dec 2019

UN (2019d) The sustainable development goals report-2019. https://unstats.un.org/sdgs/report/2019/The-Sustainable-Development-Goals-Report-2019.pdf. Accessed 15 Nov 2019

UNFCCC (2006) Technologies for adaptation to climate change. https://unfccc.int/resource/docs/publications/tech_for_adaptation_06.pdf. Accessed 18 Nov 2019

UNFCCC (2009) Risk management approaches to address adverse effects of climate change—economic diversification. https://unfccc.int/cooperation_support/response_measures/items/5003txt.php. Accessed 16 Nov 2019

UNFCCC (2019) What do adaptation to climate change and climate resilience mean? https://unfccc.int/topics/adaptation-and-resilience/the-big-picture/what-do-adaptation-to-climate-change-and-climate-resilience-mean. Accessed 19 Nov 2019

UNICEF (2019) The faces of malnutrition. https://www.unicef.org/nutrition/index_faces-of-malnutrition.html. Accessed 14 Oct 2019

Van Enst WI, Driessen PPJ, Runhaar HAC (2014) Towards productive science-policy interfaces: a research agenda. J Environ Assess Policy Manag 16(01):1450007. https://doi.org/10.1142/s1464333214500070

Vijaya Bhaskar AV, Nithya DJ, Raju S, Bhavani RV (2017) Establishing integrated agriculture-nutrition programmes to diversify household food and diets in rural India. Food Sec 9 (5):981–999. https://doi.org/10.1007/s12571-017-0721-z

Vogel E, Meyer R (2018) Climate change, climate extremes, and global food production—adaptation in the agricultural sector. In: Zommers Z, Alverson K (eds) Resilience. Elsevier, Amsterdam, pp 31–49

Vorley B, Lançon F (2016) Urban food consumption, urbanisation and rural transformation: the trade dimensions. IED, London. http://pubs.iied.org/10767IIED. Accessed 16 Nov 2019

Walters S, Bouharroud R, Mimouni A, Wifaya A (2018) The deterioration of Morocco's vegetable crop genetic diversity: an analysis of the Souss-Massa region. Agri 8(4):49. https://doi.org/10.3390/agriculture8040049

Weinberger K, Lumpkin TA (2007) Diversification into horticulture and poverty reduction: a research agenda. World Dev 35(8):1464–1480. https://doi.org/10.1016/j.worlddev.2007.05.002

WHO (2018) The double burden of malnutrition. https://www.who.int/nutrition/double-burdenmalnutrition/infographic_print.pdf?ua=1. Accessed 14 Oct 2019

World Bank (2015) Nutrition and agriculture: Bridging the gap. https://blogs.worldbank.org/health/nutrition-and-agriculture-bridging-gap. Accessed 14 Oct 2019

World Bank (2017a) Un charted waters: the new economics of water scarcity and variability. Washington, DC. https://openknowledge.worldbank.org/bitstream/handle/10986/28096/9781464811791.pdf?sequence=21&isAllowed=y. Accessed 2 Nov 2019

World Bank (2017b) Agriculture 2.0: how the Internet of Things can revolutionize the farming sector. http://blogs.worldbank.org/digital-development/agriculture-20-how-internet-things-can-revolutionize-farming-sector. Accessed 15 Nov 2019

World Bank (2019a) Agriculture and food. https://www.worldbank.org/en/topic/agriculture/overview. Accessed 10 Oct 2019

World Bank (2019b) Climate-smart agriculture. https://www.worldbankorg/en/topic/climate-smart-agriculture. Accessed 24 Nov 2019

Wright KR (2016) Machu Picchu. In: Selin H (ed) Encyclopaedia of the history of science, technology, and medicine in non-western cultures. Springer, Dordrecht, pp 2581–2590. https://doi.org/10.1007/978-94-007-7747-7_9941

Xiaowei Y, Zhanqi W, Hua W (2015) Impact of urbanization and land-use change on surface climate in middle and lower reaches of the Yangtze river, 1988–2008. Adv Meteorol 2015:1–10. https://doi.org/10.1155/2015/395094

Yadava DK, Choudhury PR, Firoz H, Kumar D (2017) Biofortified varieties: sustainable way to alleviate malnutrition. Indian Council of Agricultural Research, New Delhi. https://icar.org.in/files/BiofortifiedEnglish_.pdf

Yahaya SM, Mardiyya AY (2019) Review of post-harvest losses of fruits and vegetables. Biomed J Sci Tech Res 13(4):10192–10200. https://doi.org/10.26717/BJSTR.2019.13.002448

Yan ZW, Wang J, Xia JJ, Feng JM (2016) Review of recent studies of the climatic effects of urbanization in China. Adv Clim Cha Res 7(3):154–168. https://doi.org/10.1016/j.accre.2016.09.003

Zougmore RB, Partey ST, Totin E, Ouedraogo M, Thornton P, Karbo N, Sogoba B, Dieye B, Campbell B (2019) Science-policy interfaces for sustainable climate-smart agriculture uptake: lessons learnt from national science-policy dialogue platforms in West Africa. Int J Agr Sustain 17(5):1–16. https://doi.org/10.1080/14735903.2019.1670934

Zuma M, Kolanisi U, Modi A (2018) The potential of integrating provitamin a-biofortified maize in smallholder farming systems to reduce malnourishment in South Africa. Int J Environ Res Public Health 15(4):805. https://doi.org/10.3390/ijerph15040805

Climate Change and Food Systems: Implications on Food Security

3

Ritambhara Singh and Vishita Khanna

Abstract

Agriculture is one of the prominent activities to contribute to greenhouse gases. Its contribution is about a third of the total emissions. It was in 2017 when agriculture was discussed in climate talks. It is an emerging concern that the changing climate will affect world's developing countries more compared to rich ones as they rely more on agriculture for livelihood. Such countries are more located in the belts, which are more susceptible to droughts, floods, heatwaves, etc., eventually impacting harvests. On the other hand, the change in consumption patterns all over the world calls for more energy-intensive production that again could cause adverse environmental impacts, if not addressed appropriately. That said, the climate and food systems affect each other, which merits a deeper analysis and suitable strategies to make sure that its implications on food and nutritional security are not detrimental. The changing climate is indeed deeply impacting the quantity and quality of food. Many regions across the globe may face acute food shortages due to fluctuations in yields. Also, the population does not seem to be static. Addressing food and nutrition security of the ever-increasing population amid a changing climate is a growing concern. There are more stomachs to feed with shrinking resources in hand. A nation is said to be secure foodwise when the citizens enjoy the access to quality food in a sufficient amount that satisfies hunger and help to maintain a healthy living. The FAO

R. Singh (✉)
Department of Agribusiness Economics and Policies, International Agribusiness Management Institute, Anand Agricultural University, Anand, Gujarat, India
e-mail: SinghR@aau.in

V. Khanna
Department of Human Resource Development and Personnel Management, International Agribusiness Management Institute, Anand Agricultural University, Anand, Gujarat, India
e-mail: vishitakhanna@aau.in

V. K. Hebsale Mallappa, M. Shirur (eds.), *Climate Change and Resilient Food Systems*, https://doi.org/10.1007/978-981-33-4538-6_3

measures food security via four parameters, viz. supply of food, accessibility of food, diversity of nutrients in the available food, and stability through time. Addressing the grave concerns, this chapter discusses in detail the effect of changing climatic conditions on food systems and vice versa. The chapter also focuses on the impact on food security and the challenges that lie ahead. It also suggests various actions to be taken by individuals, by nations and by the world as a whole to overcome these challenges.

Keywords

Climate change · Food security · Resilience · Sustainability

3.1 Introduction

In recent times, the biggest challenge before of the farmers all over the world is to produce food in uncertain climate conditions. Increasing incomes and population is among the drivers of food demand. This further pressurizes production systems to yield more. However, several studies suggest that higher temperature, uneven rainfall pattern, soil degradation, etc. will affect the yield of several crops negatively in certain regions, exceptionally low-income countries. Not just this, the result of elevated levels of carbon dioxide (CO_2) in the atmosphere is the reduction of zinc, iron and other essential nutrients in crops, making them less nutritious. Thus both food availability and quality are at stake due to changing climate.

On the other hand, as demand for food increases and the consumption patterns change, our food production systems have become more aggressive emitting more greenhouse gases in the atmosphere. We are living in an era wherein we have immense challenges in the production systems, while the changing climate makes things rather difficult. Many nations are still food-insecure, and their vulnerability gets exposed further as climate changes. The challenge, therefore, is to produce qualitative food in more and more quantity without causing damage to the climate. This paper discusses the systems approach to assess and elaborate on the effects of food systems on changing climate and vice versa. The chapter also discusses the critical issues in food security and possible ways to resolve these.

3.2 Changing Climate and Changing Food Systems

3.2.1 Climate Is Changing, and That Is for Real

Climate change is not new. As per the National Aeronautics and Space Administration, in the last 650, 000 years, there have been seven catastrophic cycles; the pace differed every time. The last of them, called an ice age, is about 11,700 years old and is characterized by the birth of human civilization. With it began the interaction of humans with the environment. As the species evolved, the activities too shaped up differently. From the wild, humans shifted to practicing agriculture. From agriculture, they started manufacturing and so on. In the evolving exercise of the human being through several activities were born the greenhouse gases (GHGs), viz. nitrous oxide, methane, carbon dioxide (CO_2), chlorofluorocarbons, etc. As time progressed, the high release of the gases has thickened the GHG layer so much, so the released gases remain trapped within the earth's atmosphere, disturbing the earth's thermostat. Studies suggest that about 45% of human activities release CO_2 that remains in the atmosphere. Carbon dioxide levels have increased markedly by deforestation, cement production, burning of fossil fuels, human respiration, agriculture, etc. Furthermore, gases from decomposed wastes produced in agricultural activities, ruminant digestion and livestock management are found to be more active than CO_2. However, their contribution is less in quantitative terms. These are nitrous oxide gases. Others are chlorofluorocarbons that originate from industrial activities/processes. Chlorofluorocarbons are also GHGs and are released from industrial equipment like a refrigerator and aerosol sprays. They have a dampening effect on the atmosphere as they deplete the ozone layer. These gases cause amplified warming and contribute significantly to the change in the climate. Increasing modernization and industrialization has led to a rise in CO_2 levels to 400 ppm from 280 ppm in just 150 years. It has been found that around 77% of all emissions are CO_2 emissions, 14% are methane, 8% are nitrous oxide and 1% is chlorofluorocarbon and others (World Resource Institute 2000). Methane may enter into the system through natural processes. However, 60% of methane is released by human activities like mining, livestock rearing, use of landfills and others. Additionally, fossil fuels are used for energy generation, and it remains a persistent problem as their combustion releases greenhouse gases. Maximum, around 70% GHGs are emitted by energy resources (electricity and heat, crude and natural gas, etc.) followed by agriculture and land use and forestry which emit around 14% and 6% of the GHGs, respectively (IPCC 2018).

While these activities are anthropogenic, volcanic eruptions and seismic activity, too, cause a substantial rise in the CO_2 levels. These occurrences are now becoming more frequent and high in magnitude. Climate change is thus natural and man-made, as well. All this has led to the rise in the average temperatures and unexplained rainfall variations. The American Meteorological Society reports that approximately 80% of the events which have been reported from 2015 to 2017 showed a significant anthropogenic influence on the event's occurrence. Furthermore, the temperature during 2016 was about 1 °C higher than the 1850–1900 levels making it the warmest year on record (World Meteorological Organization 2019). Most of the warming has occurred in the last 35 years. The block 2015–2019 was the warmest 5-year period when the temperature rose by 0.2 °C over 2011–2015. The WMO also reveals that the CO_2 concentration levels in the atmosphere continue to rise, ocean acidification continues and the Arctic ice was below 1981–2010 levels.

FAO (2015) coordinated the global forest resources assessment and found that forest land is reduced by almost 1% between 1990 and 2015. As stated by the UN biodiversity report, one million species are already on the verge of extinction. The pace at which the changes are occurring and the unpredictability since 1950 is a matter of concern globally. As the damage controlling exercise, the global leaders have decided to commit and control the warming under 2 °C above 1850–1900 levels. The Impact of change in climate is visible, and it may aggravate over a while. It is, however, true that different regions of the world witness it differently. According to the Paris Climate Change Agreement 2015, climate change can cost heavily to India: almost 2.8% of its GDP by 2050 (World Bank 2018). It can also increase the poverty levels, threaten food security and reduce the living standards of the Indian population. Other countries are facing similar threats from the changing climate.

Prabhakar (2010/2007) found that The World leaders have risen to this concern and commitments for emission reduction have been made under the Kyoto Protocol. While nations have been trying to reduce their emissions individually, there are certainly many conflicts between the status that the developed and the developing countries hold on account of reduction in carbon emissions. India, however, has been prosperous in its commitments to reduce GHGs emissions. Friedman et al. (2018) found that between 1850 and 2014, as compared to India, the USA and China were the most significant contributors to climate change as these two countries emit the maximum of GHGs.

3.2.2 Food Systems Are Changing Too

Humans tend to schedule their food according to lifestyle, culture, society, religion and occasions. For example, in India, food habits differ as per the region and rituals. There is a wide variation in the consumption pattern of households as we move across the length and breadth of this country. South Indians consume fish and rice, whereas North Indians prefer bread made from whole wheat flour along with meat, chicken or vegetables. Food is not only the source of energy, but it also brings in the pleasure.

Interestingly, the food can be broadly categorized into three types:

(a) Food that provides satiety like wheat, rice, etc.
(b) Food that provides nutrition like meat, fruits and vegetables
(c) Food that enhances taste like salt, sugar, spices and condiments

It has been found that the tastes and preferences of people are changing continuously with time. Additionally, globalization has made the world a smaller place. People have been travelling more than ever until the global movement was restricted due to the COVID-19 lockdown. In the process of travelling beyond the local boundaries, people get exposed to and accustomed to the food habits of each other, thereby throwing enormous opportunities for trade. Therefore, diets do modify over a period of time. A study, 'What the world eats', conducted by the National Geographic, reveals that the dietary patterns have changed all over the world, including India. The per capita calorie intake has increased, and the consumption

Table 3.1 Changing daily diets and meat consumption in gram (g) in selected countries vis-à-vis world between 1960 and 2011

	Meat consumption(g)			Daily diet (g)		
Countries	1961	2011	% Change	1961	2011	% Change
US	285	381	34	2339	2729	17
China	20	254	1170	872	2368	171
Hong Kong	203	695	242	1160	2143	84
India	17	29	71	878	1317	38
South Korea	45	339	592	955	2167	127
Vietnam	71	260	266	876	1418	62
Australia	343	423	23	2340	2551	9
World	93	173	86	1357	1878	38

Source: What the world eats: National Geographic (2019)

Table 3.2 Global consumption of meat and meat products

	Livestock (kcal/person/day)			Beef and mutton (kcal/person/day)		
Region	2006	2050	% Change	2006	2050	% Change
European Union	864	925	7	80	75	−6
Canada and the USA	907	887	−2	117	95	−19
China	561	820	46	41	89	117
Brazil	606	803	33	151	173	15
Former Soviet Union	601	768	28	118	156	32
Other OECD	529	674	27	64	84	31
Latin America[a]	475	628	32	59	86	46
Middle East and North Africa	303	416	37	59	86	46
Asia	233	400	72	24	43	79
India	184	357	94	8	19	138
Sub-Saharan Africa	144	185	29	41	51	24
World	413	506	23	50	65	30

Source: What the world eats: National Geographic (2019)
[a]Latin America excludes Brazil

basket is driven mainly by more of meat, dairy and eggs (Table 3.1). Meat consumption has increased sharply all over the world.

Global projections for 2050 for livestock and beef and mutton (Table 3.2) reveal that except for Canada and the USA, the consumption of livestock products will further increase. It is estimated that the global consumption of these products will rise by 23%, while, for Asia, the rise is projected at 72%. The maximum growth for livestock products (94%) is estimated for India, wherein the milk and milk products continue to gain prime importance in vegetarian diets.

Further, the calorie intake from beef and mutton is estimated to decline in the European Union (−6%), Canada and the USA (−19%). The maximum rise in the case of beef and mutton, like livestock, again is estimated for India (138%) and

China (117%). This should preferably be looked at in absolute numbers. In 2006, the per capita calorie gain from beef and mutton in India was 8 kcal/day. This is expected to increase by 19 kcal/person/day, which comparatively is far less than several countries on the planet. In nations like Canada and the USA, Brazil, Former Soviet Union, European Union and others, where non-vegetarian diets are more popular, the consumption of livestock, beef and mutton is several times higher than India (Devi et al. 2014).

Such dietary modifications conglomerate to change the demand of the food, which drives the supply side and affects the complete food system. Moreover, a different trend is visible in the consumption of processed products. Countries like India and China, where people prefer to consume fresh fruits and vegetables, are investing hugely to promote processing. This is an effort to make food sustainable and reduce food wastages, which are more than 40% globally. Food processing again is an energy-intensive process. Developed nations, on the other hand, now urge to consume less processed products. There the focus is shifting to consume fresh products now. Changing consumption patterns are driving change in food systems. The systems approach is a holistic approach that takes into account all aspects from the input supply to the consumption of a food product. There is a new challenge in the modern world, and that is about mitigating climate change amid changing food systems. Many factors are responsible for such changes in food systems, and they may be attributed to:

(a) *Rise in income levels*: An increase in real income leads to a rise in consumption levels or changes in consumption patterns. With an increase in the income levels, people tend to spend more on luxurious food items like ready to eat foods, viz. chips, biscuits, juices, etc., and ready to cook products like noodles, pasta, etc., which are pre-processed. The target segment for such types of products is high-end consumers who have high spending capabilities. This kind of shift is more observable in developing countries.
(b) *Rise in nuclear families*: Due to the rise in the number of nuclear families and the increase in the number of working women, preferences increase for frozen meals and food products that are minimally processed and take minimal time to be cooked. Such type of food products are also called as convenience foods.
(c) *Changing lifestyles or increasing health consciousness*: Go vegan and go organic is the new mantra of calorie- and health-conscious people. Such people tend to incorporate more fruits and vegetables into their diets and exclude carbohydrate-rich foods and fried foods from their meals. If they follow an exercise routine, their diet also includes protein shakes, salads, green tea and other nutrient fortified products. This is a type of cognitive factor that influences food preference.
(d) *Globalization*: As already discussed, the globe has turned down to a small home and is easily connected now, more than ever. People travel to different countries and tend to consume different products, creating demand for such products in their own country. Furthermore, free trade agreements make trade simpler. Globalization has opened doors to the creation of new markets for

multinationals in fried foods, burgers, pizzas and sugary drinks. Millennials crave such foods visible on the supermarket shelves and end up buying and consuming them. Such foods have high palatability, which makes these challenging to resist, and the consumer purchases it repeatedly to satisfy the taste buds.

(e) *Urbanization*: The social circle in which people stay affects their food habits to a large extent. People living in villages have comparatively simpler food habits, and they eat home-cooked meals daily, whereas those who live in metropolitan cities tend to visit restaurants more often. The elite classes of society have more social thrust to celebrate with family and friends, which increases their consumption of cakes, pastries, chocolates, snacks and other exotic processed products. So, when people get relocated from villages to cities in search of jobs, their eating habits change.

3.3 Changing Food Systems Cause Change in Climate

A food system includes all activities right from the input supply, manufacturing/production, harvesting, processing, packaging, marketing/distribution and consumption, i.e., from farm to fork. The food system includes essential elements like how the food is grown in the field and through how many interactions it travels to the consumers' plate. The systems approach also consists of the minor components like infrastructure used and the processes involved in making food reach the consumers' plate. Furthermore, it also encompasses the wastes in the food supply chain and dietary patterns of the population (Mbow et al. 2019). While several researches have been conducted on knowing the effects of changing climate on agricultural production/yields of crops, there are not many studies that have tried to explore the impact of changing food systems on the climate. To understand this, one must question—what drives the food systems? The main drivers of the food system are socio-economic (Kearney 2010), natural resources and environmental (Ericksen 2008) in nature. Besides this, the drivers of food systems could also be research, development and innovation in food systems (Flores et al. 2010). The food systems, as driven by these factors, are linked with the climate system as various greenhouse gases are emitted at different stages of food systems through activities like production, processing, marketing and distribution. Hence, the focus should be made on the greener technologies, which will improvise the food system without deteriorating our climate. The food systems are also linked with the ecosystems as land and water are utilized at every stage to produce food. The increased greenhouse gas emissions have also been responsible for the loss of biodiversity. The quest for food for the burgeoning population further puts thrust on ecological systems and their utilization. Thus, food systems challenge the ecosystem and vice versa. The food that is produced by utilizing ecological resources produces waste too, which is disposed back into the land and water bodies and emits greenhouse gases, also polluting water bodies, resulting in soil degradation. The food systems are also linked with socio-economic systems. Socio-economic systems affecting the food system for different consumer behaviour is responsible for the demand for different foods and their production and marketing patterns. Several socio-economic constraints also challenge food systems. All components of the food system need to function efficiently, so the climate, food system, the ecosystem and socio-economic systems remain balanced (Mbow et al. 2019). The food systems must adapt to changing climate conditions, and appropriate mitigation strategies should be carved to reduce GHG emissions from the food systems. Timely adaptation and mitigation measures help in a healthy establishment of subsystems.

3.3.1 Emissions During Various Stages of Food Systems

Food systems are estimated to contribute around 19–29% to the global GHG emissions. Of this, around 80–86% is estimated from agricultural production activities (Vermeulen et al. 2012). As discussed earlier, the food system is used for umbrella activities that are core components of backward and forward linkages in the food supply chain. At all stages, the food systems require an enormous amount of energy throughout. Transportation is an activity that connects the entire food supply chain. Besides transportation, several other activities like processing and packaging, which again are energy-intensive, utilize electricity and heat. Fertilizer manufacturing is also an energy-intensive process. Food emits GHGs during its production, harvesting, processing, consumption, transportation and disposal. In the case of poultry, livestock and fishery, almost 90 million tonnes of CO_2 per year during the rearing activity (Steinfeld et al. 2006). Vermeulen et al. (2012) suggested that inputs are transported to the farms and there they get converted into outputs. The transportation of inputs from supply centres to the farms and other land-use changes release about one-fourth of the global GHG emissions. The farm output then heads for slaughtering, manufacturing or processing units, and from there, it either heads for restaurants/schools or in supermarkets/shops/other markets, etc. Consumers purchase the final products from either of the channels, and then there are some waste realizations too in the process of consumption. These post-production components of the food system yield about 5–10% of the GHG emissions. This sums up to the fact that emissions are at all stages of the food system. It is thus highly relevant to understand and work on the entire food systems instead of focusing on a single component. The systems approach is also required to frame appropriate strategies to mitigate climate change.

3.3.2 Dietary Patterns and Greenhouse Gas Emissions

The consumption activities contribute positively to the food systems. As discussed earlier, consumption patterns are changing all over the world, with a huge jump in the consumption of meat. India's consumption of meat increased by 71% in a period of 50 years between 1961 and 2011 (Table 3.1). Globally, this rise is to the tune of

Table 3.3 Contribution of food to annual greenhouse gas emissions in kg

	Frequency		
Food items	Once a week to twice	Three to five times a week	Once a day
Apple	2	5	12
Banana	5	14	25
Avocado	15	41	72
Berries	9	25	44
Oranges	2	6	11
Beans	7	20	36
Potato	3	9	16
Tomato	13	34	60
Bread	4	12	21
Wheat	2.5	7	10
Rice	26	69	121
Milk	10	35	68
Nuts	1.1	2	5
Tea	3	8	15
Coffee	33	89	155
Beef	604	1611	2820
Chicken	106	284	497
Eggs	43	115	202
Fish	146	390	683
Lamb	339	904	1582
Pork	140	375	656
Prawns	269	718	1256

Calculated from BBC Food Calculator (2019)

86%. Studies suggest (Pathak et al. 2010) that non-vegetarian diets have more emissions than vegetarian diets.

The contribution of food to annual greenhouse gas emissions is given in Table 3.3. These emissions were calculated using BBC's food calculator and were calculated for three frequency levels: (1) when the food is consumed one to two times a week, (2) when the food is consumed three to five times a week and lastly (3) when the food is consumed once a day. The standard portion was assumed the same for the population across the globe. The standard portion size was taken from the British United Provident Association and British Dietetic Association. Furthermore, the information regarding the food distance (the distance the food travels) was also included while calculating these emissions. The distance data was taken from the European Environment Agency. The agency estimates that driving a conventional diesel vehicle produces 220 g of CO_2 equivalent per kilometre over its entire life cycle, including emissions from vehicle production, fuel production and exhaust emissions per km. The average distance was based on the survey of 40,000 farms and 16,000 processors to get global average estimates.

Table 3.4 Contribution of processed food to greenhouse gas emissions in kg

	Frequency		
Food items	Once a week to twice	Three to five times a week	Once a day
Potato chips	7.5	22.5	45
Whole wheat flour	2.4	10	19
Refined wheat flour	3	16	24
Oil	21	70	153
Frozen food	88	330	665
Soft drink	24	80	125
Juices	30	100	200
Cheese	75	201	352
Tofu	12	33	58
Beer	52	139	243
Wine	24	65	114
Packaged milk	49	131	229

Calculated from Carbon Emission Calculator (2015). The average distances were taken between the significant producer and consumption centres. Consumer waste percentages were also calculated

It was found that the GHG emissions from vegetarian food items were far lower than the non-vegetarian diets. Beef, lamb and prawns emit the maximum of GHGs, while fish, pork, chicken and egg follow. Similarly, the GHGs were also calculated for processed food products (Table 3.4). In the case of processed foods, the distance data was based on the average distance the food travels from the manufacturing centres to the main consumption markets. The wastage was also considered during the processing of food items. It was found that processed foods have a high level of emissions as compared to raw products. For example, between raw milk and packaged milk, raw milk consumption if consumed every day emits around 68 kg/year of GHGs, whereas as the same amount of packaged milk (UHT) emits up to 229 kg/year due to the processes involving heat treatments, processing, packaging, transportation, etc. Of all the commonly consumed processed food items, the frozen foods emit the maximum GHGs, slightly more than that emitted by pork consumption annually.

Studies have revealed that homemade sandwiches produce only half the carbon footprints than packaged ready to eat sandwiches. This could be attributed to lower food wastages at home against commercial production. Various reports suggest around 20% of the food is wasted in industrial production processes.

3.3.3 Food Wastage and CO_2 Emissions

Global estimates suggest that close to 1.3 billion food is wasted/lost every year. The global food loss/wastage generates approximately 4.4 gigatonnes of CO_2, which is almost equal to 87% of the emissions through transportation. The entire supply chain of cereals generates about 24% as wastages, which contribute 37% to the carbon

Table 3.5 Major commodities' contribution to carbon footprint and food wastage

Commodities	Carbon footprint (per cent)	Wastage (per cent)
Cereals	37	24
Vegetables	23	25
Meat	20	4
Milk and eggs	7	7
Fruits	6	16
Roots and tubers	5	19
Fish and sea foods	5	3
Oil crop and pulses	2	4

Source: FAO (2011)

Table 3.6 Contribution of stages of a supply chain in carbon footprint and food wastage

Stage	Carbon footprint (per cent)	Wastage (per cent)
Agricultural production	16	31
Postharvest handling and storage	16	22
Processing	13	10
Distribution	14	12
Consumption	36	20

Source: FAO (2011)

footprint, the highest among all the commodities (Table 3.5). The wastages follow it in the supply chain of vegetables and meat. The wastages of these two commodity groups contribute around 23 and 20% to the carbon footprint, respectively. The least wastages are reported for the oil crops and pulses, for which the contribution to the carbon footprint is 2%. Throughout the supply chain, each added activity of processing, transportation, storage, distribution, etc. emits carbon, which is added to the total carbon footprint. During the production activities, the wastages are highest, around 31%, and these wastages contribute 16% to the carbon footprint. The wastages from post-production activity or postharvest handling and storage are next higher (22%) and contribute again around 16% to the carbon footprint. The processing and distribution add relatively lesser to the carbon footprint and low wastages, as against the consumption activity (Table 3.6). During consumption, 20% of the food is wasted, which adds maximum to the carbon footprint (36%). This suggests that if 1 kg of a commodity gets spoiled on the farm, it has less carbon footprint than 1 kg of processed product wasted by the consumers, as the amount of GHGs emitted increases cumulatively after each step of processing. Also, the average per capita carbon footprint due to food wastage in high-income countries like those in North American continent is 860 kg CO_2 per person per year which exceeds twice the amount in South and South-east Asian countries where it was 350 kg carbon dioxide per person per year (FAO 2011).

3.4 Changing Climate Has Also Affected Our Food Systems

The earlier section highlighted the activities in the food systems and their impact on climate. Interestingly, the cycle is vicious. The population demands food. The consumption patterns change, and new items get added to the food basket, trade and transportation across the borders makes sure that the countries trade to fulfil demand. So, the emissions do travel as well. The activities in the food system have added to the greenhouse gases; their accumulation over time has resulted in global warming, which further causes variations in the temperature and rainfall. Prolonged wet or dry spells have been reported in several regions of the world. One very recent example has been the Australian bushfires that the climate experts predicted long ago. The change in climate might lead. The frequency and force of bushfires have increased due to global warming, and around 100,000 square kilometres of area in Australia was burnt in the 2019 year-end bushfires. About half of the animal population lost its life, and biodiversity was disturbed. Experts believe that a few plant and animal species may have gone extinct in the bushfires. The bushfires have further resulted in warming of the continent, as the emissions from bushfires alone in 2019 were about half of the total emissions of the country in 2018. Increased incidences of the flood are reported in India now than ever, causing massive damage to human, plant and animal life and also destroying the infrastructure and GDP of the country. The interrupted climatic variables make their impact on food systems by impacting yields, causing distress, increasing disease and pest incidences, etc. Hence, there lies a vicious cycle. Food systems are chasing climate and the climate is tracking food systems. The climate affects crop productivity in two ways:

1. Direct—changing temperature, carbon dioxide and oxygen imbalance, precipitation, etc.
2. Indirect—due to changes in soil, mutation among insects, pests, diseases or weeds

The majority of the effect of changing climate has been researched and recorded for production activities or agriculture/farming. Several studies have suggested a negative impact of rising temperature and irregular rainfall on the yield of crops. The effect, however, is not found even for all commodities/crops. Some crops are also expected to make some gains from the changing climate. In totality, the impact is negative. The projected reduction in the world food supplies may build up pressure on available resources, especially in low-income countries, and thus may lead to migration in huge numbers, which in turn may lead to extreme pressure on the resources making life difficult for everyone and creating socio-economic unrest. An increase in temperature causes fewer rains with a heavy flow rate, which in turn reduces soil fertility. As a result of low fertility, the nutrition level of soil goes down, causing fluctuations in yield/output. Increased temperatures also reduce the activity level of the human population, affecting their way of life and earnings. In other words, higher temperatures have ill effects on human health. It summarizes that global warming leads to a decline in the activity level of both humans and soil, eventually leading to declining output levels, rising prices, reduced purchasing powers, increasing poverty, hunger and deteriorating health.

Climate change, therefore, is going to challenge our food systems and how we produce our food. Consequently, the food systems need to evolve and adapt to changing climate, at every level of the supply chain. However, as mentioned earlier, most of the impact is expected to be on the low-income countries, which do not have ample resources and measures to mitigate the effect of it. The developed nations, on the other hand, possess the strength to alleviate the impacts of changing climate.

The IPCC projects that by 2070, the average temperature for India will increase between 0.4 °C to 2 °C in Kharif season and 1.1 °C to 4.5 °C in Rabi season. The maximum impact is likely on maize. The yield of maize crop may reduce by 7–12% in North India (Chatterjee 1998).

Ray et al. (2019) conducted subnational analysis and projected crop yields under changed climate conditions and found that the crop yield has already been affected. The authors regressed the yield of ten globally consumed crops—wheat, rice, maize, soybean, oil palm, sorghum, rapeseed, sugarcane, barley and cassava, on weather variables. They found negative yields ranging from 13.4% in oil palm to 3.5% in soybean. Declining yields were observed in the European continent, Southern Africa and Australia, while the positive effect on yield was observed on selected crops in Latin America. Mixed trend was observed in Asia and North and Central America. The negative impact on the yield of the selected ten crops was also estimated to bring a reduction in their consumable calories (−1% average reduction) across the globe. The findings for the world's two largest populated countries were found contrasting. For China, the mean change in climatic variables benefitted crop yield and increased calorie intake for the ten selected crops. There were some provinces wherein the yields of rice and wheat registered a decline.

On the other hand, for India, some states (especially the core states of Green Revolution, viz. those in north of the country) showed a settled pattern of declining yields of all major crops. These findings were consistent with the results of Birthal et al. (2014) and Singh et al. (2017). Birthal et al. (2014) studied the effect of changing climate on the yield of nine major crops in India, including cereals, pulses and oilseeds. They found a significant negative impact of the rise in maximum temperature on the yield of all crops under study. The marginal effect analysis by Birthal et al. (2014) revealed that 1 °C increase in maximum temperature leads to decline in yield of rice and pigeon pea by 12%, sorghum by 11%, barley by 10%, groundnut and wheat by 9%, maize by 8%, rapeseed mustard by 7% and chickpea by 4%. A 1 °C rise in minimum temperature was found to have a significant positive impact on all crops except pigeon pea where the effect was non-significant and in chickpea where it was negative. The net effect of temperature was found negligible for maize and rapeseed/mustard and positive for wheat and neutral for chickpea. Singh et al. (2017) studied climate variables and their effect on the yield of cotton, wheat, rice, maize, groundnut and pearl millet in the Gujarat state of India for 32 years between 1980 and 2011. The study found that rice was the worst-hit crop due to rising temperature and will continue to register reduced yields under increasing temperature scenarios. For the state, with every 1 °C rise in maximum temperature, wheat yields were found decreasing significantly by 10%, in rice by 13% and in maize and groundnut by 9%; a rise of one degree in minimum temperature was also found negatively affecting rice, maize and groundnut yield. Rainfall was found to have a positive impact on the yield of crops in the state more so because the state is relatively dry as compared to the northern food-producing states of India. However, excessive rainfall was found posing threats to the yield of all crops. Pearl millet was the only crop that was found surviving the increase in temperature. The yield of pearl millet was found to increase by 5% with a marginal rise in both minimum and maximum temperatures during the study period. Mishra et al. (2017) studied the effect on pigeon pea in the state of Gujarat and found that 1 °C rise in the minimum temperature will reduce the yield by 4%. However, a similar surge in maximum temperature will lead to yield decline by 20%. Rainfall was found to have a significant positive impact on the yield of pigeon pea, but it was very low as against the temperature effects on yield. The study suggested that hardy varieties of the crop that can sustain high temperature should be introduced. Another study at the Indian Agricultural Research Institute found that a 2 °C rise in temperature reduces grain yield of wheat. The reduction in yield varies with the productivity of the region. The regions with higher yields tend to be less affected by the temperature change, whereas the regions with low productivity were found more affected (DEFRA 2016). Also, it was found that rain-fed areas were more sensitive to temperature change as compared to irrigated regions. In the case of rice, a temperature rise by 2 °C to 4 °C was found to reduce the yield, but the increase in radiations nullified the effect as carbon dioxide increases the yield. Thus, Indian agriculture is highly susceptible to changing climate conditions. The state of Rajasthan has been found most vulnerable, followed by Karnataka, Maharashtra and Gujarat. A few regions in Madhya Pradesh and Uttar Pradesh were also found vulnerable (Rao et al. 2013).

Niles et al. (2017) pointed out that the increase in GHGs has likely resulted in the reduction of yield of rice, wheat, soybean and maize, especially in India and China. A likely reduction in yield of wheat by 0.45 ton per hectare was found for India (IPCC 2007) if the minimum temperature rises by 0.5 °C. IPCC (2007) also reported that rice productivity would be severely affected by acute water shortages and high temperature.

As far as the productivity of vegetables is considered, the potato's productivity is expected to decline by 18.68% by 2020 (Koundinya et al. 2014). The nutritional requirements of our body can be fulfilled to a reasonable extent, with the inclusion of vegetables in our daily diet. However, these crops are more susceptible to change in the climate, which may lead to many enzymatic and physiological changes along with the occurrences of various pests and diseases (Koundinya et al. 2014). Not only yields but climate change may also alter the availability of inputs, especially water for crops. Alexandratos and Bruinsma (2012) stipulated that due to climate change, it may not be possible to harness the irrigation potential in many regions as the availability of water resources in abundant areas may not be so in the future. This is possible as precipitation and evapotranspiration may get altered due to climate change. The author duo projected the increase in irrigation potential basis of these limitations and suggested that one-third of the expansion will be in India and China.

India alone has around half of the irrigated area among developing nations. Since the beginning of the industrial era, the ocean waters have acidified more by almost 26% (IPCC 2014a; Jewett and Romanou 2017). Barange et al. (2018), in one of the technical papers, discussed the impacts of acidification of oceans as a result of anthropogenic activities and its impact on fisheries and aquaculture. Since the beginning of the industrial era, the ocean waters have acidified more, by almost 26% (IPCC 2014a; Jewett and Romanou 2017).

Ocean warming is also not even; for example, the Arctic is warming and acidifying more than the Atlantic because the cold waters have high potential to absorb CO_2. This could lead to shifts in fish distribution and migration behaviour and may affect the livelihoods of millions of small and marginal fishers and fish workers, reducing their level of income and making them more vulnerable to these changes. Furthermore, it may also hamper their consumption levels. Asia has a whopping 66% share in the global inland fishery production. China leads in Asia, contributing 20% to the production, followed by India, which contributes 13%. Low oxygen concentration as a result of changing climatic conditions may hamper the production of tunas in the Arabian Sea and the Bay of Bengal (Mislan et al. 2017). A sharp decline for hilsa shed and Bombay duck, two most important commercial species, is also expected due to such conditions (Fernandes et al. 2016).

Significant fluctuations were observed in the milk yield of cows (Das 2017) due to a change in climate variables. Das found that the average daily, weekly, monthly and daily total milk yield reduces by 0.886, 1.868, 2.471 and 4.375 kg, respectively, with a unit high temperature humidity index (THI). High temperature, along with high humidity levels, impacts the reproductive ability of cattle and buffaloes. Also, conception rates decline with higher THI (Sinha et al. 2018).

Climate change affects both the quantity and quality of food (Cramer et al. 2014; Zhu et al. 2018). Fluctuations in rainfall, groundwater contamination, higher incidences of pests and insects, more residues of contaminants and metals, higher pesticide residues, hazardous wastes, etc. hamper the food quality and further increase the threats on food and nutritional security and may have implications on international trade and incomes generated (Vermeulen et al. 2012). The FAO (2008) noted that increasing temperatures would strain the electricity grids as demand for air conditioning and refrigeration will increase. Storage costs will also double as the refrigeration costs rise. There will be a high risk for perishable foods. The countries with inadequate infrastructural facilities will face bigger risks to control the food distribution systems during floods or other calamities (Ingram 2011). The transport infrastructure is already a problem in low-income countries, so the situation will deteriorate further if there are events like floods, drought, etc.

3.5 How Food Security Is Affected Under Changing Climate Scenario

The second goal under the Sustainability Development Goals focuses on food security and promoting sustainable agriculture. Changing climate has already thrown several challenges, which have made it harder to accomplish this goal. Food security is one of the principal components to define sustainable development. A nation is food-secure if the population residing there has access to the food, which is not only sufficient but safe and accessible too. It is essential to sustain healthy living. Thus the nation which is secure foodwise is considered a healthy society. The FAO (2006) recognizes four major components of food security. These are:

(a) Food supply and production
(b) The accessibility of food
(c) Stability through time—successive agriculture
(d) Diversity of nutrients in the available food, i.e. food utilization

It is predicted that the per capita availability of food will reduce under changing climatic conditions (Funk and Brown 2009). FAO estimates reveal that food security is negatively affected due to the recent wrath of events that support climate change. The seasonal rains are delayed in Southern Africa, and as a result of prolonged dry periods, regional cereal output is estimated to decline by 8% below the 5-year average. As a result of it, 12.5 million people are expected to suffer severe food insecurity/shortages up to March 2020, which again is about 10% more than the previous year. Ethiopia, Somalia, Kenya and Uganda are facing extreme food shortages due to poor distribution of rainfall. This has left around 12.3 million people food-insecure in this region. Somalia was affected by flooding between October and November 2019. Moving to Asia, Afghanistan witnessed the worst floods of the decade in March 2019. FAO (2018) estimates suggest that as a result of this incidence, 13.5 million people are food-insecure in Afghanistan. Out of its 34 provinces, 22 are still recovering from severe drought conditions faced in 2018.

Agriculture accounts for around 14% share in India's GDP and engages approximately 60% of the country's workforce. Agricultural exports contribute between 14 and 18% of the total exports from India. This shows the significance of this sector in the economy. More than 85% of the farms are marginal (<1 ha) to small (between 1 and 2 ha) in size. About 62% of the cropped area is rain-dependent. It shows the susceptibility of Indian agriculture to the changing climate conditions. With the changing climate, rainfall has turned out to be more unpredictable in recent years. There have been increased incidences of heatwaves, drought, floods and dry spells that may affect production systems, as discussed previously, thereby weakening the foundation of the very first pillar of food security, i.e. food availability. At a temperature rise by 2 °C, there are more substantial risks of reduced yield both globally and regionally, especially in those which lie at the low latitudinal levels (Rosenzweig et al. 2013; Porter et al. 2014; Rosenzweig and Hillel 2015). Lower yields will eventually affect the overall availability of food to the population, thereby imbalancing the production vs. demand and increasing the prices of food products. IPCC (2014b) projects an increase in global food prices by 3–84% by 2050, due to changing temperature and rainfall conditions. Disturbed production cycles may increase the dependency on imports. Lower production may also result in increased unemployment and lower purchasing power of people causing variations in their consumption levels and pushing them more towards the line of undernourishment and weakening the second and third important pillars of food security, which are food accessibility and stability. Climate change may, therefore, become a barrier in the socio-economic development of the nation, creating rural distress, widening rural-urban income and consumption divide and pushing countries towards food insecurity. The fourth pillar of food security is all about the utilization of food in a way that all nutritional requirements, along with drinking water and sanitation services, for people of all ages are met. Clean drinking water availability is considered one of the critical parameters to measure food utilization, a component of food security. Cape Town in South Africa ran out of water in 2018. India is facing an acute water shortage. The per capita availability of water is estimated to decline to 1465 cubic metres by 2025 and 1235 cubic metres by 2050. If it declines further to around 1000–1100 cubic metres, then India could be declared as a water-stressed country. Chennai, one of the biggest cities in India, ran dry due to prolonged heatwaves and severe drought in 2019. Elsewhere, Eastern Australia has been facing water crises since September 2019. Many towns and cities like Queensland are approaching day zero. Due to a visible water crisis, farmers are deciding not to plant anything and shut down the business in the short run. If this continues, one can think about the harsh reality of climate crises and food problems. Around 2.2 billion people globally do not get clean drinking water, as per the World Health Organization (WHO 2018). This reflects in terms of goal six of the Sustainability Development Goals and has long-lasting repercussions on food and nutritional security of the country, for water is majorly used in growing crops, manufacturing, drinking, cleaning and several activities throughout the day.

Usually, it is believed that at 2 °C rise in global temperature, a threat to food and water availability and sustainability is severe as compared to 1.5 °C (Cheung et al. 2016; Betts et al. 2018). Regions like African Sahel, the Mediterranean, Central Europe, the Amazon and Western and Southern Africa were found more susceptible (Sultan and Gaetani 2016; Lehner et al. 2017; Betts et al. 2018; Byers et al. 2018; Rosenzweig et al. 2013). Addressing the nutritional concerns, accumulation of CO_2 leaches out the nutrients from the crops. It reduces the availability of minerals like zinc, iron and magnesium and also has adverse effects on some other vitamins and proteins in several pulses and grains. This has enormous implications not only on humans but also on animal species that depend on plants for their food. Researches have shown that cereal crops have shown a reduction in the levels of zinc, iron and proteins by 3–15% globally. This may not seem an alarming situation for the developed nations, but for the underdeveloped countries and developing countries, nutrient deficiency is a more significant challenge. The deficiency of zinc causes a reduction in immunity and lead to increased incidences of malaria, diarrhoea and pneumonia. Children below 5 years of age are more susceptible to such diseases.

According to WHO (2018), zinc deficiency causes approximately 1,76,000 diarrhoea deaths, 4,06,000 pneumonia deaths and 2,07,000 malaria deaths, worldwide, every year. Around 30–60% of women in developing countries are anaemic, and this situation will likely worsen with the non-availability of food under changing climatic conditions. The effects of protein deficiency are even more severe. It is said to affect the IQ level and physical as well as mental growth in children. Pelletier et al. (1993) have documented that protein deficiency increases the mortality rate in children due to low immunity from infections. Almost 56% of child deaths in developing countries are due to protein-energy malnutrition. Higher CO_2 concentration was found to have a negative impact on the protein concentration of wheat, barley, rice and potato, reducing it by 10–15% in these crops. It minimizes the protein concentration in soya bean by smaller yet significantly by 1.4% (Taub et al. 2008). If unaddressed, by 2050, the impact on protein availability may turn 150 million protein-deficient (Medek et al. 2017). We already have an alarming situation concerning food security. This issue will be elevated if left unaddressed. FAO has evaluated the indicators of food security region-wise, countrywide and for the world as a whole. Income-wise, low-income and less developed economies are found more prone to damage due to food insecurity. Region-wise, Africa, Central America, Central and South Asia are more susceptible to the loss arising out of food insecurity. In this section, a comparative account of food security indicators is provided, and the inferences are drawn for India and the world. Below is the list of all the indicators that were taken into consideration. These indicators were released by the FAO on October 11, 2019, and are as follows (Table 3.7).

Tables 3.8, 3.9, and 3.10 present the indicators of food security for India and the world since 2000. Data for certain years was not reported for some indicators for India, in case of some indicators for the world as a whole. India did not report the data for certain indicators at all. In such a case, the number for Southern Asia, the region of which India is a part, was taken into consideration. The availability of food does not seem to be a problem in India (Table 3.8). Foodgrain production is increasing every year. However, the concern is that our land is limited and, with the bulging population, the production is almost reaching a phase of getting stagnant, and so the people remain hungry. Besides, the problems in India are not related to production, and it's more of marketing and distribution.

To assess the food accessibility, rail line density (Table 3.9) is one of the indicators as the connecting network is necessary for the distribution of food along the length and breadth of the country. While the global figures are not available, India's rail network is one of the strongest in the world. GDP per capita is also an indicator of accessibility to food as it is generally considered that the increase in income levels leads to an increase in consumption levels and a change in the consumption from mediocre to a moderate and better diet. The GDP per capita of the Indian population increased 70% in 2018 over 2000. Even after this, the prevalence of undernourishment is high in Southern Asia (data for India was not reported), higher than the global average. This indicator is FAO's traditional hunger indicator and reveals that even though the hunger is decreasing in Southern Asia, there is a need to thrust on removing undernourishment among people. The number

Table 3.7 FAO: food security indicators

A. Food availability
a. Average dietary energy supply adequacy
b. Average value of food production
c. Share of dietary energy supply derived from cereals, roots and tubers
d. Average protein supply
e. Average supply of protein of animal origin
B. Food accessibility
a. Rail lines density
b. Gross domestic product per capita (in purchasing power equivalent)
c. Prevalence of undernourishment, 3-year averages
d. Prevalence of undernourishment, yearly estimates
e. Prevalence of severe food insecurity in the total population, 3-year averages
f. Prevalence of severe food insecurity in the total population, yearly estimates
g. Prevalence of moderate or severe food insecurity in the total population, 3-year averages
h. Prevalence of moderate or severe food insecurity in the total population, yearly estimates
C. Food stability
a. Cereal import dependency ratio
b. Per cent of arable land equipped for irrigation
c. Value of food imports over total merchandise exports
d. Political stability and absence of violence/terrorism
e. Per capita food production variability
f. Per capita food supply variability
D. Utilization
a. People using at least basic drinking water services
b. People using safely managed drinking water services
c. People using at least basic sanitation services
d. People using safely managed sanitation services
e. Percentage of children under 5 years of age affected by wasting
f. Percentage of children under 5 years of age who are stunted
g. Percentage of children under 5 years of age who are overweight
h. Prevalence of obesity in the adult population (18 years and older)
i. Prevalence of anaemia among women of reproductive age (15–49 years)
j. Prevalence of exclusive breastfeeding among infants 0–5 months of age
k. Prevalence of low birthweight

Source: FAO (2019)

of moderate to severely food-insecure people is rising in India over the years (Table 3.9). About 202 million people (2014–2016 average FAO estimates) are undernourished in India. This means that all do not access the available food or the distribution is skewed. There is a need to set up a food distribution network in a way that everyone gets access to food. The third indicator is food stability—India is the net exporter of cereals. India's 45% of arable land is equipped for irrigation, and it shows how the country could be affected by the water stress or droughts. Of the total irrigated agricultural area, 60% is irrigated via groundwater resources. According to

Table 3.8 India and the world: food availability and food stability

Food availability													Food stability			
	Average dietary energy supply adequacy (per cent)		Average value of food production (at constant 2004–2006 dollars per capita)		Share of dietary energy supply derived from cereals, roots and tubers (in per cent)		Average protein supply (gm/capita/day)		Average supply of protein of animal origin (gm/capita/day)		Cereal import dependency ratio (per cent)		Per cent of arable land equipped for irrigation (per cent)			
	W	I	W	I	W	I	W	I	W	I	W	I	W	I		
1999–2001	115	106	260	150	54	63	72.9	56.3	26.7	9.0	−0.5	−1.5	20.6	37.6		
2000–2002	114	104	261	145	53	63	73.1	55.3	27.0	9.0	−0.5	−3.2	20.9	38.4		
2001–2003	114	102	263	145	53	62	73.3	54.7	27.2	9.0	−0.4	−4.3	21.3	39.1		
2002–2004	114	101	267	142	53	62	73.4	54.0	27.5	9.0	−0.4	−4.7	21.6	39.6		
2003–2005	114	100	272	147	52	62	73.7	53.7	27.8	9.3	−0.5	−4.0	21.9	40.1		
2004–2006	115	101	277	149	52	62	74.1	54.0	28.2	9.7	−0.6	−2.3	22.1	40.6		
2005–2007	115	102	282	156	51	61	74.9	55.0	28.8	10.3	−0.5	−2.1	22.3	41.1		
2006–2008	116	104	288	162	51	60	75.7	56.7	29.3	10.7	−0.4	−2.2	22.5	41.6		
2007–2009	117	105	292	164	51	60	76.3	57.3	29.7	11.0	−0.6	−3.1	22.7	42.2		
2008–2010	117	105	296	165	50	59	76.9	58.0	29.9	11.0	−0.7	−2.7	22.9	42.7		
2009–2011	118	105	299	170	50	59	77.5	58.7	30.2	11.3	−0.7	−3.2	23.2	43.1		
2010–2012	119	105	303	176	50	58	78.1	59.3	30.5	11.7	−0.8	−5.6	23.4	43.6		
2011–2013	119	106	306	181	50	58	78.4	59.7	30.7	11.7	−1.0	−8.6	23.3	44.2		
2012–2014	120	106	309	184	–	–	–	–	–	–	–	–	23.3	44.8		
2013–2015	121	107	312	185	–	–	–	–	–	–	–	–	23.3	45.0		
2014–2016	121	108	313	186	–	–	–	–	–	–	–	–	23.3	45.0		
2015–2017	122	108	207	122	–	–	–	–	–	–	–	–	–	–		
2016–2018	122	109	–	–	–	–	–	–	–	–	–	–	–	–		

Source: FAO (2019), *W* World, *I* India

Table 3.9 India and the world: food accessibility and food stability

Food accessibility											Food stability					
	Rail line density (per 100 sq. km of land area)		GDP per capita (Purchasing power equivalent in $)		Prevalence of undernourishment (per cent)		Prevalence for severe food insecurity in the total population, yearly estimates (in per cent)		Prevalence for moderate or severe food insecurity, yearly estimates (in per cent)		Political stability and absence of violence/ terrorism (index)		Per capita food production variability (constant 2004–2006 thousand dollar per capita)		Per capita food supply variability (kcal/ capita/day)	
	W	I	W	I	W	W	I	W	I	South Asia	W	I	W	I	W	I
2000	0.8	1.9	10393.4	2710.3	14.8	–	–	–	–	18.2	–	−1	1.9	1.4	4	27
2001	0.8	1.9	10504.8	2792.3	14.9	–	–	–	–	19.8	–	–	1.2	1.9	4	20
2002	0.8	1.9	10662.5	2850	15.1	–	–	–	–	20.9	–	−1.21	2.2	1.9	6	23
2003	–	1.9	10,927	3023.6	15.1	–	–	–	–	21.8	–	−1.51	3.2	7.3	11	43
2004	0.8	1.9	11361.4	3210.9	14.9	–	–	–	–	21.9	–	−1.28	3.5	7.2	14	52
2005	0.8	1.9	11752.8	3411	14.5	–	–	–	–	21.5	–	−1.01	2.6	6.5	13	49
2006	0.8	1.9	12225.6	3629.4	13.8	–	–	–	–	19.9	–	−1.06	1.5	5.9	7	30
2007	0.9	1.9	12,719	3848.9	13.1	–	–	–	–	18.4	–	−1.15	1.6	4.6	3	25
2008	–	1.9	12922.8	3910.1	12.6	–	–	–	–	17.6	–	−1.11	1.7	5.9	8	49
2009	–	1.9	12721.2	4158.4	12.3	–	–	–	–	17.3	–	−1.35	2.7	6.9	10	60
2010	–	1.9	13220.5	4451.2	11.8	–	–	–	–	17.2	–	−1.28	2.6	5.6	9	49
2011	–	2	13584.9	4624.6	11.6	–	–	–	–	17.2	–	−1.33	2.3	4.5	7	25
2012	–	2	13872.7	4817.2	11.3	–	–	–	–	17.1	–	−1.29	1.8	4.5	6	13
2013		2	14179.4	5064.6	11.1	–	–	–	–	16.8	–	−1.23	1.9	4.4	6	30
2014	–	2	14506.1	5377.9	10.8	8	13.7	23.2	31.4	16.3	–	−1	1.7	4.6	–	–
2015	–	2	14825.7	5743.4	10.6	7.7	12.4	23.2	30.8	15.7	–	−0.95	1.6	1.7	–	–
2016	–	2	15149.9	6145.3	10.7	8	10.6	24.1	30.3	15.1	–	−0.95	2.2	5.1	–	–

(continued)

Table 3.9 (continued)

	Food accessibility										Food stability					
	Rail line density (per 100 sq. km of land area)		GDP per capita (Purchasing power equivalent in $)		Prevalence of undernourishment (per cent)		Prevalence for severe food insecurity in the total population, yearly estimates (in per cent)		Prevalence for moderate or severe food insecurity, yearly estimates (in per cent)		Political stability and absence of violence/ terrorism (index)		Per capita food production variability (constant 2004–2006 thousand dollar per capita)		Per capita food supply variability (kcal/ capita/day)	
	W	I	W	I	W	W	I	W	I	South Asia	W	I	W	I	W	I
2017	–	2	15543.4	6516.2	10.8	8.7	10.9	25.6	28.1	14.8	–	−0.83	–	–	–	–
2018	–		15940.9	6899.2	10.8	9.2	14.4	26.4	34.3	14.7	–	–	–	–	–	–

Source: FAO (2019), *W* World, *I* India

Table 3.10 India and the world: food utilization in per cent

	People using at least basic drinking water services		People using safely managed drinking water services	People using at least basic sanitation services		People using safely managed sanitation services	Percentage of children under 5 years of age who are stunted	Percentage of children under 5 years of age who are overweight	Prevalence of obesity in the adult population (18 years and older)		Prevalence of anaemia among women of reproductive age (15–49 years)		Prevalence of low birth weight
	W	I	W	W	I	W	W	W	W	I	W	I	W
2000	80.4	79	61.3	55.5	16.4	28.2	32.5	4.9	8.3	1.5	31.6	53.3	17.5
2001	80.8	79.5	61.7	56.3	18.8	28.4	–	–	8.5	1.6	31.3	53.3	17.3
2002	81.5	80.4	62.8	57.4	21.3	28.7	–	–	8.8	1.7	31	53.3	17.1
2003	82	81.2	63.6	58.5	23.8	29.6	–	–	9	1.8	30.9	53.3	16.9
2004	82.6	82	63.9	59.5	26.3	30.6	–	–	9.3	1.9	30.7	53.3	16.6
2005	84	82.8	64.2	60.3	28.8	31.5	29.3	5.1	9.5	2	30.6	53.2	16.4
2006	84.5	83.7	64.6	62.2	31.3	32.7	–	–	9.8	2.1	30.4	53	16.2
2007	85	84.5	64.9	63.2	33.8	33.8	–	–	10.1	2.3	30.2	52.7	15.9
2008	85.5	85.3	65.2	64.3	36.4	34.9	–	–	10.4	2.4	30	52.4	15.7
2009	86	86.1	65.9	65.4	38.9	35.9	–	–	10.7	2.5	29.9	52	15.5
2010	86.5	86.9	66.5	66.4	41.5	37.1	26.2	5.4	11	2.7	29.9	51.7	15.3
2011	86.9	87.8	67.1	67.4	44	38.2	25.6	5.5	11.4	2.9	30	51.5	15.1
2012	87.4	88.6	67.7	68.4	46.6	39.3	25	5.5	11.7	3	30.3	51.3	15
2013	87.8	89.4	68.3	69.4	49.2	40.4	24.4	5.6	12.1	3.2	30.7	51.2	14.8
2014	88.3	90.2	68.9	70.5	51.8	41.6	23.9	5.7	12.4	3.4	31.3	51.1	14.7
2015	88.8	91	69.5	71.5	54.3	42.7	23.3	5.7	12.8	3.6	32	51.2	14.6
2016	89.2	91.9	70.1	72.5	56.9	43.9	22.8	5.8	13.2	3.8	32.8	51.4	–
2017	89.6	92.7	70.6	73.4	59.5	45	22.4	5.8	–	–	–	–	–
2018	–	–	–	–	–	–	21.9	5.9	–	–	–	–	–

Source: FAO (2019), *W* World, *I* India

Table 3.11 India water card: water availability 2019

Water stress index	Total water resources available to a population of a region
Water stress	153,663,296 people
Water scarcity	201,170,756 people
Absolute scarcity	262,640,492 people
Percentage of people living in water-scarce area	33%
Total population	1,41,42,22,266

Source: Water Scarcity Clock (2019)

the World Resource Institute (2019), three-fourth of the India is under high to extremely high water distress. This could hamper agricultural productivity and further deteriorate the food security.

The last indicator of food security is reflected through food utilization (Table 3.10) and exposes a worrisome situation for India. Data suggested that more than half of the Indian women between 15 and 49 years are anaemic. About 33% people still live in a water-scarce area (Table 3.11), and 40% people do not use basic sanitation services (Table 3.10, refer year 2017).

India has generally been considered as a politically unstable country, and the internal unrest by different segments on the name of democracy makes the country more vulnerable to the food-insecure situations. The fluctuations in these parameters make the food systems unstable in the country. Despite the policies and efforts of government and local bodies, India is way behind the global averages. This indicates that the food is also not wisely utilized. The nutrition component is missing in the food. This is going to worsen in the climate change scenario and should seriously deal with it. In other low-income and more poverty-ridden countries, the food crisis has already emerged. With climate change, poor and low-income countries of the world may get affected as the people residing there may migrate for a better life elsewhere. This could further pressurize food systems and create political and internal unrests with nation-specific policies for migrants. However, migration is what has defined humanity. The human beings migrated in search of water, food and other non-food products in the past. They are relocating for a better standard of living now, and they will keep migrating for 'survival of fittest' which is not just a theory but a common phenomenon when the population is under distress.

Studies suggest that to increase nutritional security globally, the yearly cereal production will have to be raised by one billion tonnes by 2050 (FAO 2009). Different ways to adapt climate change with respect to food and nutritional sustainability could be through:

1. *Plant breeding*: Breeding new varieties of crops that are climate resistant and are more nutritious.
2. *Soil management*: This can be done by sequestering carbon in soil. A study by Thin Lei (2017) estimated that global croplands can store extra carbon up to 1.85 gigatonnes each year that exceed the carbon emission by the transport sector

annually throughout the world. Practicing proper crop rotation that includes legumes, using compost, and minimizing soil disturbances can make soil healthier to absorb more carbon.

3. *Biofortification*: Fortification of the food to recover the nutrient lost due to climate change and exposure to carbon dioxide to balance the overall intake.
4. *Changing our food consumption pattern*: Our traditional Indian vegetarian diet, which consists of cereals, pulses and millets, lacks many pro-vitamins and minerals, which we can get from green leafy vegetables and fruits like papaya. These fruits and vegetables can be made available at affordable rates to people by including it in the public distribution system.
5. *Gandhian Approach of self-sustainability*: Mr. Mohandas Karamchand Gandhi, or Baapu as Indians fondly remember, always emphasized nutritional security at the village level. He motivated the villagers to become self-sufficient and grow all the fruits, vegetables, cereals, pulses, etc. that are required to have a balanced diet. He also motivated them to eat what is grown locally so that even the poor can afford it. In this way, each plate would have a portion of food rich in vitamins, minerals, protein, carbohydrates, fats and micronutrients. Issues related to food security can also be addressed by reducing the losses at the production side and also minimizing the postharvest losses by investing in market infrastructure, processing, reefer vans, cold chain methodologies, etc. (Islam and Karim 2019). The food processing industry has traditionally been emitting a lot of greenhouse gases for relying on conventional methods. There is a need for innovative and green practices to curb emissions throughout the food chain. Some of the ways that can be addressed via this systems approach are:
   (a) *Green Processing technology*: Many technologies like ultrasound technology, ohmic heating, enzyme-assisted food preservation, high-pressure homogenization, irradiation, pulsed electric field, etc. are green by nature. These technologies require less temperature and time for cooking to produce better quality, eventually reducing carbon footprints, adding to the greener environment and preserving nutritional security (Chemat et al. 2017).
   (b) *Green logistics*: It is an effort to minimize the damage to the environment due to transportation, storage, distribution, inventory management and warehousing process. It also encompasses the logistics required for waste management. Use of electric vehicles, improvisation of operation space by designing customized packages accordingly, moving closer to the target customer, reusing, recycling and reprocessing the waste at the consumer end can be considered as noticeable efforts towards sustainable and green logistics (McKinnon et al. 2013).
   (c) *Green packaging*: The packaging is the core problem for environmental sustainability as it uses bubble wraps, thin films, tapes, etc. to ensure the safety of the material inside. These plastics take almost 10,000 years to decompose. Green packages made of paper, cloth and other biodegradable material have been actively encouraged by organizations to reduce their carbon footprint contribution and contribute to efforts towards a sustainable world. Multinational companies have invested in innovative concepts like

edible packaging. For example, KFC has launched a comestible cup called as Scoff-ee Cup. It is made of biscuit coated with white chocolate and glazed with sugar to make marketing lucrative. This package is entirely edible and attracts more and more customers, enhancing their brand value (Strom 2015). Air New Zealand in order to reduce onboard wastage, introduced vanilla flavoured edible coffee cups. This also brings down the costs effectively and reduces carbon footprint.

(d) *Green consumption*: It is more of an attitude where the consumers are willing to buy the products that pose no harm to the environment. Many campaigns have been conducted to promote green consumption and biodegradable packaging, but as soon as the consumers get to learn the higher costs of the finished goods, the feeling of care for Mother Nature gets subsided. The efforts need to be made in such a way that the costs to the consumer are low, and they can afford consuming green. Plastic bags available in the market were priced at 0.50 INR per bag earlier that are replaced by paper bags at 5.00 to 10.00 INR per bag, which is not welcomed by a large set of population and acts as a barrier in choosing greenways.

3.6 Global Strategies to Alleviate the Effect of Climate Change

Climate change mitigation means reducing the amount of GHGs in the environment. Several countries have adopted using cap and trade strategy to reduce carbon emissions to fulfil their pledge to the Paris Agreement to keep the rise in temperature under 2 °C against the over pre-industrial levels. Expanding carbon sinks through zero deforestation and harnessing and utilization of alternate green sources of energy, viz. wind and solar, are also followed by many. However, the irony is that though all of us are alerted to the call, not everyone is working towards controlling the damage (IPCC 2014b). Here are a few examples from the nations who have adopted various ways to decrease the carbon footprint.

Australia To control GHG emissions, Australia has introduced a carbon taxation system. Under this the country identifies 500 worst polluters who are then taxed for emitting more carbon. The tax is on per tonne of carbon basis and is effective since July 2012. Australia plans to reduce total emissions by 28% by bringing down per capita emissions 50% between 2005 and 2030 (Australian Department of Environment Analysis 2019). However, Australia has heavily been criticized for not taking the call of destruction due to changing climate during the 2019 bushfires.

China China's economy was heavily dependent on coal-based energy generation until 2010. Since 2010, China has adopted a proactive climate mitigation policy as the country felt the need to shift to renewable energy almost a decade before the Paris Agreement. China reduced the production and consumption of coal and shifted to solar, hydro and other renewable sources of energy. Between 2013 and 2017, China reduced the consumption of coal at a rate of 1.1% per year and has shifted to other sources like gas, nuclear and hydropower. China has now emerged as one of the leading solar and wind power technology producers in the world. China's focus lies on three fronts, namely, energy security, controlling air pollution and strengthening the existing economic model. The Government of China laid regulations for the electricity companies to buy a certain amount of power generated from clean energy sources and has also declared subsidies to promote clean energy sectors (RE100 The climate group 2015).

European Union (EU) The European Union pledged to enhance renewable energy consumption in the total energy requirements and reduce the emissions of GHGs by 40% over 1990. The EU has already achieved a 22% reduction in emissions up to 2017. To achieve these targets, the EU adopted policy initiatives like putting up a ban on the single-use plastics and emphasizing on waste management and recycling. The EU has guidelines for all member countries which cap the carbon amount to be emitted. The companies that emit less compared to the allowable limits can sell it to other firms that exceed the permissible limits. The EU has also proposed a 'Green Deal' package of regulations to curb GHGs. A 'carbon border tax' is also offered under this deal for importing polluting goods from countries that have less strict policies to tackle climate change (National Public Radio 2011). This, however, remains debatable as it looks like violating the most favoured nation criteria under the principle of free trade of the World Trade Organization. Individually, the countries in the EU are putting their best foot forward. Paris is promoting green transport and is building more bike lanes. The number of people cycling to work is still low (3%), but the French Government is encouraging people to bike to the workplace too. Around half of the electricity in Sweden is generated through renewable energy sources. Sweden has one of the lowest carbon emissions in Europe. Only 1% of the waste goes to landfill in Sweden as it uses advanced incineration and recycling techniques. Denmark, too, is focusing on becoming carbon-neutral by promoting all renewable energy sources, replacing coal stations. The people prefer cycling and are contributing significantly to the energy saving. Spain, too, is marching towards renewable energy. The country has proposed a $53 billion climate fund. On the other hand, Norway aims to cut its GHG emissions by 40% by 2030. The country promotes electric vehicles. Europe plans to be the first carbon-neutral continent by 2050.

Bhutan Bhutan is the first carbon-negative country. Bhutan's constitution makes it mandatory to cover 70% of its land by forests. It has created green corridors so that the animals are free to move all along the country. A landlocked and small country in the Himalayas, Bhutan, primarily practices subsistence agriculture. The country produces a surplus of hydropower that is exported to India and is one of the major sources of foreign reserves. Rising temperatures have caused several glaciers to melt in Bhutan. Floods have caused disasters. All this reflects that the actions of one country may cause havoc in another country. Yet, the committed carbon sink (70% forest cover) has helped Bhutan to remain carbon-negative. Bhutan's primary thrust is on hydropower generation, development of the industrial sector and agricultural prosperity to assure sustainable development of the country.

3.7 The Indian Way

Climate change sustainability, mitigation and adaptation have been India's focus for long. The country promotes clean energy and the use of clean technology to a large extent. The market for energy efficiency is estimated at US$ 22.81 billion. In 2001, India enacted the Energy Conservation Act to reinforce energy efficiency. The act established the Bureau of Energy Efficiency (BEE). The BEE has energy efficiency schemes for each of the sectors. To manage the demand side, the BEE had launched various schemes for the agriculture sector, municipal bodies and distribution companies. To strengthen the efficiency in the agriculture sector, the BEE also entered into a memorandum of understanding (MoU) with the Indian Council of Agricultural Research. Through this, the BEE aims at creating awareness about the energy-efficient pumps and their low-cost use on farms. Furthermore, India has explored three scenarios to estimate its energy-saving potential. The first scenario is the least effort scenario, i.e. without any technological and policy intervention. In this scenario, no change in fuel mix was proposed, i.e. the country's situation at the time of proposing the policy. Scenario II is about slight changes at the technological and political levels. Technology-wise the scenario calls for a mix of renewable energy and electricity-based energy use over fossil fuel use in various demand sectors. This is the scenario in which India expects to achieve the targets set for different programs for efficient utilization of energy by several sectors. In the end is scenario three, which calls for aggressive technological and policy push. For technology, this scenario calls for aggressive fuel mix and a shift towards renewable sources of energy. This is the scenario in which the program targets will be overachieved (Ministry of Finance, Economic Survey 2018–2019). Under each of these three scenarios, the energy-saving potential for various sectors was estimated. It was found that even with the most aggressive efforts, India's potential for energy saving is enormous. The projected energy saving in the three scenarios is shown in Table 3.12.

The focus of India's strategy is on the cleaning of exploited natural resources and resource efficiency policies. India has invested heavily in renewable energy resources. The cumulative wind power capacity has exceeded 36 GW. Under the National Solar Mission, the solar power-installed capacity was increased significantly. By March 2019, India's hydro potential was only 31% utilized, and it leaves a bigger room for the country to harness the remaining hydropower potential, which is more climate-friendly against the conventional ways of power generation. Not only this but the electric vehicles are also making their way in India. Not just this, in 2008 the country devised its own National Action Plan on Climate Change (NAPCC), which is all about reducing the emission intensity of the gross domestic product by 20–25% between 2005 and 2020. In 2014, India had already brought it down by 21%. The estimates from India's Ministry of Environment, Forests and Climate

Table 3.12 India's energy-saving potential in various demand sectors in 2031

Sector	Least effort	Moderate savings		Aggressive savings	
	Mtonne	Mtonne	Per cent	Mtonne	Per cent
Agriculture	64.4	5.7	9	9.9	15
Commercial	232.9	15.8	7	23.8	10
Domestic	98.6	12.1	12	15.1	15
Commercial	29.5	4.9	7	6.4	22
Municipal	8.0	0.9	12	1.5	19
Industries	443.4	47.5	11	72.3	16
Total (Mtonne)	876.8	86.9	10	129.0	15
Total (TWh)	10,198	1010	10	1500	15

Note: TWh = terawatt hour
Source: Ministry of Finance, Economic Survey (2019)

Change (Economic Survey 2019) reveal that the country emitted 2.607 billion tons of CO_2 equivalents, out of which natural carbon sinks offset 12%. The maximum emissions were from the energy sector (72%), followed by agriculture (16%). Waste also emits around 3% of CO_2 equivalent. Remaining 8% is emitted by industrial processes and product use.

The NAPCC plan also dealt with adaptation requirements and scientific planning to combat climate change. On similar lines, the states and Union Territories of India framed State Action Plans on Climate Change (SAPCC). Overall, India has 33 SAPCCs in force. In 2014, the Climate Change Action Programme (CCAP) was launched to build capacity for climate change assessment, create a suitable institutional framework and implement all plans on the ground for sustainable development. The 3-year budget outlay (2017–2020) for the scheme was 132.40 crores. Furthermore, in 2015, the country created a National Adaptation Fund on Climate Change to help vulnerable regions and to cover their cost of adaptation to the changing climate. The scheme is to continue until March 2020 and includes sectors such as agriculture, forestry, eco-diversity, animal husbandry and water.

India has ratified the Paris Agreement, an essential feature of which is the contribution of each country to mitigate the risks of climate change and reduce carbon footprint, to the best capacity. This is called a Nationally Determined Contribution (NDC). India has prepared its NDC. However, climate funding remains a challenge as the country has so much diversity, and several technological advancements are going on. According to some studies, by mid-century, the cost of adaptation and mitigation to climate change may reach US$ 1 trillion per annum. The multilateral climate change funds that were pledged during the Paris Agreement are still falling short. India's NDC states need of US$2.5 trillion (at 2014–2015 prices) between 2015 and 2030. Climate financing thus remains a more significant concern.

The world aims to end global hunger by 2030, and India's role is crucial in it as we are home to more than a billion people out of the total 7 billion on earth. While the numbers suggest that more than 200 million in India still remain undernourished

(2014–2016 average) and people suffering from acute food insecurity are around 10% of the population, eliminating hunger is a tedious task that too when we face the challenges arising out of climate change. The shrinking Himalayan glaciers which source the Ganges, the major lifeline of India, and many other rivers could turn out to be a bane as this may result in an expansion of deserts, retreating rivers, droughts, floods, etc. which will hamper the food security. The Government of India has various schemes and programmes to attain and sustain food security. The programmes are related to the improvement of soil health (Soil Health Card Scheme), expansion and improvement of irrigation facilities to expand cultivable land and enhance productivity, schemes to provide food to low-income population (Annapurna Scheme, Antyodaya Anna Scheme, Food Security Act) and the schemes to cover risks related to crop failure (Prime Minister Crop Insurance Scheme). India also has a midday meal programme to promote enrollments in primary schools and, at the same time, provide food security to all kids. There are several programmes in place. However, effective implementation at each stage is needed to make sure that the food and nutritional security of the population can be improvised.

3.8 Suggestions at Different Levels

India is a developing nation, and with development, there are apparent costs to be borne. The nation's economic growth is directly proportional to its per capita emissions. As the standard of living increases, the emissions tend to increase due to the haphazard usage of coal and other natural resources. Climate change has already caused irreversible changes to the environment, and now when we have realized this, we need to have a strategy so that further damage can be controlled.

Many mitigation strategies have been suggested by the United Nations, IPCC (2014c), European Union and International Financial Institutions, which have been discussed in this chapter. These strategies cannot be successful unless implemented at three levels: individual level, national level and global level.

3.8.1 Individual Front

We all are responsible for this scenario of global warming by exploiting the earth's resources for our comfort and development. Now, it is our job to create a sustainable plan for the survival of our future generations. Our efforts will be like a drop in the ocean and may not yield immediate results, but a population of 7.8 billion people taking steps towards climate resilience may help us to maintain global warming below 2 °C.

There is a need for climate education among the people and making them responsible for the outcomes. Creating awareness among the public could change their way of thinking towards the non-renewable resources. Children should be educated to respect forest and rivers and promote forestation. Many researchers have suggested that minor steps like reduction in travelling, growing our own food,

Table 3.13 Technical mitigation potential of changing diets by 2050 according to a range of scenarios examined in the literature

Diet	Composition	GHG mitigation potential (GtCO2-eq yr–1)	Researcher
Vegan	No animal source	7.8–8.0	Springmann et al. (2016), Stehfest et al. (2009)
Vegetarian	Grains, vegetables, fruits, sugars, oils, eggs and dairy, and generally at most one serving per month of meat or seafood	4.6–7.2	Springmann et al. (2016), Tilman and Clark (2014), Stehfest et al. (2009)
Flexitarian	75% of meat and dairy replaced by cereals and pulses; at least 500 g per day fruits and vegetables; at least 100 g per day of plant-based protein sources; modest amounts of animal-based proteins and limited amounts of red meat (one portion per week), refined sugar (<5% of total energy), vegetable oils high in saturated fat and starchy foods with relatively high glycaemic index	5.2–5.4	Springmann et al. (2018), Hedenus et al. (2014)
Healthy diet	Based on global dietary guidelines for consumption of red meat, sugar, fruits and vegetables and total energy intake	2.8–6.4	Springmann et al. (2018a), Bajželj et al. (2014)
Fair and frugal	Global daily per capita calorie intake of 2800 kcal/cap/day (11.7 MJ/cap/day), paired with relatively low level of animal products	0.7–7.3	Bajželj et al. (2014)
Climate carnivore	75% of ruminant meat and dairy replaced by other meat	3.4	Hedenus et al. (2014)

Source: Mbow et al. (2019)

going vegan, using more renewable energy like natural gas and solar-based equipment, the use of fresh and local food, etc. should be practiced to reduce carbon emissions at the individual level (Table 3.13). However, the question remains how can we make sure that lowering emissions at an individual level will also help us in staying food-secure. Mbow et al. (2019) presented the mitigation potential of several diets, as researched by several scientists. It was found that the vegan diet has a reduction potential (Springmann et al. 2016) of the reduction potential of 8 GtCO2-eq yr–1, maximum of all possible diet combination. Stehfest et al. (2009) suggested that under the vegan food for all scenario, sufficient food could be produced in 2050, that too from relatively lesser land than what is available today. However, it is possible only when forests are regenerated, and GHG emissions are reduced to about a third of the 'No Change or Business as Usual' scenario. This will reduce the reductions to around 7.8 Gt CO2-eq yr–1. FAO (2018), however, suggests that all diets should include nutrients, including micronutrients, as complementary to make sure that nutritional security is not disturbed. Besides this, on the individual front, the scientists and researchers have been working on improving the performance of equipments since long to curb the issue of carbon emissions and inventing energy-efficient refrigerators and cars, but the overall carbon emissions are still increasing as the number of vehicles per household is increasing, and people are commuting longer distances for work and leisure. The demand for such luxuries and immature spending of resources has to be governed in order to regulate the productions.

We need to opt for greener options for all our needs. Shifting near to the workplace and commuting on bicycles or in public transport instead of personal cars could be incentivized. Preference could be given to the local products, and the concept of healthy kitchen gardens should be promoted. Consumers should focus on having a balanced diet that would emit less carbons per calorie of food, as discussed before. Also, a one-child policy should be implemented to reduce the consumption of the resources and control population growth further.

3.8.2 National Level

Since 1750, developed nations have contributed to about 70% of the carbon emissions on account of uncontrolled and unregulated industrialization (Friedrich and Damassa 2014). A developing nation, on the other hand, faces enormous challenges towards their growth and development that encompasses all three—agriculture, manufacturing and services sectors of the economy. As the economy grows, disposable incomes rise, and development is unavoidable due to an increase in disposable incomes, living standards and globalization, but the real challenge is development without contributing to climate change or operating under the permissible limits. However, developing nations do not have sufficient alternate technologies to mitigate climate change. Also, food sustainability is a more significant threat as the existing ways of making food sustainable for the growing population are all going to exhaust energy resources leading to climate change. So, we need a more significant course of innovation here.

India has promised in the Paris Agreement that by 2030 it will cut down its emission per unit of GDP by 33–35% as compared to 2005 (Climate Action Tracker 2019). To achieve this target, the biggest contributors have to be identified, and their usage has to be limited or banned. National-level policies can be made to levy extra charges on the usage of non-renewable sources like coal, petroleum, etc. The three major contributors to carbon emission in a developing nation are the energy generation sector, industrial sector and agriculture and food sector. Countries have to revisit their policies for the same and waste management practices.

For energy generation, countries should focus on hydro-based power, solar energy and wind energy and prohibit the usage of wood and coal. The second most crucial sector is the industrial sector. This sector has lifted millions of people out of poverty and provided them employment. However, the sector has substantially contributed to GHG emissions globally. The sector needs to develop ways to promote climate resilience for sustainable growth. This sector should focus on the vulnerable sections by providing them the necessary infrastructure, food security and service to increase their resilience towards climate change. Public-private partnership models need to be assured to ensure efficient technology transfer, attract investments in various climate-resilient sectors and apply the best industrial practices to become a responsible corporate. The complete value chain has to be reformed from raw material quality to waste management practices.

In the case of agriculture and food sector, the existing subsidies that the government is giving should be redirected towards the usage of green technologies. Fertilizer subsidy and minimum support price on certain crops in India promote farmers to grow water-intensive crops and increase productivity (ET Bureau 2019). These subsidies should be diverted towards drought- and temperature-resistant crops and encourage the use of solar power for farming practices (GOI 2008).

3.8.3 Global Front

There are many fronts on which global countries should work, like energy creation, infrastructure, transportation, land, industry and finance. Adoption of renewable sources up to 30% of the global electricity generation in 2020, green infrastructure like decarbonized buildings with zero emissions and the electricity-based common transportation system should replace the existing system, and motivation should be given to the usage of public transport and reduction of deforestation with a focus on sustainable crops that help binding carbon dioxide to the soil. For all these efforts, funds are required, and so IPCC (2014c) has suggested collecting a Green Climate Fund by introducing various tariffs and market-based financial schemes to increase investment in clean energy. Approximately one trillion US $ per year will be required to be invested in climate mitigation solutions.

Every country has its strategy to resolve the issues coming up with the change in climate, but this is a global issue, and the whole world is required to come together to manage this challenge effectively. There is a mean carbon budget of 600 gigatonnes left to emit in order to reach a situation where the average temperature rise would be more than 2 °C. All the countries undertook the targets during the Paris Agreement which are still not fulfilled by many countries. We are still far away from achieving a net-zero carbon emission state. The countries who have achieved their targets and are having even better reductions as promised in the agreement receive carbon credits, and these credits can be traded off with countries who are having a shortfall and get financial or technological help in return. This carbon trading could be a win-win situation for both parties and for the world as we will achieve our overall target of reducing carbon emission without compromising the development part. To achieve these targets, we need substantial technological as well as behavioural transformation.

3.9 Conclusion

According to the United Nations, only 10 years are left to tackle climate change. If strong actions to reverse climate change are not taken by 2030, it would rather be difficult to prevent the earth from the climate-related downfall. Studies suggest that anthropogenic activities have contributed positively to climate change. Various reports indicate that humans have caused massive destruction to plant and animal life and have disturbed the ecological diversity by harming the land, air and water resources. But humans need other species—plants and animals—to thrive on. Biodiversity helps establish the earth's equilibrium by balancing different food chains. While food systems have changed climate, changing climate has affected food systems too. All that is needed is to adopt a systems approach to the problem and address all aspects. A systems approach will also help to resolve food insecurity. Agriculture's contribution to the GHGs could be mitigated by the adoption of proper livestock and crop management practices. Soil conservation and carbon sequestration ways should be considered. Efforts should also be made to predict the occurrence of droughts, storms, earthquakes, etc. to minimize the destruction and resulting crop loss. This, alongside, the efforts, should also be made to save energy across supply chain activities. Resources, including water, should be used very efficiently. People should be educated about healthy diet and economic costs of an impoverished diet and unhealthy living style. Diets that are low on carbon footprint should be preferred. This is recommended as livestock rearing is highly water consuming and greenhouse gas emissions are much higher. Approximately 15,000 L of water are used to produce 1 kg meat, whereas only 1250 L of water is sufficient to produce 1 kg of grain. Hence, cereal-based diet reduces the impact on climate by 7 to 12 times as compared to non-vegetarian food. Also, the emphasis should be given to the consumption of food in natural form as mass processing leads to wastages, which increase the carbon footprint. Reduction in wastage of food at each level of value chain like harvesting, storage, processing, transportation, etc. can increase the global food and nutritional security ratio and can feed millions of people in an economically viable manner.

Wastages should instead be brought down to zero. In order to save the food systems from the changing climate, biotechnology and other innovative ways are required to develop crop varieties that are hardy, drought-tolerant and temperature-resistant. Furthermore, the diversification should be promoted, especially on small farms in countries like India, wherein around 85% of farm holdings are <2 hectares. Good agricultural practices should be followed, and universities, industries and the government should adopt a collaborative approach to resolve the issues of yield destruction under changing climate scenarios. The contingency plans must be prepared given the climate challenges and should be promoted under contingency situations.

References

Alexandratos N, Bruinsma J (2012) World agriculture towards 2030/2050: The 2012 revision. ESA Working paper no. 12-03. FAO, Rome

Australian Department of Environment Analysis (2019). https://www.environment.gov.au/system/files/resources/f52d7587-8103-49a3-aeb6-651885fa6095/files/summary-australias-2030-emissions-reduction-target.pdf. Accessed 19 Dec 2019

Bajželj B, Richards K, Allwood J, Smith P, Dennis J, Curmi E, Gilligan C (2014) Importance of food-demand management for climate mitigation. Nat Clim Change 4:924–929. https://doi.org/10.1038/nclimate2353

Barange M, Bahri T, Beveridge M, Cochrane K, Funge-Smith S, Poulain F (2018) Impacts of climate change on fisheries and aquaculture: synthesis of current knowledge, adaptation and mitigation options. FAO Fisheries and Aquaculture Technical Paper No. 627. Rome, FAO, 628 pp

Betts R, Alfieri L, Bradshaw C, Caesar J, Feyen L, Friedlingstein P, Gohar L, Koutroulis A, Lewis K, Morfopoulos C, Papadimitriou L, Richardson K, Tsanis I, Wyser K (2018) Changes in climate extremes, fresh water availability and vulnerability to food insecurity projected at 1.5°C and 2°C global warming with a higher-resolution global climate model. Philos Trans R Soc A Math Phys Eng Sci 376(2119):20160452. https://doi.org/10.1098/rsta.2016.0452

Birthal P, Khan M, Negi D, Agarwal S (2014) Impact of climate change on yields of major food grain crops in India: Implications of food security. Agric Econ Res Rev 27(2):145–155

Byers E, Gidden M, Leclère D, Balkovic J, Burek P, Ebi K, Greve P, Grey D, Havlik P, Hillers A, Johnson N, Kahil T, Krey V, Langan S, Nakicenovic N, Novak R, Obersteiner M, Pachauri S, Palazzo A, Parkinson S, Rao N, Rogelj J, Satoh Y, Wada Y, Willaarts B, Riahi K (2018) Global exposure and vulnerability to multi-sector development and climate change hotspots. Environ Res Lett 13(5):055012. https://doi.org/10.1088/1748-9326/aabf45

Carbon Emission Calculator (2015). https://www.foodemissions.com/foodemissions/Calculator.aspx. Accessed 12 Nov 2019

Chatterjee A (1998) Simulating the impact of increase in carbon dioxide and temperature on growth and yield of Maize and Sorghum. M.Sc. thesis, Division of Environmental Sciences, Indian Agricultural Research Institute, New Delhi

Chemat F, Rombaut N, Meullemiestre A, Turk M, Perino S, Tixier A, Vian M (2017) Review of green food processing techniques. Preservation, transformation, and extraction. Innov Food Sci Emerg Technol 41(1):357–377. https://doi.org/10.1016/j.ifset.2017.04.016

Cheung W, Frölicher T, Asch R, Jones M, Pinsky M, Reygondeau G, Rodgers K, Rykaczewski R, Sarmiento J, Stock C, Watson J (2016) Building confidence in projections of the responses of living marine resources to climate change. ICES J Mar Sci 73(5):1283–1296. https://doi.org/10.1093/icesjms/fsv250

Climate Action Tracker (2019). https://climateactiontracker.org/climate-target-update-tracker

Cramer W, Yohe G, Auffhammer M, Huggel C, Molau U, Dias M, Solow A, Stone D, Tibig L (2014) Detection and attribution of observed impacts. In: Field CB, Barros VR, Dokken DJ, Mach KJ, Mastrandrea MD, Bilir TE, Chatterjee M, Ebi KL, Estrada YO, Genova RC, Girma B (eds) Climate change 2014: impacts, adaptation, and vulnerability. Part A: global and sectoral aspects. Contribution of Working Group II to the fifth assessment report of the intergovernmental panel on climate change. Cambridge University Press, Cambridge, New York, pp 979–1037

Das S (2017) Impact of climate change on livestock, various adaptive and mitigative measures for sustainable livestock production. Approaches Poult Dairy Vet Sci 1(4):64–70. https://doi.org/10.31031/APDV.2017.01.000517

Department for Environment, Food and Rural Affairs (2016) Climate change impacts on agriculture in India. http://www.indiaenvironmentportal.org.in/files/Defra-india-cc-agri.pdf

Devi S, Balachandar V, Lee SI, Kim IH (2014) An outline of meat consumption in the indian population - a pilot review. Korean J Food Sci Anim Resour 34:507–515

Ericksen P (2008) What is the vulnerability of a food system to global environmental change? Ecol Soc 13(2):14

ET Bureau (2019) Budget 2019: farmers need to be incentivised to conserve water. https://economictimes.indiatimes.com/news/economy/agriculture/farmers-need-to-be-incentivised-to-conserve-water-economic-survey/articleshow/70072687.cms

FAO (2006) Food security policy brief 02. FAO, Rome

FAO (2008) Climate change and food security- A framework document. Food and Agriculture Organization of the United Nations, Rome

FAO (2009) How to feed the world in 2050. Food and Agriculture Organization of the United Nations, Rome

FAO (2011) Food wastage footprint & climate change. Food and Agriculture Organization of the United Nations, Rome

FAO (2015) Climate change and food systems- global assessments and implications for food security and trade. Food and Agriculture Organization of the United Nations, Rome

FAO (2018) The state of world's Forest. Food and Agriculture Organization of the United Nations, Rome

FAO (2019) The state of food and agriculture. Moving forward on food loss and waste reduction. FAO, Rome

Fernandes J, Kay S, Hossain M, Ahmed M, Cheung W, Lazar A, Barange M (2016) Projecting marine fish production and catch potential in Bangladesh in the 21st century under long-term environmental change and management scenarios. ICES J Mar Sci 73(5):1357–1369. https://doi.org/10.1093/icesjms/fsv217

Flores M, Macias N, Rivera M, Lozada A, Barquera S, Dommarco J, Tucker K (2010) Dietary patterns in mexican adults are associated with risk of being overweight or obese. Nutrition and Health Research Center, National Institute of Public Health, Colonia Santa Maria Ahuacatitlan and Department of Health Sciences, Northeastern University, Mexico, Boston, MA

Food Calculator (2019) What's your diet's carbon footprint? BBC News. https://www.bbc.com/news/science-environment-46459714. Accessed 19 Nov 2019

Friedman L, Popovich N, Fountain H (2018) Who's most responsible for global warming? https://www.nytimes.com/2018/04/26/climate/countries-responsible-global-warming.html

Friedrich J, Damassa T (2014) The history of carbon dioxide emissions. https://www.wri.org/blog/2014/05/history-carbon-dioxide-emissions

Funk C, Brown M (2009) Declining global per capita agricultural production and warming oceans threaten food security. Food Secur 1:271–289

Government of India (GOI) (2008) National action plan on climate change prime minister's council on climate change

Hedenus F, Wirsenius S, Johansson D (2014) The importance of reduced meat and dairy consumption for meeting stringent climate change targets. Clim Chang 124:79–91. https://doi.org/10.1007/s10584-014-1104-5

Ingram J (2011) A food systems approach to researching food security and its interactions with global environmental change. Food Secur 3:417–431

IPCC (2007) Climate change 2007: impacts, adaptation and vulnerability. In: Parry ML, Canziani OF, Palutikof JP, van der Linden PJ, Hanson CE (eds) Contribution of Working Group II to the fourth assessment report of the intergovernmental panel on climate change. Cambridge University Press, Cambridge and New York

IPCC (2014a) Climate change: synthesis report. In: Core Writing Team, Pachauri RK, Meyer LA (eds) Contribution of Working Groups I, II and III to the fifth assessment report on the intergovernmental panel on climate change. Intergovernmental Panel on Climate Change, Geneva, 151 pp. http://www.ipcc.ch/report/ar5/syr/

IPCC (2014b) Climate change: impacts, adaptation and vulnerability. Part A: global and sectoral aspects, IPCC. http://www.ipcc.ch/report/ar5/wg2/. Accessed 15 Dec 2019

IPCC (2014c) Climate change: mitigation of climate change. http://www.ipcc.ch/report/ar5/wg3/. Accessed 15 Dec 2019

IPCC (2018) Summary for policymakers of IPCC special report on global warming of 1.5°C approved by governments — IPCC. Available at Ipcc.ch. https://www.ipcc.ch/2018/10/08/summary-for-policymakers-of-ipcc-special-report-on-global-warming-of-1-5c-approved-by-governments/. Accessed 25 Dec 2019

Islam S, Karim Z (2019) World's demand for food and water: the consequences of climate change [online first]. Intechopen 1:1–28. https://doi.org/10.5772/Intechopen.85919

Jewett L, Romanou A (2017) Ocean acidification and other ocean changes. In: Wuebbles DJ, Fahey DW, Hibbard KA, Dokken DJ, Stewart BC, Maycock TK (eds) Climate science special report: fourth national climate assessment, vol 1. Washington, DC, U.S. Global Change Research Program, pp 364–392. https://doi.org/10.7930/J0QV3JQB

Kearney J (2010) Food consumption trends and drivers Phil. Trans R Soc B Biol Sci 365 (1554):2793–2807. https://doi.org/10.1098/rstb.2010.0149

Koundinya A, Palash S, Pandit M (2014) Impact of climate change on vegetable cultivation - a review. Int J Agric Environ Biotechnol 7(1):145–155

Lehner F, Coats S, Stocker T, Pendergrass A, Sanderson B, Raible C, Smerdon J (2017) Projected drought risk in 1.5°C and 2°C warmer climates. Geophys Res Lett 44(14):7419–7428. https://doi.org/10.1002/2017gl074117

Lei T (2017) Better soil could trap as much planet-warming carbon as transport produces: study. https://www.reuters.com/article/us-climate change-agriculture-soil/better-soil-could-trap-as-mush-planet-warming-carbon-as-transport-produces-study

Mbow C, Rosenzweig C, Barioni L, Benton T, Herrero M, Krishnapillai M, Liwenga E, Pradhan P, Rivera-Ferre M, Sapkota T, Tubiello F, Xu Y (2019) Food Security. In: Shukla PR, Skea J, Buendia EC, Masson-Delmotte V, Pörtner H-O, Roberts DC, Zhai P, Slade R, Connors S, van Diemen R, Ferrat M, Haughey E, Luz S, Neogi S, Pathak M, Petzold J, Pereira JP, Vyas P, Huntley E, Kissick K, Belkacemi M, Malley J (eds) Climate change and land: an IPCC special report on climate change, desertification, land degradation, sustainable land management, food security, and greenhouse gas fluxes in terrestrial ecosystems. IPCC, Geneva

McKinnon A, Browne M, Whiteing A (2013) Green logistics: improving the environmental sustainability of logistics, 2nd edn. Kogan Page, London

Medek D, Schwartz J, Myers S (2017) Estimated effects of future atmospheric CO2 concentrations on protein intake and the risk of protein deficiency by country and region. Environ Health Perspect 125(8):087002. https://doi.org/10.1289/ehp41

Ministry of Finance, Government of India (2019) Economic survey 2018–19. OUP catalogue. Oxford University Press, Oxford

Mishra S, Singh R, Kumar R, Kalia A, Panigarhy S (2017) Impact of climate change on pigeon pea. Econ Aff 62(3):455–457

Mislan K, DeutschC BR, Dunne J, Sarmiento J (2017) Projections of climate driven changes in tuna vertical habitat based on species-specific differences in blood oxygen affinity. Glob Chang Biol 23(10):4019–4028. https://doi.org/10.1111/gcb.13799

National Geographic (2019) What the world eats. https://www.nationalgeographic.com/what-the-world-eats/. Accessed 25 Dec 2019

National Public Radio (2011) What countries are doing to tackle climate change. https://www.npr.org/2011/12/07/143302823/what-countries-are-doing-to-tackle-climate-change

Niles MT, Ahuja R, Esquivel J, Mango N, Duncan M, Heller M, Tirado C (2017) Climate change & food systems: assessing impacts and opportunities. Meridian Institute. http://bit.ly/2oFucpe

Pathak H, Jain N, Bhatia A, Patel J, Aggarwal P (2010) Carbon footprints of Indian food items. Agric Ecosyst Environ 139:66–73

Pelletier D, Frongillo E, Habicht J (1993) Epidemiologic evidence for a potentiating effect of malnutrition on child mortality. Am J Pub Health 83:1130–1133

Porter J, Xie L, Challinor A, Cochrane K, Howden S, Iqbal M, Lobell D, Travasso M (2014) Food security and food production systems. In: Field CB, Barros VR, Dokken DJ, Mach KJ, Mastrandrea MD, Bilir TE, Chatterjee M, Ebi KL, Estrada YO, Genova RC, Girma B (eds) Climate change 2014: impacts, adaptation, and vulnerability. Part A: global and sectoral aspects.

Contribution of Working Group II to the fifth assessment report of the intergovernmental panel on climate change. Cambridge University Press, Cambridge and New York, pp 485–533
Prabhakar S (2010/2007) Promoting adaptation and disaster risk reduction in the post-kyoto climate regime. In: Shaw R, Pulhin J, Pereira J (eds) Climate change adaptation and disaster risk reduction: asian perspective, vol 4. Climate Change Adaptation and Disaster Risk Reduction: Issues and Challenges (Community, Environment and Disaster Risk Management), Emerald Group Publishing, Bingley, pp 77–98
Rao C, Raju B, SubbaRao A, Rao K, Rao V, Ramachandran K, Venkateswarlu B, Sikka A (2013) Atlas on vulnerability of indian agriculture to climate change. Central Research Institute for Dry land Agriculture, Hyderabad, p 116
Ray D, West P, Clark M, Prishchepov A (2019) Climate change has likely already affected global food production. PLoS One 14(5):e0217148. https://doi.org/10.1371/journal.pone.0217148
RE 100 The Climate Group (2015) China's fast track to a renewable future, China analysis, 2015. https://www.theclimategroup.org/sites/default/files/archive/files/RE100-China-analysis.pdf
Rosenzweig C, Hillel D (eds) (2015) Handbook of climate change and agroecosystems. Imperial College Press, London, 1160 pp. https://doi.org/10.1142/p970
Rosenzweig C, Jones W, Hatfield J, Ruane A, Boote K, Thorburn P, Antle J, Nelson G, Porter C, Janssen S, Asseng S, Basso B, Ewert F, Wallach D, Baigorria G, Winter J (2013) The Agricultural Model Intercomparison and Improvement Project (AgMIP): Protocols and pilot studies. Agric For Meteorol 170:166–182. https://doi.org/10.1016/j.agrformet.2012.09.011
Singh R, Devi G, Parmar D, Mishra S (2017) Impact of rainfall and temperature on the yield of major crops in Gujarat state of India: a panel data analysis (1980-2011). Curr J Appl Sci Technol 24(5):1–9
Sinha R, Lone S, Ranjan A, Rahim I, Devi I, Tiwari S (2018) The impact of climate change on livestock production and reproduction- ameliorative management. Int J Livest Res 7(6):1–8. https://doi.org/10.5455/ijlr.20170417042102
Springmann M, Godfray H, Rayner M, Scarborough P (2016) Analysis and valuation of the health and climate change cobenefits of dietary change. Proc Natl Acad Sci 113(15):4146–4151. https://doi.org/10.1073/pnas.1523119113
Springmann M, Wiebe K, Mason-D'Croz D, Sulser TB, Rayner M, Scarborough P (2018) Health and nutritional aspects of sustainable diet strategies and their association with environmental impacts: A global modelling analysis with country-level detail. Lancet Planet Heal 2:e451–e461. https://doi.org/10.1016/S2542-5196(18)30206-7
Springmann M, Clark M, D'Croz D, Wiebe K, Bodirsky B, Lassaletta L, Vries W, Vermeulen S, Herrero M, Carlson K, Jonell M, Troell M, DeClerck F, Gordon L, Zurayk R, Scarborough P, Rayner M, Loken B, Fanzo J, Godfray H, Tilman D, Rockstrom J, Willett W (2018a) Options for keeping the food system within environmental limits. Nature 562:519–525. https://doi.org/10.1038/s41586-018-0594-0
Stehfest E, Bouwman L, Vuuren D, Elzen M, Eickhout B, Kabat P (2009) Climate benefits of changing diet. Clim Chang 95:83–102. https://doi.org/10.1007/s10584-008-9534-6
Steinfeld H, Gerber P, Wassenaar T, Castel V, Rosales M, Haan C (2006) Livestock's long shadow: environmental issues and options. FAO, Rome
Strom S (2015) Edible coffee cup in KFC tests in Britain. https://www.nytimes.com/2015/02/26/business/the-chicken-at-kfc-is-ok-but-the-cups-are-delicious.html. Accessed 25 Dec 2019
Sultan B, Gaetani M (2016) Agriculture in west Africa in the twenty-first century: climate change and impacts scenarios, and potential for adaptation. Front Plant Sci 7:1262. https://doi.org/10.3389/fpls.2016.01262
Taub D, Miller B, Allen H (2008) Effects of elevated CO2 on the protein concentration of food crops: a meta-analysis. Glob Change Biol 14:565–575
Tilman D, Clark M (2014) Global diets link environmental sustainability and human health. Nature 515:518–522
Vermeulen S, Campbell B, Ingram J (2012) Climate change and food systems. Annu Rev Environ Resour 37:195–222

Water Scarcity Clock (2019) Worldwater.io. https://worldwater.io/. Accessed 28 Dec 2019

WHO (2018) Double burden of malnutrition. https://www.who.int/nutrition/double-burden-malnutrition/en

World Bank (2018) Bringing the concept of climate-smart agriculture to life: insights from CSA country profiles across Africa, Asia, and Latin America. World Bank, Washington, DC

World Meteorological Organization (2019) Global climate in 2015–2019: climate change accelerates. https://public.wmo.int/en/media/press-release/global-climate-2015-2019-climate-change-accelerates. Accessed 20 Dec 2019

World Resource Institute (2000) World GHG emissions flow chart. World Resource Institute, Climate Analysis Indicators Tool (CIAT). http://cait.wri.org/figures.php?page=/World-FlowChart

World Resource Institute (2019) Reduce growth in demand for food and other agricultural products (synthesis). https://wrr-food.wri.org/course/reduce-growth-demand-food-and-other-agricultural-products-synthesis

Zhu C, Kobayashi K, Loladze I, Zhu J, Jiang Q, Xu X, Liu G, Ebi K, Drewnowski A, Fukagawa N, Ziska L (2018) Carbon dioxide (CO2) levels this century will alter the protein, micronutrients, and vitamin content of rice grains with potential health consequences for the poorest rice-dependent countries. Sci Adv 4(5):eaaq1012. https://doi.org/10.1126/sciadv.aaq1012

Resilient Measures in Face of Climate Change to Strengthen Food and Nutritional Security

4

D. Vijayalakshmi and Mrunal D. Barbhai

Abstract

A distinct but complex relationship exists between climate, food, and nutritional security of human beings. Climate change impacts agriculture, livestock, fisheries, etc. reducing its productivity and yield leaving an adverse effect on the economic condition of the people, which in turn affects food availability, accessibility, and utilization. With increased food and nutritional insecurity, the nation is at risk of higher malnutrition. To address these insecurities, all sectors like agriculture, food, environment, health should join hands. Various solutions can be taken up for tackling the problems like diversification of agriculture, leading to diet diversification and introduction of innovative alternative nutritious food sources. Bringing into limelight the underutilized fruits, vegetables, cereals, and grains could help reach food and nutritional security. Blending indigenous knowledge and scientific understanding can pave ways to improve the utilization of existing resources. Value addition of empty-calorie food with food wastes obtained during processing such as vegetable and fruit peels, cereal brans, etc. can increase the nutritional quality of such food products contributing to nutritional security. To remove food and nutritional insecurities, nutrition education will render a helping hand alo with the capacity building of people.

D. Vijayalakshmi (✉)
Department of Food Science and Nutrition, University of Agricultural Sciences, GKVK, Bengaluru, India

M. D. Barbhai
Department of Foods and Nutrition, Professor Jayashankar Telangana State Agricultural University, Hyderabad, India

V. K. Hebsale Mallappa, M. Shirur (eds.), *Climate Change and Resilient Food Systems*, https://doi.org/10.1007/978-981-33-4538-6_4

Keywords

Climate change · Nutrition security · Malnutrition · Diet diversification · Indigenous knowledge

4.1 Introduction

Climate change significantly influences agriculture, forestry, livestock, health, etc. Climate changes have not been defined universally in a concrete way. It is referred to as "change in the state of the climate that can be identified (e.g., by using statistical tests) by changes in the mean and/or the variability of its properties and persisting for an extended period, typically decades or longer" (IPCC 2014). According to UNFCCC, "climate change is directly or indirectly attributed to human activities in addition to natural climate variability that alters the composition of the global atmosphere over comparable periods." This separates climate variability associated with natural causes from anthropogenic activities caused by climate change. A general term by WMO (World Meteorological Organization) encompasses all the inconsistencies in climate despite their statistical nature and physical cause. A more restricted term used by WMO (1992), updated in 2005, suggests climate change as "significant change, with important economic, environmental and social effects, in the mean values of a meteorological element (particularly temperature or amount of precipitation) during a certain period, where the means are taken over periods of a decade or longer" (FAO 2008). WMO uses an average period of 30 years that helps to eliminate the climate variations within year to year.

Human activities fuel climate change significantly. It has accelerated the process of global warming by releasing greenhouse gases (GHGs) through fossil fuel ignition, changing earth's average temperature. Industrial revolution escalated quantities of GHGs like carbon dioxide, methane, nitrous oxide, etc., creating a blanket effect trapping the heat on earth's surface. Global average temperature rise recorded by the Goddard Institute for Space Studies (GISS) is up to 1 °C since 1880 (NASA 2020). Influences of climate change are already obvious, like a rise in surface temperatures, melting glaciers leading to rising sea level, heat and cold waves, drought, and declining quality of water and air (Riebeek 2010). All these factors affect agriculture as it is closely related to and depends on weather and climate. Agricultural production is adversely affected by climatic conditions like drought, floods, heat waves, cold waves, etc. (Niles et al. 2017). Agriculture is deeply intertwined with food systems. Both agriculture and food system have impacted climate change and vice versa. Food systems contribute approximately 19–29% to the world's GHGs out of which 80–86% is contributed by agricultural activities (Vermeulen et al. 2012). Soil erosions and deforestation also contribute to climate change (Niles et al. 2017). Food systems include all the activities related to production, processing, distribution, preparation, and consumption of food and their interactions within and between biogeophysical and human environments, and they enclose the effect of these acts on food security, viz., (1) food availability related to production, distribution, and exchange; (2) food access related to affordability, allocation, and preferences; and (3) food utilization associated to nutritional and social value and food safety (GECAFS 2005). Vermeulen et al. (2012) reported that the food systems affect climate through its pre- and post-processing, storage (chilling, cold storage), transport, and wastage (wastage during production, processing, storage, transport, and also plate waste). Food security is defined as "all people, at all times, have physical and economic access to sufficient, safe, and nutritious food to meet their dietary needs and food preferences for an active and healthy life" (FAO 1996) and is supported by food systems. Thus, it can be understood that food security can be reached by fulfilling four dimensions: availability, accessibility, utilization, and stability.

Food insecurity can lead to compromising the nutritional status of individuals leading to nutritional insecurity. Food security and nutrition security go hand in hand and share a complex relationship (Hwalla et al. 2016) emphasizing on nutritional and health components of individual or community with access to a wholesome and adequate diet and clean drinking water, combined with hygienic and excellent health-care services for all individual members of the community (FAO, WFP and IFAD 2012). In this era of climate change, where the burden of undernutrition and overnutrition exists, it is the need of the hour to adapt agricultural practices and diets for achieving food and nutritional security.

4.2 The Linkage Between Climate Change, Agriculture, Food, Nutrition, and Health

Agriculture, food, nutrition, and health go hand in hand and share a complex relation. If food production is insufficient, then it may lead to hunger and malnutrition. The linkage between agriculture, climate change, food system, nutrition, and health is illustrated in Fig. 4.1. Anthropogenic actions are one of the major reasons for climate change that are rapidly affecting the composition of the global atmosphere. Emission of GHGs has resulted in increased temperature, which in turn has accelerated melting of glaciers, rise in sea levels, and extreme weather conditions, viz., frequent and intense floods, dry spells, storms, etc. Agricultural activities and allied sectors like livestock production, fisheries, etc. get affected and add up to climate change. The extensive application of synthetic fertilizers is one of the prime contributors of GHGs from the pre-production activities. Animal feed production from livestock fields accounts for 45% of emissions (Vermeulen et al. 2012). Agricultural activities are dependent and deeply intertwined with local climate and weather, e.g., rainfall, temperature, and winds affect crop production, productivity, and cultivation. Elevated temperatures and resulting heat stress, elevated CO_2 levels, extreme weather events, pests, changed rainfall patterns, etc. are affecting agriculture and food production (FAO 2008; Otieno et al. 2013; Thornton et al. 2018). All the abovementioned changes are adversely affecting food supply locally and globally. Agriculture not only is the main provider of food, which satisfies the basic human need, but also generates employment for 36% world population and 40–50% in Asia

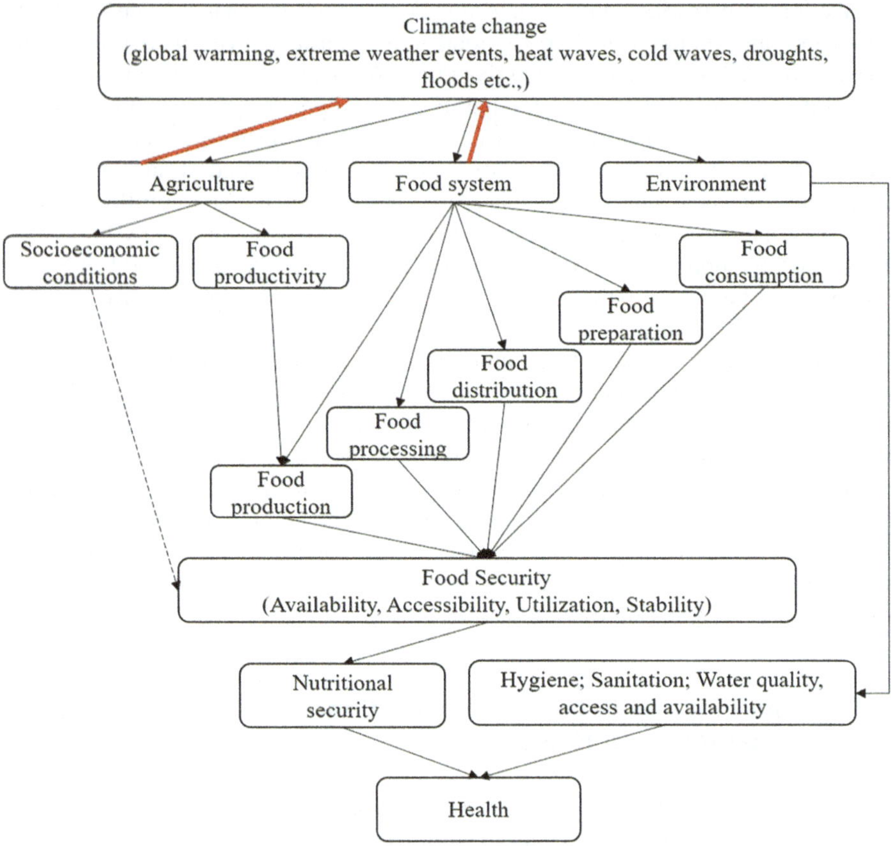

Fig. 4.1 Climate change linkages between agriculture, nutrition, and health

and the Pacific (FAO 2008). Around 78% of the world's poverty-struck people reside in rural sectors depending majorly on farming (World Bank 2014). Thus, the negative consequences of climate change on farming also affect a population's socioeconomic status, which further reduces their purchases and buying ability resulting in inadequate consumption. Reduced food supply, consumption, purchasing power, and accessibility results in hunger and malnutrition. The food system and environmental conditions are also affected by climate change, further risking human health as a whole. Post-production activities in food system such as processing, distribution, cold chain, and marketing also pitch into climate change. These account to long-term effects of climate change on all the four dimensions of food security (Vermeulen et al. 2012; Masipa 2017). Increased frequencies of rainfall, floods, heat stress, etc. may damage the roadway and railway infrastructure, compromising the transport of food and food distribution. This will reduce access to food to certain groups of the population (Rao et al. 2017). Environmental health conditions related to sanitation and hygiene are at risk with changing climate. Changes in rainfall pose a threat to water availability. A shift in rainfall may cause the risk of water scarcity and unavailability of good quality water. Climate change may also increase incidences of food-, vector-, and water-borne diseases affecting health and diminishing nutrient absorption leading to a vicious circle of malnutrition. Vector such as mosquitoes, flies, and ticks can carry pathogenic bacteria and viruses causing various diseases and infections. Altering air composition due to climate change might lead to increased allergies, asthma, respiratory tract problems, and also cardiovascular problems in the population (USGCRP 2016).

4.3 Urbanization Influence on Climate, Agricultural Production, Lifestyle, and Consumption Pattern

The concept of "urbanization" has been explained by the theory of modernization of infrastructure (Srivastava 2017). It is referred to as a shift of population from village areas to metropolitan cities (McGranahan and Satterthwaite 2014). The process of urbanization started with the industrial revolution. It is estimated that 6 billion will dwell in urban areas due to urbanization by 2050 (McCarthy et al. 2010). Urbanization is associated with economic growth (Regmi and Dyck 2001) and better-centralized infrastructure, educational systems, and health facilities (Knorr et al. 2018). It has a significant effect on lifestyle both in positive and negative ways. On one hand, it is associated with economic growth, but on the other hand, it also poses challenges of income inequalities, environmental burden, health issues, and increase in urban poor population. It affects the rural-urban dynamics. Urbanization has always posed a risk on the environment due to pollution, overburden on resources, unplanned expansion of metropolitan cities, urban heat islands, etc. Land use changes with urbanization resulting in the creation of heat islands (Zhou 2004). Regions with a high growing population in urban areas also have the potential to cause urban heat island causing a dramatic effect on the health of the people. These urban heat islands in populous cities have to project been shown a similar or greater climate change effect as compared to that of GHG emissions (Stone 2007; Campbell-Lendrum and Corvalan 2007). Thus, urbanization can be associated with warming, increased CO_2 emission, and extreme heat events resulting from increased energy consumption, use of electrical appliances at home and in industries, ever-growing vehicle congestion on roads, etc. (Pigeon et al. 2007; McCarthy et al. 2010). The health of the urban population is also at risk due to floods and storms resulting because of housing construction with a poor quality, unplanned, inadequate, and improper drainage system. In such situations, heavy rains, resulting in flash floods and water logging, make people vulnerable to water- and vector-borne infections (Campbell-Lendrum and Corvalan 2007).

Urbanization also decreases the direct interaction of people with food production as they are not involved with farmers or in any food production activities (Jennings 2015). Urbanization also has affected the food demands, consumption patterns, lifestyle due to increased working hours, and per capita income of the population (Regmi and Dyck 2001; Jennings 2015; Knorr et al. 2018). Convenience ready-to-

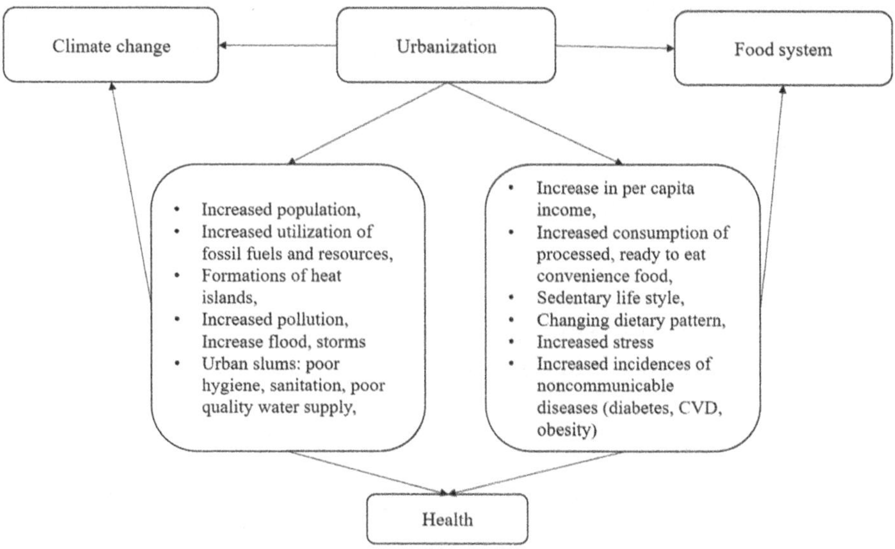

Fig. 4.2 The linkage between urbanization, climate change, and health

eat food and processed foods high in fat, salt, sugars, etc. have become parts of the diets accompanied with less physical activity, sedentary lifestyle, and stress increasing the risk of non-communicable diseases like metabolic syndrome X, diabetes, cardiovascular diseases, increased obesity, etc. (Fig. 4.2).

4.3.1 Effect on Agriculture and Allied Sectors

Agriculture faces a negative impact due to climate change in low-latitude and tropical areas especially affecting major staple cereals (wheat, rice, etc.). In contrast, there might be some beneficial effects for crop production in high-latitude regions. The decline in agricultural productivity is estimated to affect more in regions like sub-Saharan Africa (9% decline) (Masipa 2017). Several studies conducted reveal agriculture being adversely affected in countries like Kenya, India, and Korea (Otieno et al. 2013; Chakrabarty 2016; Cenacchi, et al. 2016) (Table 4.1). Cereals are a staple diet for the majority, and changes or decrease in cereal crop production affects availability and distribution. Smallholding farmer groups are highly at risk, especially those depending on rainfed farming due to the changing seasonality exerted by climate change. A rise in CO_2 levels in the atmosphere might indirectly affect the nutritional composition of crops such as cereals, pulses, legumes, and beans, reducing nutrients, viz., protein, iron, zinc, and calcium (Tirado et al. 2013; Myers et al. 2014; Thornton et al. 2018; Hummel et al. 2018; Beach et al. 2019). This will indirectly affect the nutritional status of human beings. Animal fodder also gets affected nutritionally, reducing milk and meat yield (Thornton et al. 2018). This will adversely affect the livestock sector. Undernutrition, coupled with obesity and

Table 4.1 Summary of direct and indirect consequences of changes in climate

Changes in climate	Effects on	End results
Agriculture and allied sectors		
Rise in temperature Extreme climatic events Altered rainfall Heat and cold waves Droughts Increased floods Storms/change in wind patterns	Reduced crop yields, crop production Reduced production of livestock Water shortage for irrigation Degradation of soil	Reduced income from agricultural sources Increased food prices Reduced affordability Decreased food access Leading to conditions of food insecurity Adverse impact on health of farmers, agricultural labor Adverse impact on fisheries
Water and sanitation		
Increased floods coupled with an improper drainage system and urbanization, releasing untreated wastewater in local water bodies Altered rainfall pattern	Unhygienic water Water shortage for drinking	Health problems associated with water-borne diseases Scarcity of clean drinking water Increase in vector- and food-borne diseases
Food systems		
Extreme climatic events Increased floods Storms/change in wind patterns, erratic rainfalls damaging connectivity	Decreased production Food wastage (from production to consumer plate waste) Improper transportation connectivity due to floods, extreme climate events damaging road and railways	Reduced food availability, increased food prices, decreased food consumption especially by poor Alerted nutritional and health status

hidden hunger, i.e., micronutrient deficiencies, still prevails even after continuous ongoing efforts with a slight reduction in malnutrition over the past few decades. In addition to this, decreased nutritional composition of plants will further contribute to malnutrition and slow down the process of attaining food and nutritional stability, especially in Asian, Middle Eastern, and North African countries. Climate change via extreme events may even have a deleterious impact on the health of farmers, agriculture, and fisheries due to extreme heat, droughts, inadequate drinking water, etc. (Meybeck et al. 2018).

Challenges Faced by Agriculture and Allied Sector

- Changes/decrease in the production of crop: e.g., production of rice, maize, wheat, and cash crops like cocoa and coffee is affected in many regions.
- Changes in nutritional content: Crops produced may be rich in carbohydrates and deficient in other nutrient content, viz., protein, iron, zinc, and calcium. Even the forages and fodder will have similar effects of reduced nutritional content.
- The decreased nutritional content of fodder, leading to reduced livestock production and milk yields.

4.3.2 Effect on Food System

A food system includes all agricultural activities, transport and intake, and these together affect food and nutrition security (Table 4.1). As discussed earlier, declining production and increasing population will cause stress on food availability that is defined by FAO (2008) as "physical quantities of food produced, stored, processed, and transported." It also includes imports and exports and is calculated using food balance sheets. It accounts for net food remaining after deducting exports from production, stock, and imports of all items included in food balance sheets (FAO 2008). The growing population increases the number of households to be fed. When this is coupled with decreased agricultural production and an increase in food demand, the food prices also tend to increase, leading to reduced food availability forcing the poor to buy and consume less than their requirements. This will, in turn, result in food insecurity by affecting the nutritional status of individual and increase malnutrition (Nelson et al. 2009). Inadequate consumption increases the risk and vulnerability to infections due to malnutrition and compromises body's ability to absorb nutrients (Crahay et al. 2010). According to FAO (2009a), food-deprived and hunger-struck population almost exceeds 1 billion globally.

On one hand, it is estimated that agriculture production should increase more than 60% for fulfilling the demand for the increased population. Still, on the other hand, it also should be noted that out of the current production, one third percent of the edible food portions is wasted or lost every year (Meybeck et al. 2018). The wastage can take place right from the production site, such as harvesting losses, processing losses due to lack of processing technology, and storage losses due to inadequate warehouses, storage facilities, and plate wastes. The wastes contribute to GHG emission, via., the landfill sites affecting climate change (Vermeulen et al. 2012). Higher temperatures can affect the shelf life of food, especially perishable items. Lack of proper transport facilities, viz., lack of road and rail facilities, may also contribute to food losses and wastage as a result of poor transport from its production site to the consumers. This may also limit the food access and increase food prices, again affecting the utilization of food by the consumers away from production sites.

4.3.3 Effect on Water and Sanitation

Safe and hygienic water, along with sanitary conditions, are needed to maintain health and nutrition. With changing climate and alteration in rainfall, the surface water resources are getting depleted (Table 4.1). It affects not only the quantity but also the quality of water (IPCC 2001; Crahay et al. 2010). Climate change, along with urbanization, adversely affects water quality with increased pollutants due to untreated drainage water, which may end up in local water bodies, leaving the water bodies polluted (USEPA 2002; Franco et al. 2018; Cullis et al. 2019). Increased frequencies of flood may pitch in unsanitary conditions leading to a spike in water-, food-, and vector-borne disease incidence, e.g., malaria, dengue, etc.

4.4 Transformation in Food Consumption Pattern and Its Consequences on Climate Change, Food, and Nutrition Security

The food consumption pattern is based on culture and food availability, determining individuals' nutritional and health status (Prabhat and Begum 2012) (Table 4.1). Diets have been changing due to increased urbanization associated with a sedentary lifestyle and increased income (Regmi and Dyck 2001; Popkin et al. 2012; Tilman and Clark 2014). Global nutrition transition with changed diets of individuals is evident over the period of several decades (Ghattas 2014). They have become more calorie-dense (Niles et al. 2017), especially in high-income countries. Some studies also suggest that the calorie intake of individuals, e.g., in Portugal, people have increased above recommended intakes (Galli et al. 2017). There is an increase in consumption of meat, processed and packaged convenience ready-to-eat products, bakery and confectionery items, energy-dense foods, etc. (Tilman and Clark 2014; Ghattas 2014) resulting in increased prevalence of non-communicable diseases like obesity, diabetes, cardiovascular diseases (CVD), and hyperlipidemia, lowering the life expectancy of the individual (Ng 2014; Popkin et al. 2012).

Urban dwellers' meat consumption pattern was found high than among the rural population in nations like China, Indonesia, Pakistan, etc. (Regmi and Dyck 2001). It has also upsurged the intake of meals outside the house, e.g., fast-food consumption at food joints, eating in restaurants, mobile food trucks, local food vendors, etc. (Knorr et al. 2018). Changes in lifestyle have also resulted in increased snacking habits like consumption of fried foods, bakery and confectionery products, and processed foods (Tefft et al. 2017). Urbanization provides a wide range of market for the purchase of food, but the urban poor still consumes limited food products due to a limited budget. Because of increased and irregular working hours of both men and women, there is limited time to prepare meals at home. Thus, they popularly prefer energy-dense foods; easy-to-prepare or ready-to-eat, convenience food; or eating outside the home (Jobbins and Henley 2015). Processed and packaged foods are gaining lots of attention due to their high palatability, advertisement promotion, and availability in various outlets, thus growing faster in low-income nations compared to high-income nations (Tefft et al. 2017).

Diets of people significantly affect GHG emission from a food system (Aleksandrowicz et al. 2016). Analyzing the effects of these changing dietary patterns and food consumption trends on food and nutrition security is vital. Diets with higher animal products have high GHG emission compared to plant-based diets. Reduction in animal-based foods and focus on vegan, vegetarian, and pescatarian diets may positively influence the environment. Vegan diets had a higher capacity to reduce emission of GHGs. More than 70% reduction in GHG emissions and land usage and a 50% reduction in water usage were observed as a result of a shift in western to environmentally suitable dietary patterns (Aleksandrowicz et al. 2016).

The varying diets and shift from nutrient-dense meal to consumption of energy-dense meals coupled with a sedentary lifestyle are increasing the burden of malnutrition with overnutrition leading to obesity and related complications. Increased consumption of processed foods and ready-to-eat foods adversely affects the climate with the emission of GHGs during processing and also affects the health of an individual, putting them at a higher risk of nutritional insecurity. Diets only focusing on energy-rich foods tend to cause micronutrient deficiencies. Other changes like reduction in consumption of leafy vegetables and fruits with increased snacking habits due to availability of more processed, ready-to-eat, deep-fried snacks in the market shelf and their access to even rural areas are having an adverse effect on health and contributing to malnutrition and nutritional insecurity. The double burden of malnutrition is experienced with increased obesity on one hand and undernourishment, stunting, wasting, and micronutrient deficiencies on the other.

In one scene, increased income due to urbanization has led to increased calorie consumption. However, micronutrient deficiencies still prevail, and, in another picture, due to insufficient purchasing power, majority rural population and urban poor are forced to decrease consumption resulting in hunger and undernutrition.

4.5 Possible Innovative Solutions/Alternatives to Strengthen Food and Nutritional Status Amid Climate Change

Malnutrition is still a major problem that needs to be addressed at the earliest. Multiple malnutrition grades, undernourishment (wasting, stunting, underweight), overnourishment, obesity, and nutritional deficiencies (especially micronutrients) are affecting all the countries. Insufficient access to a balanced and nutritious diet affects the health of the population (Govender et al. 2017). Globally, almost 45% of children from middle-income and low-income countries die due to undernutrition, but there is also a rise in childhood obesity. Among all the adults, 1.9 billion are above normal weight or obese, and 462 million are underweight (WHO 2018). Along with these prevailing conditions, climate change through increased heat stress, droughts, floods, storms, pollution, and changes in air composition has further worsened the situation by growing threats of diseases, hunger, malnutrition, and food and nutritional insecurity. This is especially more conspicuous among the urban poor and majority rural population (Watts et al. 2015). Tackling climate change to maintain individual's nutritional status remains the biggest challenge of twenty-first century. A multi-sectorial adaptation and mitigation approach is required to solve this problem. Adaptation is the process to adjust with existing climate or expected changes and its effect. It also includes facilitating interventions for adapting to expected climate and using opportunities for reducing harm. Mitigation measures in terms of climate change are human interventions to reduce sources of emissions and increase the sinks of GHGs (IPCC 2014). The following section highlights the possible adaptations and mitigation solutions that can be taken up to combat the adverse effects of climate change on human health and food and nutritional security.

4.5.1 Possible Solutions Through Acclimatization and Mitigation in Farming and Food System

The food system, as discussed earlier, includes activities right from production to food consumption, i.e., reaching the consumer's plate. In addition to agriculture that deals with food production, all the activities related to storage, transport, and food handling also come under the food system. Various pathways on climate change can be adopted to make agriculture sustainable, resilient for combating its harmful effect.

Sustainable, nutrition-sensitive, and climate-smart farming could be a possible solution to be adopted against climate change. Sustainable agriculture aims at conventional and organic production practices helping to improve resource efficiency and minimize waste. This system unifies plant and animal production, which aims at producing human food, animal feed, and fuel for satisfying the need of the ever-growing population. Sustainable agriculture intends to save the environment and maintain soil fertility. It also ensures food and nutritional security for everyone in present and future (Velten et al. 2015; HLPE 2016).

Nutrition-sensitive agriculture directs at providing affordable, nutritious, safe food in both quality and quantity to meet the nutritional requirements of the population, thus improving health and nutritional status of an individual. Adding a nutrition lens to agriculture requires coordination and cooperation of multiple stakeholders from agriculture and health and nutrition sectors. Nutrition sensitivity in agriculture includes approaches such as diversification of agriculture, improving horticulture productions, production of nutrient-dense crop varieties, and improving the quality of livestock and fisheries (Harvey et al. 2014; FAO 2017). Agriculture caters to food needs, and it is an integral part of reducing hunger and malnutrition. Only increasing the food production is not enough to improve the nutritional status of the population; thus, nutrition-sensitive agriculture programs focusing on diversification of agriculture, rearing livestock, and dairy and promoting micronutrient-rich crops may lead to dietary diversity (Ruel et al. 2017). For making agriculture nutrition-sensitive, some steps can be taken like incorporating nutrition objectives in agricultural projects and timely assessment of indicators to ensure achievements of nutritional goals. Following up impact of the project from farm to plate (production to consumption) for checking complete implementation and nutritional benefit of the project is essential. Collaborating and involving all the stakeholders in a project also make it more successful (Garrett and Kennedy 2015).

Climate-smart agriculture includes a set of approaches aiming at sustainable increase in productivity, resilient agriculture and food security system, reduction or removal of GHGs, and promotion of food and nutritional security (FAO 2013a). In many developing nations like India, agricultural productivity needs improvement to cater to the needs of a growing population (Chakrabarty 2016). Climate-smart agriculture also focuses on poverty reduction and economic upliftment of farmers (Steenwerth et al. 2014).

4.5.2 Water Management

Water is the main resource for agriculture as it is mainly dependent on rainfall and climate. Thus, scarcity of water in the era of climate change is the biggest challenge to be faced, and climate-resilient practices need to be adopted. Irrigation facilities should be focused on managing water properly. The wise use of irrigation facilities and promoting micro-irrigation could be possible solutions for water resource management. Efficient irrigation practices like micro-irrigation, drip irrigation, sprinkle irrigation, and water harvesting according to field situations will be beneficial strategies (Dickie et al. 2014; FAO 2015; Rao et al. 2017). Similarly, drought-, pest-, and disease-resistant, temperature-tolerant crop varieties can also be developed and adopted. This may help to reduce crop losses due to extreme climatic conditions and improve productivity.

4.5.3 Reduce Deforestation

Forests are one of the best examples of carbon sink. Forest management via ensuring agroforestry and reducing deforestation will improve resilience to climate change. Planting trees will not only help in removing carbon, but it will also contribute to food security (FAO 2015). Deforestation resulting from transforming the forest into cultivable lands is one of the sources of emission of GHGs, which can be mitigated through sustainable agricultural intensification and conservation of forests.

4.5.4 Diversification of Agriculture

Diversification of agriculture could contribute to dietary diversity as it aims at increasing the availability and accessibility of diverse foods. Thus, agriculture extension services can promote crop diversity and integrated approaches for improved nutrition. Rice production is a major contributor to GHG emissions from agriculture (Saxena et al. 2018). Thus, efforts to reduce emissions from rice fields, popularizing and expanding no-tillage farming, and water management are essential. New varieties of rice with enhanced yields and nutritional qualities and less water requirement need to be developed. Consultative Group for International Agricultural Research (CGIAR) and FAO are promoting soil, water, and nutrient management through Rice Integrated Crop Management Systems (RICMS) (FAO 2008). Growing non-paddy crops and focusing on the production of other cereal grains can also help to reduce emissions and increase dietary diversity (Rao et al. 2017). Focusing on horticulture and olericulture can also add variety to the diets by including various vegetables and fruits, thus improving vitamins, antioxidants, and micronutrients in the diet.

4.5.5 Blending Traditional Knowledge with Modern Technology and the Preservation of Resources

Traditional expertise or indigenous knowledge stems from an accumulation of experience throughout generations, which is passed on orally from generation to generation. Indigenous knowledge has a cultural base specific to particular tribes, locality, region, etc., and it is learned from observation of nature, thus being nature friendly. Combining indigenous knowledge with newer technology can provide a powerful tool to promote food and nutritional security during climate changes (Gyampoh et al. 2009; Porter et al. 2014). Understanding the importance of indigenous practices, FAO promotes and develops innovative projects supporting the use of traditional knowledge (FAO 2009b). Traditional knowledge regarding agricultural practices are the local experience of thousands of years of the folk. Cropping systems like mixed cropping where farmers grow more than two crops at a time increase the diversity of crops and also reduce the risk from failure of a single crop (Singh and Singh 2017). Agroforestry is blending agriculture with forestry. It benefits the farmers by growing trees in fields and improving soil fertility, creating a favorable microclimate for crops, providing fruits, wood, etc. (Mbow et al. 2014; Meybeck et al. 2018). Integrated farming with conservation goals and wildlife-friendly and land-sharing practices is one way of protecting biodiversity (Garnett et al. 2013). Some, e.g., of agriculture from different ethnic-cultural groups of Vietnam, include "rock pocket agriculture" that can be adopted in other places where similar conditions exist. In rock pocket agriculture, rock walls are created on a slope which protects the soil inside it from erosion for almost 6–7 years, and maize or other such crops are cultivated there (Vien 2003). Complementing and supplementing agriculture with livestock, aquaculture, and horticulture in the integrated farming system can be useful in dealing with current conditions (Descheemaeker et al. 2016; Meybeck et al. 2018), e.g., an integrated farming system with fish farming, where the farmers can benefit from synergism of both conventional farming and fish cultivation. This approach can be utilized by various smallholding farmers. It will also help in improving food and nutritional security. Such integrated agriculture-aquaculture is practiced in various countries in Asia, such as Vietnam, Indonesia, Malaysia, Bangladesh, China, India, Thailand, etc. (Phong 2010).

4.5.6 Promotion of Millets Through Increasing Their Production

Millets are termed as "nutri-cereals" as they are nutrient-rich and provide various health benefits. Minor cereals and millets have a lower carbon footprint compared to rice and wheat. Millets are also drought-resistant crops and can survive in conditions like less water and higher temperature (Saxena et al. 2018). Major millets are popularly cultivated, and minor millets, viz., kodo, proso, barnyard, foxtail, little, and browntop millets, are also gaining importance due to their nutritional content. All these millets are stress and drought tolerant and can be grown in regions with water scarcity, as they require low inputs. Both their nutritional benefits and ability to withstand lack of water and increased temperature coupled with reduced GHG emission make them the most suitable solution for food stability during climate change. Agrarian-nutritional importance of millets calls for the attention of promoting millets through various agriculture and nutritional programs. These underutilized grains have wide natural diversity and can be further explored to have high yielding varieties with more research attention (Kumar et al. 2018; Upadhyaya and Vetriventhan 2018).

4.5.7 Soil, Crop, and Livestock Management

Soil restoration and maintenance of soil health and grasslands are some potential mitigation solutions having low to moderate costs. Regular soil testing will help to make an appropriate decision in nutrient management. Crops grown in healthy soils will produce higher yields, thus increasing availability of better and nutritious foods. Crop yields can be managed or increased by changing crop dates, especially for cereals and oilseeds, to prevent from drought. Early sowing and transplanting may improve crop yields (Porter et al. 2014). Livestock management can also help in the reduction of GHGs. Improving feeds and feeding practices for livestock can help in the reduction of methane from animal production and increase productivity in terms of milk, eggs, and meat. This can lead to increased availability, thus making diets diversified and nutritious (Dickie et al. 2014; FAO 2015).

4.5.8 Management of Fertilizers and Its Better Application

This can be achieved through genetically modified plant breeds that provide the same or increased yields even with the application of lesser fertilizers. Training on the precise and appropriate use of fertilizers and their optimum dosage for particular crops can reduce the burden on soil and environment. Excessive use of synthetic fertilizers poses a serious threat on the crop as well as soil. It also has an adverse impact on climate (Dickie et al. 2014). Organic fertilizers should be used to enrich soil nutrients.

4.5.9 Use of ICT (Information and Communication Technology) and Remote Sensing Tools

Remote sensing can help in providing a location-specific solution for the problems. Once the site-specific personalized issues are identified, personalized messages can be sent through mobile SMS. Various such pilot studies are being taken up globally. Efforts are also taken in developing countries. One such pilot study conducted during 2015 in India at Anantapur district is *Harita Priya*, a precision agricultural initiative by the government of Telangana in collaboration with the Centre for

Development of Advanced Computing (C-DAC) providing personalized agro-advisories based on information collected from farms to farmers in regional language (Telugu), viz., SMS (CDAC 2016).

Technological advancement in agriculture should help in establishing, monitoring, and detection systems to give precautionary warnings about climate change and pest attacks. Warning system providing early detection of pest and diseases outbreak can help in taking quick preventive measures (Rao et al. 2017).

4.5.10 Decreasing Food Wastage and Adopting Nutritious Diets

Nutrient-sensitive sustainable climate-smart agriculture should be coupled with a nutrition-smart food system, and for it to be nutrient-sensitive, focus should be on all its stages right from producing to consuming. Avoiding food losses and wastage is a core concern to achieve sustainability in the food system (SDSN 2013). Food losses occur at crop production stage due to deficient food supply chain; improper infrastructure, technology, knowledge, and skills; and inadequate access to a market, whereas food waste occurs at consumer stage due to discarding foods unfit for human consumption (e.g., packaged food crossing expiry date, plate waste, etc.). It compromises the availability of food and adds to the burden of agriculture, demanding an increase in production (FAO 2013b; Garnett et al. 2013). It also poses an additional burden on water, land, and emission of GHGs. Food wastage is estimated to be one third of that associated with consumption. If we can reduce the food losses and wastage at each stage of the food chain, then we may help in reducing the burden on agricultural production (Dickie et al. 2014). Post-harvest storage and processing play an essential role in maintaining the quality of the food. Inadequate storage facilities, transport facilities, and processing plants may increase the food wastage. Thus, adequate infrastructure should be developed for the same.

Along with reduced food wastage, shifting to nutritious diets is most important. With urbanization there is increased consumption of refined carbohydrates, sugars, and fats leading to declining in dietary diversity. Unhealthy diets are a global burden in promoting diseases (Caron et al. 2018).

4.5.11 Increasing Focus on Underutilized Crops (Fruits, Vegetables, Pseudo-Cereals, Grains, Etc.)

Diversification of agriculture has been discussed in the sections earlier. Diversification, in its extended form, will focus on underutilized crops, leading to increased options of a variety of nutritious food items in the shelf. This, in turn, may reduce food prices. Thus, scaling up and focus on diversification of underutilized crops are required at a local, regional, national, and global level (FAO 2017). Improving yields of underutilized crops and wild foods will help in providing numerous options, especially for the poor. Various underutilized cereals, fruits, and vegetables are available, which can add to food and nutritional security, such as quinoa, drumstick leaves (*Moringa oleifera*), karonda (*Carissa carandas*), jujube, etc. For example, in India drumstick leaves—with abundant antioxidants, protein, and minerals especially iron—have the potential to be used as a source of functional food for value addition of baked goods. It can also be used in preparing *chutney* (Durst and Bayasgalanbat 2014). Quinoa—an underutilized pseudo-cereal also called as golden grain of Andes, with ample amount of proteins, amino acids, minerals, and phytochemicals such as phenols, saponin, etc.—also helps in diversifying diet and making it nutrient-rich (Farinazzi-Machado et al. 2012). Backyard home gardening of some indigenous, underutilized food can be promoted to increase the nutrition of family diet.

4.5.12 Biofortification, Fortification, and Value Addition of Available Food Products to Enhance Their Nutritional Composition

There are still nutritional deficiencies existing in population. Thus, biofortification of foods will help to improve nutritional value of foods. Promoting biofortified food will improve food and nutritional security (Garnett et al. 2013). Efforts are being taken by various organizations like the ICAR-Central Research Institute for Dryland Agriculture (CRIDA) to grow biofortified crops. ICAR-Indian Institute of Rice Research (IIRR), Hyderabad, and ICAR-Central Rice Research Institute (CRRI), Cuttack, are genetically modifying rice to make it more nutritious. The ICAR-Directorate of Mushroom Research and ICAR-Central Institute of Post-Harvest Engineering and Technology, Ludhiana, and CSIR- CFTRI, Mysore, are developing products from biofortification of different crops. Developing low-cost nutrient-rich foods will improve health status. This can be done especially for the children as they are the most vulnerable. Involving government agencies, NGOs, and women SHGs in the development of nutritious value-added food can help to combat malnutrition, and involvement of women SHGs will make them financially stronger. ICRISAT is developing agribusiness platforms to support such ventures where entrepreneurs can be encouraged and linked to market facilities (Satarupa et al. 2018).

Food fortification is a cheap source to combat micronutrient deficiencies. Food fortification is an addition of essential nutrient or trace element to food for preventing or correcting the nutrient deficiency of a population. It includes restoration and enrichment, i.e., addition of nutrients lost during manufacturing (Allen et al. 2006). Popular examples of fortification are iron and iodine fortified salt, vitamin A and D fortified oils, fortified wheat flours, etc. Fortification can be mandatory/compulsory or voluntary and market-driven or mass (for the population as a whole) or target fortification (for specific groups).

Various processing methods such as dehulling, dehusking, polishing, milling, etc. are responsible for the removal of bran and bran fractions from cereals and millets. Processing techniques used to prepare fruit juices, pulps, jams, jellies, etc. are responsible for removal of peel and pomace from fruits. All these processing techniques result in improved palatability, but reduced nutritional qualities of the product. Removal of bran and peel results in removal of fiber component. It is also evident from some researches that these components are a rich source of antioxidants, minerals, etc. These by-products generated during processing are discarded as waste or some time used in animal feeds. Researchers are finding ways to incorporate such food wastes as novel food ingredients in an edible food product for value addition (O'Shea et al. 2012; Ahmad et al. 2018).

Brans of rice, wheat, oats, barley, foxtail millets, barnyard, sorghum, etc. are identified as a potential source of nutrients and are used in the value addition of products. These brans obtained as waste from processing industry could be exploited as nutraceuticals to enhance the value of products having less nutrients such as bakery items (Patel 2015). Various researches are taken up to develop products like high-fiber pasta incorporated with wheat bran (Sudha et al. 2011), sorghum bran-supplemented bread (Dahlberg et al. 2004), wheat bran-enriched muffins (Romjaun and Prakash 2013) and cookies (Ertas 2015), biscuits developed using rice bran protein concentrates (Yadav et al. 2011), etc. Phytochemical-, polyphenol-, carotenoid-, and dietary fiber-rich mango peel biscuits were developed by another researcher (O'Shea et al. 2012; Baddi et al. 2015). Peels and pomace of some fruits like apple and citrus lime are used alternatively as animal feed (O'Shea et al. 2012). Apple peels and pomace having high concentrations of antioxidants and phenolic compounds can be used in the preparation of sausages, jams, muffins, cereal bar, fruit bars, fruit leathers, etc. (Wolfe and Liu 2003; Henríquez et al. 2010; O'Shea et al. 2012). Grape seeds, skins, and stems, lime, orange, peach peels, and vegetable peels from carrot, onion, tomato, etc. can also be used as functional foods (O'Shea et al. 2012). Fruit seeds from avocado, jackfruit, longan, mango, and tamarind—rich source of antioxidants—can also be used as additives in the functional food industry (Soong and Barlow 2004). Banana pseudo-stem contributes to a lot of waste generated from banana cultivation every year, which is rich in sugar and minerals. It has many therapeutic benefits and can be used in value addition of foods. Research has been conducted to develop fermented beverages (Puranik 2017) which also contribute toward generating additional income for banana growers through waste utilization. Scaling up and popularization of such value-added foods can provide a helping hand in promoting food and nutritional security. It may also reduce some GHG emission that is contributed from food wastes.

4.5.13 Alternative Innovative Food Products

Innovative alternatives should be promoted and implemented to alleviate malnutrition. Some such alternatives include consumption of edible insects, lab-grown meat, seaweeds, etc. Edible insects are consumed traditionally in various parts of Asia and Africa. They are rich sources of proteins, amino acids, fat, vitamins, and minerals, nearly similar to that of consumed livestock. Some insects popularly consumed are crickets, silkworm pupae, beetles, larvae, bees, ants, etc. (Tao and Li 2018). Insect production/harvesting/rearing is a low-cost affair. Thus, in the face of growing malnutrition and climate change, edible insects could prove as super-food as they are nutrient-rich and also emit less GHGs (Huis et al. 2013). With all the nutritional benefits spoken, insects are not part of the majority of diets. This is due to the disgusting concept associated with the consumption of insects and also some section of the population being vegetarians and vegans. For the other section of the population who merely reject edible insect due to disgust factor, innovative processing techniques can be applied to convert insects into flours and pastes to be value-added to other products. Lab-grown meat or cultured meat is one such innovative approach to deal with malnutrition in an era of climate change as it can help to cut down GHG emission. But there are a lot of factors to be considered before it popularly hits the market floor, such as safety regulations; cultural, social, economic, and political background; consumer acceptance; etc. (Stephens et al. 2018). Seaweeds are another such alternative with a highly nutritive profile containing vitamins, proteins, minerals, dietary fiber, and essential fatty acids (Ortiz et al. 2006). Seaweeds form part of traditional diets in China, Japan, and Korea with wide range of products that can be prepared like sushi, noodles, pickles, soup, tea, wine, jams, jellies, cheese, chocolates, salad, curry, etc. (Kaliaperumal 2003).

4.5.14 Empowerment and Capacity Building

Training and capacity building will help to improve their skills in adjusting to climate changes (Suchiradipta and Saravanan 2018). Making communities aware about climate change through various empowerment and capacity building programs, teaching them regarding the causes, impacts, and adverse effects, will make them more aware of climate change and help in adopting new ideas and experimenting with new crop varieties to adopt climate change (Kindra 2010). Empowering women socially and economically and politically should be focused, as they are more at risk of malnutrition and undernutrition as they lack access to many facilities. It is also observed that if a woman is controlling the household, there is an improvement in child nutrition (Chung 2012). In many countries, women eat last after feeding the whole family. They also have to face secondary status compared to men in decision-making. Agriculture extension service can help in building a more robust program with integrating gender and nutrition in agriculture. This will help in enhancing the contribution of women to household income, reducing gender gaps, and improving the nutritional status of families. It has been observed in many studies that involvement of females in decision-making in agriculture, family expenditure, and food distribution can improve the nutritional status of the family (Satarupa et al. 2018). Women can be involved in activities like the value addition of food products. Further market opportunities can be created for these products to increase their economic stability that will contribute to improving household practices of preservation and nutrition improvement.

4.5.15 Nutrition Education Through Educational Institutions, Farmer Field Schools, and Mass Media

Nutrition education helps in acquiring nutrition-related knowledge and skills through instruction and training (ADA 2011). It assists the population to make an informed decision regarding their food choices, health, and well-being. Nutrition education is a good solution to tackle malnutrition and climate change (FAO 2017). Nutrition, agriculture, environment, and extension departments should join hands to tackle the problem of food and nutrition security coupled with climate change. Strengthening extension services is one step toward addressing these issues (Garnett et al. 2013).

The integration of nutrition intervention within agricultural platforms is a must to reach a larger section of the population. Nutrition education through already existing agriculture programs as complementary services will fetch larger participation at the village level. Thus, imparting nutrition education to people via agricultural extension systems as they already have established infrastructure, close connection with farmers, and community trust will prove beneficial (Babu et al. 2015; Fanzo 2015). Collaboration is required within government staff, health-care workers, agriculture subject matter specialists, and extension workers, to create awareness and adopt high-impact nutritional practices (SPRING Bangladesh 2017; Aakesson et al. 2014).

Using technologies with the participation of the villagers, health workers, and women self-help groups (SHGs) while preparing and disseminating the nutrition-related videos will help in educating more people in a rural area as they well accept the information (Kadiyala et al. 2016).

Farmer field schools (FFS) are based on experiential learning principles, where farmers with same interest join together for a weekly meeting to learn about agricultural practices. A similar approach can be used to promote nutrition education. The concept of FFS can be used as a base to develop Farmer Nutrition School (FNS), which will help to reduce malnutrition (SPRING Bangladesh 2016).

School-based approach nutrition education can help in improving child nutrition (Tirado et al. 2013). Including nutrition education and backyard gardening exposure to children at an early age can help in developing healthy eating habits. Connecting people and farms is one way to diversifying the diet; based on the concept of connecting farms, farmers' and consumers' efforts can be taken to involve the non-farming community in agriculture production via exposure visits to farms, especially for the preschool and school-going children. This will create awareness about locally grown foods and understanding their importance in the diet. An excellent example of nutrition education based on such concept is created by Rutgers Cooperative Extension (RCE) for 3–8-year-old children called "From Our Farms" (Hughes 2002).

4.5.16 Policy Changes and Support

Policy-level changes and support should be given to adopting climate-smart technologies to achieve sustainable agriculture, reduced GHGs, effective and efficient food system, and equal access to food for all (Suchiradipta and Saravanan 2018). Encouraging research on linkages between food, nutrition, health care, and climate change for developing models providing holistic solutions is needed. More investment on researches can help to improve monitoring and surveillance systems and provide a better, realistic, implementable solution (Watts et al. 2015).

The collaboration of various research agencies, a multidisciplinary approach on research, will throw light on all angles of the problem. Coordinating government and non-governmental organizations will also improve the reach and access to extend the climate-smart interventions toward food and nutritional security. Dealing with climate change alone is not going to fetch any positive results; thus, the multidisciplinary, multi-sectorial approach is required. Thus, collaborating and empowering different government ministries and departments, viz., environment, agriculture, women and child development, and health, will help to fight climate change with a holistic approach (Watts et al. 2015; Meybeck et al. 2018).

Developing national and international cooperation such as South-South cooperation, where expertise is exchanged between the developing countries helping each other with knowledge, technical assistance, and/or sometimes investments, will help to get a wider vision of the problem and access to new solutions or technologies used by other countries to deal with similar problems (Meybeck et al. 2018).

Public and private sector tie-ups and collaboration between all stakeholders, viz., scientists in various fields (agriculture, water, environment, health, and food research), farmers, policymakers, technology providers, and engineers, were suggested in Extreme Weather and Resilience of the Global Food System Report (2015).

Developing a reliable monitoring system to observe and analyze weather and climate will enable providing early warnings (FAO 2008) and help in better utilization of available weather information and management tools. Other policies include promoting soil testing and soil health card schemes and strengthening crop management system. Insurance schemes can be implemented to deal with climate change risks.

Promoting agrobiodiversity and effective use of natural resources is an approach that assists in tackling and increasing adaptability toward deleterious impacts of climate change. Integrated policies should be formed and implemented to boost agroforestry. Funds should be provided for such initiatives. Various trained groups at the local level can be formed to promote, educate, and train people about agroforestry (Meybeck et al. 2018).

Poverty can be a root cause for undernutrition. Thus, poverty eradication should be focused by policymakers, and initiatives should be taken for the same. Policies promoting access and availability for the consumption of nutritious foods should be taken up to reduce food and nutritional insecurities (Ghattas 2014).

Health-related issues are still increasing due to climate change. Thus, policies providing maternal and child health care should be strengthened, and more investment and planning should be done to formulate such policies. Strengthening surveillance systems giving early alerts of infectious diseases and environmental risk factors can help in the prevention of various food-, water-, and vector-borne diseases (Tirado et al. 2013).

Policies should be established to promote, protect, and utilize underutilized foods and traditional knowledge and practices. Protecting traditional knowledge and practices with the help of intellectual property rights can safeguard them. Policies should be formulated to develop standards for underutilized foods. The promotion of traditional knowledge can be done through its inclusion in school and college curriculum. Incentives should be given for the people who help in documenting and following traditional practices based on indigenous knowledge as this is a way to deal with climate change (Durst and Bayasgalanbat 2014).

4.6 Summary and Conclusion

With the majority relying on farming, and extreme stress on natural resources, the influence of climate change on agriculture is visible globally, especially in low-latitude and tropical regions. These forecasted impacts may further decrease yields of staple cereals in many countries contributing to food insecurity. Livelihood security of poor will be worst hit in the long run due to difficulty in maintaining agricultural productivity and the consequences of climate change.

All the four dimensions of food security, important in meeting dietary needs, can be affected by changes in agricultural production resulting from climate change, thus lowering food production. Some short-term impacts on many food crops include a reduction in yield which can be offset by adaption measures based on available technologies like altering planting dates, heat-tolerant varieties, efficient water management, conservation agriculture, protected cultivation, etc. But addressing long-term impacts requires investments in strategic research, extension, and helping farmers in risk management through innovative insurance services. Droughts, floods, and extreme weather events decrease crop yields and nutritional properties and also affect incomes and livelihoods of farmers and farm laborers due to loss of wages. The nutritional quality of foods can be compromised in terms of protein and mineral content due to depletion of micronutrients in the soils.

Existing statistics on malnutrition suggest high rates with 39% of adults globally are overweight or obese and approximately 20 million babies have low birth weight thus are more prone to malnutrition. Anemia is also one of the major prevailing problems in women and girls. It is also a well-known fact that deficiency of micronutrients like iron and zinc can cause hidden hunger. Climate change, coupled with the current malnutrition status, may worsen the situation further. There are few technology and policy direction, which need attention in planning adaptation toward climate change, i.e., prioritization and focus on vulnerable hotspots, investments on technology generation and dissemination, prudent use of natural resources, leveraging the ongoing schemes and missions, livelihood diversification and risk transfer, financing adaptation costs, and capacity building. Investment plans and policies in agriculture, rural infrastructure, and water resources need to be tailored according to the requirements of regions and communities. Apart from climate change, rapidly changing food preferences and habits are challenging agriculture to produce quality food across the seasons. Preferences for proteins, carbohydrates, specific fatty acids, vitamins, minerals and other secondary metabolites, fiber, etc. have increasing trends for a complete diet. Growing fruit trees at the farm, kitchen garden, and along farm borders and roads provides easy access to fruits, which will help in sequestration and achieving ecological balance. Food basket has to be expanded by exploiting edible neglected or underutilized species. These can encourage the farmers toward diversification. Nutrition education, along with the diversification of crops and kitchen gardening, can help in improving the dietary habits leading to better nutritional status among the population.

A multidisciplinary perspective is necessary for dealing with climate change and food and nutritional security. Measures need to be taken for ensuring the food quality at production and consumption levels. Collaborations within various departments linked with climate, agriculture, food, health, and NGOs should be strengthened, and interconnections between social activists, health workers, extension personnel, etc. will help to combat climate change more effectively.

References

Aakesson A, Pinga V, Titus S (2014) Using agriculture extension agents to promote nutrition: a process review of three feed the future activities in Ethiopia. Arlington, VA: USAID/ strengthening partnerships, results, and innovations in nutrition globally (SPRING) project

ADA (2011) International dietetics and nutritional terminology-reference manual: standard language for the nutrition care process, 3rd edn. American Dietetic Association, Chicago

Ahmad F, Pasha I, Saeed M, Asgher M (2018) Biochemical profiling of Pakistani sorghum and millet varieties with special reference to anthocyanins and condensed tannins. Int J Food Prop 21(1):1586–1597

Aleksandrowicz L, Green R, Joy EJM, Smith P, Haines A (2016) The impacts of dietary change on greenhouse gas emissions, land use, water use, and health: a systematic review. PLoS One 11 (11):e0165797. https://doi.org/10.1371/journal.pone.0165797

Allen L, Benoist B, Dary O, Hurrell R (2006) Guidelines on food fortification with micronutrients. World Health Organization and Food and Agriculture Organization of the United Nations, Geneva. https://www.who.int/nutrition/publications/guide_food_fortification_micronutrients.pdf

Babu SC, Havimo T, Pehu E (2015) Fostering agriculture–nutrition links, recommendations for agriculture extension curriculum reforms in India. World Bank, Washington, DC

Baddi J, Vijayalakshmi D, Durgannavar NA, Chandru R (2015) Mango peel: A potential source of natural bioactive phyto-nutrients in functional food. Asian J Dairy Food Res 34(1):75–77. https://doi.org/10.5958/0976-0563.2015.00016.0

Beach RH, Sulser TB, Crimmins A, Cenacchi N, Cole J, Fukagawa NK, Mason-D'Croz D, Myers S, Sarofim MC, Smith M, Ziska LH (2019) Combining the effects of increased atmospheric carbon dioxide on protein, iron, and zinc availability and projected climate change on global diets: a modelling study. Lancet Planet Health 3:307–317

Campbell-Lendrum D, Corvalan C (2007) Climate change and developing-country cities: implications for environmental health and equity. J Urban Health 84(1):109. https://doi.org/10.1007/s11524-007-9170-x

Caron P, Loma-Osorio GFYD, Nabarro D, Hainzelin E et al (2018) Food systems for sustainable development: proposals for a profound four-part transformation. Agron Sustain Dev 38:41. https://doi.org/10.1007/s13593-018-0519-1

CDAC (2016). Center for Development of Advanced Computing (C-DAC). https://www.cdac.in/index.aspx?id=pdf_annualreport_15-16. Accessed 25 Sept 2019

Cenacchi N, Lim Y, Sulser TB, Islam S, Mason-D'Croz D, Robertson RD, Kim C, Wiebe KD (2016) Climate change, agriculture, and adaptation in the Republic of Korea to 2050: an integrated assessment. IFPRI discussion paper 1586. International Food Policy Research Institute (IFPRI), Washington, DC. http://ebrary.ifpri.org/cdm/ref/collection/p15738coll2/id/130999

Chakrabarty M (2016) Climate change and food security in India. ORF issue brief no. 157. New Delhi, pp 1–12. http://cf.orfonline.org/wp-content/uploads/2016/09/ORF_IssueBrief_1571.pdf

Chung K (2012) An introduction to nutrition-agriculture linkages. MINAG/DE research report 72E. Directorate of Economics, Ministry of Agriculture, Maputo, Mozambique. https://www.spring-nutrition.org/sites/default/files/chung_msu_2012_ag-nutrition_linkages.pdf

Crahay P, Mitchell A, Gomez A et al. (2010) The threats of climate change on undernutrition—a neglected issue that requires further analysis and urgent actions SCN, 2010. Newsletter #38—climate change: food and nutrition security implications. http://www.unscn.org/files/Publications/SCN_News/SCN_NEWS_38_03_06_10.pdf

Cullis JDS, Hornb A, Rossouwa N, Fisher-Jeffes L, Kunnekeb MM, Hoffman W (2019) Urbanisation, climate change and its impact on water quality and economic risks in a water scarce and rapidly urbanising catchment: case study of the Berg River catchment. H2Open J 2 (1):146–167. https://doi.org/10.2166/h2oj.2019.027

Dahlberg JA, Wilson JP, Snyder T (2004) Sorghum and pearl millet: health foods and industrial products in developed countries. CFC and ICRISAT. Alternative uses of sorghum and pearl millet in Asia: proceedings of the expert meeting, ICRISAT, Patancheru, Andhra Pradesh, India, 1–4 July 2003. CFC technical paper No. 34. Amsterdam, Common Fund for Commodities, International Crops Research Institute for the Semi-Arid Tropics, Andhra Pradesh, India. p 364, ISBN 92-9066-471-1

Descheemaeker K, Oosting SJ, Tui SH, Masikati P, Falconnier GN, Giller KE (2016) Climate change adaptation and mitigation in smallholder crop-livestock systems in sub-Saharan Africa: a call for integrated impact assessments. Reg Environ Chang 16:2331–2343. https://doi.org/10.1007/s10113-016-0957-8

Dickie A, Streck C, Roe S, Zurek M, Haupt F, Dolginow A (2014) Strategies for mitigating climate change in agriculture: abridged report. Climate focus and California environmental associates, prepared with the support of the climate and land use alliance. Report and supplementary materials. www.agriculturalmitigation.org

Durst P, Bayasgalanbat N (2014) Promotion of underutilized indigenous food resources for food security and nutrition in Asia and the Pacific. Food and Agriculture Organization of the United Nations, Rome

Ertas N (2015) Effect of wheat bran stabilization methods on nutritional and physico-mechanical characteristics of cookies. J Food Quality 38:184–191

Extreme Weather and Resilience of the Global Food System (2015) Final project report from the UK-US taskforce on extreme weather and global food system resilience. The Global Food Security Programme, London

Fanzo J (2015) Integrating nutrition into rural advisory services and extension. Note 9. GFRAS good practice notes for extension and advisory services. GFRAS, Lindau, pp 1–4

FAO (1996) Report of the world food summit. FAO, Rome. http://www.fao.org/3/w3613e/w3613e00.htm

FAO (2008) Climate change and food security: a framework document. Food and Agriculture Organization of the United Nations, Rome. http://www.fao.org/forestry/15538-079b31d45081fe9c3dbc6ff34de4807e4.pdf
FAO (2009a) The state of food insecurity in the world. In: Economic crisis—impacts and lessons learned. Food and Agriculture Organization, Rome. http://www.fao.org/tempref/docrep/fao/012/i0876e/i0876e_flyer.pdf
FAO (2009b) FAO and traditional knowledge: the linkages with sustainability, food security and climate change impacts. Food and Agriculture Organization of the United Nations, Rome. http://www.fao.org/3/a-i0841e.pdf
FAO (2013a) Climate-smart agriculture source book. Food and Agriculture Organization of the United Nations, Rome. ISBN 978–92–5-107720-7 (print) E-ISBN 978–92–5-107721–4 (PDF). http://www.fao.org/3/i3325e/i3325e.pdf
FAO (2013b) Food wastage footprint: impacts on natural resources. Food and Agriculture Organization of the United Nations, Rome
FAO (2015) Review of climate change adaptation and mitigation in agriculture in the United Republic of Tanzania mitigation of climate change in agriculture (MICCA) Programme background report 7. Food and Agriculture Organization of the United Nations, Rome
FAO (2017) Nutrition-sensitive agriculture and food systems in practice options for intervention. Food and Agriculture Organization of the United Nations, Rome. http://www.fao.org/3/a-i7848e.pdf
FAO, WFP and IFAD (2012) The state of food insecurity in the world 2012. Economic growth is necessary but not sufficient to accelerate reduction of hunger and malnutrition. Rome, FAO. http://www.fao.org/3/a-i3027e.pdf
Farinazzi-Machado FMV, Barbalho SM, Oshiiwa M, Goulart R, Pessan O (2012) Use of cereal bars with quinoa (Chenopodium quinoa W.) to reduce risk factors related to cardiovascular diseases. Ciênc Tecnol Aliment Campinas 32(2):239–244. https://doi.org/10.1590/S0101-20612012005000040
Franco S, Gaetano V, Gianni T (2018) Urbanization and climate change impacts on surface water quality: enhancing the resilience by reducing impervious surfaces. Water Res 144:491–502. https://doi.org/10.1016/j.watres.2018.07.058
Galli A, Iha K, Halle M, El Bilali H, Grunewald N, Eaton D, Capone R, Debs P, Bottalico F (2017) Mediterranean countries' food consumption and sourcing patterns: an ecological footprint viewpoint. Sci Total Environ 578:383–391. https://doi.org/10.1016/j.scitotenv.2016.10.191
Garnett T, Appleby MC, Balmford A et al (2013) Sustainable intensification in agriculture: premises and policies. Science 341(6141):33–34. https://doi.org/10.1126/science.1234485
Garrett J, Kennedy S (2015) Nutrition-sensitive agriculture and rural development. International Fund for Agricultural Development, Rome
GECAFS (2005) Science plan and implementation strategy. Earth system science partnership (IGBP, IHDP, WCRP, DIVERSITAS) report no. 2. p 36, Wallingford. http://www.ihdp.unu.edu/docs/Publications/GECAFS/GECAFS_Report_1_Science_Plan_and_Implementation_Strategy.pdf
Ghattas H (2014) Food security and nutrition in the context of the nutrition transition. FAO, Rome. http://www.fao.org/economic/ess/ess-fs/voices/en/
Govender L, Pillay K, Siwela M, Modi A, Mabhaudhi T (2017) Food and nutrition insecurity in selected rural communities of KwaZulu-Natal, South Africa—linking human nutrition and agriculture. Int J Environ Res Public Health 14(17):1–21. https://doi.org/10.3390/ijerph14010017
Gyampoh BA, Amisah S, Idinoba M, Nkem J (2009) Using traditional knowledge to cope with climate change in rural Ghana. In: Perlis A, Unasylva (eds) Adapting to climate change, No. 231/232 60:1–2. Food and Agriculture Organization of the United Nations, Rome
Harvey M, Dangour A, Lambert R et al. (2014) Can agriculture interventions promote nutrition? Agriculture and nutrition evidence paper march 2014. UK Department for International Development. https://pdfs.semanticscholar.org/5531/e74654e5cec01ec2d02dc9b58301afb966a5.pdf

Henríquez C, Speisky H, Chiffelle I, Valenzuela T, Araya M, Simpson R, Almonacid S (2010) Development of an ingredient containing apple Peel, as a source of polyphenols and dietary Fiber. J Food Sci 75(6):172–181. https://doi.org/10.1111/j.1750-3841.2010.01700.x

HLPE (2016) High level panel of experts sustainable agricultural development for food security and nutrition: what roles for livestock? High Level Panel of Experts on Food Security and Nutrition of the Committee on World Food Security, Rome

Hughes JL (2002) Using agriculture as the foundation for an extension nutrition education program. J Extension 40(6):1–5

Huis AV, Itterbeeck JV, Klunder H, Mertens E, Halloran A, Muir G, Vantomme P (2013) Edible insects: future prospects for food and feed security. Food and Agriculture Organization of the United Nations, Rome. http://www.fao.org/3/i3253e/i3253e.pdf

Hummel M, Hallahan B, Brychkova G et al (2018) Decline in nutritional quality of common bean under climate change induced drought stress in Africa. Sci Rep 8:16187. https://doi.org/10.1038/s41598-018-33952-4

Hwalla N, Labban SE, Bahn RA (2016) Nutrition security is an integral component of food security. Front Life Sci 9(3):167–172. https://doi.org/10.1080/21553769.2016.1209133

IPCC (2001) In: JJ MC, Canziani OF, Leary NA, Dokken DJ, White KS (eds) Climate change 2001: impacts, adaptation, and vulnerability contribution of working group ii to the third assessment report of the Intergovernmental Panel on Climate Change. Intergovernmental Panel on Climate Change, Cambridge University Press. https://library.harvard.edu/collections/ipcc/docs/27_WGIITAR_FINAL.pdf

IPCC (2014) Annex II: glossary. In: Mach KJ, Planton S, von Stechow C, Pachauri RK, Meyer LA (eds) Climate change 2014: synthesis report. Contribution of working groups I, II and III to the fifth assessment report of the intergovernmental panel on climate change. IPCC, Geneva, pp 117–130. https://www.ipcc.ch/site/assets/uploads/2018/02/AR5_SYR_FINAL_Annexes.pdf

Jennings S, Cottee J, Curtis T, Miller S (2015) Food in an urbanised world: the role of city region food Systems in resilience and sustainable development. FAO, Rome, pp 1–85

Jobbins G, Henley G (2015) Food in an uncertain future: the impacts of climate change on food security and nutrition in the Middle East and North Africa. Overseas Development Institute/World Food Programme, London/Rome

Kadiyala S, Morgan EH, Cyriac S, Margolies A, Roopnaraine T (2016) Adapting agriculture platforms for nutrition: a case study of a participatory, video-based agricultural extension platform in India. PLoS One 11(10):e0164002. https://doi.org/10.1371/journal.pone.0164002

Kaliaperumal N (2003) Products from seaweeds. In: Proceedings of the national seminar on "Marine biodiversity as a source of food and medicine", SDMRI, 26–28 September 2002. SDMRI Research Publication No. 3, pp 33–42

Kindra J (2010) Empower people to become resilient. In: SCN, 2010. Newsletter #38—climate change: food and nutrition security implications. http://www.unscn.org/files/Publications/SCN_News/SCN_NEWS_38_03_06_10.pdf

Knorr D, Khoo CSH, Augustin MA (2018) Food for an urban planet: challenges and research opportunities. Front Nutr 4:73. https://doi.org/10.3389/fnut.2017.00073

Kumar A, Tomer V, Kaur A, Kumar V, Gupta K (2018) Millets: a solution to agrarian and nutritional challenges. Agric Food Secur 7:31. https://doi.org/10.1186/s40066-018-0183-3

Masipa TS (2017) The impact of climate change on food security in South Africa: current realities and challenges ahead Jàmbá. J Disaster Risk Stud 9(1):a411. https://doi.org/10.4102/jamba.v9i1.411

Mbow C, Noordwijk MV, Luedeling E et al (2014) Agroforestry solutions to address climate change and food security challenges in Africa. Curr Opin Environ Sustain 6:61–67. https://doi.org/10.1016/j.cosust.2013.10.014

McCarthy MP, Best MJ, Betts RA (2010) Climate change in cities due to global warming and urban effects. Geophys Res Lett 37:L09705. https://doi.org/10.1029/2010GL042845

McGranahan G and Satterthwaite D (2014) Urbanisation concepts and trends. IIED Working Paper. IIED, London. http://pubs.iied.org/10709IIED

Meybeck A, Laval E, Lévesque R and Parent G (2018) Food security and nutrition in the age of climate change. In: Proceedings of the international symposium organized by the government of Québec in collaboration with FAO. Québec City, September 24–27, 2017. FAO, Rome. p 132

Myers SS, Zanobetti A, Kloog I et al (2014) Increasing CO_2 threatens human nutrition. Nature 510:139–142. https://doi.org/10.1038/nature13179

NASA Earth Observatory (2020). World of change: global temperatures. https://earthobservatory.nasa.gov/world-of-change/global-temperatures. Accessed 26 Apr 2020

Nelson GC, Rosegrant MW, Koo J et al (2009) Climate change impact on agriculture and costs of adaptation. Food policy report. International Food Policy Research Institute, Washington, DC. https://doi.org/10.2499/0896295354

Ng M, Fleming T, Robinson M et al (2014) Global, regional, and national prevalence of overweight and obesity in children and adults during 1980-2013: a systematic analysis for the global burden of disease study 2013. Lancet 384(9945):766–781. https://doi.org/10.1016/S0140-6736(14)60460-8

Niles MT, Ahuja R, Esquivel MJ et al (2017) Climate change and food systems: assessing impacts and opportunities. Meridian Institute, Washington, DC. https://pdfs.semanticscholar.org/4ead/7c6212d181350cdc7d2527bec8a4b8cf3995.pdf. Accessed 04 Sept 2019

Ortiz J, Romero N, Robert P et al (2006) Dietary fiber, amino acid, fatty acid and tocopherol contents of the edible seaweeds *Ulva lactuca* and *Durvillaea Antarctica*. Food Chem 99:98–104

O'Shea N, Arendt EK, Gallagher E (2012) Dietary fibre and phytochemical characteristics of fruit and vegetable by-products and their recent applications as novel ingredients in food products. Innov Food Sci Emerg 16:1–10. https://doi.org/10.1016/j.ifset.2012.06.002

Otieno AC, Mugalavai V, Kamau EH (2013) Study on impact of climate change on the food and nutrition security in children of small-scale farmer households of Migori area. Glob J Biol Agric Health Sci 2(1):67–71

Patel S (2015) Cereal bran fortified-functional foods for obesity and diabetes management: triumphs, hurdles and possibilities. J Funct Foods 14:255–269. https://doi.org/10.1016/j.jff.2015.02.010

Phong LT (2010) Dynamics of sustainability in integrated agriculture aquaculture systems in the Mekong Delta. PhD thesis, Wageningen University, Wageningen, ISBN: 978–90–8585-662-7. https://pdfs.semanticscholar.org/3b3a/75471ee81f8090227ba72c844e8aef470ebf.pdf

Pigeon G, Legain D, Durand P, Masson V (2007) Anthropogenic heat release in an old European agglomeration (Toulouse, France). Int J Climatol 27:1969–1981. https://doi.org/10.1002/joc.1530

Popkin BM, Adair LS, Ng SW (2012) Global nutrition transition and the pandemic of obesity in developing countries. Nutr Rev 70(1):3–21. https://doi.org/10.1111/j.1753-4887.2011.00456.x

Porter JR, Xie L, Challinor AJ, Cochrane K, Howden SM, Iqbal MM, Lobell DB, Travasso MI (2014) Food security and food production systems. In: Field CB, Barros VR, Dokken DJ, Mach KJ, Mastrandrea MD, Bilir TE, Chatterjee M, Ebi KL, Estrada YO, Genova RC, Girma B, Kissel ES, Levy AN, MacCracken S, Mastrandrea PR, White LL (eds) Climate change 2014: impacts, adaptation, and vulnerability. Part A: global and sectoral aspects. contribution of working group II to the fifth assessment report of the intergovernmental panel on climate change. Cambridge University Press, Cambridge, pp 485–533

Prabhat A, Begum K (2012) Food consumption pattern and nutritional status of women laborers from coastal areas of Karnataka. Nat J Commun Med 3(2):321–325

Puranik S (2017) Microbial processing of banana pseudo stem core for development of fermented beverage. MSc thesis, University of Agricultural Sciences, GKVK, Bengaluru

Rao RK, Sharma PK, Raghuraman M (2017) Impact of climate change on food security. Indian J Agri Allied Sci 3(2):29–40

Regmi A, Dyck J (2001) Effects of urbanization on global food demand. Changing structure of global food consumption and trade/WRS-01-1 Economic Research Service, pp 23–30

Riebeek H (2010) Global warming. https://earthobservatory.nasa.gov/features/GlobalWarming. Accessed 9 Sept 2019

Romjaun ZZ, Prakash J (2013) Development and assessment of fibre-enriched muffins. Adv Food Sci 35(4):159–165

Ruel MT, Quisumbing AR, Balagamwala M (2017) Nutrition-sensitive agriculture what have we learned and where do we go from here? IFPRI Discussion Paper 01681. http://www.indiaenvironmentportal.org.in/files/file/Nutrition-sensitive%20agriculture.pdf

Satarupa M, Suchiradipta B, Saravanan R (2018) Integrating gender and nutrition in agriculture and advisory services. MANAGE discussion paper 2, MANAGE—Centre for Agricultural Extension Innovations, Reforms and Agripreneurship (CAEIRA), National Institute of Agricultural Extension Management, Hyderabad

Saxena R, Vanga SK, Wang J et al (2018) Millets for food security in the context of climate change: a review. Sustainability 10:2228. https://doi.org/10.3390/su10072228

SDSN (2013) Sustainable development solutions network. Thematic Group on Sustainable Agriculture and Food Systems of the Sustainable Development Solutions Network Solutions for Sustainable Agriculture and Food Systems. Technical report for the post-2015 development agenda. http://unsdsn.org/wp-content/uploads/2014/02/130919-TG07-Agriculture-Report-WEB.pdf

Singh R, Singh GS (2017) Traditional agriculture: a climate-smart approach for sustainable food production. Energ Ecol Environ 2(5):296–316. https://doi.org/10.1007/s40974-017-0074-7

Soong YY, Barlow PJ (2004) Antioxidant activity and phenolic content of selected fruit seeds. Food Chem 88:411–417. https://doi.org/10.1016/j.foodchem.2004.02.003

SPRING/Bangladesh (2016) Reducing malnutrition through farmer nutrition schools: key elements for implementation based on the SPRING experience in Bangladesh. https://www.spring-nutrition.org/sites/default/files/publications/briefs/spring_bangladesh_reducing_malnutrition_farmer_nutrition_schools.pdf

SPRING/Bangladesh (2017) Engaging government agriculture staff to promote nutrition at the community level. A retrospective analysis from Bangladesh. https://www.spring-nutrition.org/publications/briefs/engaging-government-agriculture-staff-promote-nutrition-community-level

Srivastava RK (2017) Climate change and urbanisation impact. Business standards. https://www.business-standard.com/article/printer-friendly-version?article_id=117102100637_1

Steenwerth KL, Hodson AK, Bloom AJ et al (2014) Climate-smart agriculture global research agenda: scientific basis for action. Agric Food Secur 3:11

Stephens N, Lucy Di Silvio LD, Dunsford I et al (2018) Bringing cultured meat to market: technical, socio-political, and regulatory challenges in cellular agriculture. Trends Food Sci Technol 78 (2018):155–166. https://doi.org/10.1016/j.tifs.2018.04.010

Stone B (2007) Urban and rural temperature trends in proximity to large US cities: 1951–2000. Int J Climatol 27:1801–1807. https://doi.org/10.1002/joc.1555

Suchiradipta DB, Saravanan R (2018) Climate smart agriculture towards triple win: adaptation, mitigation and food security. MANAGE discussion paper 5, MANAGE-Centre for Agricultural Extension Innovations, reforms and Agripreneurship (CAEIRA), National Institute of agricultural extension management, Hyderabad

Sudha ML, Ramasarma PR, Venkateswara RG (2011) Wheat bran stabilization and its use in the preparation of high-fiber pasta. Food Sci Technol Int 17(1):47–53

Tao J, Li Y (2018) Edible insects as a means to address global malnutrition and food insecurity issues. Food Qual Saf 2:17–26. https://doi.org/10.1093/fqsafe/fyy00

Tefft JF, Jonasova M, Ramziath ATOA, Morgan AM (2017) Food systems for an urbanizing world: knowledge product (English). World Bank Group, Washington, DC. http://documents.worldbank.org/curated/en/454961511210702794/Food-systems-for-an-urbanizing-world-knowledge-product

Thornton P, Dinesh D, Cramer L et al (2018) Agriculture in a changing climate: keeping our cool in the face of the hothouse. Outlook Agric 47(4):283–290. https://doi.org/10.1177/0030727018815332

Tilman D, Clark M (2014) Global diets link environmental sustainability and human health. Nature 515:518–522

Tirado MC, Crahay P, Mahy L et al (2013) Climate change and nutrition: creating a climate for nutrition security. Food Nutr Bull 34(4):533–547

Upadhyaya HD, Vetriventhan M (2018) Underutilized climate-smart nutrient rich small millets for food and nutritional security. In: Regional expert consultation on underutilized crops for food and nutritional security in Asia and the Pacific – thematic, strategic papers and country status reports. Asia-Pacific Association of Agricultural Research Institutions (APAARI), Bangkok, pp 109–120

USEPA (2002) What can you do to protect local waterways? United States Environmental Protection Agency. https://www3.epa.gov/npdes/pubs/centralized_brochure.pdf

USGCRP (2016) In: Crimmins A, Balbus J, Gamble JL, Beard CB, Bell JE, Dodgen D, Eisen RJ, Fann N, Hawkins MD, Herring SC, Jantarasami L, Mills DM, Saha S, Sarofim MC, Trtanj J, Ziska L (eds) The impacts of climate change on human health in the United States: a scientific assessment. U.S. Global Change Research Program, Washington, DC, p 312. https://doi.org/10.7930/J0R49NQX

Velten S, Leventon J, Jager N et al (2015) What is sustainable agriculture? A systematic review. Sustainability 7:7833–7865. https://doi.org/10.3390/su7067833

Vermeulen SJ, Campbell BM, Ingram JSI (2012) Climate change and food systems. Annu Rev Env Resour 37:195–222. https://doi.org/10.1146/annurev-environ-020411-130608

Vien TD (2003) Culture, environment, and farming Systems in Vietnam's Northern Mountain region. S Asia 41(2):180–205

Watts N, Adger WN, Agnolucci P et al (2015) Health and climate change: policy responses to protect public health. Lancet 23:2015. https://doi.org/10.1016/S0140-6736(15)60854-6

WHO (2018) Malnutrition. https://www.who.int/news-room/fact-sheets/detail/malnutrition

WMO (1992) International meteorological vocabulary, 2nd edn. WMO, Geneva. https://doi.org/10.1002/qj.49709339524

Wolfe KL, Liu RH (2003) Apple peels as a value-added food ingredient. J Agric Food Chem 51 (6):1676–1683. https://doi.org/10.1021/jf025916z

World Bank (2014). worldbank.org/en/news/feature/2014/11/12/for-up-to-800-million-rural-poor-a-strong-world-bank-commitment-to-agriculture

Yadav RB, Yadav BS, Chaudhary D (2011) Extraction, characterization and utilization of rice bran protein concentrate for biscuit making. Brit Food J 113(9):1173–1182

Zhou L, Dickinson RE, Tian Y, Fang J, Li Q, Kaufmann RK, Tucker JC, Myneni RB (2004) Evidence for a significant urbanization effect on climate in China. Proc Natl Acad Sci (PNAS) 101(26):9540–9544

Climate-Smart Agriculture: An Integrated Approach for Attaining Agricultural Sustainability

5

Abhilash, Alka Rani, Arti Kumari, Ram Narayan Singh, and Kavita Kumari

Abstract

Impacts of climate change (CC) and climate variability (CV) are felt throughout the world as it's getting hotter. As a result, irregular distribution of precipitation, a progressive rise in the oceanic levels, and more frequent occurrence of extreme weather events (EWE) have become a common phenomenon in recent years. Extended episodes of drought, floods, and shifts in the agroclimatic areas are threatening agricultural crop production (ACP) throughout the globe. However, their severity is especially felt and perceived more in developing and least developed nations from southern Asia and African regions. Their impacts are expected to become further severe as the average global temperature of Earth is projected to rise by another 1.1 °C to 6.4 °C till the end of the twenty-first century. In addition to being affected by CC and CV, agriculture also exacerbates it *via* emitting a large amount of greenhouse gases (GHGs) in the atmosphere. The GHG is reported to be intensive in usual production methods, such as conventional cultivation and plant nutrient and irrigation management systems by the farmers. In this context, climate-smart agriculture (CSA) can bring adaptation and

Abhilash (✉)
CCS Haryana Agricultural University, Hisar, Haryana, India
e-mail: abhilashaanu92@hau.ac.in

A. Rani
ICAR-Indian Institute of Soil Science, Bhopal, MP, India

A. Kumari
ICAR -Research Complex for Eastern Region, Patna, Bihar, India

R. N. Singh
National Institute of Abiotic Stress Management, Baramati, Maharashtra, India

K. Kumari
ICAR – National Rice Research Institute, Cuttack, Odisha, India

V. K. Hebsale Mallappa, M. Shirur (eds.), *Climate Change and Resilient Food Systems*, https://doi.org/10.1007/978-981-33-4538-6_5

mitigation strategies to sustain ACP. For instance, CSA contributes to the development of climate-resilient agricultural systems by increasing soil characteristics and the efficiency of water and nutrient use and by providing more stable yields and reducing emissions of GHGs. Although the advantages of CSA are broadly recognized, there is still a limited and dispersed holistic assessment of adaptation and the mitigation potential of CSA techniques.

Keywords

Climate-smart agricultural practices · Climate change · Climate variability · Climate change mitigation and adaptation · Sustainable agricultural development

5.1 Introduction of Climate-Smart Agriculture (CSA)

Human interferences like industrialization, deforestation, etc. have augmented the concentrations of GHGs in the atmosphere. As stated by the Intergovernmental Panel on Climate Change (IPCC (Intergovernmental Panel on Climate Change) 2007a), the atmospheric CO_2 concentration has been amplified more than 50% concerning to the pre-industrial concentration. The contribution of GHG emissions varied in different sectors like energy (57.8%), industry (21.7%), agriculture (17.6%), and waste areas (3%). Significant sources of GHG emissions are the cultivation of rice (20.9%), soils (13%), and burning of crop residues in fields (2%), which adds 35.9% of total emissions from agriculture (INCCA (Indian Network for Climate Change Assessment) 2010). All these factors have given rise to the problem of CC and CV. The impacts of CC and CV are felt as the global temperature is increasing and there are irregular precipitation pattern, gradual rise in sea levels, and increased frequency of EWE. Their negative impacts are expected to deepen shortly. Not only is agriculture affected by the impacts of CC and CV, but it also intensifies this problem by emitting GHGs through various farming practices. In this context, the CSA can bring adaptation and mitigation to sustain ACP. As identified and addressed by the FAO at the 2010 Hague Conference on Agriculture, Food Security, and CC, CSA corresponds to the accomplishment of sustainable development objectives. This combines three elements (financial, environmental, and social) of sustainable agricultural development (SAD) by tackling food, nutrition, and climate security challenges together.

CSA is a strategy for improving scientific rule and investment setting to achieve SAD to ensure food availability in the light of CC (FAO 2012). CSA is a strategy to design a framework of policy, capital, and technology to accomplish SAD for ensuring FNS in the context of changing climatic scenarios. CSA also seeks to improve the livelihoods and food safety, particularly of small and marginal farmers, by enhanced management and utilization of natural endowments and employing suitable production, development, storage, processing, and marketing strategies for agricultural commodities. CSA takes account of the socioeconomic and environmental conditions in which the potential gains are to be maximized and trade-offs minimized. Impacts on available regional resources and energy are also evaluated. The holistic landscape strategy, which embraces the concepts of environmental conservation, is centered on balanced land and water usage, which is an essential component here.

CSA is not a standard, specifically relevant agricultural technology or procedure. It is a methodology that calls for site-based evaluations to determine the correct techniques and methods for ACP. This approach:

a. Attempts to recognize and examine interconnected alternatives that establish synergetic advantages and reduce trade-offs in complex and closely related problems of SAD, FNS, growth, and CC
b. Identifies the choices sculpted by specific situations and capability of each nation and by the particular socioeconomic and environmental condition to which they are used
c. Evaluates the relationships among industries and the demands of various stakeholders engaged
d. Recognizes obstacles to adoption, particularly between farmers, and offers effective policy, strategy, intervention, and opportunity strategies and solutions
e. Strives for the integration of strategies, financial investment, and institutionalized structures to build supporting environments
f. Endeavors to accomplish several goals by recognizing the need to prioritize CSA practices and to agree together on its various potential advantages and compromises
g. Ought to give priority to improve living standards, particularly livelihoods of small farmers by enhancing access to information, awareness, knowledge and capital, resources, financial products, and marketing sector of economies
h. Improves resilience and enhances adaptability to shocks, particularly that associated with CC, as weather impacts have severe consequences on rural and agricultural development
i. Acknowledges the mitigation of CC as a probable secondary advantage, particularly for poor farming people
j. Strives to define the possibilities and recognize transparency gaps for climate-related funding and incorporate them into conventional agricultural investment financing outlets

CSA puts together and incorporates methods, strategies, and organizations which are not precisely firsthand but which are not familiar to peasants, shepherds, farmers, growers, or fishermen in the light of changing climate. What's unique and innovative is also the idea that the integrated and comprehensive solution to several challenges concurrently and internationally experienced by agriculture and food processes is approached concomitantly and comprehensively, which prohibits detrimental and ineffective strategies, regulation, funding, and investment. CSA primarily focuses on the following three objectives:

a. A sustainable rise in income and productivity

b. Enhanced adaptability to CC and CV
c. Contribution toward mitigation of CC

5.2 Climate Change and Climate Variability

It is imperative to understand the concept of CC and CV before gaining insights into CSA. According to IPCC, "Climate change is defined as a change in the state of the climate that can be identified (e.g., by using statistical tests) by changes in the mean and/or the variability of its properties and that persists for an extended period, typically decades or longer. CC may be due to natural internal processes or external forcings such as modulations of the solar cycles, volcanic eruptions, and persistent anthropogenic changes in the composition of the atmosphere or in land use" (IPCC (Intergovernmental Panel on Climate Change) 2018).

CC is a long-term gradual shift (rise or fall) in normal (e.g., normal temperature) or the range of weather conditions (e.g., frequency and severity of EWE). It is slow and steady and is quite hard to observe without scientific records of statistical data of climatic variables as it progresses gradually, dissimilar from year-to-year variability. It is attributable to both natural variabilities and human activity. It takes place as a result of variations in the earth's atmosphere, such as deviations in its orbit around the sun or alteration in the atmosphere due to human activities. There's nothing fundamentally wrong with CC. It was experienced in the past and will happen in the future again. The present problem comes from the rate of change – how rapidly changes are happening. There is strong evidence enough to prove that human-driven anthropogenic activities are causative of this exceptional, unprecedented rate of global temperature rise. Since the pre-industrial era in the mid-twentieth century, anthropogenic actions such as fossil fuel (FF) burning and land clearing for intensive agricultural practices and industrialization have enhanced to drastic GHG emissions in the atmosphere, which is responsible for trapping more thermal energy to raise the surface temperature of the earth.

CV is more harmful to agriculture and livelihood as compared to CC. According to IPCC, "Climate variability is defined as variations in the mean state and other statistics (such as standard deviations, the occurrence of extremes, etc.) of the climate on all spatial and temporal scales beyond that of individual weather events. Variability may be due to natural internal processes within the climate system (internal variability), or to variations in natural or anthropogenic external forcing (external variability)" (IPCC (Intergovernmental Panel on Climate Change) 2018). Scientists believe CV in the manner that it oscillates around the long-term statistical value of climatic normal on seasonal, annual, or decadal time scale. For easy understanding, this scenario can be divided into two parts: average and range. Working out the range gives a rough idea about average, and the converse is also true. So, both are complementary to each other.

5.2.1 Average

The World Meteorological Organization (WMO), Geneva, Switzerland, defines climate as the average weather conditions over 30-year time intervals. Such 30-year averages are termed as climatological normal, which are used to calculate, observe, and define the climate—or a particular subset of the climate—at a specific area. A 30-year data is sufficient enough to estimate the average, which is not affected by year-to-year fluctuations. Normal is determined using the data collected from weather stations for a number of climate parameters, like temperature and rainfall, in the area of concern. Normal is computed for many weather parameters, such as temperature or precipitation, depending upon available data from weather stations of the concerned region. In these 30-year averages, significant year-to-year variations can be notable. For example, in Fig. 5.1, the normal annual maximum temperature (T_{max}) for India is 29.61 °C from 1981 to 2010. But the annual T_{max} is not precisely 29.61 °C for each year in that range (e.g., in 1995 it was 30.18 °C; in 1997 it was 29.05 °C). This year-to-year instability around the statistically derived normal is CV.

5.2.2 Range

Only half of the story of CV is covered by average; the fluctuations around the average are another half, i.e., the range. To calculate the average, the variety within data is "smoothed," but there is so much to learn by studying this variation,

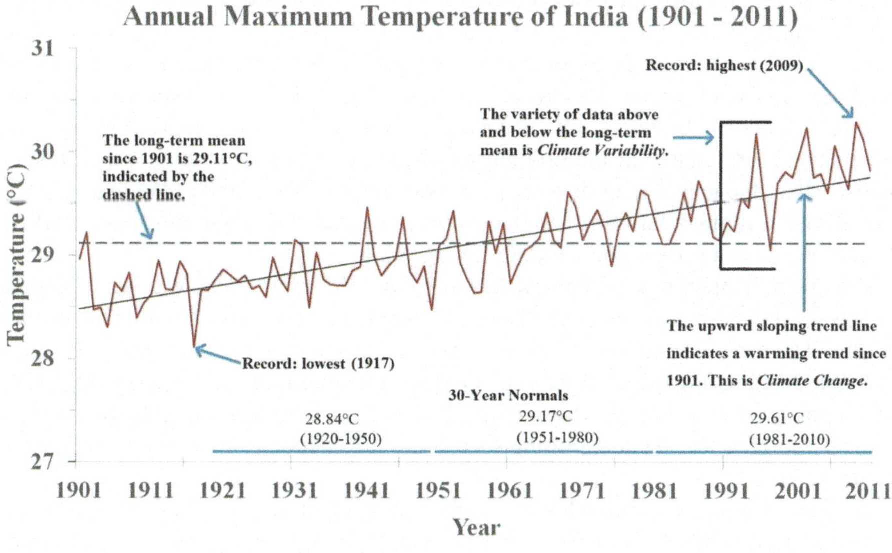

Fig. 5.1 Representation of all India long-term annual maximum temperature (T_{max}) (Data source: https://data.gov.in/resources/annual-and-seasonal-maximum-temperature-india)

particularly the extremes. For example, in Fig. 5.1, the long-term average for annual T_{max} in India is 29.11 °C from 1901 to 2011. But during that time frame, the lowest T_{max} was recorded in the year 1917 (28.11 °C), and the highest T_{max} was recorded in the year 2009 (30.29 °C), which shows the range of annual T_{max} around the long-term average. So, CV is the unanticipated and sporadic monthly or seasonal or periodic fluctuation in climate or its components without following any explicit trend with the past events. Both CC and CV have a significant effect on the ecosystem, agriculture, livelihood, and FNS. These problems are becoming severe due to a continuous increase in GHG emission owing to anthropogenic activities.

5.3 Impacts of Climate Change and Climate Variability on Agriculture

The latest documented fifth assesment report of IPCC (2014) shows that the atmosphere of the earth was exposed to extraordinary heating in recent decades, due to which hydrological cycle and rainfall trends across the planet will be significantly affected. All through the period from 1880 to 2012, global temperatures escalated by 0.85 °C. Weather and climate are the primary driving force that is thoroughly interlinked with ACP (Selvaraju et al. 2011). The amount, intensity, and nature of precipitation distribution are crucial to assess the features of rainy season, irrigation scheduling, and agricultural and livestock production systems. The inter-annual and intra-seasonal fluctuations in precipitation hinder ACP in tropical and subtropical ranges. The consequences of CC would have a remarkable impression on ACP, by decreasing productivity in many regions and increasing uncertainty in production to the degree that improvements would be required to sustain the ACP in the geographical region in which food is grown. Economically fragile tropical and subtropical nations that are already uncertain of their FNS will be severely altered by a decline in ACP and livestock production because of CC (IPCC (Intergovernmental Panel on Climate Change) 2007a). This will lead to insignificant trades, market shifts, cost fluctuations, and net food imports among the different nations, which, as a result, will escalate the gap among developed and developing economies (Padgham 2009). Small and marginal farmers will face these complicated, regional effects (IPCC (Intergovernmental Panel on Climate Change) 2007a).

Frequent happenings of natural hazards, like droughts, heat waves, wildfires, floods, cyclones, storms, etc., are also the characteristic traits of CC in the upcoming future. Tackling the effects of climate shocks in the current scenario may offer valuable insights for future situations. Overall, the length of growing period (LGP) in cropping season during which crop plant is subjected to temperatures above a critical point has lengthened, putting several crops under heat stress in different areas of the world.

Higher evaporative demand of atmosphere along with unpredictable precipitation patterns on the soils of semiarid tropical areas with high runoff and reduced water retention capability leads to an elevated risk of moisture deficit at every point during crop growth (Lipiec et al. 2013). Soil moisture deficits during initial plant growth stages lead to increased mortality of seedlings, disrupted plant growth, and diminished ACP. During late plant growth stages, several cases are observed in which moisture deficit is evident. Throughout the monsoon season, wet spells and occurrences of heavy precipitation produce waterlogging in the vicinity of roots, decrease crop growth, and impede operations in the farmland.

The length of the rainy season is among the key drivers, which significantly influence the opportunities for ACP. At any particular location, the onset and withdrawal of rainfall in the monsoon season differ immensely from 1 year to another. Along with the volatility in regional precipitation, higher temperatures during the crop growth period can have drastic effects on crop yield, agricultural income, and FNS (Battisti and Naylor 2009). In the course of crop-growing period, the temperature sometimes crosses the optimum limits ideal for metabolic, biochemical, and physiological processes like attaining different phenological stages, expansion of the leaf area, assimilation of photosynthates, and filling of grains. Higher ambient temperature prevailing near the flowering stage of a crop may decrease the viability of pollen grains, hence reducing the overall grain setting as well as ACP in most of the cereals grown in tropical areas (paddy, maize, pearl millet, jowar, etc.). Throughout plant development, the prevalence of elevated soil temperature is indeed a risk in semiarid and arid ecosystems like Indian, sub-Saharan African, and Australian regions where a failure of seedling emergence or thermal damages to the crop are common.

Current variability and extremes being observed in different climates are putting pressure on ACP systems, but additional CC-related issues that will arise in upcoming years will affect agricultural systems in the respective regions (Karl et al. 2008). In the current century, the severity, strength, and magnitude of tropical cyclones and extreme precipitation events are likely to rise over different regions of the globe (Knutson et al. 2010). Simultaneously, the fractions of arid acreage are expected to rise (Bates et al. 2008), which can be attributed to long dry summers, particularly in low- and mid-latitude subtropical areas. The occurrences of drought are also expected to escalate because of reduced rainfall and intensified evapotranspiration in these regions (IPCC (Intergovernmental Panel on Climate Change) 2012).

CC is also jeopardizing the biodiversity of agriculture. If the global mean temperature exceeds 1.5–2.5 °C as compared to what it was in 1980–1999, then roughly 25–30% of flora and fauna species evaluated this far are prone to the excessive danger of extinction (IPCC (Intergovernmental Panel on Climate Change) 2007a). The spectrum of crop weeds, pests, insects, and diseases is expected to shift to higher latitudes (Lamichhane et al. 2015). Coastal areas, especially fishing industries, are vulnerable to potential risks of rising oceanic level, change in concentrations of sea salinity, cyclones, and a reduction in fishing stocks because of rising oceanic water temperatures (Hall-Spencer et al. 2008).

Farming has been tremendously affected by CC, which is further projected to, directly and indirectly, impact the ACP (Lobell et al. 2011). Intensifying global temperature; variations and unpredictability in temperature and precipitation patterns; fluctuations in the availability of moisture; severity and intensity of EWE; rise in oceanic level and salinity; and disruptions in the agro-ecological system will have dramatic effects on livestock, agriculture, forestry and fishery (FAO 2008). The scale of such disruptions not only is determined by the frequency, intensity, and periodicity of these impacts but also depends on their combined effects depending upon the regional factors.

ACP in developing nations is projected to decline, but the scenario in developed nations is probably the opposite (Fischer et al. 2001; IPCC (Intergovernmental Panel on Climate Change) 2007b). The rising temperatures will extend the LGP in cropping seasons, favoring farming, at higher latitudes where plenty of developed nations are situated. However, this impact will not be noticed in developing economies, which are primarily in the tropical regions. Investment potential among various agricultural industries should be recognized if the decline of yield from any sector is to be balanced (Crosson 1997). However, a reduction in yield would lead to an escalation of agricultural commodity prices, and its significant effects will be observed in food-insecure tropical developing nations (Rosenzweig and Parry 1994).

An unarguable rise of air temperatures across India, of around 0.6 °C, has been noticed in the previous century from 1901 to 2010 (Krishnan and Sanjay 2017). In a broad sense, while keeping everything else constant, CC impacts on ACP are projected to be pervasive throughout the earth, as researchers have found that agricultural productivity in mid- to high-latitude temperate regions may see a rise. In contrast, tropical and subtropical areas like the Indian subcontinent and sub-Saharan African regions are prone to the most drastic impacts owing to their high dependency on low-input rainfed agriculture and small adaptation potential (IPCC (Intergovernmental Panel on Climate Change) 2007a; Mertz et al. 2009). Overall, CC is going to affect agricultural productivity by fluctuations in growing environmental conditions explicitly and secondarily by shifting of geographical location, the spread of agricultural pest and disease endemics, related effects on soil, irrigation water quality, biological activity, and linked biological diversity in agricultural ecosystems. Although many projections of the effect of CC appear to be detrimental to ACP, increased CO_2 will add to improved fertilization effects—however, there are lots of discussion regarding the degree to which it will improve ACP.

In the changing climatic scenario, India has witnessed various types of EWE like heat and cold waves, hefty rain, cyclone, etc. During the year 2015, following the flooding in February and March attributed to unseasonal precipitation, one of the lethal and worst heat waves claimed around 2,300 lives in the month of May, succeeded by a below-normal rainfall during monsoon season (NOAA 2015). Further, rains during August, and during December in Chennai, had huge repercussions (Livemint 2015).

These occurrences have become more severe, and their frequency has increased in recent decades, with catastrophic consequences on human health and life, agriculture, and natural resource endowments. High intra-seasonal variability increases the number of EWE that directly influence crop productivity and, ultimately, the farm income. For instance, an unexpected 49% shortfall in mean precipitation of India for July in 2002 inevitably resulted in a significant drought. This resulted in a drop in agricultural productivity, causing a loss of around more than 10% of total food grain production, specifically because of mid-season breaks in monsoon (Samra and Singh 2002).

CC puts added pressure on a country's FNS by threatening ACP in several ways. For example, there may be a dramatic expansion in the range of inter-annual and intra-seasonal monsoon precipitation variability. Based on the current policy scenario of the International Energy Agency and other energy sector economic models, the World Bank's figures suggest that in a 4°C global warming situation, annual average monsoon precipitation intensity and inter-annual precipitation variability will rise by 10% and 15%, respectively. It forecasts a higher probability of drought conditions in northwest India whereas increased wetness in the southern peninsula (World Bank 2013).

CC implications on ACP are influenced by the sensitivity of the crop to average temperature and its fluctuations, the physiological reaction of the plant to rising levels of CO_2, the dynamics between moisture stress and CO_2, and the relationships between multiple factors and their relative adjustments (Challinor et al. 2009). In this changing climate scenario, an elevated concentration of CO_2 is expected to benefit many crops. Still, an associated rise in temperature and variability in precipitation would pose a significant threat to the production of food grains. The latest IPCC assessment reports and several other global research findings suggest a possibility of a 15–40% reduction in ACP in the Indian subcontinent because of the rise in temperature by 2080–2100. Accelerated melting of glaciers in the Himalayas will alter the availability of irrigation, particularly the ACP in the Indo-Gangetic Plains, which is a region of utmost importance to sustain FNS, as it significantly contributes to ensuring food buffer stocks in India. Also, the rising temperatures would probably decrease the efficacy of fertilizer used. To meet the future food grain requirement of the rising population of the country, the demand for fertilizer will also rise. Consequently, the excessive use of fertilizers will lead to increased GHG emissions. In the context of rising global warming due to excessive GHG emissions, this might be an issue of significant concern during international negotiations in the future.

Rising oceanic temperature is expected to influence the breeding, migration, and harvesting of fish. The coral ecosystem is also expected to collapse significantly in the Indian seas to a great extent within the next few decades. Because ACP is tightly linked with its environment and ecosystem, principally the local climate, most of the CC- and CV-related prediction would have a tremendous effect on ACP and, therefore, on the FNS. It is, therefore, essential to evaluate the effects of CC on agricultural field crops to address the sensitivity of the system to CC and its variability.

Extreme temperature, especially in the northern and southern regions of India, is a crucial factor influencing the production of annual crop plants. However, deviations in the average temperature and its effects on the rates of crop growth may be more significant as compared to the heating stress caused by various events of extreme temperatures in the same season. However, this relies upon the sensitivity of the distinct genotypes (Challinor and Wheeler 2008).

CC affects the agricultural sector by altering the dynamics of pests and diseases such as weeds, insects, nematodes, fungi, bacteria, and viruses. Fluctuations in weather variables amend the dynamics of their population that consequently leads to a decline in ACP. Increased temperatures and deviations in rainfall, humidity, as well as other abiotic variables influence the heterogeneity, diversity, and sensitivity and broaden the spectrum of various pests and diseases across different agroclimatic regions (Estay et al. 2009). Among all the climatic variables affecting ACP because of agricultural pests and diseases, the temperature is considered to be the most crucial weather variable to influence insect ecology, epidemiology, and dissemination. In contrast, humidity and precipitation trends determine the sensitivity of pathogens toward crop plant (Hatfield et al. 2008).

Global warming may escalate the water, shelter, and energy needs for farm animals to fulfill the predicted milk requirements. CC is going to amplify the problem of hyperthermia in cattle that will negatively affect their milching yield and reproductive efficiency. Therefore, the overall area where high-yield dairy animals can be economically raised is shrinking. In this way, we can realize that CC and CV may result in having severe impacts on all components of agriculture.

5.4 Need for CSA in the Present Scenario

CC and CV are adversely affecting agriculture, and their impact is projected to become more severe shortly. CC and CV are going to pose a serious risk to the FNS in the near future. It is estimated that ACP in Asia will decline by 2.5–10% from 2020 and 5–30% from 2050, with the most horrific decline in South and Central Asia (Cruz et al. 2007). The FNS of India is also going to be influenced by it. Low crop productivity is among one of the greatest challenges faced by the agricultural sector in India. The productivity of cereals in India is significantly lesser as compared to developed nations/regions like North America, South Africa, East Asia, the European Union, etc. (see Fig. 5.2). Apart from lower global productivity, there is a wide variation in the yield among different states of India (see Fig. 5.3). This condition is going to become worse in the near future due to CC and CV.

Global research findings show the statistical likelihood that ACP in India will drop from 10 to 40% because of rising temperature by 2080–2100. There is a high probability of a loss in wheat production up to 4–5 Mt with each 1 °C increase of temperature all through the crop-growing season even when carbon fertilization is taken into account, excluding other adaptation measures (Aggarwal 2008). This decrease in wheat production is expected to hit 19 Mt and 27.5 Mt, respectively, with a rise of 3 °C and 5 °C. The General Large-Area Model (GLAM) for annual crops was used to forecast the groundnut production where it was concluded that the production of groundnut could decline as much as 65–70%, based on genotype responses of the cultivars used and resource management practices on the field (Challinor et al. 2007).

Some recent Crop Simulation Model (CSM) studies have advocated that the production of maize during monsoon season is expected to experience adverse

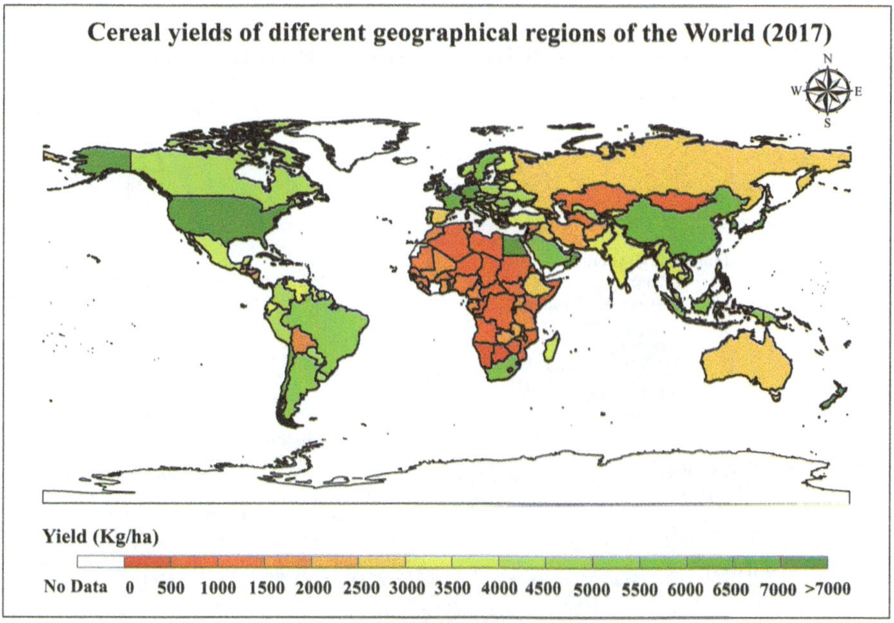

Fig. 5.2 Cereal yields of different geographical regions of the world for the year 2017 (Source: World Bank Database)

effects of increased global average ambient temperature. Increased precipitation, however, can partially offset these losses, and spatial fluctuations in predicted temperature and precipitation will result in distinct effects in respective areas. Investigations on sorghum also stated that a projected rise in precipitation could counterbalance a decline in crop production because of increasing temperature. But, even if the amount of precipitation got doubled, a complete improvement to offset the loss of crop yield over 2 °C cannot be achieved. CROPGRO soybean model-based studies predicted a rise of 50% in ACP in central India because of the doubling of the concentration of CO_2. However, an escalation of 3 °C in mean ambient temperature may nullify these beneficial effects of increased CO_2. Assessment of the scenarios of future CC has shown that production of mustard is likely to decline in irrigated as well as rainfed areas, but these shortfalls in production have spatial variability in the distinct mustard-producing regions of the country. It is evident that CC and CV are going to influence approximately all kinds of cultivated crops, which provide carbohydrates, proteins, fats, vitamins, and minerals to us. Thus, there is a risk to our FNS. The calorie and protein intake of Indians is depicted in Figs. 5.4 and 5.5, respectively. This calorie and protein consumption is projected to increase marginally by the year 2023 (see Fig. 5.6). So, there is a need to achieve sustainable ACP under CC scenario to attain FNS in India, which is possible through the adoption of CSA practices.

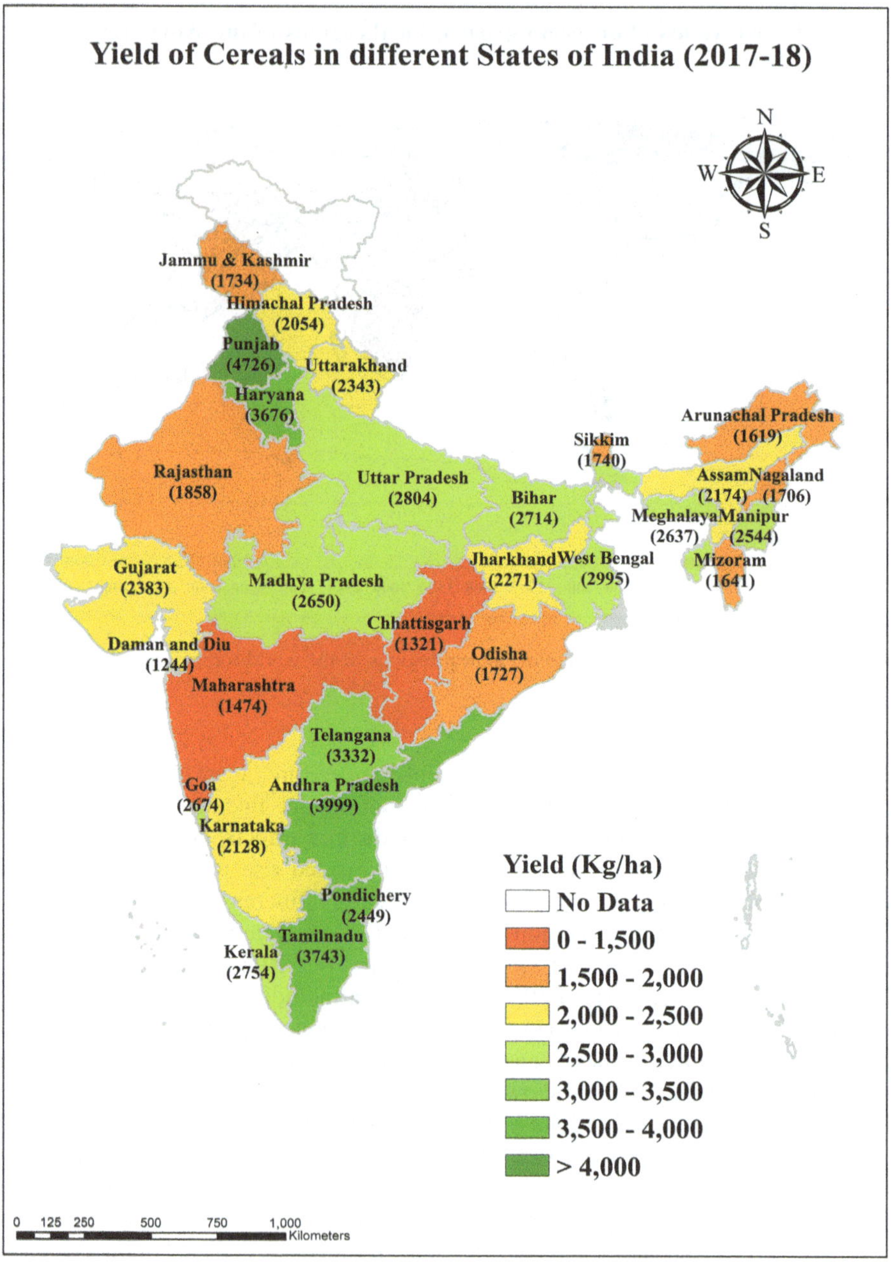

Fig. 5.3 Yield of cereals in different states of India for the year 2017–2018 (Source: Directorate of Economics & Statistics)

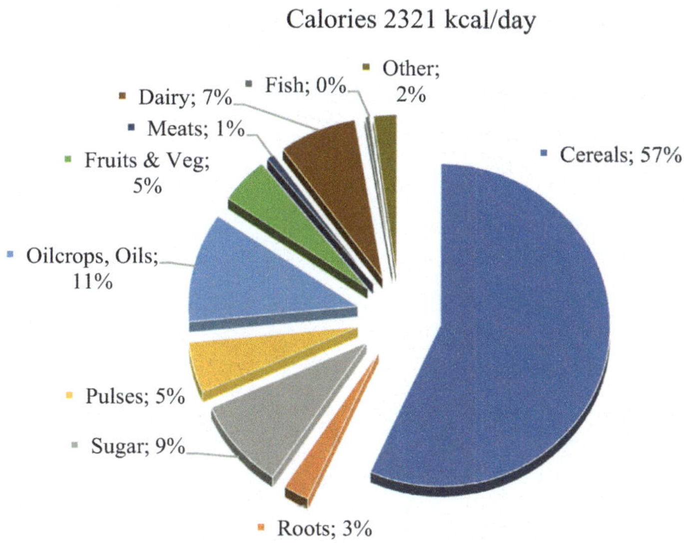

Fig. 5.4 Estimated daily calorie intake by food items in India (2009). Data source: FAO (2013), FAOSTAT (database), http://faostat.fao.org/

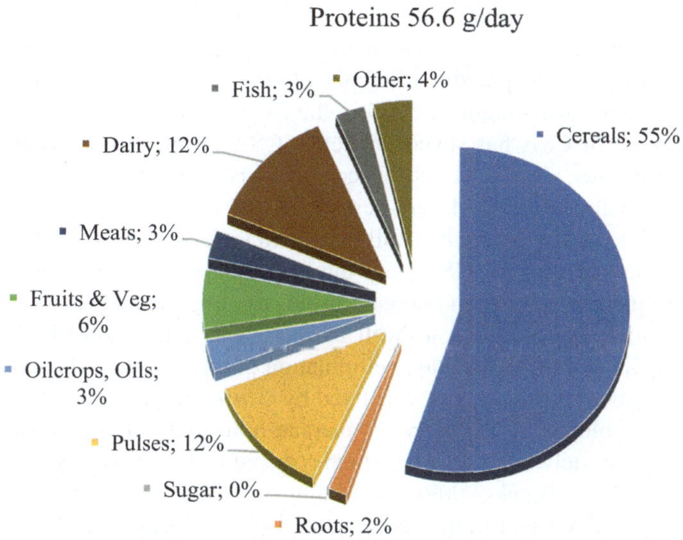

Fig. 5.5 Estimated daily protein intake by food items in India (2009). Data source: FAO (2013), FAOSTAT (database), http://faostat.fao.org/

A minor fluctuation in temperature at the critical crop growth stage may compromise the ACP. Because fluctuations in variability are easier to grasp and understood by farmers, the early adaptation measures to tackle the variability can be the priority

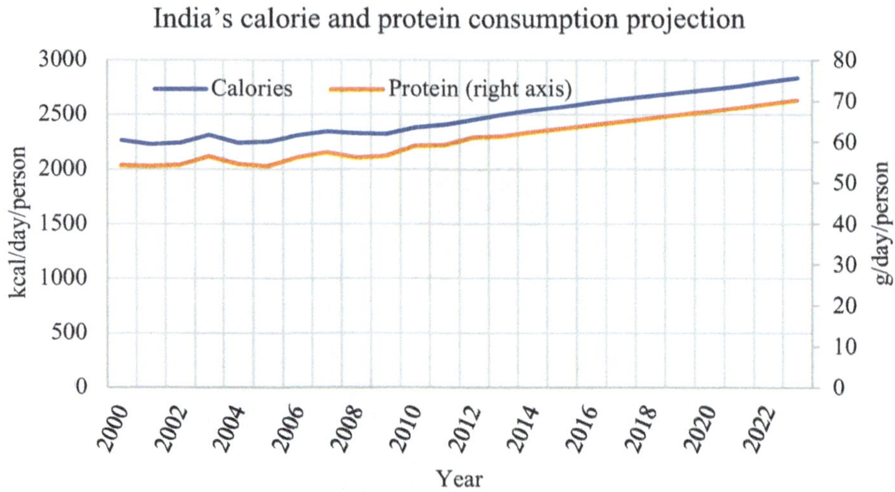

Fig. 5.6 Projection of India's calorie and protein consumption (Data source: OECD and FAO Secretariats (https://doi.org/10.1787/888933099333))

of farmers (Padgham 2009). Thus, it is essential to differentiate and classify the two impact categories to illustrate two ways of adaptation, both with different time frames: improving resilience in case of more significant variability and improving adaptive capacity and preparedness in case of slow-onset changes. In this context, CSA is important for maintaining SAD in the CC scenario.

The adoption of CSA has several benefits. CSA ensures diversification of land use and crop species, which can reduce ecological risks associated with uniform crop covers such as vulnerability to insect-pest attack and unexpected EWE (Scherr et al. 2012). Thus, it leads toward higher resilience with lower risk. Kim (2012) conducted a study on one approach of CSA, i.e., intercropping of *Gliricidia* and maize. He reported that there was a net annual gain in the soil organic carbon (SOC) of about 3.5 t/ha through sequestration, as well as there was mitigation of nitrous oxide equivalent to 3.5–4.1 t CO_2/ha due to elimination of the use of synthetic nitrogenous fertilizers as atmospheric nitrogen was fixed by *Gliricidia* plant. In terms of energy, the adoption of biogas in Nepal by an average household of six to seven persons saved 2–3 tons of fuelwood per year, which reduced GHG emissions of about 5 tons of CO_2 equivalent (Blank 2009). According to IPCC estimates, the no-tillage intervention of CSA has a mitigation potential of 0.17 tons of carbon per hectare per year under cool-dry conditions.

In contrast, under warm-humid conditions, it is 0.72 tons of carbon per hectare per year (Smith et al. 2014). Therefore, it is evident from many studies that the adoption of CSA is quite relevant in the present scenario for mitigating CC and CV and promoting resilience in agriculture for future scenarios. This would ultimately reduce the impacts of CC on agriculture, which would ensure FNS and enhance ecosystem services.

5.5 Pillars of Climate-Smart Agriculture

CSA deals with the three aspects of sustainable development (economic, social, and environmental) to tackle the challenges of FNS and CC (FAO 2010). CSA has the following three main pillars:

5.5.1 Productivity

First pillar deals with the sustainable enhancement of agricultural productivity and livelihood. CSA is a way of achieving both sustainable development and green economy objectives. This aims to achieve FNS as well as the protection of natural resources (Amin et al. 2015). Productivity enhancement can be achieved through the use of renewable energy, efficient resource management, resource conservation technologies, land use management, etc. which are discussed in the further section.

5.5.2 Adaptation

Second pillar deals with adapting and developing resilience toward the impacts of CC. Adaptation aims "to reduce the vulnerability of human or natural systems to the impacts of CC and climate-related risks, by maintaining or increasing adaptive capacity and systems resilience" (OECD-DAC 2011). Adaptation initiatives tackle CC impacts by reducing the vulnerability of human and natural systems toward it. CC adaptation in the agricultural system can be achieved either by specific actions like cultivating the improved crop variety or by systemic changes involving livelihood diversification, better resource management through institutional reforms, etc.

5.5.3 Mitigation

Mitigation deals with the reduction of atmospheric GHG concentration by tackling its emission sources. The strategies for CC mitigation involve the assimilation of those technologies which reduce GHG emissions and inputs per unit of output. Agriculture and deforestation activities contribute to 30% global emissions of GHGs, which provides adequate potential for mitigation. The three ways to mitigate CC in agriculture are given below:

a. Reducing GHG emissions: The emission of CO_2, CH_4, or N_2O can be minimized by efficient management of carbon and nitrogen flows in the agricultural ecosystems.
b. Avoiding or displacing emissions: There is much scope in improving energy efficiency in the agricultural sector. For instance, the use of biofuels in place of FF in agricultural operations can avoid or displace a significant amount of GHG emissions.

c. Removing emissions: The GHGs or their precursor can be removed and sequestered from the atmosphere through CSA operations.

All these pillars can be achieved by embracing various interventions of CSA.

5.6 Interventions for Climate-Smart Agriculture

There is a need for integrated, scientific, multidisciplinary, and diversified approaches that concentrate on natural resource management. It is the most prominent of a paradigm shift that will transform traditional agriculture into modernization under CSA. Therefore, many interventions in the areas of land, water, soil, ACP, energy, livestock, etc. are essential for making agriculture climate-smart. The holistic and integrated adoption of these approaches based on site-specific resources is the need of the hour.

5.6.1 Land Use Management

Land acts as a storage house of GHGs and has a remarkable role in the interchange of materials between soil and atmosphere. Land bio-networks are most susceptible to continuing CC as well as EWE, to various extents. This calls for eco-friendly land management, which contributes to decreasing the adverse effects of CC and variability. Land use management includes widespread choices of farm practices and techniques, including the restoration of degraded and wastelands, which ultimately enhances the soil's nutrients and its water holding capacity. Land use management requires developing comprehensive land cover (LC) database and expert systems, which help in providing a baseline for natural resource management and land use planning. Various farm practices and technologies like minimum/zero tillage, efficient management of irrigation and nutrients, as well as the incorporation of residue can enhance crop productivity, water use efficiency (WUE), and nutrient use efficiency (NUE) and decrease the release of GHGs from various agricultural operations (Branca et al. 2011; Jat et al. 2014). Several activities which will help in increasing productivity sustainably along with contributing to CC mitigation and adaptation are mentioned below:

- Land use management options like afforestation, agroforestry, reforestation, SOC management, and biochar addition to soil contribute to delivering carbon sequestration (CS) in soil or vegetation and help in mitigating the negative effect of CC.
- Conservation and restoration of natural bio-networks such as wet and coastal lands, peatlands, and forests and biodiversity conservation, minimizing antagonism for land management as well as disaster risk management, also help in reducing GHG emissions and ultimately contribute to sustainable development under CSA.

- Soil degradation through agricultural activities can be tackled via sustainable land use management options like green manuring and cover crop production, retention of crop residue, minimum/zero tillage, and improved grazing management with co-benefits for CC adaptation. Mass production of forage legumes such as lablab, cowpea, pigeon pea, lucerne, and sesbania as sources of protein to enhance feed conversion efficiency ultimately decreased methane emissions by about 25–33%.
- Forest and crop diversity, appropriately diversified crop rotations, range and pasture management, etc. can also maintain or enhance forest carbon stocks, which also help in lowering GHG emissions and can contribute to CC adaptation. This diversification will also enhance the nutritional value of the food intake. Such type of interventions successively raises overall herd productivity as well as its adaptability to CC through the diverse use of land resources.
- The most commonly used farm practices like adding soil organic matter (SOM), controlling soil erosion, improved fertilizer, crop management (fallow rice management), and cultivation of drought and flood-tolerant varieties, etc. can significantly contribute toward attaining CC adaptation and mitigation.
- Spatial and integrated landscape planning and land use zoning can achieve positive adaptation and mitigation outcomes under CC.
- The restoration of wastelands for forestry, grassland, agroforestry, and ACP by the management of water and nutrients.

Altering land use practices like cropping pattern, livestock production, and shifting crop or livestock production from vulnerable areas, altering the frequency of application of inorganic fertilizers and chemical pesticides, capital, and labor can assist in reducing the risks posed by CC to ACP. Apart from it, land use management options like trees grown in fields as windbreaks, live fences, fodder banks, alley cropping, or improved fallows can sequester atmospheric carbon in biomass as well as soil and provide fuelwood and other forest-based commodities. This also helps in preventing the destruction of natural forests and helps in adaptation and mitigation under CC (Awazi and Tchamba 2019). In areas prone to drought and heat under CC, the agroforestry system can play a crucial role in CS and making effective carbon sinks. Keeping in view, sustainable land management with co-benefits for CC mitigation helps in integrating land, water, biodiversity, and environmental management to meet the FNS of burgeoning populations.

5.6.2 Crop Production Management

ACP with appropriate management practices is very significant for attaining sustainability under CC scenarios. Sustainable crop production intensification (SCPI) is essential in this aspect. SCPI is a method of ACP that conserves and enhances natural resources and reduces the harmful impacts on the environment by using natural biological inputs and processes. It contributes to making agricultural systems resilient concerning CC. SCPI is formulated on the management practices,

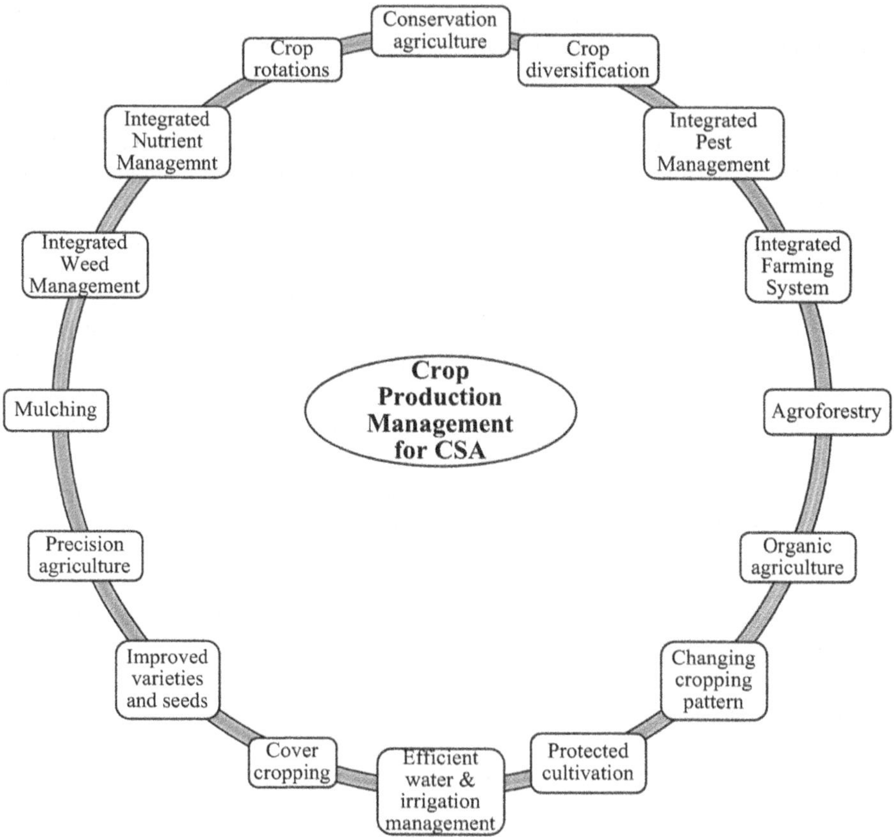

Fig. 5.7 Crop production management strategies for CSA

which maintain soil health, avoid monocropping, and cultivate well-adapted, high-yielding varieties with the use of good quality seeds or planting materials, integrated management of nutrients, pest, weeds, and diseases, as well as efficient management of water. Thus, climate-smart ACP is the sustainable production of crops in the context of CC such that crops grown are less vulnerable to CV. The examples of approaches for CC adaptation and mitigation are shown in Fig. 5.7 (FAO 2011).

These approaches are discussed below:

- Conservation agriculture (CA), which is formed on the dogma of less soil disturbance, residue retention, and crop rotation, should be adopted. With the adoption of zero tillage, the farmers can sow wheat crops soon after the harvesting of paddy or cotton. This practice saves the wheat crop from terminal heat stress during the grain-filling period (Pathak 2009). No-tillage reduces the emission of GHGs and sequesters SOC (Zingore 2010). Resource conservation technologies (RCTs) consist of techniques that enhance efficiency in resource management or input application and thus use direct, identifiable, and comprehensive economic advantages such as declining cost of production; reducing the use of fuel, labor, and water; and timely crop seeding, resulting in enhanced productivity (Amin et al. 2015).

CA technologies involve no or minimal tillage with direct seeding and planting of crops, residue management, and crop diversification. It has the potential to improve productivity and soil quality, primarily by building up SOM. Conservation farming systems seem to be a viable option for achieving safe and intensive ACP in various agro-ecological environments because they allow effective use of available resources and preserve soil fertility (Amin et al. 2015).

- Integrated nutrient management, which supplies nutrients to the crops from all possible sources like inorganic fertilizers, green manure, biofertilizers, and organic manure, can drastically reduce the demand for chemical fertilizers, thereby reducing GHG emission. Site-specific, demand-driven, optimum, and balanced utilization of fertilizers and nutrients is required. Application of microbes for augmenting soil fertility and crop productivity is also useful in the form of biofertilizers.
- Mulch cropping and cover cropping are very useful in maintaining soil health and favorable soil conditions for crop growth. These practices also protect the soil from erosion. The emergence of weeds is suppressed, which eliminates the use of weedicides.
- Changing cropping patterns and crop rotations, particularly with legume crops and millets, can avoid the infestation of specific insects, pests, diseases, and weeds in a particular area along with maintaining proper nutrient balance in the soil. Inclusion of millets in the cropping system can increase the CC adaptability as millets can adapt to a wide range of agro-ecological conditions, require less external inputs for their cultivation, and have less vulnerability to stress. Also, millets are nutritionally rich than major cereal crops (Bandyopadhyay et al. 2017). Therefore, the adoption of legumes and millets in CSA is essential for attaining nutritional security.
- Diversification of crops can build resilience in several ways, such as creating the capacity to control pest outbreaks, dampening pathogen spread, protecting ACP from the harmful effects of increased CV as well as EWE, and rising household income. Diversification of crop as well as livestock, including replacement of crop types, varieties, and hybrids, along with animal breeds possessing improved drought or heat tolerance, is recommended as they may have the potential to enhance productivity under heat and moisture stress. Diversification of the seed's genetic structure and composition is also identified as a potent defense against outbreaks of pest and disease and climate hazards. A substitution from the rice-wheat system to high-value cash crops will augment farmers' income, leading to the minimization of water use and chemical fertilizers (Sapkota et al. 2015). This diversification will enhance the nutrient value of the food.

- High-quality seeds or planting materials should be used for sowing the crops for obtaining a better germination rate and vigor of the crops. Resistant or resilient cultivars toward the adverse effect of CC like insect-pest attack, drought, high temperature, etc. should be planted for increasing adaptation and reducing vulnerability.
- Integrated pest management, which combines various physical, chemical, and biological techniques for controlling pests like biocontrol agents, traps, mulches, soil sterilization, pesticides, resistant varieties, etc., should be practiced.
- Integrated weed management is an approach for managing weeds by cultural, mechanical, biological, and chemical methods to minimize the use of weedicides for avoiding its harmful effects to the environment and, thus, can contribute to CSA.
- Efficient water and irrigation management through a sprinkler or drip irrigation, avoiding water conveyance losses, reducing water losses through evaporation, runoff and drainage, application concerning the demand of the crops, and so on should be practiced in the agricultural fields.
- Organic agriculture, which supplies nutrients to the crops through organic sources of nutrients and eliminates the use of chemicals, can also be an alternative.
- The nutrient-use efficient crop varieties should be cultivated to minimize the need for external application of fertilizers, thereby reducing GHG emissions.
- Integrated farming system by the involvement of crops, poultry, dairy animals, and fishery is a better option for CSA along with sustaining livelihood, especially for small and marginal farmers. This integrated farming system provides nutrients from various food sources and, thus, plays an important role in achieving nutritional security.
- Energy crops required for the production of biofuels should be cultivated, which tends to eliminate the use of FF.
- GHG emission can be significantly decreased by reducing fuel consumption in mechanical farm operations.
- Improved rice cultivation techniques as cultivation of rice crops under flooded conditions leads to the emission of methane gas, which is a GHG with high global warming potential. Its cultivation can reduce methane gas emission from rice crop through the System of Rice Intensification (SRI) technique. In the SRI technique, soil condition is kept moist, 8–15-day-old rice seedlings are transplanted at wider spacing, chemical inputs with organic matter are applied, and frequent weeding is done. Another method is the cultivation of aerobic irrigated rice for CSA adaptation (Friedrich and Kassam 2009).
- Agroforestry is also an efficient method for CSA. It is estimated that agroforestry system sequesters 50–75 Mg C ha^{-1} as compared to row crops with less than 10 Mg C ha^{-1} (Verchot et al. 2007).
- The release of methane gas from the rice crop can be minimized by the application of properly fermented organic waste like compost made from biogas slurry rather than unfermented waste (Pathak and Wassmann 2007).
- It is important to distinguish agroclimatic regions and crops, which are highly sensitive to the adverse effects of CC to reposition them in more appropriate

areas. For instance, a rise in temperature will severely influence the growth of crops like tea and Basmati rice. So, suitable regions for their growth need to be identified to protect the crop from yield loss. Another alternative is to change the crop sequence by modifying the time of crop sowing, applying pesticides, irrigation, fertilizers, and harvesting crop produces for obtaining optimum yield under the altered LGP and the associated change in the heat and humidity levels (Amin et al. 2015).
- Precision agriculture, which involves the application of inputs at the appropriate time, by the appropriate method, and at the appropriate location through the use of RS and GIS, sensors, and variable rate applicators, should be adopted for increasing productivity, increasing input use efficiency, and protecting the environment.
- Protected cultivation of crops through various technologies like polyhouses, greenhouses, shade houses, hotbeds, cold frames, etc. for providing favorable environmental conditions for crop growth by controlling temperature, moisture, insects, pests and diseases, etc. can also be practiced.

In this way, it is clear that appropriate crop management is a very crucial component for attaining the objectives of CSA. The farmers can adopt these practices for sustainable agriculture along with CC mitigation and adaptation. These practices will help to ensure FNS in the long run under CC scenarios.

5.6.3 Soil Management

Soil is a crucial natural resource for achieving sustainability through CSA. Soil acts as a medium for cultivating plants and providing various nutrients and water for their growth. It supports soil biodiversity and regulates carbon, oxygen, and many other nutrient cycles. So, proper soil management is a crucial practice in the CSA. CC impacts soil through the following ways:

- Erratic rainfall patterns and frequent drought deplete the water and nutrients supplying capacity of soil to the plants.
- Increased risk of soil erosion by higher rainfall intensity.
- Increased rate of mineralization of SOM due to rise in temperature.

SOC sequestration has the tendency to decrease the CO_2 content in the atmosphere and, thus, contributes to CC mitigation. According to Lal (2004a, 2015), there is the global potential of SOC sequestration of 0.9 ± 0.3 Pg/year *via* adopting recommended practices like residue retention, CA, rational utilization of farm inputs, etc., which may counterbalance 25–33% of the annual increment in the atmospheric CO_2. Through recommended management practices, the SOC sequestration rate fits a sigmoid curve in which the maximal level of SOC sequestration rate is attained in 5–20 years (Lal 2004b) that further continues at diminishing rates till SOC stock reaches new equilibrium level in 20–30 years (IPCC (Intergovernmental Panel on

Climate Change) 2007b). The management of the soil for CSA can be done in the following ways:

- Before adopting any CSA practice for soil management, various physical, chemical, and biological characteristics of the soil influencing soil health and SOC sequestration should be assessed by in situ inspection through soil testing kits or by taking soil samples and analyzing it in the laboratory. The CSA practice should be adopted by these analyzed properties (Faurès et al. 2013).
- The practice of minimum or no-tillage substantially decreases the runoff rate, increases the soil water infiltration, and avoids the formation of plough pan in the subsoil. CA also augments the SOM and decreases SOC mineralization rate, which further facilitates the process of SOC sequestration. In this way, CA contributes to the reduction of GHG emission (Faurès et al. 2013). Crop rotation in CA, especially with leguminous, will reduce the infestation of pests and diseases along with boosting the soil nutrients. CA reduces soil erosion by making protective cover over the soil, which reduces runoff and increases soil water infiltration. Altieri et al. (2011) reported a 90% decrease in soil erosion in no-till plots in comparison to conventionally tilled plots in a study conducted in Parana, Brazil. CA based on rice-wheat systems emits 10–15% less GHG as compared to conventional systems by creating more aerobic soil environments (Sapkota et al. 2015).
- In areas having steeper slopes, the soil erosion can be prevented in many ways, like by plantation of the vegetation across the slope or by the construction of soil and water conservation structures like tied ridges, bunds, terraces, trenches, etc. The excess runoff water can be safely disposed of the slopes by using grassed waterways, chute spillway, drop inlet spillway, etc. (Faurès et al. 2013).
- Many agronomic practices like agroforestry, mixed cropping, cover cropping, contouring, strip cropping, etc. also help in decreasing soil erosion and increasing SOC sequestration.
- In arid and semiarid regions, there is a problem of wind erosion, which may either take away the fertile topsoil or deposit the blown away sand dunes on the productive soils. This can be prevented by planting drought-resistant species, rotational grazing, and planting windbreaks in the direction perpendicular to the prevailing wind.
- Mulching by crop residues is also an essential practice for buffering soil temperature, decreasing soil water evaporation and nutrient loss, and increasing SOM, which further enhances the soil moisture content, soil biodiversity, soil structure, and soil water infiltration. This practice also reduces soil erosion by avoiding the dispersion of soil particles by raindrops or runoff. The risk of development of salinity or waterlogged condition in the soil can also be reduced by it (Faurès et al. 2013).
- The proper nutrient management in the crop is very important in CSA as improper nutrient management leads to an increase in GHG emission from the soil. Among all nutrients, proper application of nitrogenous fertilizer is crucial as it may convert to nitrous oxide gas under the anaerobic condition, which has a global

warming potential of 300. The nitrate ions produced by mineralization of nitrogenous fertilizers move freely with soil water, which increases their leaching from the soil through drainage. The placement of nitrogenous fertilizers in the reduced zone of the soil also reduces the emission of nitrous oxide. These fertilizers should be applied in the zone of active root uptake. In CSA, the proper recommended amount of fertilizers and manures should be applied at the recommended time as required by the crop to increase their efficiency and reduce their wastage.

- Integrated soil fertility management should be performed in CSA as it provides nutrients to the plants through various sources like compost, organic manure, green manure, crop rotations, intercropping, and inorganic fertilizers in the desired amount along with conservation of soil and water in order to achieve decreased nutrient losses, increased soil CS, enhanced water storage, reduced soil erosion, increased NUE, and reduced GHG emissions.
- Precision nutrient management can improve fertilizer use efficiency by application of nutrients in the appropriate form, in the optimum amount, at the appropriate time synchronizing with crop demand, and at the correct place. The tools for precision nutrient management of nitrogenous fertilizers are Leaf Color Chart, chlorophyll meter, and optical sensors like GreenSeeker (Singh et al. 2015). Decision support systems, which are computer or android mobile-based, such as Nutrient Expert and Crop Manager, can also facilitate the farmers for precise management of nutrients (Pampolino et al. 2012).
- The agroforestry system with leguminous trees or shrubs can also be practiced.
- Integrated and site-specific nutrient management should be done as it stores SOC and reduces GHG emissions. It prevents the wastage of inputs, which improves the quality of soil and water.

5.6.4 Water Management

Water is the finite natural resource, which is diminishing at a faster pace due to haphazard and unscrupulous exploitation. Due to the ever-increasing population, water resources are under enormous pressure, which emphasizes the need for sustainable use of water resources. Presently, agriculture consumes about 70% of the total water withdrawn, primarily for irrigation purposes. Irrigation is not viable if provisions of water resources are not sufficient. Particularly in water-scarce regions, the prerequisite for irrigation development is to reduce losses caused by percolation, evaporation, seepage, etc. The impacts of CC on agricultural water is estimated to be through the increased rainfall variability, higher temperatures, and life-threatening EWE like floods and droughts. In this scenario, sustainable water management is essential under CC mitigation. In the water management practices, more attention is given to irrigation scheduling, precision irrigation, efficient drainage systems, in situ moisture conservation, and rainwater harvesting structures to contribute toward three pillars of CSA (Fig. 5.8). However, both irrigation methods and scheduling are interrelated.

Fig. 5.8 Water management options for climate change adaptations

The adaptation strategies under CC for smart water management are given below:

- Micro-irrigation systems (micro-sprayers, trickle, or drip irrigation) are efficient methods for irrigating crops in which water is directly applied to the root system of plants. Micro-irrigation system saves 20–48% irrigation water, 10–17% energy, 30–40% labor cost, and 11–19% fertilizer and enhances the ACP from 20 to 38% (PMKSY (Pradhan Mantri Krishi Sinchayee Yojana) 2019). These localized irrigation options are not only vital for conserving water but also pertinent to save energy (Shah 2009) and decreasing carbon emissions. Moreover, the adoption of a solar-powered drip system can help in saving water and energy as well as ultimately reduce carbon emissions. Due to the initiatives of the central government through Pradhan Mantri Krishi Sinchayee Yojana (PMKYS), the area under micro-irrigation is increasing in the various states of India, as depicted in Fig. 5.9 which is a positive step toward CSA.

- Pressurized micro-irrigation systems with sensor-based irrigation scheduling have great potential for enhancing crop water productivity by maintaining soil moisture content at field capacity in the root zone of crops and helps in solving problems related to manual irrigation. Keeping in view, various soil moisture sensors like tensiometers, gypsum or resistance blocks, frequency domain reflectometry (FDR), time domain reflectometry (TDR), etc. have been used for irrigation scheduling to monitor in situ soil moisture (Cardenas and Dukes 2012). Apart from it, wireless sensor array is also used for irrigation scheduling on a real-time basis. For monitoring plant water stress, plant-based sensors like sap flow, infrared thermometer, trunk diameter variation, and leaf turgor

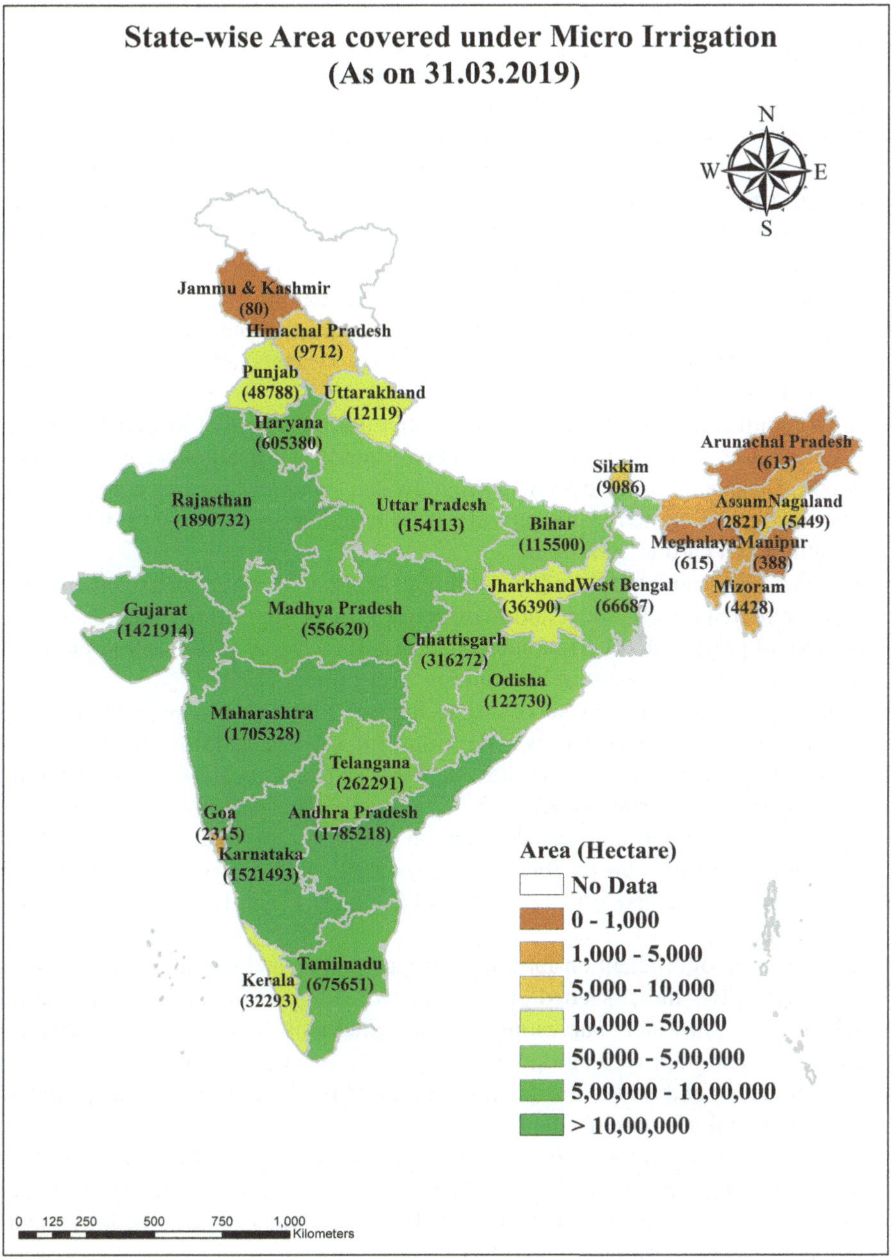

Fig. 5.9 Area covered under micro-irrigation in various states of India (Source: Department of Agriculture, Cooperation & Farmers Welfare)

pressure-related sensors are used. These are the utmost adaptation strategies to mitigate CC.

- Latest advances in information and communication technology (ICT) support development of irrigation scheduling and soil water balance softwares like BEWARE (Chartzoulakis et al. 2008), IrriSatSMS (John et al. 2009), IrriSat (Urso et al. 2013), IRRISA (Boyer and Campagnaud 1996), CROPWAT (Smith 1991), PILOTE (Khaledian et al. 2009), or SIMDualKc (Rolim et al. 2007) which help in irrigation scheduling of different crops. Apart from it, under a CC scenario, regulated deficit irrigation (RDI) or subsurface irrigation (SSI) can be adopted under CC, especially in water-scarce regions.
- To minimize the water losses during conveyance, distribution, and application networks, advanced technologies like telemetry systems, RS, and GIS are very beneficial under CC.
- Solar pumps can serve as an ideal alternative for the usage of available groundwater with less emission in an area where groundwater availability is at shallow depth. Despite it, solar-operated groundwater pumping system coupled with micro-irrigation systems also provides an alternative in minimizing diesel consumption and carbon emissions.
- RS and GIS can be applied in different aspects of water management like command area development and management, mapping of cropping pattern and crop yield projection, flood monitoring and hazard mapping, as well as environmental impact assessment in interlinking river project.
- Remote sensing is the new surveying technique, which also helps in the identification of groundwater occurrence, development, storage, and flow direction. This technique is also helpful in aquifer mapping and identification of the potential area for groundwater recharge.
- Similarly, ICT-based automated irrigation systems, crop and agro-meteorology advisories, and insurances for crops and livestock can also assist farmers in minimizing the adverse effects of CC and CV (Altieri and Nicholls 2013; Mittal 2012).
- Development of efficient on-farm infrastructures like land levelling, minimum or no-tillage, SRI, direct-seeded rice, crop diversification, appropriate irrigation scheduling, rainwater harvesting, site-specific soil and water conservation structures, and improved agronomic practices helps in enhancing on-farm irrigation efficiency and arrests the declining water table which ultimately helps in CC mitigation and adaptation.
- The promotion of rainwater harvesting in rainfed areas is very beneficial and also acts as a drought proofing strategy. Apart from it, integrated watershed management and the construction of artificial recharge structures can also be the adaptation approaches to mitigate CC.
- For the same water efficiency, many high-pressure drip irrigation structures have lower-energy efficiency than gravity-based irrigation. Therefore, in order to achieve sustainability, adjustments must be addressed to attain a balance between increased energy efficiency and water efficiency.

Sustainable water management helps in solving problems regarding erosion, drainage, irrigation, flood, drought, etc. In a technological era, the Government of India has also given the urgency for conservation and management of water resources by launching PMKSY (Pradhan Mantri Krishi Sinchayee Yojana) (2019) for fulfilling the dream of "more crop per drop" and "Her khet ko Pani." In this context, more crop per drop is achieved by familiarizing modern irrigation like drip and sprinkler as well as sensor- and IoT (Internet of Things)-based irrigation and "Her khet ko pani" by on-farm development, integrated farming, as well as integrated watershed management.

5.6.5 Livestock Management

The livestock sector contributes to the CC as well as it is influenced by its effects (FAO 2006). Eighteen percent of GHG emissions are from livestock (Steinfeld et al. 2006). Livestock mainly emits methane gas through an enteric fermentation process, which has 25 times more heat-trapping capacity than CO_2. This methane emission from livestock can be decreased by improving their diet. The diet can be improved by feeding them with good quality feed additives as well as by replacing feeds having low digestibility with that of high digestibility. Inclusion of higher fats and protein and providing antioxidants, vitamins, and mineral supplementations in the livestock diet are beneficial in CC adaptation and mitigation (Hristov et al. 2013; Havlik et al. 2013). Grazing management through rotational grazing helps in the restoration of degraded grasslands, improves soil health, and enhances climate resilience. Improved varieties of pastures should be cultivated in grasslands for cattle grazing. Research developments are in progress for developing vaccines against methanogens present in the rumen of livestock, which would eventually minimize the release of methane gas (Wright and Klieve 2011). Manure also releases GHGs in the atmosphere. Therefore, better manure management techniques like composting should be adopted. The adoption of these measures has a total mitigation potential of 417,000 Gg CO_2 eq. (Thornton and Herrero 2010; Herrero et al. 2016).

Livestock diversification is an approach in CSA for increasing resistance against CC-related pests and diseases (Batima et al. 2005). The breeds having a higher tolerance to temperature and humidity, resistance toward diseases, and the ability to survive under low input conditions should be reared (Pankaj et al. 2013). Many managemental strategies can help in building CC adaptation in livestock (Pankaj et al. 2013). A regular supply of clean and cool water to animals is one such strategy. Splashing cool water on animals during a hot period at regular intervals can reduce heat stress. The stocking density of animals should be reduced during the hot period. Animals should be kept under proper shade as well-designed shades can reduce the heat load of about 30–40% on animals. Roofs of cattle sheds composed of hay or corrugated steel sheets are good for shading purposes. Ventilation and air circulation should be increased in animal sheds by the use of fans and open housing system or by increasing the height of buildings. Plantation of trees around cattle sheds can provide long-term cooling effect (Das 2017).

5.6.6 Genetic Approaches for Building Climate Resilience

The genetic makeup of the crops determines how a particular crop responds to the external environmental conditions and cope with the various abiotic and biotic stresses resulting from extreme temperatures, flood, drought, attack of pests and diseases, etc. The genetic makeup also regulates the LGP and phenology of the crop as well as influences the efficiency of the crops to utilize inputs like fertilizers and water. Conservation of genetic resources of various crops is essential for their contribution to breeding crops having better adaptation under CC conditions. For CC adaptation, the following traits are mainly required in the crops (Faurès et al. 2013):

- Tolerance to water and temperature stress
- Tolerance or resistance to pests and diseases
- Ability to efficiently utilize scarce nutrient supply
- Capacity to grow under poor soil conditions
- Phenotypic plasticity

Conservation of genetic resources by both in situ and ex situ methods can be done. However, ex situ conservation in gene banks and botanical gardens is more cost-effective and easily accessible to the users. Genetically modified organisms can be created by the insertion of genes from a foreign organism or deletion of the existing genes by the use of specific enzymes.

Emission of methane gas from the rice crop can be minimized by the cultivation of varieties having higher root oxidative activity, few unproductive tillers, and higher harvest index (Nagargade et al. 2017).

There is a need to cultivate varieties of crops that can reduce the emission of GHGs (Barfoot and Brookes 2014). Genetically modified crops can help in alleviating the emission of GHGs by reducing fuel use for farm operations and by increasing the uptake of atmospheric CO_2 gas and their conversion to oxygen (Nagargade et al. 2017). The NUE of the crops can be improved by genetic engineering or by breeding methods. By increasing nitrogen use efficiency of the crops, the amount of nitrogenous fertilizers applied to the crops will be reduced, which thereby minimizes the emission of GHGs. The target gene for improving NUE in crops for breeding purpose is the alanine aminotransferase gene found in barley (Shrawat and Good 2008). There are some genes in plants like *Brachiaria humidicola* and *Leymus racemosus*, which release such compounds from their roots that inhibit the nitrification process in the soil by suppressing the activity of *Nitrosomonas* bacteria (Subbarao et al. 2007). The results of research on transferring the nitrogen-fixing genes from atmospheric nitrogen-fixing leguminous plants to non-nitrogenous fixing plants may enable them to fix nitrogen on their own without application of synthetic nitrogenous fertilizers (Lutz 2013).

Methane gas released from ruminants can be minimized by breeding the livestock for high feed efficiency through genetic selection as it provides a permanent solution (Alford et al. 2006). Bentley and Hegarty (2008) found that the development of composite cattle breeds by crossbreeding of locally adapted breeds with good quality breeds has improved heat tolerance, disease resistance, health, and reproductive traits as evident from the study in tropical grasslands of Northern Australia. So, crossbreeding can be an option for building climate resilience in the livestock.

Transfer of stress-tolerant or resistance genes through genetic engineering or molecular plant breeding to the target crop can make the crops adaptable to CC. The tolerance to high temperatures in the plant is possible by various mechanisms such as maintaining membrane stability or through ion carriers, osmoprotectants, proteins, and anti-oxidants (Hasanuzzaman et al. 2013). There are few successful examples of building heat stress tolerance in crops by transgenics like the insertion of ROB5 genes from bromegrass to canola and potato enhanced their performance under high temperatures. Similarly, a heat shock protein gene "hsp101" was transferred from *Arabidopsis* to Basmati rice, which improved its growth under heat stress (Katiyar-Agarwal et al. 2003).

Metabolites like trehalose, glycinebetaine, and mannitol or indirectly controlling gene expression via transcription factor and kinases in signal transduction could provide enhanced tolerance to drought (Bhardwaj and Yadav 2012). *Arabidopsis* plant has a drought- and salt-tolerant "HARDY" gene, which on transferring to the transgenic rice crop improved its WUE by increasing photosynthetic assimilation and minimizing transpiration (Karaba et al. 2012). Castiglionit et al. (2008) found that the insertion of cold shock protein B (CspB) encoding gene derived from a soil bacterium *Bacillus subtilis* enhanced the adaptation of transgenic maize plant under drought stress through interrupting the disruption of trehalose sugar by trehalase enzyme, thereby slowing growth rate of the plant and conserving water for the essential processes. Few rice varieties are also released in India, which are tolerant to droughts, such as Sahbhagi Dhan (Dar et al. 2012) and IR64-Drought1 (Haefelea et al. 2016).

There is an increase in the development of salinity due to CC. Hence, there is a requirement to develop salt-tolerant crop varieties through either conventional breeding or genetic engineering for improved CC adaptation. Genes improve plant resistance to salinity by using many mechanisms like reducing the absorption rate of salts from the soil and inhibiting further transportation of salts within the plant system, controlling the leaf growth as well as plant senescence, and modifying the ionic and osmotic cell balance in roots as well as shoots (Chinnusaamy et al. 2005). He et al. (2005) developed transgenic cotton plants having an expression of an *Arabidopsis* vacuolar sodium/proton antiporter gene "AtNHX1" and found that these transgenic plants yielded more fiber content and biomass when subjected to salt stress under greenhouse condition.

Developing new crop varieties having higher yield potential as well as resistance or tolerance to multiple stresses such as drought, salinity, flood, etc. is essential for maintaining yield stability. Under the CC scenarios, it is crucial to develop varieties with tolerance toward multiple abiotic stresses and a variety of responsive inputs. The need for the hour is to move from C3 to C4 plant to enhance the efficiency of radiation use and water use. Improving the root efficiency for water mining and nutrient absorption is critical. The use of genetic engineering for "gene pyramidization" has become necessary to consolidate all the desirable traits within one plant to obtain the "ideal plant form" (Pathak et al. 2012).

5.6.7 Energy Management

Energy is very crucial for the agriculture sector and non-renewable energy sources like FF responsible for emissions of GHGs, which are interrelated to CC. Therefore, there is a prerequisite to address these challenges by introducing renewable energy sources. The huge demand and exhaustive use of energy in agriculture call for prudent management of both renewable and non-renewable energy sources. Energy management mainly focuses on optimization of energy in context to sustainability. The main focus of energy management is on energy conservation and energy efficiency.

Globally, agricultural energy is dependent mainly on FF due to the marked shift from human and animal power to tractors, electricity, and diesel power. In this regard, Smith et al. (2014) reported that the agriculture sector adds 5.0–5.8 Gt CO_2 eq./year, which forms approximately 10–12% of total anthropogenic GHG emissions. But, the average contribution of agriculture-based emissions is about 35% of global emissions, which are generated in developing countries. The highest growth rate of 10.4% energy use in the agriculture sector, 3.6% in industry, and 3.2% in the transport sector has been reported during the past years (Jha et al. 2012). The energy consumption in agriculture can be mostly divided into two categories, viz., direct and indirect use of energy. In agriculture, basically, energy is directly consumed for pumping and mechanization (tractors, power tiller, etc.) and indirectly through fertilizers and pesticides. The electricity consumption in agriculture is mainly due to greater irrigation demand for various crops and subsidized electricity given to farmers. This calls for efficient use of a non-renewable source of energy and their substitution with economically feasible renewable sources. In the context of the Indian agriculture sector, most of the energy used, particularly electricity, is in pumping water for irrigation purposes. So, it is of utmost importance to save energy in all possible ways in water pumping, which can be achieved by enhancing the efficiency of pump sets (Fig. 5.10).

The activities that will help in managing energy efficiently in a sustainable manner under CC are given below:

- There is a need to increase bio-based product inventories for replacing petroleum-based products to reduce GHG emissions. Biomass like wood, animal dung, agricultural waste, etc. can be decomposed by anaerobic microorganisms to produce biogas. This biogas can be utilized for the purposes of heating and lighting. The slurry left behind biogas production is rich in nitrogen and phosphorous and also used as manure for crops.
- Laser-aided land levelling has also been proven one of the promising technologies to mitigate CC, which saves water up to 40%, improves fertilizer use efficiency, and enhances crop yields. This technique decreases GHG

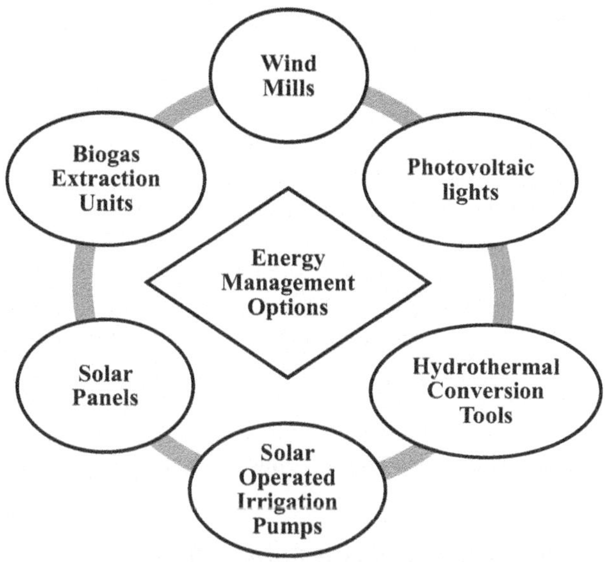

Fig. 5.10 Energy management options for climate-resilient agriculture

emissions through many farm operations, particularly by minimizing irrigation water demand resulting in the reduction of energy needs for pumping water.

- The micro-irrigation system, which reduces pumping energy requirement, should be promoted to save energy. Irrigation during hot sunny and windy days should be avoided.
- It is essential to promote CA machinery like zero till, Happy Seeder, seed-cum-fertilizer drill, raised bed planter, and laser-guided land leveller as they have already proved to be a better option for saving energy during various farm operations.
- It is very crucial to choose the right capacity of pumps to meet the irrigation requirement and their matching pump sets with the source of water—canal or well for saving energy under a CC scenario.
- Installation of variable speed drives (VSDs) on pumps is also an important energy-saving measure that provides variable speeds through which pumps can operate at the optimal rate. It was also reported that reducing the speed of a motor by just 20% can cause an energy saving of up to 50%.
- Regular pump maintenance not only prolongs pump life but also provides optimum energy efficiency of pumping machine.
- IoT-based smart irrigation system can help to irrigate the field at optimal times and under optimal conditions, which is also beneficial under CC adaptation measures.
- For energy-smart food systems, a number of novel technologies that can be very relevant consist of windmills, photovoltaic lights, solar panels, power generators, units for biogas extraction, bio-oil mining, and purification tools, pyrolysis units, ethanol extraction by fermentation and distillation processes, tools for hydrothermal conversion, bio energy-operated water pumps, renewable energy-powered

vehicles, ICT, and monitoring systems, cooking stoves, tools for water purification, supply and distribution.

- Methods, like replacing synthetic artificial fertilizers with manure and farm residues that need less external inputs and improve crop yields may together lead to improved energy efficiency as well as sustained productivity, farm income, and profits.
- In situ renewable energy production may enable farmers to raise their incomes sustainably by selling solar power to the electricity grids or biogas to the regional marketplace or minimal purchases of FF.
- In situ generation of biogas may enable the use of its by-products as liquid organic fertilizer that will augment the crop yields and minimize environmental pollution.

Keeping in view, under CC, there is a need to promote energy-smart technologies, which can help in reducing energy consumption during farming operations and also help in reducing crop water requirements.

5.6.8 Modelling and Forecasting

CC and mainly the related EWE are presently one of the major concerns of the scientific and farming fraternity. For achieving sustainability of the ACP system and to feed the burgeoning population under prevailing CC, crop modelling and forecasting help in mitigation of CC. Crop modelling has played a crucial role in simulating the relationship between plant and environment and in predicting future CV. To understand the effect of CC on crop growth, development, and yield, crop simulation models played a significant role. Crop models also integrate different components and help in understanding genotype and environmental interactions. Crop models help the policymakers to address the complex problems in agricultural livelihoods. They can be used in crop management and evaluating the weather risks in crop planning under CC. They also play an important role in deciding the best management practices under certain cropping systems. Apart from that, they also help in comprehending the spatial and temporal variation in agricultural farming systems. There are various kinds of models available for assessment and prediction of crop growth and yield like deterministic and stochastic, dynamic and static, empirical, statistical, mechanistic, etc. Generally, modelling helps in simulating the behavior of the system. Presently, crop models like APSIM, AquaCrop, CERES-Wheat, CropSyst, DSSAT, EPIC, GLAM, ORYZAv3, SORKAM, SorModel, SWAP, and WOFOST, etc. are most commonly used. The minimum input data required for any crop models are shown in Fig. 5.11.

The WOFOST, CropSyst, EPIC, and CERES-Maize CSMs are mainly used for simulating maize crop, whereas SorModel, SORKAM, and SORGF CSMs are used to deal with the problems related to sorghum crop management. DSSAT is also a CSM with different modules that not only simulate the crop growth, development, and yield but also calculate the water, soil heat, and nutrient dynamics, which helps in economic and weather risks associated with CC. Despite it, EPIC (Williams et al.

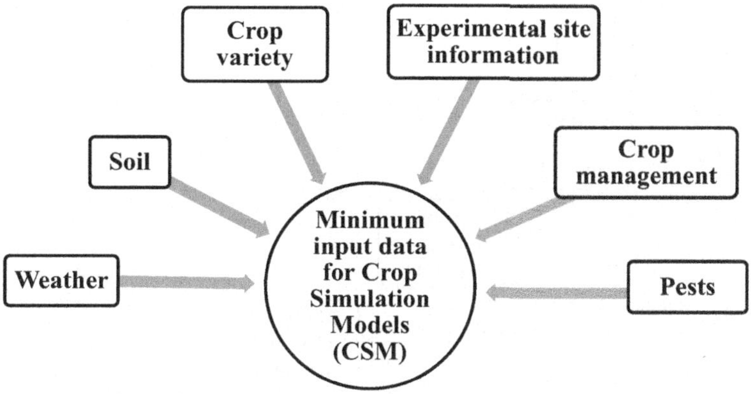

Fig. 5.11 Minimum input data for crop models

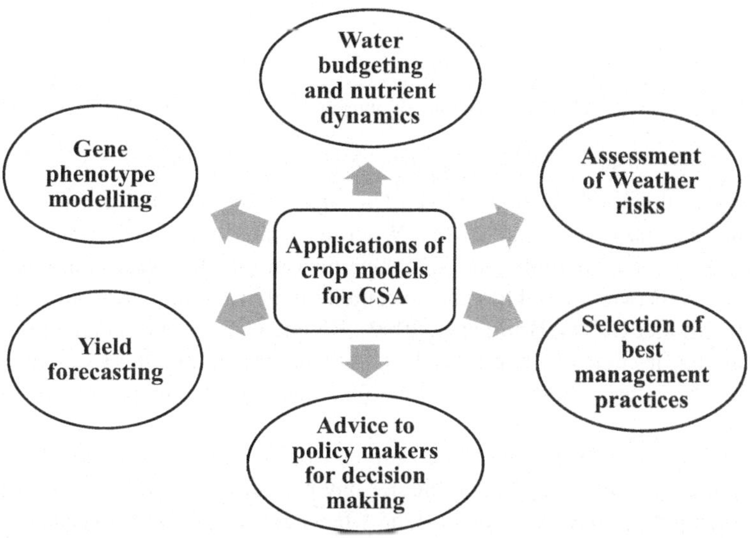

Fig. 5.12 Applications of crop models for CSA

1989) also simulates 80 different crops through one generic crop growth module and also consists of soil erosion module. APSIM model with a plant, soil, and management module is able to simulate several different crops along with some trees and weeds under climate risk (Keating et al. 2003). The major limitations of CSMs are availability, quality, and accuracy of input data. In spite of it, crop models also require multidisciplinary knowledge, technical skills, and a large amount of input data. Therefore, proper calibration is needed before predicting the future scenario of agricultural systems. The application of crop models that are supporting CC adaptation decisions is given below in Fig. 5.12.

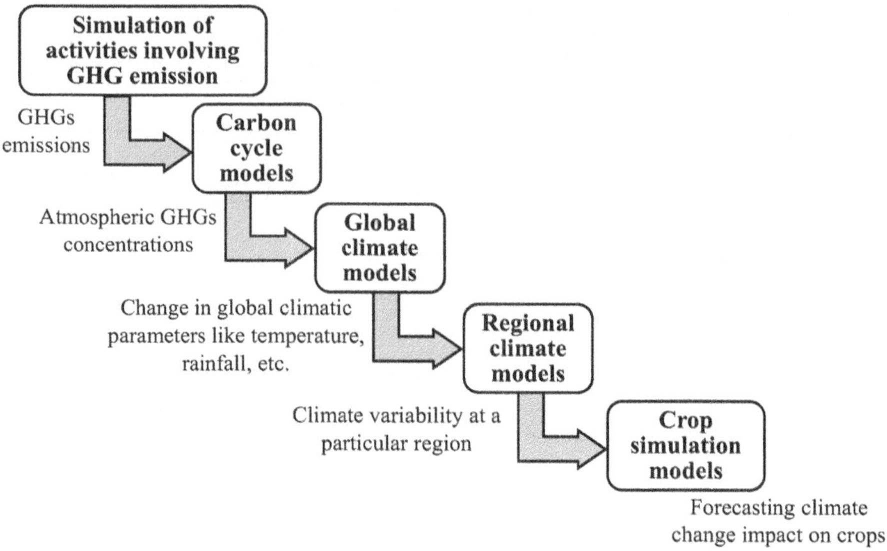

Fig. 5.13 Generalized schematic diagram of integrated assessment model for CC impact on agriculture

Early warning and weather forecasting systems also play a crucial part in decreasing the risks of climate losses. When integrated with ICT, these may assist the policymakers in planning contingency measures. The new technique Multi-criteria Analysis (MCA) tool has also executed for alerting mitigation and adaptation alternatives under CC (de Bruin et al. 2009). RS and GIS also help in the preparation of risk and hazard maps, which played a significant role in disaster risk management under CC. Keeping in view, these adaptation strategies must be promoted under CC scenarios.

The integration of CSM with the outputs of simulation models of the climate, carbon cycles, etc. in a sequential manner having cause and effect relationship can provide an appropriate way to forecast the effects of CC and CV on a specific crop in a particular region. This is possible by integrated assessment models. According to IPCC (Intergovernmental Panel on Climate Change) (1996), a full-scale integrated assessment model has several sub-models which simulate the activities responsible for the emission of GHGs, carbon cycle and processes to yield atmospheric GHG concentrations, CC and CV due to changes in atmospheric GHG concentrations through global and regional climate models, and finally the impact of CC on crops through CSM. The schematic diagram of the integrated assessment model for simulating the effects of CC on agriculture is given in Fig. 5.13.

Thus, in an integrated assessment model, the output of one model is used as input to another model. In this way, these models incorporate scientific knowledge from different areas for forecasting the effects of CC and CV on agricultural crops. These models serve as an effective medium for understanding the complex interactions between various components at different scales. They can help policymakers for better decision-making. An example of an integrated assessment model is MOSAICC (Modelling System for Agricultural Impacts of Climate Change) developed by FAO, which integrates five components, i.e., climate, crops, hydrology, forests, and economy. Thus, modelling and forecasting are a significant intervention for enhancing understanding and making decisions about CC mitigation and adaptation in CSA.

5.6.9 Development of Sustainable and Inclusive Food Value Chains

Food wastage is nearly 1/3 of the food produced (Gustavsson et al. 2011). According to FAO (2011), approximately 38% of the total energy used by the entire food chain is consumed in annual food losses. All components involved in food value chains, i.e., agriculture, transportation, processing, storage, cooking, and consumption, are possible fields for improvement of energy efficiency concerning CSA. Therefore, there is a need to develop a sustainable and inclusive framework for a food value chain that paves the path toward CSA. FAO (2013) defines sustainable and inclusive food value chain as "the full range of farms and firms and their successive coordinated value-adding activities that transform raw agricultural materials into food products that are sold to final consumers and disposed of after use, in a manner that is profitable throughout the chain, has broad-based benefits for society and does not permanently deplete natural resources."

Food wastage occurs at every stage ranging from food production to final disposal of food (Amin et al. 2015). At the food production stage, food wastage is due to poor production practices, pest and disease damage, poor water management, and so on. This can be reduced by improving production, cultivation, and harvesting practices. In the case of livestock, wastage occurs due to poor nutrition, diseases, and bad milking techniques. This can be managed by proper maintenance of housing, feed, diet, water, milking hygiene, and animal health. In the next stage of post-harvest handling and food storage, food losses are due to poor storage and inadequate transportation infrastructure. This can be minimized by investing in proper food storage, cooling, and transportation facilities. Food wastage at the processing stage takes place through substandard processing and packaging techniques, which need to be counteracted by enhancing their standards. The losses at the distribution phase occur due to poor transport and marketing infrastructure, which need to be improvised through policy interventions.

The reasons for food losses at the retail stage of developing nations are different from that of developed nations. In developing countries, the main factors for food wastage at retail markets are food handling through unhygienic methods, inadequate food storage, and improper cooling conditions. In contrast to this, the food losses in developed countries are due to the display of food commodities in large quantities and varieties at supermarkets, packaging in large quantities, and high-quality safety regulations for food products. For combating this problem in developing countries, appropriate infrastructure development in the storage, transportation, and processing of food products is required. The steps for reducing this wastage in developed

Table 5.1 Improving energy efficiency by direct or indirect technological and social measures across the food chain

In situ	Ex situ
• Precise irrigation methods like drip and sprinkler • Adopt and maintain motor engines with higher fuel efficiency • Site-specific application of fertilizers • Adopt zero and minimum tillage techniques • Microclimate Controlled building • Thermal heat control in greenhouses • Design and model of fishing vessel propeller • Low-input-requiring cultivars and animal species • Less input demand of crop varieties and animal breeds • Soil erosion control • Lower moisture demand and losses • Use of organic manures and biofertilizers • Economical manufacture of efficient machinery	• Design and operation of transport vehicles • Electric motors of adjustable speed • Effective lighting and temperature control • Cold storage insulation • Minimization of food packaging • Improve cooking and space heating efficiency • Improvement of highway infrastructure • City planning to lessen distance and time for distribution and purchase of food • Reduce the loss of crop production at all stage • Changing food habits by curtailing animal-based products in the diet • Reduce the levels of obesity • Identification of food products by labelling

countries include smaller packages, fewer varieties of food products on shelves, increasing shelf life by improving packaging material, coordinated and coherent regulatory frameworks for food safety, labelling and legislation, and alternative markets and food uses. The ultimate consumer of the food causes food wastage at the food consumption stage, which can be tackled by spreading awareness among consumers. The food value chain can be strengthened at the final stage of disposal of food by improving the waste separation and management practices. In this way, a sustainable and inclusive food value chain needs to be adopted for CSA (Table 5.1).

5.7 Policy Instrumentation

Interventions of CSA should be aligned with and promoted by governmental policies along with a strong legal, regulatory framework. The policies and legislation have considerable influence over each component of CSA, which could facilitate or obstruct their adoption by the stakeholders. In this context, the Indian government has launched the National Action Plan on Climate Change (NAPCC) in 2008 with the following eight national missions which directly or indirectly contribute toward the attainment of CSA objectives:

1. National Solar Mission: This mission aims to promote the production and utilization of solar energy for electricity and power generation in order to make solar

power competitive with other alternatives of FF-based energy. It also involves setting up a solar research center, enhanced global technology development cooperation, increasing domestic production capability, and increased funding by government agencies and international support. India generates more than 74 GW of energy through renewable resources, out of which about 25 GW is contributed by solar energy (Press Information Bureau (PIB), Government of India, Ministry of Environment, Forest and Climate Change 2019). Recently, the government of India launched a 750 megawatt solar project in Rewa, Madhya Pradesh, on July 10, 2020, which will reduce carbon emission equivalent to approximately 15 lakh ton of CO_2 per year along with providing electricity (Press Information Bureau (PIB), Prime Minister's Office 2020).

2. National Mission for Enhanced Energy Efficiency: This mission endorses that specific energy consumption reductions be required in broad energy-intensive industries, with a framework for businesses to exchange energy-saving certificates, and funding for public-private collaborations to decrease energy consumption via demand-side management systems in the industrial, construction, and agricultural sectors, along with energy incentives, constituting minimal taxes on energy-efficient devices. Schemes such as UJALA for the distribution of LED bulbs have covered more than 320 million households, whereas UJJWALA for the distribution of clean cooking stoves has reached more than 63 million families of women below poverty line (Press Information Bureau (PIB), Government of India, Ministry of Environment, Forest and Climate Change 2019).
3. National Mission on Sustainable Habitat: The goal of this mission is to promote energy efficiency as a crucial part in urban planning by expanding the current Energy Conservation Building Code, improving the implementation of regulations for automotive fuel economy, and applying price policies to facilitate the procurement of efficient vehicles and public transport opportunities. It also focuses on the management and recycling of waste.
4. National Water Mission: This mission places a target of a 20% increase in WUE by pricing and other steps to tackle water scarcity resulting from CC. Assessment of 138 blocks of Punjab and 128 blocks of Haryana was carried out, out of which 109 and 78 blocks were categorized as "Over-exploited," respectively (Central Ground Water Board (CGWB) 2017). This is primarily because of the cultivation of rice as these areas use two to three times more irrigation water for producing 1 kilogram of rice than West Bengal and Bihar. To tackle such issues, National Water Mission launched a campaign named "Sahi Fasal" in November 2019 so that the income of farmers can be enhanced with appropriate ACP, using less water on a sustainable basis (National Water Mission (NWM), Ministry of Water Resources, River Development, and Ganga Rejuvenation 2019).
5. National Mission for Sustaining the Himalayan Ecosystem: This mission aims at preventing the melting of the Himalayan glaciers and protecting the Himalayan region's biodiversity.
6. Green India Mission: This mission aims at afforestation of degraded forest lands of 6 M ha area and increasing the forest cover from 23% to 33%. The forest and tree cover of India has increased by 1 percent as compared to the assessment in

2015 (Press Information Bureau (PIB), Government of India, Ministry of Environment, Forest and Climate Change 2019).

7. National Mission for Sustainable Agriculture: This mission focuses to promote climate adaptation in agriculture by developing climate-resilient crops and expanding mechanisms for weather insurance, along with improved agricultural operations. Because of readiness and introduction of climate-resilient varieties, total food grain production has significantly increased from 208.60 Mt in 2005–2006 to 284.95 Mt in 2018–2019, as well as horticulture production has increased from 116.9 Mt in 2004–2005 to 313.85 Mt in 2018–2019 (Press Information Bureau (PIB), Government of India, Ministry of Agriculture, and Farmers Welfare 2020). It also includes interventions like Soil Health Card (SHC) where around 115 million SHC were distributed under Soil Health Card Scheme from 2017 to 2019 among different states of India (Fig. 5.14) and more than 267.8 million soil samples were taken under the scheme to assess the nutrient status and health of soil (Directorate of Economics Statistics (DES) 2020). This will enable the farmers to improve farm productivity and farm economics by applying fertilizers, only to the particular patch, instead of the whole field, following the site-specific nutrient management approach. Indian Council of Agricultural Research (ICAR) has also developed 45 models for climate-resilient integrated farming systems (IFS) that are promoted by Krishi Vigyan Kendras (KVKs) by demonstrating and extending them to farmers through the Rainfed Area Development (RAD) program (Press Information Bureau (PIB), Government of India, Ministry of Environment, Forest and Climate Change 2019).
8. National Mission on Strategic Knowledge for Climate Change: This mission envisages a new Climate Science Research Fund, better climate modelling, and improved international collaboration to attain a good knowledge of climate science and its impacts and constraints. It also supports private sector projects to use venture capital funds to develop adaptation and mitigation technologies.

In this way, NAPCC mainly aims at the enhancement of understanding of CC and its adaptation and mitigation, improving energy efficiency, and conservation of natural resources contributing toward the attainment of CSA. Apart from NAPCC, several other schemes were initiated by the Indian government for the promotion of CSA. These schemes include National Project on Organic Farming (NPOF), National Project on Promotion of Balanced Use of Fertilizers (NPPBUF), National Food Security Mission (NFSM), and Accelerated Irrigation Benefit Programme (AIBP). Crop insurance scheme named as Pradhan Mantri Fasal Bima Yojana (PMFBY) was launched to provide financial insurance to farmers in case of crop losses caused by natural calamities like hailstorm, cyclone, flood, drought, etc. as well as by attack of pests and diseases. This scheme also insures for post-harvest losses caused within 14 days of the harvested crop. PMFBY alone has benefitted around 44.2 million farmers in 2018–2019, and around 51.9 M ha, that is, around 26.35% of gross cropped area of India (Directorate of Economics Statistics (DES)

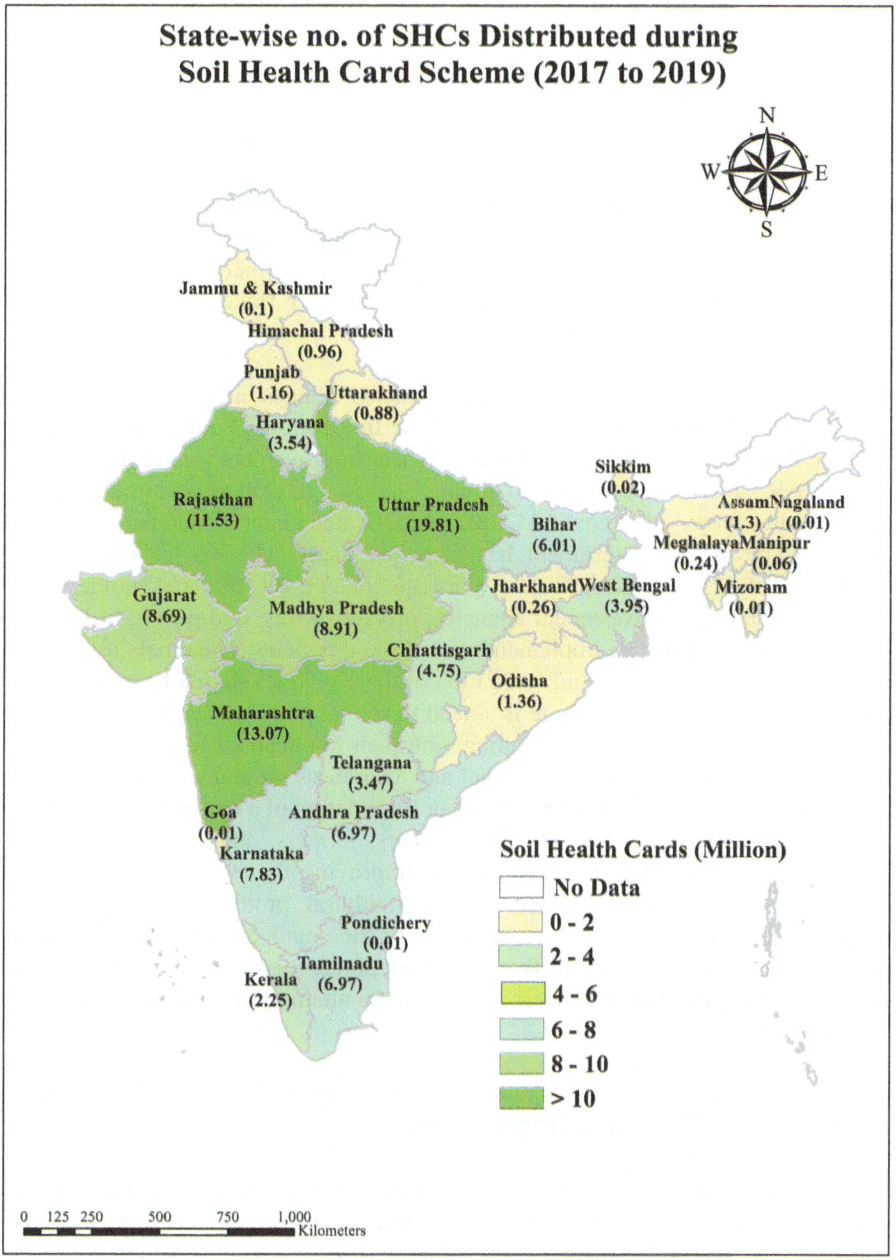

Fig. 5.14 State-wise number of SHCs distributed during Soil Health Card Scheme (2017–2019) (Source: Department of Agriculture, Cooperation & Farmers Welfare)

2020). The state-wise gross cropped area under all kinds of insurance schemes is depicted in Fig. 5.15.

National Initiative on Climate Resilient Agriculture (NICRA) program is initiated by the ICAR in February 2011 with the following objectives (Nagargade et al. 2017):

a. Increase the resilience of Indian agriculture to CC and CV by development and implementation of production and risk management strategies.
b. Demonstration of site-specific technologies on farmer's field.
c. It is enhancing the capacity of scientists and other stakeholders for climate-resilient agriculture research and its application.

Under the NICRA project, climate-resilient villages are developed in the 151 districts of India. The collaboration of the International Food Policy Research Institute (IFPRI), International Maize and Wheat Improvement Center (CIMMYT), and CGIAR's research program on Climate Change, Agriculture and Food Security (CCAFS) has proposed to cover 237,000 hectares area under climate-smart villages in India. The Indian government has launched many schemes for fulfilling the objectives of CSA. The effective participation of all the stakeholders like bureaucrats, farmers, government agencies, private sectors, and people in business is required for the effective implementation of these policies. The timely assessment of their impact, monitoring, and evaluation of these policies should be conducted to quantify their effectiveness. There is a need to spread proper awareness about these policies and about CSA practices, along with their implications among the farming community through extension personnel and awareness programs. Proper feedback mechanism for improvising these policies or formulation of a new policy should also be established in conjunction with researchers, farmers, and other stakeholders. There is a need to increase investment for improving the storage, transportation, processing, and marketing facilities of agricultural produce at ground level. Incentives or subsidies should be provided to the small or marginal farmers for adopting CSA practices. An effective and strong policy for promotion and successful adoption of CSA by all stakeholders is essential for attaining sustainability in AP and FNS under CC scenario.

5.8 Constraints and Opportunities

For the successful adoption of CSA practices, there is a requirement to identify constraints in its adoption by various stakeholders and harnessing of the available opportunities. The main constraints for adoption of CSA practices by Indian farmers are mentioned below (Deepika and Saravanan 2018):

a. Farmers with small landholdings don't have space for the installation of water harvesting structures, specifically in rainfed regions.
b. Less availability of labors for carrying out CSA practices.

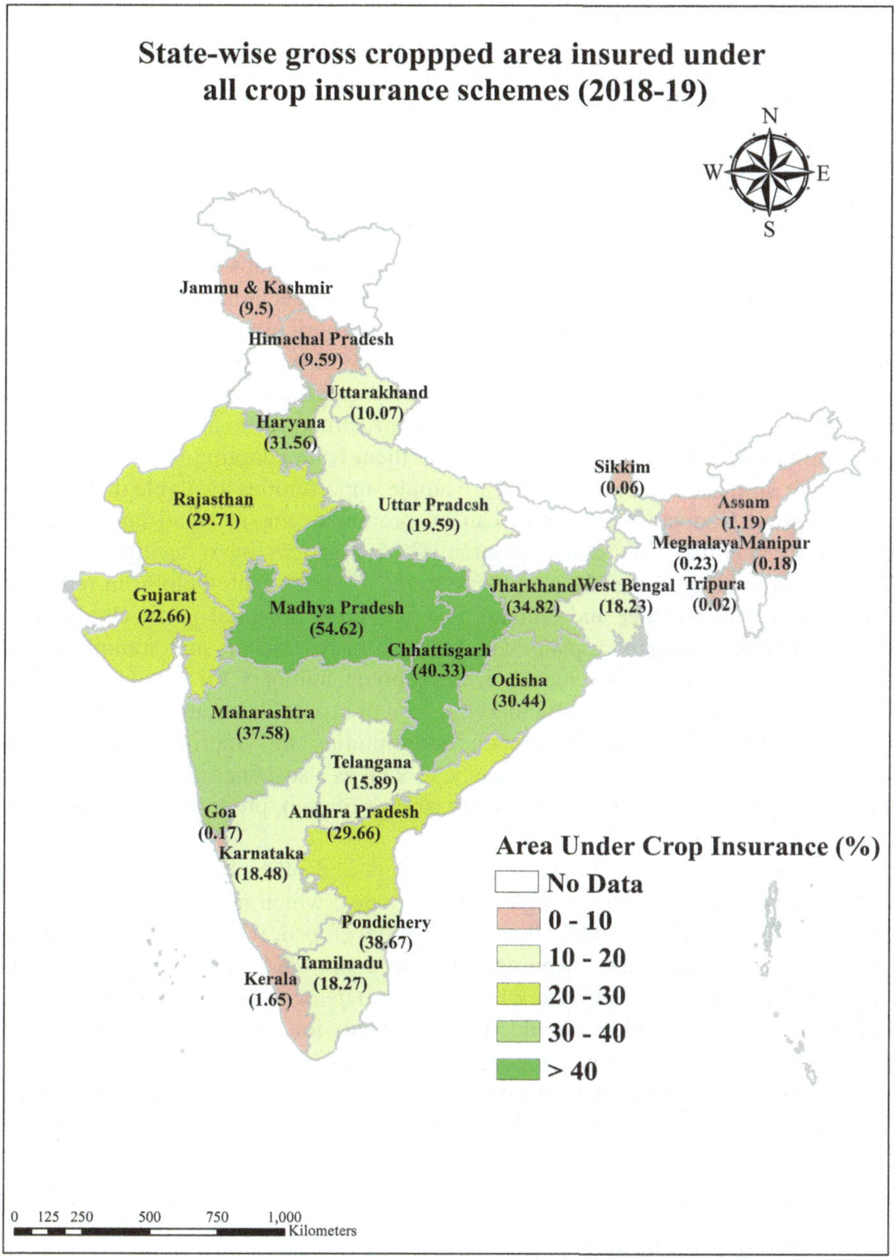

Fig. 5.15 State-wise gross cropped area insured under all crop insurance schemes (2018–2019) (Source: Department of Agriculture, Cooperation & Farmers Welfare)

c. Unavailability of the good quality inputs to the farmers by the local dealers or traders.
d. Lack of awareness and willingness among farmers for adopting CSA practices.
e. Unavailability of proper marketing infrastructures.
f. Inaccessibility of farmers to credit facilities for insurance of their crops.
g. Conflicts among farmers while conducting extension activities like demonstrations, training programs, etc.

Other constraints include lack of proper awareness about CC and its mitigation and adaptation by CSA practices among the various stakeholders. Lack of proper storage, processing, and transportation facilities for agricultural products is another barrier. Unavailability or lack of timely availability of machineries, implements, or various inputs required for practicing CSA also creates an obstruction. The poor financial condition of most of the farmers and unwillingness to change their attitude toward innovative CSA practices also hinder them from adopting it.

Despite all these constraints, there are ample opportunities available that need to be harnessed. Cooperative farming can be promoted among small and marginal farmers so that they can easily adapt to CSA practices by reducing their risk. Weather forecasting and early warning systems could be quite helpful in reducing risks linked with weather and climate. ICT has great potential in developing and communicating contingency plans with the help of researchers and administrators. The seed banks should be created in appropriate numbers for timely providing quality seeds to the farmers, particularly in case of crop failure caused by unwanted weather events. There are enormous opportunities for improved post-harvest technologies that aids in energy efficiency, as well as a sustained rise in productivity and revenue generation, like better quality storage of crop, produce, its packing, and supply.

The awareness can be spread among farmers and other stakeholders by using voice messages and videos through mobile phones, which may bridge the knowledge gap. Climate-smart model villages can be developed, which can act as a model village among farmers and encourage them to adopt CSA practices. The rewards should be provided to incentivize innovative farmers or agencies involved in CSA, and their success stories should be promoted for sensitizing farmers and stakeholders. Appropriate financing and funds should be provided to various sectors for the development of proper infrastructure and spreading awareness, which may facilitate the adoption of CSA. So, there is a need to increase investment in promoting CSA adoption.

5.9 Future Perspectives

Reliable and accurate weather forecasts for several locations can help to develop contingency plans for various crops and cropping systems. There is a need to refine the outputs of CSM for better decision-making for the policymakers and for providing suitable and timely advisories to the farmers. There is a need for conducting further research on precise and site-specific management of inputs suitable for small and marginal farmers. There is also a need to research nanotechnology for improving input use efficiency in ACP.

Further research should be conducted to study the long-term effects of CA. Integrated farming system models should be developed based on the location and resources available to the farmers. More crop varieties should be developed, having more climate resilience. Research should be conducted to develop sensors which can monitor the performance of CSA practices and real-time condition of crops while using green energy for its operation. These sensors should be able to notify farmers through text messages about field operations. Robotics has great scope in the mechanization of CSA practices. Research is required to evaluate the application of drones in CSA. Innovative extension systems should be developed for convincing farmers and stakeholders toward the adoption of CSA. There is a requirement to develop a proper integration framework of policymakers, private partners, farmers, researchers, bureaucrats, traders, and governmental agencies for successful and effective implementation of CSA at field level.

5.10 Conclusions

CC and CV are affecting agriculture, and its adverse effects are projected to become more grave in the future. CSA is very pertinent for mitigation and building resilience or adaptation of agricultural system along with increasing or sustaining production in CC scenarios. There are various CSA practices which should be adopted based on the farmers' condition and availability of resources. These practices include cultivation of improved varieties resistant to insect, pest, diseases, high temperature, drought, or salinity; CA; using energy-smart technologies like biofuels, solar energy, etc.; efficient water management by micro-irrigation, rainwater harvesting, drainage structures, etc.; precision farming, integrated farming system; integrated pest, weed, and nutrient management; agroforestry, crop, and livestock insurance; improved breeds of livestock; modelling and forecasting for appropriate decision-making; and so on. The effective adoption of these CSA interventions can be possible by a holistic and integrated approach of all stakeholders involved in agriculture. CSA has huge potential in combating CC and CV. Appropriate policies and their effective implementation at the field level are essential for the success of CSA. The policies should focus on spreading awareness by ICT-based technologies and demonstration of climate-smart villages along with providing financial assistance for the adoption of CSA practices. Appropriate and energy-efficient storage, transport, processing, and marketing facilities should also be developed for the promotion of CSA. Thus, CSA has huge potential for attaining agricultural sustainability, providing FNS, and improvement of livelihood and income in an environment-friendly way.

References

Aggarwal PK (2008) Global climate change and Indian agriculture: impacts, adaptation and mitigation. Ind J Agri Sci 78:911–919

Alford AR, Hegarty RS, Parnell PF, Cacho OJ, Herd RM, Griffith GR (2006) The impact of breeding to reduce residual feed intake on enteric methane emissions from the Australian beef industry. Aus J Exp Agri 46:813–820

Altieri MA, Nicholls CI (2013) The adaptation and mitigation potential of traditional agriculture in a changing climate. Clim Change. https://doi.org/10.1007/s10584-013-0909-y

Altieri MA, Marcos LA, Bittencourt HV, Kieling AS, Comin JJ, Lobsyo PE (2011) Enhancing crop productivity via weed suppression in organic no-till cropping systems in Santa Catarina, Brazil. J Sustainable Agri 35(8):1–15. https://doi.org/10.1080/10440046.2011.588998

Amin A, Mubeen M, Hammad HM, Nasim W (2015) Climate smart agriculture: an approach for sustainable food security. Agri Res Communication 3:13–21

Awazi NP, Tchamba MN (2019) Enhancing agricultural sustainability and productivity under changing climate conditions through improved agroforestry practices in smallholder farming systems in Sub-Saharan Africa. Afr J Agric Res 14(7):379–388

Bandyopadhyay T, Muthamilarasan M, Prasad M (2017) Millets for next generation climate-smart agriculture. Frontiers in Plant Science 8:1266

Barfoot P, Brookes G (2014) Key global environmental impacts of genetically modified (GM) crop use 1996-2012. GM Crops and Food 5(2):149–160

Bates BC, Kundzewicz, ZW, Wu S, Palutikof JP (eds) (2008) Climate change and water. Technical Paper of the Intergovernmental Panel on Climate Change, IPCC Secretariat, Geneva

Batima P, Bat B, Tserendash L, Bayarbaatar S, Shiirev-Adya S, Tuvaansuren G, Natsagdorj L, Chuluun T (2005) Adaptation to climate change, vol 90. ADMON Publishing, Ulaanbaatar

Battisti DS, Naylor RL (2009) Historical warnings of future food security with unprecedented seasonal heat. Science 323:240–244

Bentley D, Hegarty R (2008) Managing livestock enterprises in Australia's extensive rangelands for greenhouse gas and environmental outcomes: a pastoral company perspective. Aus J Exp Agric 48:60–64

Bhardwaj J, Yadav SK (2012) Genetic mechanisms of drought stress tolerance, implications of transgenic crops for agriculture. In: Agroecology and strategies for climate change. Springer, New York, pp 213–235

Blank D (2009) PoA cdm manual. Mini biogas plants for households. GFA Envest, Hamburg

Boyer F, Campagnaud M (1996). Synthesis of experimentation on software IRRISA. Chamber of agriculture dordogne, p 18

Branca G, Lipper L, Neves N, Lopa D, Mwanjioka I (2011) Payments for watershed services supporting sustainable agriculture development in Tanzania. J Environ Dev 20(3):278–302

Cardenas LB, Dukes MD (2012) Soil moisture sensor landscape irrigation controllers: a review of multi-study results and future implications. American Society of Agricultural and Biological Engineers 55(2):581–590

Castiglionit P, Warner D, Bensen RJ, Anstrom DC, Harrison J, Stoecker M, Abad M, Kumar G, Salvador S, D'Ordine R, Navarro S, Back S, Fernandes M, Targolli J, Dasgupta S, Bonin C, Luethy MH, Heard JE (2008) Bacterial RNA chaperones confer abiotic stress tolerance in plants and improved grain yield in maize under water–limited conditions. Plant Physiology 147:446–455

Central Ground Water Board (CGWB) (2017) Ministry of Jal Shakti, Department of Water Resources, River Development and Ganga Rejuvenation. Dynamic Ground Water Resources of India. http://cgwb.gov.in/GW-Assessment/GWRA-2017-National-Compilation.pdf. Accessed 04 April 2020

Challinor AJ, Wheeler TR (2008) Use of a crop model ensemble to quantify CO2 stimulation of water-stressed and well-watered crops. Agric For Meteorol 148:1062–1077

Challinor AJ, Wheeler TR, Craufurd PQ, Ferro CAT, Stephenson DB (2007) Adaptation of crops to climate change through genotypic responses to mean and extreme temperatures. Agri Ecosys Env 119:190–204

Challinor AJ, Ewert F, Arnold S, Simelton E, Fraser E (2009) Crops and climate change: progress, trends, and challenges in simulating impacts and informing adaptation. J Exp Bot 60:2775–2789

Chartzoulakis, K, Kasapakis I, Tzobanoglou (2008) Improving water efficiency: the irrigation advisory service of Crete, Greece. In: The 3rd international conference on water resources and arid environments and the 1st Arab Water Forum

Chinnusaamy V, Jagendorf A, Zhu JK (2005) Understanding and improving salt tolerance in plants. Crop Sci 45:437–448

Crosson P (1997) Impacts of climate change on agriculture. Climate issues brief No. 4. Resources for the Future, Washington, DC

Cruz RV, Harasawa H, Lal M, Wu S, Anokhin Y, Punsalmaa B, Honda Y, Jafari M, Li C, Huu N (2007). Asia climate change 2007: impacts, adaptation and vulnerability, pp 469–506

Dar MH, Singh S, Zaidi NW, Shukla S (2012) Sahbhagi Dhan: science's answer to drought problems. STRASA News 5:1–13

Das S (2017) Impact of climate change on livestock, various adaptive and mitigative measures for sustainable livestock production. Appro Poult Dairy Vet Sci 1(4):517. https://doi.org/10.31031/APDV.2017.01.000517

de Bruin K, Dellink RB, Ruijs A, Bolwidt L, van Buuren A, Graveland J, de Groot RS, Kuikman PJ, Reinhard S, Roetter RP, Tassone VC, Verhagen A, van Ierland EC (2009) Adapting to climate change in The Netherlands: an inventory of climate adaptation options and ranking of alternatives. Climatic Change 95:23–45

Deepika, Suchiradipta B, Saravanan R (2018) Climate smart agriculture towards triple win: adaptation, mitigation and food security. MANAGE discussion paper 5, MANAGE-Centre for Agricultural Extension Innovations, Reforms and Agripreneurship (CAEIRA), National Institute of Agricultural Extension Management, Hyderabad, India

Directorate of Economics Statistics (DES) (2020) Pocket book of Agricultural Statistics 2019 Government of India, Ministry of Agriculture & Farmers Welfare. https://eands.dacnet.nic.in/PDF/Pocket%20Book%202019.pdf. Accessed 04 April 2020

Estay SA, Lima M, Labra FA (2009) Predicting insect pest status under climate change scenarios: combining experimental data and population dynamics modelling. J Appl Ento 133(7):491–499

FAO (2006) Livestock's long shadow: environmental issues and options. FAO, Rome

FAO (2008) Climate change and food security: a framework document. FAO, Rome

FAO (2010) Climate-smart agriculture: policies, practices and financing for food security, adaptation and mitigation. Report prepared for The Hague Conference on Agriculture, Food Security, and Climate Change 31 October to 5 November 2010 Rome, Italy

FAO (2011) Save and grow: a policymaker's guide to the sustainable intensification of smallholder crop production. FAO, Rome

FAO (2012) Towards the future we want: end hunger and make the transition to sustainable agricultural and food systems. FAO, Rome

FAO (2013) Principles of sustainable food value chain development with a meta guide to value development resource materials. Draft

Faurès JM, Bartley D, Bazza M, Burke J, Hoogeveen J, Soto D, Steduto P (2013) Climate smart agriculture sourcebook. FAO, Rome, p 557

Fischer G, van Velthuizen H, Shah M, Nachtergaele F (2001) Global agro-ecological assessment for agriculture in the 21st century. International Institute for Applied Systems Analysis (IIASA), Laxenburg, Austria and Food and Agriculture Organization, Rome

Friedrich T, Kassam A (2009) Adoption of conservation agriculture technologies: constraints and opportunities. Invited paper at the IV World Congress on Conservation Agriculture. New Delhi, India

Gustavsson J, Cederberg C, Sonesson U, van Otterdijk R, Meybeck A (2011) Global food losses and food waste: extent, causes and prevention. Study conducted for the International Congress

SAVE FOOD! At Interpack 2011, Düsseldorf, Germany. Food and Agriculture Organisation (FAO) of the United Nations, Rome, Italy, p 29

Haefelea SM, Kato Y, Singh S (2016) Climate ready rice: augmenting drought tolerance with best management practices. Field Crops Res 190:60–69

Hall-Spencer JM, Rodolfo-Metalpa R, Martin S (2008) Volcanic carbon dioxide vents show ecosystem effects of ocean acidification. Nature 454:96–99

Hasanuzzaman M, Nahar K, Alam MM, Roychowdhury R, Fujita M (2013) Physiological, biochemical, and molecular mechanisms of heat stress tolerance in plants. Int J Mol Sci 14:9643–9684

Hatfield J, Boote K, Fay P, Hahn L, Izaurralde C, Kimball BA, Mader T, Morgan J, Ort D, Polley W, Thomson A, Wolfe D (2008) Agriculture. In: The effects of climate change on agriculture, land resources, water resources, and biodiversity in the United States. A Report by the U.S. Climate Change Science Program and the Subcommittee on Global Change Research. Washington, DC., USA, p. 362. Available at: https://www.usda.gov/oce/climate_change/SAP4_3/CCSPFinalReport.pdf. Accessed 31 Oct 2019

Havlik P, Valin H, Mosnier A, Obersteiner M, Baker JS, Herrero M, Rufino MC, Schmid E (2013) Crop productivity and the global livestock sector: implications for land use change and greenhouse gas emissions. Am. J. Agric. Econ 95:442–448

He C, Yan J, Shen G, Fu L, Holaday AS, Auld D, Blumwald E, Zhang H (2005) Expression of an Arabidopsis vacuolar sodium/proton antiporter gene in cotton improves photosynthetic performance under salt conditions and increases fiber yield in the field. Plant Cell Phys 46:1848–1854

Herrero M, Henderson B, Havlík P, Thornton PK, Conant RT, Smith P, Wirsenius S, Hristov AN, Gerber P, Gill M, Butterbach-Bahl K, Valin H, Garnett T, Stehfest E (2016) Greenhouse gas mitigation potentials in the livestock sector. Nature Climate Change 6:452–461. https://doi.org/10.1038/nclimate2925

Hristov AN, Oh J, Firkins JL, Dijksta J, Kebreab E, Waghorn G, Makkar HPS, Adesogan AT, Yang W, Lee C, Gerber PJ, Henderson B, Tricarico JM (2013) Mitigation of methane and nitrous oxide emissions from animal operations: I a review of enteric methane mitigation options. American Society of Animal Science 91:5045–5069

INCCA (Indian Network for Climate Change Assessment) (2010) Assessment of the greenhouse gas emissions: 2007, the Ministry of Environment & Forests, Govt. of India, New Delhi.

IPCC (Intergovernmental Panel on Climate Change) (1996) Climate change 1995: economic and social dimensions of climate change. In: Bruce J, Lee H, Haites E (eds) Contribution of Working Group III to the second assessment of the intergovernmental panel on climate change. Cambridge University Press, Cambridge

IPCC (Intergovernmental Panel on Climate Change) (2007a) Climate change 2007: impacts, adaptation and vulnerability. In: Parry ML, Canziani JP, Palutikof PJ, Hanson CE (eds) Contribution of Working Group II to the Fourth Assessment Report of the IPCC. Glossary. Cambridge University Press, Cambridge, pp 869–883

IPCC (Intergovernmental Panel on Climate Change) (2007b) Climate change 2007: Mitigation. In: Metz B, Davidson OR, Bosch PR, Dave R, Meyer LA (eds) Contribution of Working Group III to the fourth assessment report of the Intergovernmental Panel on Climate Change. Cambridge University Press, Cambridge, pp 497–540

IPCC (Intergovernmental Panel on Climate Change) (2012) Managing the risks of extreme events and disasters to advance climate change adaptation. In: Field CB, Barros V, Stocker TF, Qin D, Dokken DJ, Ebi KL, Mastrandrea MD, Mach KJ, Plattner GK, Allen SK, Tignor M, Midgley PM (eds) A special report of Working Groups I and II of the Intergovernmental Panel on Climate Change. Cambridge University Press, Cambridge

IPCC (Intergovernmental Panel on Climate Change) (2014) In: IPCC (ed) Climate change 2014: Synthesis report. Contribution of Working Groups I, II and III to the fifth assessment report of the Intergovernmental Panel on Climate Change, Geneva

IPCC (Intergovernmental Panel on Climate Change) (2018) Annex I: glossary. Masson-Delmotte,-V, Zhai P, Pörtner HO, Roberts D, Skea J, Shukla PR, Pirani A, Moufouma-Okia W, Péan C,

Pidcock R, Connors S, Matthews JB, Chen Y, Zhou X, Gomis MI, Lonnoy E, Maycock T, Tignor M, and Waterfield T (eds.). Global Warming of 1.5°C. An IPCC Special Report on the impacts of global warming of 1.5°C above pre-industrial levels and related global greenhouse gas emission pathways, in the context of strengthening the global response to the threat of climate change, sustainable development, and efforts to eradicate poverty. IPCC, Geneva

Jat RK, Sapota TB, Singh RG, Jat ML, Kumar M, Gupta RK (2014) Seven years of conservation agriculture in a rice-wheat rotation of eastern Gangetic Plains of South Asia: Yield trends and economic profitability. Field Crops Res 164:199–210

Jha GK, Pal S, Singh A (2012) Changing energy-use pattern and the demand projection for Indian agriculture. Agricultural Economics Research Review 25(1):61–68

John WH, Nicholas JC, Evan WC, Thomas MS, Bill W (2009). IrriSatSMS irrigation water management by satellite and SMS - a utilization framework, CSIRO Land and Water Science

Karaba A, Dixit S, Greco R, Aharoni A, Trijatmiko KR, Marsch-Martinez N, Krishnan A, Nataraja KN, Udayakumar M, Pereira A (2012) Improvement of water use efficiency in rice by expression of HARDY, an Arabidopsis drought and salt tolerance gene. Proceedings of the National Academy of Sciences of the United States of America 104:15270–15275

Karl TR., Meehl GA, Miller CD, Hassol SJ, Walpe AM, Murray WL (2008) Weather and climate extremes in a changing climate. Synthesis and assessment product 3.3. Report by the US Climate Change Science Program (CCSP) and the Subcommittee on Global Change Research, Department of Commerce, NOAA National Climate Data Center, Washington, DC

Katiyar-Agarwal S, Agarwal M, Grover A (2003) Heat-tolerant basmati rice engineered by over-expression of hsp101. Plant Mol Biol 51:677–686. https://doi.org/10.1023/A:1022561926676

Keating BA, Carberry PS, Hammer GL, Probert ME, Robertson MJ, Holzworth D (2003) An overview of APSIM, a model designed for farming systems simulation. Euro J Agron 18:267–288

Khaledian MR, Mailhol JC, Ruelle P, Rosique P (2009) Adapting PILOTE model for water and yield management under direct seeding system: the case of corn and durum wheat in a Mediterranean context. Agric Water Manage 96:757–770

Kim DG (2012) Estimation of net gain of soil carbon in a nitrogen-fixing tree and crop intercropping system in sub-Saharan Africa: results from re-examining a study. Agrofor Syst. https://doi.org/10.1007/s10457-011-9477-1

Knutson TR, McBride JL, Chan J, Emanuel K (2010) Tropical cyclones and climate change. Nat Geosci 3:157–163

Krishnan R, Sanjay J (2017) Centre for Climate Change Research (CCCR) in the Indian Institute of Tropical Meteorology (IITM) Pune. Climate Change over INDIA: an interim report. Available at http://cccr.tropmet.res.in/home/docs/cccr/climate-change-report-2017.pdf. Accessed 27 October, 2019

Lal R (2004a) Soil carbon sequestration to mitigate climate change. Geoderma 123(1-2):1–22

Lal R (2004b) Soil carbon sequestration impacts on global climate change and food security. Science 304:1623–1627. https://doi.org/10.1126/science.1097396

Lal R (2015) Sequestering carbon and increasing productivity by conservation agriculture. J Soil Water Conserv 70(3):55A–62A. https://doi.org/10.2489/jswc.70.3.55A

Lamichhane JR, Barzman M, Booij K, Boonekamp P, Desneux N, Huber L, Kudsk P, Langrell SR, Ratnadass A, Ricci P, Sarah JL (2015) Robust cropping systems to tackle pests under climate change. A review. Agronomy for Sustainable Development 35(2):443–459

Lipiec J, Doussan C, Nosalewicz A, Kondracka K (2013) Effect of drought and heat stresses on plant growth and yield: a review. International Agrophysics 27:4

Livemint (2015). Available at http://www.livemint.com/Politics/4SbC4kGC9cYt2oxtrYS6eP/A-year-of-extreme-weather-conditions-for-India.html. Accessed 27 October, 2019

Lobell D, Schlenker W, Costa-Roberts J (2011) Climate trends and global crop production since 1980. Science 333(6042):616–620. https://doi.org/10.1126/science.1204531

Lutz D (2013) Creating plants that make their own fertilizer. Washington University in St. Louis News Release. 22 August 2013. Available at: http://news.wustl.edu/news/Pages/25585.aspx. Accessed 27 October 2019

Mertz O, Halsnaes K, Olesen JE, Rasmussen K (2009) Adaptation to climate change in developing countries. Environmental Management 43:743–752

Mittal S (2012) Modern ICT for agricultural development and risk management in smallholder agriculture in India. CIMMYT

Nagargade M, Tyagi V, Singh MK (2017) Climate smart agriculture: an option for changing climatic situation. Plant Engineering 143:88

National Water Mission (NWM), Ministry of Water Resources, River Development & Ganga Rejuvenation (2019). http://nwm.gov.in/node/542. Accessed 04 April 2020

NOAA (2015). Available at https://www.climate.gov/news-features/event-tracker/india-heat-wave-kills-thousands. Accessed 27 October 2019

OECD-DAC (2011) Available at Tracking aid in support of climate change mitigation and adaptation in developing countries. Available at: https://www.oecd.org/dac/stats/47477193.pdf. Accessed 27 October 2019

Padgham J (2009) Agricultural development under a changing climate: opportunities and challenges for adaptation. World Bank, Washington, DC

Pampolino MF, Witt C, Pasuquin JM, Johnston A, Fisher MJ (2012) Development approach and evaluation of the Nutrient Expert software for nutrient management in cereal crops. Comput Electron Agric 88:103–110

Pankaj PK, Ramana DBV, Pourouchottamane R, Naskar S (2013) Livestock management under changing climate scenario in India. World Journal of Veterinary Science 1(1):25–32

Pathak H (2009) Agriculture and environment. In: Hand book of agriculture, Directorate of Information and Publication Agriculture. ICAR, New Delhi, pp 62–92

Pathak H, Wassmann R (2007) Introducing greenhouse gas mitigation as a development objective in rice-based agriculture: II. Cost-benefit assessment for different technologies, regions and scales. Agric Syst 94:826–840

Pathak H, Aggarwal PK, Singh SD (2012) Climate change impact, adaptation and mitigation in agriculture: methodology for assessment and applications. Indian Agricultural Research Institute, New Delhi, p 302

PMKSY (Pradhan Mantri Krishi Sinchayee Yojana) (2019). Available at: https://pmksy.gov.in/. Accessed 31 Oct 2019

Press Information Bureau (PIB), Government of India, Ministry of Agriculture & Farmers Welfare (2020). https://pib.gov.in/PressReleseDetailm.aspx?PRID=1606803. Accessed 04 April 2020.

Press Information Bureau (PIB), Government of India, Ministry of Environment, Forest and Climate Change (2019). https://pib.gov.in/Pressreleaseshare.aspx?PRID=1564033. Accessed 04 April 2020.

Press Information Bureau (PIB), Prime Minister's Office (2020). https://pib.gov.in/PressReleasePage.aspx?PRID=1637549. Accessed 09 July 2020.

Rolim J, Godinho P, Sequeira B, Paredes P, Pereira LS (2007) Assessing the SIMdualkc model for irrigation scheduling simulation in Mediterranean environments. Opt Mediterr 56:49–61

Rosenzweig C, Parry ML (1994) Potential impact of climate change on world food supply. Nature 367:133–138

Samra JS, Singh G (2002) Drought management strategies. Indian Council of Agricultural Research, New Delhi, p 68

Sapkota TB, Jat ML, Aryal JP, Jat RK, Chhetri AK (2015) Climate change adaptation, green-house gas mitigation and economic profitability of conservation agriculture: some examples from cereal system of Indo-Gangetic Plains. J Integrative Agri 14(8):1524–1533

Scherr SJ, Shames S, Friedman R (2012) Agric Food Secur 1:12. https://doi.org/10.1186/2048-7010-1-12

Selvaraju R, Gommes R, Bernardi M (2011) Climate science in support of sustainable agriculture and food security. Clim Res 47:95–110

Shah T (2009) Climate change and groundwater: India's opportunities for mitigation and adaptation. Environment Research Letters Journal 4(03):1–13

Shrawat AK, Good AG (2008) Genetic engineering approaches to improving nitrogen use efficiency. ISB News Report, pp 1–5. Available at https://pdfs.semanticscholar.org/f9f6/7c56c7e0d8a5dd843aedf3c20c9b73365815.pdf. Accessed 31 Oct 2019

Singh B, Singh V, Purba J, Sharma RK, Jat ML, Singh Y, Thind HS, Gupta RK, Choudhary OP, Chandna P, Khurana HS, Kumar A, Singh J, Uppal HS, Uppal RK, Vashistha M, Gupta RK (2015) Site-specific nitrogen management in irrigated transplanted rice (Oryza sativa) using an optical sensor. Precision Agriculture 16:455–475. https://doi.org/10.1007/s11119-015-9389-6

Smith M (1991) CROPWAT: manual and guidelines. FAO of UN, Rome

Smith P, Bustamante M, Ahammad H, Clark H, Dong H, Elsiddig EA, Haberl H, Harper R, House J, Jafari M, Masera O, Mbow C, Ravindranath NH, Rice CW, Robledo Abad C, Romanovskaya A, Sperling F, Tubiello F (2014) Agriculture, forestry and other land use (AFOLU). Climate change 2014: mitigation of climate change. Contribution of Working Group III to the fifth assessment report of the Intergovernmental Panel on Climate Change. Cambridge University Press, Cambridge, pp 811–922

Steinfeld H, Gerber P, Wassenaar T, Castel V, Rosales M, de Haan C (2006) Livestock's long shadow: environmental issues and options. Food and Agriculture Organization of the United Nations, Rome

Subbarao GV, Ban T, Kishii M, Ito O, Samejima HY, Wang SJ, Pearse S, Gopalakrishnan K, Nakahara AKM, Zakir Hossain H, Tsujimoto WL, Berry W (2007) Can biological nitrification inhibition (BNI) genes from perennial Leymus racemosus (Triticeae) combat nitrification in wheat farming? Plant Soil 299:55–64

Thornton PK, Herrero M (2010) Potential for reduced methane and carbon dioxide emissions from livestock and pasture management in the tropics. Proceedings of the National Academy of Sciences 107(46):19667–19672. https://doi.org/10.1073/pnas.0912890107

Urso GD, Michele CD, Bolognesi SF (2013) "IRRISAT: The Italian on-line satellite irrigation advisory service". EFITA-WCCA-CIGR Conference "Sustainable agriculture through ICT innovation", Turin, Italy

Verchot L, Van Noordwijk M, Kandji S, Tomich T, Ong C, Albrecht A, Mackensen J, Bantilan C, Anupama K, Palm C (2007) Climate change: linking adaptation and mitigation through agroforestry. Mitig Adapt Strat Glob Chang 12:901–918. https://doi.org/10.1007/s11027-007-9105-6

Williams JR, Jones CA, Kiniry JR, Spanel DA (1989) EPIC crop growth model. Trans Am Soc Agric Eng 32:497–511

World Bank (2013) "Turn down the Heat: climate Extremes, regional impacts, and the case for resilience". Available at https://www.worldbank.org/content/dam/Worldbank/document/Full_Report_Vol_2_Turn_Down_The_Heat_%20Climate_Extremes_Regional_Impacts_Case_for_Resilience_Print%20version_FINAL.pdf. Accessed 31 Oct 2019

Wright AG, Klieve AV (2011) Does the complexity of the rumen microbial ecology preclude methane mitigation? Animal Feed Science Technology 166–167:248–253. https://doi.org/10.1016/j.anifeedsci.2011.04.015

Zingore S (2010) SOC sequestration in farming systems in Africa: potential, opportunities and challenges. In: 2nd meeting of the round table on organic agriculture and climate change, RTOACC.

Climate Change and Its Impact on Rice Productivity and Quality

6

Mangal Deep Tuti, Mahender Kumar Rapolu, and Brajendra

Abstract

Most of the population of Asian and African countries completely depend on rice for their daily food, and 90% of world rice comes from Asia. Nearly four and half billion of the total world population completely depend on rice as a primary source of their food. In India and other parts of the world, rice is taken as a staple food. Considering its importance, the United Nation declared the year 2004 as the International Year of Rice. The sustainable goals of the United Nation are to reduce hunger, poverty, malnutrition, yearning, and ailing health of the world. A 100 g of white, short-grain, and cooked rice contains 130 calories, 28.7 g carbohydrate, 2.36 g protein, and 0.19 g fat. Agriculture, horticulture, agroforestry, and all agriculture-related ecosystems are very much closely linked with climatic variables. So climate change and its impact on all these agro-ecosystems have been the prime agenda for research in recent times. Environmental changes and extreme biotic and abiotic stresses are posing genuine hurdles for rice production which affects badly to farmers' livelihood. There is an earnest need to devise and outline systematic procedures against these extremes, to adapt against these negative effects of climate change. The current chapter gives an outline of the ongoing studies on climate change and its likely effects on rice productivity and quality. Further, it suggests the mitigation strategy through crop improvement and crop management technologies.

Keywords

Climate change · Ecosystems · Livelihood · Rice quality · Rice productivity

M. D. Tuti (✉) · M. K. Rapolu · Brajendra
ICAR-Indian Institute of Rice Research, Hyderabad, Telangana, India
e-mail: mangal.tuti@icar.gov.in

V. K. Hebsale Mallappa, M. Shirur (eds.), *Climate Change and Resilient Food Systems*, https://doi.org/10.1007/978-981-33-4538-6_6

6.1 Introduction

In India, 40% of the total cropped area is irrigated, and rest 60% of the total cropped area is rainfed and completely dependent on an uncertain monsoon. This shows the dependency of Indian agriculture on climate. The worldwide population at the current pace will be in excess of 9 billion by the middle of the twentieth century and will prompt a significant increase in demand for rice. Approximately 20 million hectares of the world's rice-growing areas especially in India and other Asian countries are directly affected by climate change. It leads to an increase in the price of rice between 32 and 37% due to a reduction in productivity. By 2050, the International Food Policy Research Institute predicted that an increase in rice price will be 10–14% in Asia and 15% in sub-Saharan Africa. The major climatic parameters that influence rice productivity in India and most of the Asian nations are (1) rise in temperature, (2) uneven distribution of precipitation, (3) emergence of new insect pest and diseases, (4) greenhouse gas emission from rice fields, and (5) natural calamities, for example, long dry spell during the cropping season, heat waves, landslides, and flooding. Climate change affects not only rice productivity but also rice quality. To begin with, more than 80% of the world's rice is consumed by people, and rest of the cereals are used as animal feed or to produce other food product (GRiSP 2013). Again, in low- and middle-income countries, other than as staple food, rice is the major source of protein and micronutrients. Further, its appearance and biochemical properties are crucial for consumer acceptance because rice is generally consumed directly (Lyman et al. 2013; Cooper et al. 2008). So, studying and exploring the relationship between different climatic parameters and grain quality is very much important. Similarly, suitable strategies need to be discovered for rice crop to adapt to these climate changes. There is an alarming increase in earth's atmospheric temperature due to an increase in greenhouse gases (methane, carbon dioxide, and nitrous oxide). Rice contributes significantly to greenhouse gas like methane. Weather pattern changes due to an increase in earth's surface temperature. The Intergovernmental Panel on Climate Change in 2001 reported that there is an increase of 0.6 °C in worldwide temperature in the past century. It was estimated that average temperatures will increase up to 5.8 °C by 2100 (Nguyen 2002) which will lead to emergence of harmful insect pest and diseases. Further, it will have a greater harmful effect on agriculture ecosystems.

6.2 Rice Productivity and Climate Change

A higher concentration of carbon dioxide in the atmosphere leads to an increase in temperatures which further leads to a decrease in rice production globally. Rice crop fixes lesser CO_2 than that of efficient C_4 crop. Rice grown in all ecosystems emits greenhouse gases (methane and nitrous oxide gases) and contributes significantly to global warming. Rice cultivation contributes more than 10% of released methane from the rice field. It was found that an increase in CO_2 level and rise in temperature decrease the productivity of rice. Temperature more than 45 °C is detrimental for rice

Table 6.1 The critical temperature for growth and development of rice

Rice growth stage	Temperature (°C)	
	Low	High
Germination	16	45
Rooting	16	35
Elongation of leaf	7	45
Active tillering	9	33
Panicle initiation	15	–
Anthesis	22	36
Ripening	12	>30

Yoshida (1978)

Table 6.2 Various symptoms of rice crop affected by heat stress

Growth stage	Symptoms
Vegetative	Reduction in height, tillering, white leaf tip, white bands and specks, chlorotic bands and blotches
Reproductive	Reduction in number of spikelet, increase in sterility
Ripening	Reduction in grain filling

(Yoshida 1981)

growth and adversely affects its productivity. The critical temperature ranges for rice growth at different phases are given in Table 6.1. Temperature plays a major role in rice growth and shows different symptoms in rice (Table 6.2). A higher temperature during the most sensitive stages of rice, i.e., reproductive and grain filling stage, results in decreased grain yield and quality.

6.2.1 Climate Change and Rice Yield

By 2039, rice yield is going to be reduced significantly (4.5–9%) due to changes in climatic parameters (Guiteras 2009). In *kharif* season, in India, not only monsoon rainfall but also its arrival and distribution pattern affect the rice yield (Cruz et al. 2007). Global panel studies indicated an increase in the frequency of hot and warmer nights which will put additional challenges to achieve higher rice yield (IPCC 2013). Climatic parameters such as rainfall, temperature, solar radiation, and atmospheric CO_2 play an important role in rice production (Nyang'au et al. 2014). An increase in rainfall and temperature variation was found to be harmful and beneficial to autumn and winter rice yield, respectively; however, these climatic parameters were found to be insignificant for summer rice (Nath and Raju 2018). Daily maximum temperature plays a crucial role in rice spikelet fertility, and increase in temperature adversely affects the yield. However, an increase in the atmospheric CO_2 level could increase the yield of rice (Dharmarathna et al. 2012). The higher temperature at critical growth stages increases spikelet sterility and reduces rice crop duration (Jagadish et al. 2008), reduces the duration of grain filling (Kim et al. 2011), and enhances the rate of respiration (Mohammed and Tarpley 2009) resulting in lower grain yield and grain quality (Fitzgerald and Resurreccion 2009). The flowering stage of rice is highly susceptible to high temperature (Jagadish et al. 2008). Similarly, the duration of solar radiation plays a crucial role in rice growth and yield (Tao et al. 2012). Increased night temperature along with global warming decreases rice yield. Higher minimum temperature (Wassmann et al. 2009) and lower solar radiation during the latter part of crop growth stages reduce rice yield (Peng et al. 2004). Studies on rice phenology revealed that warmer climate over the past 30 years had shortened the rice crop duration in China (Zhang et al. 2013). Similarly, higher temperature during the flowering stage of rice results in more spikelet sterility (Nakagawa et al. 2003).

The yield of rice crop may increase by 10–15% due to increase in CO_2 level from 340 to 680 ppm (Allen 1990; Cure and Acock 1986); at the same time, it will decrease the incidence of photosynthetically active radiation by 1% (Hume and Cattle 1990). It was revealed that physiological parameters (membrane stability index, relative water content, chlorophyll content, photosynthetic rate) increase under elevated CO_2; however, these traits were negatively affected due to elevated temperature (Dwivedi et al. 2015). The similar study confirmed that the panicle initiation stage of the rice crop was severely affected due to water stress and resulted in lower grain yield (Zaman et al. 2018).

6.2.2 Photoperiod and Temperature

Emergence to flowering stage of rice is generally responsive to photoperiod duration. Among the three stages of the pre-flowering period, only inductive phase is affected by photoperiod duration (Roberts et al. 1987; Yin et al. 1997). The pre-inductive phase which is also called as vegetative phase duration was extended due to increased photoperiod duration (Vergara and Chang 1976). These studies further validate that simply ambient air temperature studies may not be enough but remodeling of climate change responses with respect temperature is crucial. A study revealed that temperature at the growing point of rice plant affects a lot (Jamieson et al. 1995). Rice plant cooling effect by transpiration is reduced due to large differences between air and tissue temperatures at higher CO_2 concentration. Hence, the timing of flowering within a season is largely controlled by responses to temperature and photoperiod. So water-saving technologies and heat- and drought-tolerant cultivars will play a major role in enhancing the rice yield under future hotter climates (Jagadish et al. 2015).

6.3 Abiotic and Biotic Stresses and Rice Crop Yield

6.3.1 Abiotic Stress

Several abiotic stresses have emerged due to climate change. These have affected negatively the rice plants and reduced their overall growth and development. Most of the abiotic stresses are elaborated below:

Drought Rice is very much sensitive to water stress. Among all abiotic stresses, drought has a devastating effect on rice crop affecting millions of hectares of cultivated land in India and Asia. In India, drought affects more than 50% of cropland and causes a huge loss in rice crop. In a few states of India, i.e., Jharkhand, Orissa, and Chhattisgarh, drought is more frequent and may cause yield loss more than 35%. Drought severely affects germination, plant height, tillering, spikelet fertility, flower sterility, seed quality, chalkiness, and time of flowering ultimately reducing the yield. Increase in temperature increases respiration rate and reduces rice yield. So in the coming days, rice crop yield will reduce significantly due to an increase in temperature. Various studies revealed that there is a positive correlation between CO_2 and drought. Moisture stress reduces growth and physiological parameters such as photosynthesis, rate of transpiration, stomatal conductance, membrane stability index, water use efficiency, relative water content, and abscisic acid content and thereby reduces its yield (Pandey and Shukla 2015). Water stress at reproductive stage reduces grain formation and inhibits pollen development and panicle exsertion which leads to 75% spikelet sterility (O'toole and Namuco 1983). Water stress also inhibits pollen germination, anther dehiscence, and pollen fertilization (Ekanayake et al. 1990).

Submergence/Flooding Flooding significantly reduces the rice grain yield especially when it is grown under rainfed lowland condition. Frequent flooding happens in major eastward river basins of India. Flooding has a severe effect on rice crop due to improper drainage facility in rice fields. Flooding can happen at any stage of the rice crop. Rice can tolerate partial submergence; however, it dies within a few days of complete submergence. Flooding in rice creates anaerobic conditions which favor more emission of CH_4 from the rice field (Sun et al. 2016). This anaerobic condition affects ethanolic fermentation pathway and reduces glycolysis promoting ethanol and lactate production (Chirkova and Yemelyanov 2018). Ethanol is toxic to plant since through diffusion it forms acetaldehyde which is an intermediate toxic substance (Rahman et al. 2001). Submergence at early vegetative stage showed varied response among the varieties. Duration of submergence had a significant effect on rice morphology (Sultana et al. 2018).

Chilling Stress The seedling stage is affected the most by chilling temperature (Buti et al. 2018). There is more accumulation of reactive oxygen species (ROS) in rice plant cells due to chilling stress (Cen et al. 2018). *Indica* cultivars are more sensitive to cold than *japonica* cultivars.

Soil Salinity Major part of cultivable land in Asia and India are not suited for rice cultivation due to higher accumulation of salt. Higher temperatures prompt high evapotranspiration rates leading to more salt accumulation in rice fields. Salt stress on rice plant has a similar effect as that of drought stress leading to lower yield. Soil salinity affects more than 20% of total worldwide cultivated area and 33% of total irrigated agricultural lands. In addition to this, the salinized areas are increasing at a rate of 10% annually. It has been estimated that more than 50% of the arable land would be salinized by the year 2050. Rice crop produces less yield in saline soils due to high osmotic stress, nutrient toxicities, and poor soil physicochemical properties. Rice is especially susceptible to salinity during the seedling stage and panicle exsertion stage. In India, nearly 6.7 million ha cultivable lands are affected by soil salinity stress. The detrimental effect of soil salinity on rice crop is further aggravated by high temperature and relative humidity (Tack et al. 2015). Moderately salt-sensitive rice crop significantly reduces yield under high-salinity conditions. Under high-salinity conditions, accumulation of reactive oxygen species damages the cell of rice plant (Apse et al. 2003).

Acid Soils Acid soils (defined as soils with pH <6.5 in the top layer) in India cover around 49 million hectares arable land of which 25 million hectares have pH below 5.5 and 23 million hectares have pH between 5.6 and 6.5. Acid soils have toxic levels of iron (Fe), aluminum (Al), and manganese (Mn) coupled with lower levels of phosphorous (P) (Kochian et al. 2004).

6.3.1.1 Water Management and Rice Yield

Water demand for rice is going to be changed due to climate change. Climate change will affect rainfall pattern, soil water balance, and rate of evapotranspiration. Water demand for rice will increase with an increase in temperatures and variability of precipitation. Rice crop may require more irrigation water in the future than now. Water stress at anthesis stage may result in the sterility of rice crop. It has been observed that total precipitation is increased in the high-latitude regions of northern hemisphere and tropics. Similarly, semi-tropical regions of the world witnessed a decrease in precipitation over the past several decades.

6.3.2 Biotic Stress

Generally, insect pest and diseases pose a major threat to rice cultivation. On an average farmers lose 37% of their rice yield due to pests and diseases. Rice yield loss may vary from 15 to 20% depending upon the pressure of insect pests. It is estimated that the average yield losses in rice vary from 21 to 51%. The major pests of rice, i.e., yellow stem borer (25–30%), plant hoppers (10–70%), and gall midge (15–60%), reduce the plant growth and cause severe yield loss. At national level, stem borers accounted for 30% of the losses, while plant hoppers 20%, gall midge 15%, leaf folder 10%, and other pests 25%. Yield losses due to the incidence of insect pests vary depending upon the stage and duration of rice crop.

6.4 Climate Change and Rice Grain Quality

There is very much urgency to study the effect of climate change on food quality (Högy and Fangmeier 2008; Porter and Semenov 2005; Loladze 2002). A large part study focused on nitrogen and protein concentrations under elevated CO_2 conditions. Furthermore, the studies on the effect of climate change on nutritional value and quality of rice grain are very few. Moreover, this insufficient information does not help us to draw firm conclusions on how the climate changes will affect rice grain quality for human consumption. It was found that few degree variation in temperature has more pronounced overall effects on quality than that of increased CO_2. To assess the effects of temperature and CO_2 interactions on rice grain quality needs understating about assimilate supply and demand (Morison and Lawlor 1999).

6.4.1 Milling Quality

6.4.1.1 Carbon Dioxide

Milling quality is directly related to the market price of rice (Cooper et al. 2008). Carbon dioxide studies on rice grain quality revealed deterioration of processing quality of rice grain under elevated CO_2, whereas a different free-air CO_2 enrichment (FACE) study in Japan confirmed that the milling quality of *japonica* variety was significantly reduced under elevated CO_2 condition (Terao et al. 2005). Similarly, decrease in milling rice percentage (3%) and head rice recovery percentage (24%) of *japonica* and IR24 variety was reported under elevated CO_2 (Yang et al. 2007). The changes in processing quality induced by elevated CO_2 influenced the yield of head rice, brown rice, and milled rice.

6.4.1.2 Ozone

Higher ozone level reduces the processing quality of rice. *Indica* hybrid also shown a decreasing percentage of brown rice and milled rice due to ozone stress (Wang et al. 2012, 2014). Shen et al. (2016) also reported that the brown rice percentage and head rice percentage of *japonica* varieties were reduced due to higher ozone level. However, the effect of higher ozone on milled rice percentage was prominent.

6.4.1.3 CO_2, Temperature, and Ozone

Grain filling stage is susceptible to a higher temperature and affects the grain quality which further reduces the milling quality of rice (Lyman et al. 2013). Madan et al. (2012) did not found any interaction between temperature and CO_2 level during short-term studies at flowering stage of rice. However, a long-term study revealed that there is an interaction between temperature and elevated CO_2. Xie et al. (2009) reported that elevated CO_2 and air temperature together affected the processing quality of *japonica* cultivar while their individual effect did not reflect the same. Usui et al. (2014) also found that the grains are damaged due to elevated temperature and CO_2 grown under many seasons. Contradictory to previous studies, interaction effects of CO_2 and O_3 did not influence the processing quality of rice (Wang et al. 2014).

6.4.2 Appearance Quality

6.4.2.1 Carbon Dioxide

Appearance quality can be defined as the physical appearance of brown rice or polished rice. It is the first sensory impression of rice consumers. Generally, elevated CO_2 increases grain size and chalkiness. There was a significant increase in rice grain chalkiness (3%) due to elevated CO_2 (Terao et al. 2005). Yang et al. (2007) also reported a similar increase in chalky area (3%), chalkiness degree (28%), and chalky grain percentage (17%) under elevated CO_2. Similarly, white-base grains increased by 8.4% in ambient plots. The white-base grain increased by 17.1% in sensitive cultivars; however, the same for heat-tolerant cultivars were 2.1% and 4.4%, respectively (Usui et al. 2014). This confirmed that the grain chalkiness varies with cultivars also apart from elevated CO_2. Milling quality deteriorates due to elevated CO_2 (Yang et al. 2007). However, no change (Madan et al. 2012) or a decrease in grain chalkiness (Xu et al. 2008) due to elevated CO_2 is also reported.

6.4.2.2 Ozone

Ozone stress increases chalkiness but reduces grain size. Higher ozone level increased the chalky grain percentage (15%), chalkiness area (42%), and chalkiness degree (61%) of *japonica* cultivars (Shen et al. 2016).

6.4.2.3 CO_2, Temperature, and Ozone

Lighter and smaller grains are formed due to high temperature during grain filling stage (Yamakawa et al. 2007). High temperature increases the percentage of milky white rice and white chalky rice by inhibiting grain filling (Tsukaguchi and Iida 2008). The formation of more chalky grains under high temperatures is attributed to less starch accumulation. Jing et al. (2016) reported that there was no significant interaction effect of CO_2 and temperature on grain weight or size. However, grain chalkiness was increased due to elevated CO_2 and temperature. The interaction effect of these two further increased the chalky grain percentage, chalkiness degree, and chalkiness area. This result confirmed the additive interactive effects of CO_2 and temperature on rice chalkiness formation (Jing et al. 2016).

6.4.3 Carbohydrate

In rice, grain quality is determined by amylose content and gelatinization. Elevated CO_2 did not affect amylose content and palatability of rice (Terao et al. 2005). However, Yang et al. (2007) found lower amylose content (3.6%), decreased grain hardness, and improved palatability under elevated CO_2. Elevated CO_2 negatively affected the milled rice percentage and head rice percentage; however, it increased the chalky grain percentage and chalkiness degree by 16.9% and 28.3%, respectively. This inconsistent finding may be due to various cultivars used in the above-cited studies. Ziska et al. (1997) reported that increased amylose content of rice grain at higher temperatures might be associated with increased stickiness.

6.4.4 Minerals

Seneweera and Conroy (1997) reported that elevated CO_2 reduced the concentrations of N (14%), P (5%), Zn (28%), and Fe (17%) but increased the concentration of Ca (32%) in rice grain.

6.4.5 Proteins and Their Fractions

Sometimes higher nitrogen fertilization and most of the times normal recommended levels can minimize protein content in rice (Stafford 2008). However, reduced protein content in rice grain was also reported under elevated CO_2 (Taub et al. 2008). Application of a higher dose of nitrogen fertilization would be unwise in terms of costs and environmental consequences. Again, a higher dose of nitrogen may affect the physical quality of rice grain as reported by Yang et al. (2007).

6.5 Mitigation Strategies

Climate change requires global attention to reduce greenhouse gas emissions from agricultural lands. This strategy needs long-term solutions to the climate change problem. Specially to control methane emission, there are several options available, e.g., technologies such as mid-season drainage and switching to alternative fertilizers and heat–/drought-tolerant rice varieties. Agronomic management practices such as adjusting sowing dates and efficient crop establishment methods are some of the measures to reduce the methane emissions from rice fields. New techniques of plant breeding and genetics can also be used to counteract the negative impact of climate change. Proactive, proper strategies and policies from different countries are very much needed to reduce emissions of greenhouse gases from rice cultivation. Refinement of water-saving technologies and adoption of more efficient water management practices are the need of the hour to mitigate drought stress.

6.6 Mitigation Strategy Through Crop Improvement

- Development of high temperature- and drought-tolerant varieties with better adaptation to other biotic stresses. Developing submergence- and flood-tolerant rice varieties suitable for various agro-ecological conditions.

- Salinity- and alkalinity-tolerant cultivars should be developed to grow at salt-affected regions of India.
- Varieties resistant to emerging insect pest and diseases need to be developed to overcome the biotic stresses.
- Holistic research is needed to convert from C_3 rice to C_4 rice.

6.7 Future Perspectives

The study of the effect of climate change on grain quality is lagging behind as compared to rice yield. Even as the study is picking pace now, many aspects in these studies are inconclusive, and not yet completely established. These uncertainties greatly influence global food security. This chapter indicated that elevated temperature and CO_2 significantly reduce the grain yield and quality. So, there is an urgent need to improve our understanding of the impacts of climate change on rice yield and quality through systematic investigation.

1. Strengthening the basic climate change research consists of all the climatic parameters. Further research on the response of rice quality to climate changes specially to study the biotic and abiotic factors. In-depth studies of these interactions will help to develop effective adaptation and mitigation strategies for climate change.
2. Strengthen the research through new biotechnological tools which include genomics, metabolomics, and proteomics and need to be introduced to study the rice quality and its response mechanism. To understand the response mechanism is very much needed (Long et al. 2006; Ainsworth et al. 2012).
3. An in-depth study on the adaptation strategy of rice under the different climatic regime is very much needed. The research focus should be on increasing the yield with better grain quality under various biotic and abiotic stresses.

References

Ainsworth EA, Yendrek CR, Sitch S, Collins WJ, Emberson LD (2012) The effects of tropospheric ozone on net primary productivity and implications for climate change. Annu Rev Plant Biol 63:637–661

Allen LH (1990) Plant responses to rising carbon dioxide and potential interactions with air pollutants. J Environ Qual 19:15–34

Apse MP, Sottosanto JB, Blumwald E (2003) Vacuolar cation/H+ exchange, ion homeostasis, and leaf development are altered in a T-DNA insertional mutant of *AtNHX1*, the *Arabidopsis* vacuolar Na+/H+ antiporter. Plant J 36:229–239

Buti M, Pasquariello M, Ronga D, Milc JA, Pecchioni N et al (2018) Transcriptome profiling of short-term response to chilling stress in tolerant and sensitive *Oryza sativa* ssp. Japonica seedlings. Funct Integr Genomics 18:627–644

Cen W, Liu J, Lu S, Jia P, Yu K et al (2018) Comparative proteomic analysis of QTL CTS-12 derived from wild rice (*Oryza rufipogon* Griff.), in the regulation of cold acclimation and de-acclimation of rice (*Oryza sativa* L.) in response to severe chilling stress. BMC Plant Biol 18:163

Chirkova T, Yemelyanov V (2018) The study of plant adaptation to oxygen deficiency in Saint Petersburg university. Biomed Commun 63:17–31

Cooper NTW, Siebenmorgen TJ, Counce PA (2008) Effects of nighttime temperature during kernel development on rice physicochemical properties. Cereal Chem 85(3):276–282

Cruz RV, Harasawa H, Lal M, Wu S, Anokhin Y et al (2007) Asia. Climate change 2007: working group II: impacts, adaptation and vulnerability. Cambridge University Press, Cambridge, pp 469–506

Cure JD, Acock B (1986) Crop responses to carbon dioxide doubling: a literature survey. Agric For Meteorol 38:127–145

Dharmarathna WR, Weerakoon SB, Rathnayaka UR, Herath S (2012) Variation of irrigated rice yield under the climate change scenarios. In: SAITM research symposium on engineering advancements, pp 31–34

Dwivedi SK, Kumar S, Prakash V, Mondal S, Mishra JS (2015) Influence of rising atmospheric CO2 concentrations and temperature on morphophysiological traits and yield of rice genotypes in sub humid climate of eastern India. Am J Plant Sci 6(14):2239–2249

Ekanayake IJ, Steponkus PL, De Datta SK (1990) Sensitivity of pollination to water deficits at anthesis in upland rice. Crop Sci 30:310–315

Fitzgerald MA, Resurreccion AP (2009) Maintaining the yield of edible rice in a warming world. Funct Plant Biol 36(12):1037–1045

GRiSP (2013) Rice Almanac, 4th edn. International Rice Research Institute, Los Baños

Guiteras R (2009) The impact of climate change on Indian agriculture. University of Maryland, College Park, pp 1–54

Högy P, Fangmeier A (2008) Effects of elevated atmospheric CO_2 on grain quality of wheat. J Cereal Sci 48:580–591

Hume CJ, Cattle H (1990) The greenhouse effect meteorological mechanisms and models. Outlook Agric 19:17–23

IPCC (2013) Working group I contribution to the IPCC fifth assessment report on climate change 2013: the physical science basis, summary for policymakers. IPCC, Geneva, pp 1–28

Jagadish SV, Craufurd PQ, Wheeler TR (2008) Phenotyping parents of mapping populations of rice for heat tolerance during anthesis. Crop Sci 48:1140–1146

Jagadish SVK, Murty MVR, Quick WP (2015) Rice responses to rising temperatures–challenges, perspectives and future directions. Plant Cell Environ 38:1686–1698

Jamieson PD, Martin RJ, Francis GS, Wilson DR (1995) Drought effects on biomass production and radiation-use efficiency in barley. Field Crop Res 43:77–86

Jing L, Wang J, Shen S, Wang Y, Zhu J, Wang Y, Yang L (2016) The impact of elevated CO_2 and temperature on grain quality of rice grown under open-air field conditions. J Sci Food Agric 96 (11):3658–3667

Kim J, Shon J, Lee CK, Yang W, Yoon Y et al (2011) Relationship between grain filling duration and leaf senescence of temperate rice under high temperature. Field Crop Res 122:207–213

Kochian LV, Hoekenga OA, Pineros MA (2004) How do crop plants tolerate acid soils? Mechanisms of aluminum tolerance and phosphorous efficiency. Annu Rev Plant Biol 55:459–493

Loladze I (2002) Rising atmospheric CO_2 and human nutrition: toward globally imbalanced plant stoichiometry? Trends Ecol Evol 17:457–461

Long SP, Ainsworth EA, Leakey AD, Nosberger J, Ort DR (2006) Food for thought: lower-than-expected crop yield stimulation with rising CO_2 concentrations. Science 312(5782):1918–1921

Lyman NB, Jagadish KS, Nalley LL, Dixon BL, Siebenmorgen T (2013) Neglecting rice milling yield and quality underestimates economic losses from high-temperature stress. PLoS One 8(8): e72157

Madan P, Jagadish SVK, Craufurd PQ, Fitzgerald M, Lafarge T, Wheeler TR (2012) Effect of elevated CO_2 and high temperature on seed-set and grain quality of rice. J Exp Bot 63 (10):3843–3852
Mohammed AR, Tarpley L (2009) Impact of high nighttime temperature on respiration, membrane stability, antioxidant capacity, and yield of rice plants. Crop Sci 49:313–322
Morison JIL, Lawlor DW (1999) Interactions between increasing CO_2 concentration and temperature on plant growth. Plant Cell Environ 22:659–682
Nakagawa T, Kitagawa H, Yasuda Y, Tarasov PE, Nishida K (2003) Asynchronous climate changes in the North Atlantic and Japan during the last termination. Science 299:688–691
Nath HK, Raju M (2018) Heterogeneous climatic impacts on agricultural production: evidence from rice yield in Assam, India. Asian J Agric Develop 15(1):23–42
Nguyen VN (2002) Productive and environmentally friendly rice integrated crop management systems. IRC Newsl 51:25–32
Nyang'au WO, Mati BM, Kalamwa K, Wanjogu RK, Kiplagat LK (2014) Estimating rice yield under changing weather conditions in Kenya using CERES rice model. Int J Agron 2:1–12
O'toole JC, Namuco OS (1983) Role of panicle exsertion in water stress induced sterility. Crop Sci 23:1093–1097
Pandey V, Shukla A (2015) Acclimation and tolerance strategies of rice under drought stress. Ric Sci 22:147–161
Peng S, Huang J, Sheehy JE, Laza RC, Visperas RM et al (2004) Rice yields decline with higher night temperature from global warming. P Natl Acad Sci USA 101:9971–9975
Porter JR, Semenov MA (2005) Crop responses to climatic variation. Philos Trans Royal Soc 360:2021–2035
Rahman M, Grover A, Peacock WJ, Dennis ES, Ellis MH (2001) Effects of manipulation of pyruvate decarboxylase and alcohol dehydrogenase levels on the submergence tolerance of rice. Funct Plant Biol 28:1231–1241
Roberts EH, Summerfield RJ, Atherton JG (1987) Manipulation of flowering: proceedings of previous easter schools in agricultural science, 17th edn. Butterworth Heinemann, London
Seneweera SP, Conroy JP (1997) Growth, grain yield and quality of rice (*Oryza sativa* L.) in response to elevated CO_2 and phosphorus nutrition. Soil Sci Plant Nutr 43(sup1):1131–1136
Shen S, Zhang D, Yang K, Wang Y, Zhu J, Yang L, Wang Y (2016) Effect of elevated surface layer ozone concentration on grain quality of two rice cultivars d a FACE study. Chin J Eco-Agric 24 (9):1231–1238
Stafford N (2008) The other greenhouse effects. Nature 448:526–528
Sultana T, Ahamed KU, Naher N, Islam MS, Jaman MS (2018) Growth and yield response of some rice genotype under different duration of complete submergence. J Agric Ecol Res Int 15 (1):1–11
Sun H, Zhou S, Fu Z, Chen G, Zou G et al (2016) A two-year field measurement of methane and nitrous oxide fluxes from rice paddies under contrasting climate conditions. Sci Rep 6:28255
Tack J, Singh RK, Nalley LL, Viraktamath BC, Krishnamurthy SL et al (2015) High vapor pressure deficit drives salt-stress-induced rice yield losses in India. Glob Chang Biol 21:1668–1678
Tao F, Zhang S, Zhang Z (2012) Spatiotemporal changes of wheat phenology in China under the effects of temperature, day length and cultivar thermal characteristics. Eur J Agron 43:201–212
Taub D, Miller B, Allen H (2008) Effects of elevated CO_2 on the protein concentration of food crops: a meta-analysis. Glob Chang Biol 14:565–575
Terao T, Miura S, Yanagihara T, Hirose T, Nagata K, Tabuchi H et al (2005) Influence of free-air CO_2 enrichment (FACE) on the eating quality of rice. J Sci Food Agric 85(11):1861–1868
Tsukaguchi T, Iida Y (2008) Effects of assimilate supply and high temperature during grain filling period on the occurrence of various types of chalky kernels in rice plants (*Oryza sativa* L.). Plant Prod Sci 11(2):203–210
Usui Y, Sakai H, Tokida T, Nakamura H, Nakagawa H, Hasegawa T (2014) Heat tolerant rice cultivars retain grain appearance quality under free-air CO_2 enrichment. Rice 7(1):6

Vergara BS, Chang TT (1976) The flowering response of the rice plant to photoperiod: a review of the literature. International Rice Research Institute, Los Baños, pp 1–66
Wang Y, Yang L, Han Y, Zhu J, Kobayashi K, Tang H, Wang Y (2012) The impact of elevated tropospheric ozone on grain quality of hybrid rice: a free-air gas concentration enrichment (FACE) experiment. Field Crop Res 129:81–89
Wang Y, Song Q, Frei M, Shao Z, Yang L (2014) Effects of elevated ozone, carbon dioxide, and the combination of both on the grain quality of Chinese hybrid rice. Environ Pollut 189:9–17
Wassmann R, Hirano T, Miyata A, Kumar A, Padre A et al (2009) CO_2/heat fluxes in rice fields: comparative assessment of flooded and non-flooded fields in the Philippines. Agric For Meteorol 149:1737–1750
Xie L, Ma Z, Han X, Lin E (2009) Impacts of CO_2 enrichment and temperature increasing on grain quality of rice. J Northeast Agric Univ 40(3):1–6
Xu CL, Li JY, Xie H, Zhu JG, Fan GZ, Cai QS (2008) Effect of free air CO_2 enrichment to rice quality of rice. Chin Agric Sci Bull 24:391–397
Yamakawa H, Hirose T, Kuroda M, Yamaguchi T (2007) Comprehensive expression profiling of rice grain filling-related genes under high temperature using DNA microarray. Plant Physiol 144 (1):258–277
Yang L, Wang Y, Dong G, Gu H, Huang J, Zhu J, Han Y (2007) The impact of free-air CO_2 enrichment (FACE) and nitrogen supply on grain quality of rice. Field Crop Res 102 (2):128–140
Yin X, Kropff MJ, Horie T, Nakagawa H, Centeno HG et al (1997) A model for photothermal responses of flowering in rice I. model description and parameterization. Field Crop Res 51:189–200
Yoshida S (1978) Tropical climate and its influence on rice. IRRI, Los Baños
Yoshida S (1981) Fundamentals of rice crop science. IRRI, Los Baños, p 269
Zaman NK, Abdullah MY, Othman S, Zaman NK (2018) Growth and physiological performance of aerobic and lowland rice as affected by water stress at selected growth stages. Rice Sci 25:82–93
Zhang T, Huang Y, Yang X (2013) Climate warming over the past three decades has shortened rice growth duration in China and cultivar shifts have further accelerated the process for late rice. Glob Chang Biol 19:563–570
Ziska LH, Namuco O, Moya T, Quilang J (1997) Growth and yield response of field-grown tropical rice to increasing carbon dioxide and air temperature. Agron J 89(1):45–53

Changing the Way We Produce Food: An Overview of the Current Agricultural Food Production Industry and Worldwide Trends for Sustainable Production

7

Rowena P. Varela, Raquel M. Balanay, and Anthony Halog

Abstract

The rapidly expanding population and the changing lifestyle amid climate change pose tremendous challenges to the way food is produced globally. Various food production strategies have been tested worldwide to guarantee adequate food supply. This chapter puts together the findings and insights of the various initiatives in the food production industry to ensure sustainable production. The shift from the green revolution to sustainable agriculture has significantly changed the agricultural production landscape by integrating the market, policy, research and innovation, and society's perspectives. Agriculture is undergoing a technology revolution, now the Agriculture 4.0, with the introduction of artificial intelligence (AI), sensors, and the Internet of Things (IoT). With the current food production and consumption patterns impacting the environment, life cycle assessment (LCA) is crucial to improve food-related supply chains. Therefore, a sustainable agricultural production system that embraces technological advancement, ecological soundness, and sociological perspectives is the way forward to ensure food supply.

R. P. Varela (✉) · R. M. Balanay
Department of Agricultural Sciences, College of Agriculture and Agri-industries, Caraga State University, Ampayon, Butuan, Philippines
e-mail: rpvarela@carsu.edu.ph

A. Halog
School of Earth and Environmental Sciences, The University of Queensland, Brisbane, QLD, Australia
e-mail: a.halog@uq.edu.au

V. K. Hebsale Mallappa, M. Shirur (eds.), *Climate Change and Resilient Food Systems*, https://doi.org/10.1007/978-981-33-4538-6_7

Keywords

Climate change · Food security · Sustainable agriculture · Agriculture 4.0 · Agri-food life cycle assessment

7.1 Overview of the Agricultural Food Production System

The changing lifestyles of people and the challenges posed by climate change call for modifications in the way food is produced. After World War II, the "Green Revolution" was the trend in agriculture where agrochemicals (fertilizers and pesticides) were applied to increase agricultural productivity. This era was marked by pollution and degradation of the environment. The extensive use of highly toxic pesticides in agriculture and forestry killed non-target organisms, which prompted Carson (1962) to publish the *Silent Spring* that documented the adverse environmental effects caused by the agrochemicals. Since then, the USA and other countries initiated agricultural programs that promote the safe use of agrochemicals, such as sustainable agriculture, ecological agriculture, and bio-intensive agriculture. This era led to more research on alternative agricultural production systems that are environment-friendly and have improved agricultural input-output efficiency.

With the current global population growth rate and the changing lifestyles of people, the Earth has been under extreme pressure. Ensuring food security to feed the increasing population with minimal environmental impacts and easing the biosphere's pressure are today's biggest challenge. Citing the UN population projections, Alexandratos and Bruinsma (2012) pointed out that an adaptive food production system is essential to meet the global population's food demand, which by 2050 is projected to balloon up to 9.15 billion. The adaptive food production system considers the repercussions of agriculture on biodiversity conservation, habitat loss, judicious use of external inputs in farming, and the use of methods that optimize water use efficiency. With the challenges posed by climate change, adaptive agriculture is necessary. In sub-Saharan Africa, the leading approach to analyzing climate change and its impact on food security is modeling the effects of future climate scenarios on food production to provide valuable information on the next production levels (Connolly-Boutin and Smit 2016). However, socioeconomic data (e.g., income, health, and assets) are necessary as food accessibility and utilization are among food security issues. The recent decisions at the micro and macro level are dictated by the use of data with the help of advances in data modeling and simulation with virtual resources.

A new revolution called Agriculture 4.0 is being advocated to optimize food production with minimal environmental impacts. Agriculture 4.0 is a new concept in the agri-food production system that considers the use of digital technology for more precise and smart decision-making to improve efficiency. This concept aligns with the Fourth Industrial Revolution, which harnesses the use of sensors, computers, and information and communication technology (ICT) in agriculture systems to increase food production efficiency. Precision agriculture (PA) and the integration of digital technology are becoming the modern approaches to food production in Europe to increase input-output efficiency with less environmental footprints (Lindblom et al. 2017). The applications of ICT are imperative for lasting and sustainable development in agriculture through computer-human interaction toward developing a decision support system (DSS). In literature, the uses of ICT in agriculture are referred to as "smart agriculture," "intelligent agriculture," and "digital farming" or "digital agriculture." These ICT-based agriculture approaches are based on the ideas of precision farming (Zambon et al. 2019). Some researchers looked at Agriculture 4.0 as a concept beyond smart agriculture (De Clercq et al. 2008; Himesh et al. 2018; Huh and Kim 2018; Yahya 2018) and, from the perspective of Agriculture 4.0, shared that establishing smart farms is essential to monitor and manage carbon emissions in real time. According to Wolfert et al. (2017), smart farming is impacted by big data required in decision-making along the whole supply chain. Hence, research is crucial in line with governance and business models.

With climate change disrupting the food production system, disturbances in the markets ensued, posing risks to the food supply. The instabilities in food production systems due to climate change impacts can be reduced. The countries all over the world need to initiate programs on enhancing the adaptive capacity and the resilience of farmers and on increasing the resource use efficiency to produce food despite the threats of climate change continually. Climate-smart agriculture (CSA) is advocated internationally to stimulate coordinated actions by all major agricultural production players. According to Lipper et al. (2014), the agri-food production system's key players need to participate actively in critical aspects of CSA. The aim is to increase institutional efficiency at the local level and ensure consistency of agricultural policies concerning climate matters and agricultural investment related to climate. This is the underpinning of CSA, as it focuses on capacity building among farmers and support groups to implement solutions to climate-linked issues. Richardson et al. (2018) cited that the vulnerability in parts of sub-Saharan Africa and South Asia to food insecurity is estimated to increase significantly under all emission scenarios. However, high levels of adaptation and mitigation in farming communities can improve food production in the future compared to present-day circumstances. This scenario then highlights the impacts of combined mitigation and adaptation to avoid the worst impacts of climate change and to make gains in dealing with food insecurity. Varela et al. (2019) assessed climate risks in agriculture toward initiating climate-resilient agricultural initiatives in Caraga Region, Philippines, and they found that communities have existing knowledge about climate-smart agriculture.

Nonetheless, the farmers and the support groups need to enhance their understanding of climate change resilience and focus on improving their adaptive mechanisms toward food security. With limited knowledge of the likelihood of food insecurity under different climate change scenarios, providing information to farmers on the impacts of climate change is crucial in achieving food security (Antón et al. 2013). Recently, there have been programs and activities in various parts of the world to offset climate change impacts in the agri-food production system. These initiatives also address the need for greater input-output efficiency as a requirement for sustainability. The current agriculture-food production system adopts the

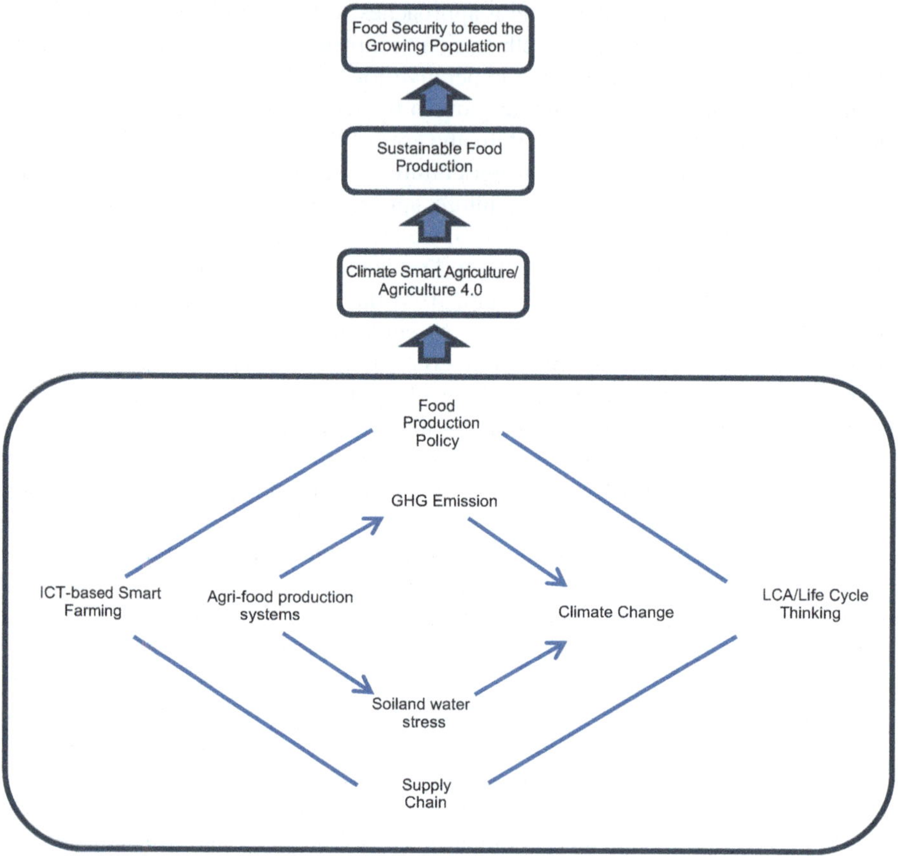

Fig. 7.1 A framework of the agricultural production system to ensure food and environmental sustainability amid climate change. The agri-food production practices result in GHG emission and soil and water stresses that contribute to climate change. Addressing these through ICT-based smart farming and understanding the supply chain, life cycle thinking, and food production policy will lead to sustainable climate-smart food production and food security

framework shown in Fig. 7.1 to achieve food security and ecosystem sustainability with the challenges posed by climate change.

7.2 The Current State of Food Production Globally

Conventional agriculture is still the foundation of modern agriculture. The usual farming and soil management principles remain a solid basis in current food production systems. However, modifications in conventional agriculture are made to adapt to the demands of time and society amid the challenges of climate change. Agriculture is identified to contribute significantly to GHG emissions. About

52–84% of global anthropogenic methane and nitrous oxide emissions are determined to come from agriculture (Scheehle and Kruger 2006). The global targets for greenhouse gas mitigation can be realized by making significant adjustments to food supply chains (Whitfield et al. 2018). Hence, they outlined the research space to make adjustments in the agricultural system to adapt to changing and unpredictable climatic conditions. As agriculture ensures food availability to spur local development and alleviate poverty without altering ecological functions and services, innovative technologies supporting climate-smart food systems must be in place to ensure sustainability.

The advent of Industry 4.0 has also affected the agriculture sector; hence it is now undergoing a new technology revolution globally. Industry 4.0 is under the Fourth Industrial Revolution that involves the digitalization of industrial processes, allowing better connectivity between customers and supply chains through real-time access to production and consumption information. In this era, the use of novel approaches (AI, sensors, and IoT) becomes trendy. While these technologies have been regarded as essential inputs in achieving agri-food system efficiency, the social implications of these technologies are being put aside. Sociological studies suggest that there may be concerns among farmers and associated communities about revolutionary agricultural technologies. Thus, Rose and Chilvers (2018) argued that agricultural innovations be further developed and tested among farming communities to be considered as promising farm technologies. They added that frameworks on ICT-based smart innovations in agriculture be field-tested to see if these can influence the agenda of agricultural innovation. Agriculture 4.0 involves smart farming (SF) for intelligent decision-making in response to the need of the growing population and the changing climate. In Agriculture 4.0, the ICT-based technologies have been integrated to facilitate decision-making concerning agriculture-food systems. In Brazil, ICT-based systems are the main limiting factors to smart farming evolution (Pivoto et al. 2018). The ability and skills of farmers to understand and handle smart farming tools are also regarded as limiting factors. They noted that countries with more research and development investments usually led to smart farming.

7.3 Environmental Issues Created by Conventional Food Production Processes

Analysis of life cycle assessments (LCA) of 742 agricultural systems shows organic systems have greenhouse gas emissions (GHGs) comparable to conventional methods (Clark and Tilman 2017). Smith and Gregory (2013) reviewed and outlined some of the likely impacts of climate change on agriculture. They looked at the mitigation measures to reduce GHG emissions and how to produce food sustainably for ten billion people. With the enormous challenge, radical alterations in the production and consumption patterns of food systems are inevitable. Increasing agricultural input efficiency would benefit both crop and livestock systems environmentally (Clark and Tilman 2017).

The production and consumption patterns in agriculture-food systems are the leading causes of environmental impacts. Therefore, these are crucial in assessing the food production systems and food supply chains to mitigate environmental footprints and gas emissions that contribute to climate change. However, it is challenging to conduct LCA due to the intricate food production systems and supply chains. Notarnicola et al. (2017) suggested that dedicated modeling approaches have to be adopted to address the critical variability of agriculture-food production assemblies. These models tackle the distinction between the technosphere and the ecosphere. Likewise, these consider the biological systems' multi-function and the emission modeling linked with LCA. Food production and consumption pose a significant challenge, considering the environmental impacts of agricultural systems (Liu et al. 2015). Also, the competing demands from the expansion of biofuel production, implications of using genetically modified organisms, and the changing lifestyle of people amplify the global food crisis. There will be trade-offs if practical solutions are not determined as these are crucial for agricultural sustainability. As a result, life cycle thinking (LCT) and life cycle assessment (LCA) are critical elements in determining more sustainable solutions to address global food challenges (Sala et al. 2017). In outlining the solutions, LCA should be used to assess agri-food supply chains and be considered a reference. They identified LCA and life cycle thinking as necessary for identifying hotspots of impacts along food supply chains, food supply chain optimization, and assessment of future scenarios, particularly in consideration of climate change.

7.4 Current and Future Sustainable Production and Consumption Trends in the Food Industry Worldwide

In producing food, the impacts of climate change are evidenced by the crop failures leading to food shortages and rising food prices. Thus Whitfield et al. (2018) discussed climate-smart agriculture (CSA), which considers strategies used by local communities and integrates technological solutions to complex issues that would set out a direction for change toward improved food production. They emphasized the need to discuss broad systemic perspectives to provide adequate healthy food for the growing population, ensure sustainable land use, and adapt and mitigate climate change. The idea is to discuss these approaches concerning the goal of designing climate-smart food production systems. In the era of Industry 4.0, decision support tools are an essential part of the pursuit of a decision support system (DSS) for agriculture to improve productivity and reduce environmental footprints. Rose et al. (2016) investigated the factors affecting the use of DSS tools by farmers and advisers in the UK. They found that there were 15 factors influencing farmers and advisers to use DSS tools. These include usability, cost-effectiveness, performance, relevance to the user, and compatibility with compliance demands. A better understanding of these factors is imperative for a more effective design of DSS tools to entice farmers to use these with CSA. The Thailand Ministry of Agriculture and Cooperatives has been promoting smart farming and upgrading farmers to "smart farmers," who can use technology in food production and marketing for better prices (Jones and Pimdee 2017). The Thailand 4.0 program of the government adopts the principle "not to leave anyone behind." However, farmers and other agriculture players must use the DSS in their transactions. In this context, human labor must not be replaced by machines; however, humans need to develop the ability to speak foreign languages and evolve to be "smart people." The importance of food innovation is valued and recognized in Thailand as the next big thing, as the human population continues to increase.

Nanotechnology and its application in the food industry have been recognized to have many contributions to food processing, food packaging, and food preservation. The introduction of nanotechnology in the food industry has made the transport of foods to different parts of the world easy and resulted in less spoilage of most food products (Hamad et al. 2018). However, nanotechnology has not been proven effective, being a burgeoning field of science; thus, more research is needed. In the aspect of food safety, nanotechnology can contribute to reducing food wastage caused by microorganisms that result in food spoilage. In Agriculture 4.0, many innovations are associated with sustainable agriculture, interacting and co-evolving as "agricultural innovation systems" (Klerkx et al. 2012). These are facilitated by the emerging smart technologies that are computer-based and the smart farmers involved in community-based agricultural innovations (Waters-Bayer et al. 2015). However, in embracing smart technologies to make food production systems efficient and ecologically sound, the community's perspectives and the various players in the system should be considered. Macnaghten and Chilvers (2014) stated that addressing sustainable agriculture problems has to incorporate public concerns about emerging science and technology. This action promotes the cultivation of shared responsibilities across innovation ecologies. With collective responsibilities, there will be better governance and coordination between key actors in directing agricultural revolutions toward more socially responsible actions.

7.5 Challenges for Implementation and Adaptation of More Sustainable Food Production Processes

The global climate unpredictability resulted in crop productivity reduction, particularly in arid and semiarid areas, causing food scarcities and food price increases. Adaptation of more sustainable food production systems and processes, therefore, has been in discussion in several forums due to the increasing demand for food and to the uncertainties brought about by climate change. However, adaptation and mitigation initiatives at the national and global levels tend to have a disconnect resulting from a range of constraints. Harvey et al. (2014) demonstrated that climate change mitigation and adaptation benefits were associated with tropical agricultural systems, considering the landscape approach. The landscape-scale agricultural systems have better mitigation and adaptation potentials resulting from synergies due to the spatial arrangements. However, this approach will require re-shaping of current policies, institutional arrangements, and funding mechanisms to enhance the adoption of climate-smart strategies in agricultural landscapes. On managing transformative processes in agriculture, Vermeulen et al. (2018) suggested providing more comprehensive and long-term approaches to climate adaptation planning. Within this framework are financial and technical packages, and incentives are given to multifunctional farming systems to improve the effectiveness of outcomes.

Agroforestry, which is advocated to make agricultural production similar to natural systems, can make a significant contribution to climate change mitigation and adaptation. The limited knowledge in agroecological principles and the need for stable finances are among the key limitations in adopting agroforestry. Hence, the indirect solutions suggested are training programs and the development of economically and ecologically feasible climate-smart agriculture investment (Hernández-Morcillo et al. 2018). The recommended solutions are adopting practices through the implementation of multifunctional hedgerows and windbreaks capable of increasing soil organic carbon for nutrient cycling. Intensive agriculture (e.g., monoculture, use of agrochemicals) has led to biodiversity loss, climate change, and groundwater stress. In the Philippines, rice production, which is usually adopting intensive agriculture, has been re-oriented to *Palayamanan*, taking the principles of agroecology to preserve agricultural biodiversity and restore the soil fertility naturally. *Palayamanan* is an agriculture-food production system highlighting the purposive integration of other components to rice farming (Corales et al. 2004). In 2011 dry and wet seasons, results showed that rice insect pests were relatively low in the Palayamanan rice field compared to those in rice monoculture (Arida et al. 2016). The agro-biodiversity in the Palayamanan rice field includes the dense populations of beneficial organisms that result in a high incidence of egg parasitism in brown plant hopper. Between the Palayamanan system and the rice monoculture system, there is a striking difference in the overall rice field productivity. However, Mariano et al. (2012) identified some aspects of the *Palayamanan* to make the farming system more adopted by farmers. This is focused on the social benefits such as technical and financial assistance to the farmers. In addition, the government policies should include intervention packages toward ensuring food production by offsetting environmental adversities during climate extremes (e.g., drought and flooding). Moreover, an adaptive management approach for creating and utilizing learning tools tailored for biodiversity-based agriculture may provide "safety nets" for climate variabilities (Duru et al. 2015).

Countries in the tropics such as Africa and Asia are more economically associated with natural resource and climate-dependent sectors. In Africa, Misra (2014) highlighted the importance of groundwater recharge by utilizing wastewater using the soil aquifer treatment (SAT) method in irrigation. He presented options under climate change to ensure water and food security as inputs in the formulation of effective adaptation and mitigation policies and strategies to reduce the impact of climate change on water resources and irrigation. Iglesias and Garrote (2015) pointed out that the implementation of climate-smart agriculture requires overhauling current water policy and providing adequate training to farmers and viable financial instruments to improve their adaptive capacity. These arrangements will assist farmers and other key players in addressing the adaptation challenge, thereby developing mechanisms to reduce the vulnerability of agriculture to climate change.

Similarly, sustainable intensification (SI) is currently a catchphrase in food security discussions as an approach to meet the demand for food while conserving land, water, and other resources. SI is emerging as an exciting approach nowadays, as it considers the human condition, nutrition, and social equity. However, the method requires a new set of metrics and indicators to track progress, assess trade-offs, and identify synergies. Smith et al. (2017) assessed indicators and metrics and concluded that some indicators require strong sets of parameters.

The introduction of ICT and digital technologies triggered an evolution in agriculture in the early 2010s leading to the era of Agriculture 4.0. The introduction of sensors and actuators, low-cost microprocessors, and other ICT-based analytics has been changing the agricultural systems' landscape. The agriculture evolution, which promotes smart agriculture, started in the more developed countries in the west, but China and Japan have also joined the trend in the east. Table 7.1 summarizes the initiatives in changing the agriculture landscape with the introduction of Agriculture 4.0.

Along with the new technologies introduced concerning Agriculture 4.0, there have been discussions about the precision farming evolution perspective, Industry 4.0 in agriculture perspective, digital technologies perspective, informed decision-making perspective, beyond the farm boundaries perspective, and ultimate goal perspective. With many innovations coming in, the concept of Agriculture 4.0 has become challenging to comprehend at the grassroots level. For farmers in developing countries, education to acquire knowledge and skills related to the use of sensors, computers, and others is essential to make them accustomed to the changing way of producing food. These are both challenges and opportunities in the age of climate change and innovations.

7.6 Moving Forward for Facing the Challenges in Ensuring Food Security

The challenges associated with changing the way we produce food in the era of climate change and agricultural innovations can be dealt with through progressive education of the various players in the food supply chain. In the past decades, if food production is intended only for household consumption, the current agriculture-food system has ballooned due to the exponentially growing population globally and with the people's changing lifestyle. Sustainable agricultural production system, which embraces ecological soundness and sociological perspectives, are, therefore, the way forward. This production system has to consider the production-processing-market linkage, gas emissions, and related environmental disturbances along the food production chain. Food production system digitalization is also crucial in managing inputs and outputs. Likewise, the policy to support farm mechanization in tropical countries in South Asia and Africa is necessary to improve the food production system's timing and efficiency.

Table 7.1 Agricultural innovations and policy interventions to make agriculture attuned to the changing climate and lifestyle of the people

Agricultural/ICT-based innovation	Brief description and highlight of the project	Author/s
Multi-perspective approach in agriculture 4.0	A systematic review to provide a clear and holistic definition of agriculture 4.0 and the related boundaries	Sponchioni et al. (2019)
Swedish project on developing an agricultural decision support system (AgriDSS)	The project was to apply a user-centered design (UCD) approach. There had been pitfalls; however, despite the challenges, ICT has been contributing to developing DSS	Lindblom et al. (2017)
Ecosystem service (ESS)-based adaptation (EbA) to climate change in Bangladesh's policy-making process	ESS is marginally considered an adaptation component, especially at the top strategic level (vertical mainstreaming). The arguments related to policy and institutional capacities have offset the EbA mainstreaming process that needs to be addressed for climate change adaptation	Huq et al. (2017)
Industry vs. agriculture in future development for small and medium enterprises (SMEs)	Analysis of the farming supply chain to permit the effective implementation of industry 4.0 guidelines. There are questions about how industry 4.0 approaches can be improved and be pertinent to the agricultural sector to progress at a much faster rate	Zambon et al. (2019)
Forecasting soil temperature to estimate monthly soil temperature (ST) at multiple depths with a hybrid multi-layer perceptron algorithm integrated with firefly optimizer algorithm (MLP-FFA)	The hybrid MLP-FFA model is drawn upon a limited set of predictors. This can be used to visualize the degree of similarity between the observed and forecasted soil moisture	Samadianfard et al. (2018)
Understanding the concepts of innovative supply chain management	Consistent supply chain management re value creation is a common approach in the industrial sector. However, the processes require reconsidering the supply chain: Empirically characterized processes, stochastic environmental conditions, mobility of the production facilities, and low division of work need to be considered	Braun et al. (2018)
Green growth engine in Thailand's food sector, to create sustainability through environmentally friendly development	The model follows the 20-years national strategic plan by building strength from within and connecting Thailand to the global community under the principle of sufficiency philosophy	Jones and Pimdee (2017)

(continued)

Table 7.1 (continued)

Agricultural/ICT-based innovation	Brief description and highlight of the project	Author/s
Smart farming using ICT	Smart farming is influencing the entire food supply chain. Smart farming may unravel into a highly integrated food supply chain or collaborative systems involving the production-processing-market linkage	Wolfert et al. (2017)
IoT and radiofrequency identification (RFID) technologies	Agricultural information is constructed with a combination of IoT and RFID. This system significantly improves the efficiency of hardware resources in the agricultural information network	TongKe (2013)
Smart agricultural model in Japan using data-on-demand information exchange based on smart agricultural machinery systems (SAMS)	Four machines were developed, namely, smart rice transplanter, smart second fertilizer applicator, yield monitor; and farm activity record management system (FARMS)	· Morimoto and Hayashi (2017)

The promise of the sustainable agricultural production system is bright for advanced economies. However, for the developing economies, more education of the various players of the agriculture-food system and policy reforms have to be undertaken to meet the food demands of the growing populations amid climate change challenges.

References

Alexandratos N, Bruinsma J (2012) World agriculture towards 2030/2050: the 2012 revision. ESA working paper No. 12-03. Agricultural Development Economics Division, Food, and Agriculture Organization of the United Nations. June 2012. www.fao.org/economic/esa

Antón J, Cattaneo A, Kimura S, Lankoski J (2013) Agricultural risk management policies under climate uncertainty. Glob Environ Chang 23(6):1726–1736. https://doi.org/10.1016/j.gloenvcha.2013.08.007

Arida GS, Punzal BS, Marquez LV, Corales RG, Settele J (2016) Impact of Palayamanan fields on conservation biological control in rice-based cropping system. Philipp Entomol 30:85–94

Braun AT, Colangelo E, Steckel T (2018) Farming in the era of Industrie 4.0 (2018). Procedia CIRP 72:979–984. https://doi.org/10.1016/j.procir.2018.03.176

Carson RL (1962) The silent spring. Houghton Mifflin Harcourt, Boston, MA

Clark M, Tilman D (2017) Comparative analysis of environmental impacts of agricultural production systems, agricultural input efficiency, and food choice. Environ Res Lett 12(6):064016

Connolly-Boutin L, Smit B (2016) Climate change, food security, and livelihoods in sub-Saharan Africa. Reg Environ Chang 16(2):385–399. https://doi.org/10.1007/s10113-015-0761-x

Corales RG, Juliano LM, Capistrano AOV, Tobias HS, Dasalla NV, Cañete SD, Casimero MC, Sebastian LS (2004) Palayamanan: a Rice-based farming systems model for small-scale farmers. Philippine J Crop Sci 29(1):21–27

De Clercq M, Vats A, Biel A (2008) Agriculture 4.0: the future of farming technology. Proceedings of the world government summit, Dubai, UAE (2018) Feb. 11–13
Duru M, Therond O, Martin G, Martin-Clouaire R, Magne MA, Justes E, Journet EP, Aubertot JN, Savary S, Bergez JE, Sarthou JP (2015) How to implement biodiversity-based agriculture to enhance ecosystem services: a review. Agron Sustain Dev 35(4):1259–1281
Hamad AF, Han JH, Kim BC, Rather IA (2018) The intertwine of nanotechnology with the food industry. Saudi J Biol Sci 25(1):27–30. https://doi.org/10.1016/j.sjbs.2017.09.004
Harvey CA, Chacón M, Donatti CI, Garen E, Hannah L, Andrade A, Bede L, Brown D, Calle A, Chara J, Clement C (2014) Climate-smart landscapes: opportunities and challenges for integrating adaptation and mitigation in tropical agriculture. Conserv Lett 7(2):77–90
Hernández-Morcillo M, Burgess P, Mirck J, Pantera A, Plieninger T (2018) Scanning agroforestry-based solutions for climate change mitigation and adaptation in Europe. Environ Sci Pol 80:44–52
Himesh S, Rao EV, Gouda KC, Ramesh KV, Rakesh V, Mohapatra GN, Rao BK, Sahoo SK, Ajilesh P (2018) Digital revolution and big data: a new revolution in agriculture. CAB Rev 13 (021):1–7. https://doi.org/10.1079/PAVSNNR201813021
Huh JH, Kim KY (2018) Time-based trend of carbon emissions in the composting process of swine manure in the context of agriculture 4.0. Processes 6(9):168. https://doi.org/10.3390/pr6090168
Huq N, Bruns A, Ribbe L, Huq S (2017) Mainstreaming ecosystem services based climate change adaptation (EbA) in Bangladesh: status, challenges and opportunities. Sustainability 9(6):926
Iglesias A, Garrote L (2015) Adaptation strategies for agricultural water management under climate change in Europe. Agric Water Manag 155:113–124
Jones C, Pimdee P (2017) Innovative ideas: Thailand 4.0 and the fourth industrial revolution. Asian Int J Social Sci 17(1):4–35. https://doi.org/10.29139/aijss.20170101
Klerkx L, van Bommel S, Bos B, Holster H, Zwartkruis JV, Aarts N (2012) Design process outputs as boundary objects in agricultural innovation projects: functions and limitations. Agric Syst 113:39–49. https://doi.org/10.1016/j.agsy.2012.07.006
Lindblom J, Lundström C, Ljung M, Jonsson A (2017) Promoting sustainable intensification in precision agriculture: review of decision support systems development and strategies. Precis Agric 18(3):309–331. https://doi.org/10.1007/s11119-016-9491-4
Lipper L, Thornton P, Campbell BM, Baedeker T, Braimoh A, Bwalya M, Caron P, Cattaneo A, Garrity D, Henry K, Hottle R (2014) Climate-smart agriculture for food security. Nat Clim Chang 4(12):1068
Liu Y, Pan X, Li J (2015) Current agricultural practices threaten future global food production. J Agric Environ Ethics 28(2):203–216. https://doi.org/10.1007/s10806-014-9527-6
Macnaghten P, Chilvers J (2014) The future of science governance: publics, policies, practices. Environ Plann C: Gov Policy 32(3):530–548. https://doi.org/10.1068/c1245j
Mariano MJ, Villano R, Fleming E (2012) Factors influencing farmers' adoption of modern rice technologies and good management practices in the Philippines. Agric Syst 110:41–53
Misra AK (2014) Climate change and challenges of water and food security. Int J Sustain Built Environ 3(1):153–165
Morimoto E, Hayashi K (2017) Design of smart agriculture Japan model. Adv Anim Biosci 8 (2):713–717
Notarnicola B, Sala S, Anton A, McLaren SJ, Saouter E, Sonesson U (2017) The role of life cycle assessment in supporting sustainable Agri-food systems: a review of the challenges. J Clean Prod 140:399–409
Pivoto D, Waquil PD, Talamini E, Finocchio CP, Dalla Corte VF, de Vargas Mores G (2018) Scientific development of smart farming technologies and their application in Brazil. Inform Process Agric 5(1):21–32. https://doi.org/10.1016/j.inpa.2017.12.002
Richardson KJ, Lewis KH, Krishnamurthy PK, Kent C, Wiltshire AJ, Hanlon HM (2018) Food security outcomes under a changing climate: impacts of mitigation and adaptation on vulnerability to food insecurity. Clim Chang 147(1–2):327–341. https://doi.org/10.1007/s10584-018-2137-y

Rose D, Chilvers J (2018) Agriculture 4.0: responsible innovation in an era of smart farming. Front Sustain Food Syst 2:87. https://doi.org/10.3389/fsufs.2018.00087

Rose DC, Sutherland WJ, Parker C, Lobley M, Winter M, Morris C, Twining S, Ffoulkes C, Amano T, Dicks LV (2016) Decision support tools for agriculture: towards effective design and delivery. Agric Syst 149:165–174. https://doi.org/10.1016/j.agsy.2016.09.009

Sala S, McLaren SJ, Notarnicola B, Saouter E, Sonesson U (2017) In quest of reducing the environmental impacts of food production and consumption. J Clean Prod 140:387–398

Samadianfard S, Ghorbani MA, Mohammadi B (2018) Forecasting soil temperature at multiple-depth with a hybrid artificial neural network model coupled-hybrid firefly optimizer algorithm. Inform Process Agric 5(4):465–476. https://doi.org/10.1016/j.inpa.2018.06.005

Scheehle EA, Kruger D (2006) Global anthropogenic methane and nitrous oxide emissions. Energy J 1:33–44

Smith P, Gregory PJ (2013) Climate change and sustainable food production. Proc Nutr Soc 72 (1):21–28. https://doi.org/10.1017/S0029665112002832

Smith A, Snapp S, Chikowo R, Thorne P, Bekunda M, Glover J (2017) Measuring sustainable intensification in smallholder agroecosystems: a review. Glob Food Sec 12:127–138. https://doi.org/10.1016/j.gfs.2016.11.002

Sponchioni G, Vezzoni M, Bacchetti A, Pavesi M, Renga FM (2019) The 4.0 revolution in agriculture: a multi-perspective definition. In Summer School Francesco Turco-Industrial Systems Engineering, pp. 143–149

TongKe F (2013) Smart agriculture based on cloud computing and IoT. J Converg Inf Technol 8 (2):26. https://doi.org/10.4156/jcit.vol8.issue2.26

Varela RP, Balanay RM, and Apdohan AG (2019) Assessment of climate risks in agriculture towards climate resilient agricultural initiatives: the case of Caraga region, Philippines. Proceedings of the international conference and workshop on basic and applied sciences (ICOWOBAS), KSL resorts, Johor Bahru, Malaysia, July 14–17, 2019

Vermeulen SJ, Dinesh D, Howden SM, Cramer L, Thornton P (2018) Transformation in practice: a review of empirical cases of transformational adaptation in agriculture under climate change. Front Sustain Food Syst 2:65

Waters-Bayer A, Kristjanson P, Wettasinha C, Van Veldhuizen L, Quiroga G, Swaans K, Douthwaite B (2015) Exploring the impact of farmer-led research supported by civil society organisations. Agric Food Secur 4(1):4. https://doi.org/10.1186/s40066-015-0023-7

Whitfield S, Challinor AJ, Rees RM (2018) Frontiers in climate smart food systems: outlining the research space. Front Sustain Food Syst 2:2. https://doi.org/10.3389/fsufs.2018.00002

Wolfert S, Ge L, Verdouw C, Bogaardt MJ (2017) Big data in smart farming–a review. Agric Syst 1 (153):69–80. https://doi.org/10.1016/j.agsy.01.023

Yahya N (2018) Agricultural 4.0: its implementation toward future sustainability. In: Green Urea. Springer, Singapore, pp 125–145

Zambon I, Cecchini M, Egidi G, Saporito MG, Colantoni A (2019) Revolution 4.0 industry vs. agriculture in a future development for SMEs. Processes 7(1):36. https://doi.org/10.3390/pr7010036

Climate Change, Soil Erosion Risks, and Nutritional Security

8

Shrikaant Kulkarni

Abstract

The climatic changes at a global level may further aggravate the risks of soil erosion due to temperature variations and precipitation in the time to come. The factors like continuously growing ocean temperature, sea level, ice meltdown from glaciers and icebergs, frequent floods, storms, cyclones in tropical areas, rainfall, and temperature pattern shifts are the indicators of climatic changes which can hamper soil stability and influence erosion dynamics. Further, there would be a rise in the rate of runoffs, particularly in high altitude and erosion due to arid areas. The extent to which wind and water erosion increase depends on the ecosystem, topography, and management and is region-centric. The developing world would be more vulnerable to the effects of climatic changes on potential risks associated with soil erosion in the form of a greater degree of eroded soils and lack or inadequate access to soil erosion remedies effectively. Rise in erosion by the wind in semiarid or arid areas is attributed to high temperature and low rate of precipitation. Wind erosion has already damaged severely cultivable lands to the extent of 25% in arid regions. The fallout of soil erosion is a substantial reduction in biomass and grain production. Reduction in vegetation cover and crop residues can exacerbate soil erosion and have a cascading effect on soil degradation and desertification. On farm experiments and modeling are the two major approaches adopted for understanding the potential effects of change in climatic conditions on soil erodibility. To simulate global warming effects, small plots of land are warmed up using underground electric cables and heaters. Soil response based on its quality in crop production subjected to conditions created

S. Kulkarni (✉)
Department of Chemical Engineering, Vishwakarma Institute of Technology, Bibwewadi, Pune, India

V. K. Hebsale Mallappa, M. Shirur (eds.), *Climate Change and Resilient Food Systems*, https://doi.org/10.1007/978-981-33-4538-6_8

artificially is monitored over a given period. This chapter reviews the effect of climatic changes on soil erosion, the cause of the soil erosion, approaches employed to monitor the impact of climatic variations and global warming on soil erosion, and measures to maintain soil fertility, crop productivity, and nutritional security.

Keywords

Climate change · Erosion risks · Ecosystem · Topography · Erodibility · Simulation

8.1 Introduction

Climate change at the global level is recognized across the board. It is a fallout of human intervention activities (anthropogenic), which in turn are major causes of greenhouse gases (GHGs) (Bergh and Linder 1999). The concentration level presently of GHGs in the atmosphere is at a peak from 650,000 years (Blanco-Canqui and Lal 2007). The potential global warming effects are ascertained by the heat-trapping capacity and the pace at which each GHG decays against carbon dioxide (CO_2). The CO_2 is the major GHG candidate and contributes 61% of the total GHG contribution (Boardman and Favis-Mortlock 1993). The pace of CO_2 liberation has grown up substantially since 1995, and the increase annually till 2005 was at the rate of 1.9 ppm/year as against 1.4 ppm from 1960 to 2005 (Buol et al. 1990). Further during 1750 to 2005, the increase in CO_2 was 31% (379 ppm from 280 ppm), methane (CH_4) by 151% (1774 ppb from 715 ppb), and nitrous oxide (N_2O) by 17% (319 ppb from 270 ppb) (Ci 1998; Gao et al. 2002).

Burning fossil fuels added substantive amounts of CO_2. Alterations in land usage patterns (e.g., deforestation, tillage of the soil, application of fertilizers, burning of biomass, extrication of crop leftovers) led to furtherance in the generation of GHGs. Agricultural operations are a major cause of CH_4 and N_2O liberation. The present pace at which CO_2 emits by the burning of fossil fuels supersedes that from a change in land usage varying from cultivating vegetation to agricultural produce. Still, the contribution in totality made by land usage to GHG liberation since the beginning of established agricultural practices is substantial (Groisman et al. 2001).

8.2 Effects of Various Factors on Climate Change

8.2.1 Greenhouse Gas Emissions

The gradual increase in GHG emissions in the atmosphere and the commensurate climatic changes have brought about an alteration in the balance in the energy of the Earth (Harte et al. 1995). Precipitation and alteration in temperature patterns are the fallouts of these changes globally, and climatic changes predicted to be severe in the time to come. The greenhouse effect is the trapping of long-wave infrared (IR) radiations by the greenhouse gases, leading to the heating up of the Earth's atmosphere, rise in oceanic temperature, ice and snow meltdown from glaciers and bergs, and rise in seawater level, which are the implications of global warming. There has been a disparity with some regions getting drier, while others are receiving increasingly heavy but uneven rains attributed to either a decrease in precipitation or an increase in evapotranspiration rates. Sharp or sudden change in patterns of temperature and rain too is expected in the time to come (IPCC 2007), which will have its own but greater manifestations than progressive changes. Predictions made based on simulations show that the rise in GHG levels in the future will have more severe effects on climatic changes.

In the twenty-first century, warm climate may, on the one hand, raise the rate of precipitation and runoff in humid regions and on the other, causing acute scarcity of water in dry areas (IPCC 2001). Water resources, both qualitatively and quantitatively, are affected both by precipitation and temperature pattern changes.

8.2.2 Temperature

The average temperature in the terrestrial and aquatic ecosystems is on the rise. During 1995–2006, a major chunk of the period was warmer on average as against earlier years. An increase in mean air temperature by 0.75 °C from 1851 to 2006 has almost doubled during the last 50 years (1955–2005) compared to the past 100 years (1855–1955). There has been a consistency in trends about the maximum and minimum soil temperature rise with air temperature. It is predicted that with the increase in temperature of about 0.21 °C/decade, the sea level will rise to about 24 mm/decade by 2100. The mounting atmospheric GHG concentration level reflects upon its own consequences in the long run. The present level of GHGs would carry on to bring about a rise in temperature of the order of 0.1 °C/decade, even if we stop its emissions immediately. It is attributed to thermal inertia that indicates the present GHG concentration levels have not still shown their impact on the climatic conditions in full.

8.2.3 Droughts

The amount of precipitation has reduced substantially in arid as well as semiarid areas across continents such as Africa, Asia, and South America. This reduction is reflected upon not only frequent droughts but also an increase in drought periods, and intensities since the 1970s attributed to high temperature, reduced amounts of precipitation, an increase in the rate of evaporation, and intense wind storms. Risk potential associated with drought and flood occurrence is vulnerable to increase with latitude. Soil yield gets lowered under both drought and higher temperatures.

8.3 Climate Change Indicators

Frequency of flood occurrences, prevailing heat wave conditions, wind storms, cyclones in tropical areas, and other extreme climate change events are on the rise in recent past. The extent of water vaporization, degree of salinity of the sea or oceans, and rise in sea level are also on the rise, over the decades. The mean sea level has increased to the extent of 1.7 ± 0.6 mm/year from 1860 to 2002 and 3.0 ± 0.6 mm/year from 1992 to 2002. The reduced ocean's ability to absorb carbon sink may decrease, which will further aggravate the global warming problem by increasing the temperature across the globe by about 4.4 °C in 2100 (Lal 2006). A rise in air temperature has resulted in a reduction in ice in the Arctic sea by about 28.82% in 2015 as per the satellite image data obtained by NASA. There have been an expansion in size and an increase in a number of glacial lakes in recent decades because of ice and snow meltdown in a sustained manner as well as soil quality degradation in arctic and boreal regions.

8.4 Impacts of Climatic Changes

8.4.1 Soil Erosion

It is predicted that climatic changes will affect increasing the potential soil erosion risks, which will further bring about soil degradation and desertification (Lal 2007). Conditions will govern the magnitude of this effect in enhancing the water and wind erosion risks at local and regional level (Larson et al. 1997). It may lead to an increase in soil erosion by 5–95% and loss in a runoff by 5–100% in cultivable lands (Lavelle et al. 1997). Precipitation pattern changes, in conjunction with the factors like land usage, vegetative cover, and soil erosion behavior that influence the soil erosion rate. Global predictions about soil erosion are given in Table 8.1.

8.4.1.1 Site-Centric Dependence of Climatic Variations on Soil Erosion

In developing nations, as the farmers are resource-poor, the impacts of climatic changes on soil erosion are expected to be greater in intensity and severity. It may be prominently due to larger areas of degraded soils and either nonexistent or inadequate erosion control practices. Heavy rainfall would cause flooding of low-lying areas due to heavy runoff. Indeed, the magnitude and erosion capacity of rain may decrease in semiarid and arid regions that are prone to further water erosion as in the humid areas (Lee et al. 1996). Thus, reduction in precipitation velocities, however,

Table 8.1 Global predictions about soil erosion

Year	Potential soil erosion quantity (Pg/year)
2001	34.9
2012	35.8
2023	36.7

may not necessarily lead to lowering soil erosion behavior. Downsloping lands and undulating terrains are responsible for a significant amount of losses of runoff water.

8.4.2 Landscape Stability

Runoff sediment is a cause of grave concern to downstream water bodies. Highly concentrated or saturated runoffs in lands sloping down are responsible for an increase in landslides, stream bank erosion, and mudflows. Landscape characteristics are affected by runoff rise as well as soil erosion and temporary and permanent gullies. Intense rainstorms and large runoff volumes can lead to severe flow erosion of farmlands. Large runoff volumes move downslope with its own field topography end up as natural swales therefore, can cause concentrated flow erosion. The erosion at large can become increasingly significant and is influenced by rainfall intensity and frequency.

8.4.3 Water Erosion

Based on available data (observed and simulated), it is found that runoffs and soil erosion rise with the amount of precipitation. The intense rainstorms give rise to water erosion due to variations in rainstorm intensity, which primarily affects soil erosion over occurrence and quantum of rainfall. The proportional rise in water erosion with the increase in precipitation takes place in a few areas because of positive feedback effect, i.e., a rise in soil erosion vulnerability. Average runoff rates and CO_2 levels are projected to grow by 31–41% in high latitude and southeast areas and reduce by 11–31% in arid and semiarid drought-prone regions like Central Asia, South America, and Africa by 2050. Coastal areas and degraded agricultural lands are vulnerable to an increase in runoffs and soil erosion risks. More substantial rainstorms and flash floods have further escalated the runoff losses in northern latitudes. Predictions base on simulation studies show that in the Midwest USA, there would be rise by about 85% over the next five decades in soil erosion rates.

8.4.4 Nutrient Losses in Runoff

It is projected that the increase in soil erosion by virtue of climatic changes can bring about contamination of water bodies with soluble and insoluble (turbidity imparting) impurities. Sediments transported by runoff to water downstream further increase the precipitation (Lukewille and Wright 1997). Soil heating triggers decay and mineralization of organics in soil; dissolved nutrients and chemicals in soil may be passed on to runoffs. The dissolved impurities may, therefore, be transferred to surface and groundwater reservoirs by leaching and runoffs, e.g., in Norway, runoff from synthetically heated fields had more concentration levels of nitrates and ammonia than those without warming (Melillo et al. 2002). Excessive nutrient supply may lead to eutrophication of water bodies. Various bio-channels, including earthworm burrows, may lead to preferentially bypassing the flow of water contaminated with hazardous chemicals into the low-lying areas (Nash and Gleick 1991). Cracks in the soil are also routing responsible for bypassing flow of the rainwater and soluble chemicals. Such flow bypassing may enhance the pollution risk potential of water both ground and underground with nutrients and agrochemicals.

8.4.5 Wind Erosion

Wind erosion may further reduce precipitation and temperature rise both in semiarid and arid regions of about 25% of cultivated lands. The increase in temperature of the air can enhance evaporation and lower soil humidity with a warming of the soil while reducing vegetative cover as well as biomass generation. These conditions are conducive for a rise in the rate and erodibility of wind, increasing wind erosion rates (Nearing et al. 2005). Water erosion substantially enhances wind erosion, particularly in dry regions. An increase in time span and intensity of dry seasons in conjunction with intense wind can improve the potential soil erosion risks in drylands. In China, severe sandstorms have raised by fivefold in the 1950s as against 20-fold/year in the 1990s (O'Neal et al. 2005). In North China, with temperature rise by 1 °C, the increase in average wind erosion takes place by 31 Mg/km^{2}year, and the shift in land use was the primary factor responsible for the wind erosion risks (Pruski and Nearing 2002). An increase in CO_2 concentrations may to inevitable extent influence wind erosion because of increase in biomass generation and vegetative residue (Rastogi et al. 2002); the reduced and increased rates of precipitation and evaporation, respectively, result into thermal and drought stress as well as a decrease in plant production.

8.4.6 Climate Change Complexity Impacts

Climate change in the future will have its impact on soil erosion, which is varying and complex in nature on ecological, managerial, climatic, and landscape fronts. The variations in the precipitation and temperature patterns exacerbate soil erosion effects. Few of the soils are vulnerable to erosion due to their typical nature. Climate change predictions show that their impacts on soil erosion are uncertain because of many other governing factors like rainfall, soil erodibility, vegetation cover, etc. In a few areas, a slight change in precipitation may bring about a substantial rise in runoffs and soil erosion because of the synergistic effects. In contrast, in other areas, it may reduce erosion rates of wind and water due to more vegetation generation with higher temperature and rainfall (Reay 2007), as shown in Table 8.2. In sandy soils in the northern part of China, an increase in temperature by 1 °C lowered the water erosion mean rate by 5 Mg km^{-2} year^{-1} (Rillig et al. 2002).

Table 8.2 Synergistic effect of various factors on climate change

Increase in temperature (°C)	Precipitation decrease (%)	Runoff reduction (%)
2 °C	10	20
4 °C	20	30

8.5 Effects of Climatic Variations on Soil Erosion

8.5.1 Rainfall Distribution

There has been a substantial change in the rainfall distribution patterns across the globe due to the global warming effect and a change in mean total precipitation too. It is predicted that a shift in rainfall distribution patterns shall be rampant to a higher degree in the twenty-first century. The 2019 annual precipitation over India at large was 109% of its long period average (LPA) value for the period 1961–2010. The monsoon season (June–September) precipitation was 110% of its LPA, while during the northeast monsoon season (October–December) over the NE monsoon core region of the south peninsula, precipitation was 109% of its LPA. From 1950 to 2018, average annual precipitation has increased in 90% of the US states analyzed (NOAA 2016). Eighteen states have recorded a rise of 5 in or more, such as New Hampshire (7.0 in), Vermont (7.0 in), and Indiana (6.6 in). The eight largest increases are from the Northeast and Midwest, where *heavy* rainfalls are intensifying the most (Blunden and Arndt 2016).

Rainstorms are far robust than before (Shipitalo and Gibbs 2000). Although the amount of rainfall is steady in a few areas, the frequency of strong storms increased in the northern and southern parts of Europe that occurred from 1970 to 1998 than from 1909 to 1971. In the USA, overall precipitation has enhanced from 1911 to 1995, of which 52% was attributed to climate-based events. The predictions show that throughout the twenty-first century, the number of times the intense storms will occur will increase by 20–60% (Soil and Water Conservation Society 2003).

Loss and transport of soil by runoff depend on the intensity and quantity of rainfall. Experimental findings show that erosion due to rains is to rise significantly with the climatic changes in the future. Rainfall intensity affects erosion profoundly over its quantity and frequency of occurrence. Soil erosion depends on rainfall erosion behavior, as rainstorm energy multiplied by peak half an hour intensity. Even a small amount of intense rainstorms can bring about substantial damage to the soil. A rise in 1% of total rainfall may cause an increase in soil erosion rate by 1.8%, with the rise in rainfall intensity to the commensurate amount.

In comparison, an increase in erosion rate by 0.85% would occur if rainfall intensity remains constant (Sarah 2005). An increase in erosive energy of rainfall increases rate of soil erosion. Rain causes both positive and negative impacts on soil resilience. It works against soil stability by enhancing the loss and transport of soil particles in runoffs. It favors by improving the water holding capacity of the soil and boosting the growth potential of plants and the vegetation cover. The more biomass generation shields the soil from erosion caused by rainfall and checks runoffs and damage to the soil. The favorable effect of rainfall on biomass generation, however, counterbalanced by rising temperatures, which may increase evaporation rate, drought stress, and lower water content available to plant. An increase in temperature of air causes snow melting with rapidity, which furthers runoffs and soil erosion as snowpacks break and melt down quickly with a rise in temperature. Transitions from snowfall to rainfall because of the increase in temperature being considered as the next possible source responsible for storm enabled runoff instrumental in soil erosion. Snowstorms are expected to be substituted by rainstorms with changes in climate (Southworth et al. 2002a). The runoff during winter is more significant than during summer because of the premature melting of snow and higher rainfall-to-snow ratio.

8.5.2 Soil Erodibility

Soil erodibility is affected by the variations in water runoffs and temperature and can substantially affect soil mechanisms and characteristics. It brings about a reduction in macroporosity and water infiltration rates. Degraded soil structures lose soil to the maximum by wind and water erosion. Alterations in the subsurface soil environments like surface plugging, crusting, and densification are the processes caused by changes in climatic conditions, which are instrumental in increasing soil erosion. Soil particles on aggregation and subsequent stability of it may reduce with the rise in temperature and reduce during rainfall because of variations in vegetation cover (Southworth et al. 2002b). Changes in land usage, crop types, tillage techniques, and species of plants in tune with the variation in climate, too, are expected to influence soil erodibility. Soil structural stability is determined by land usage and management, soil type, organic matter contents, and rates of water infiltration that will decide the sensitivity of soil to erosion. Climatic changes make an impact on surface runoff, draining of water through the soil. Change in climate causes changes in vegetation cover due to variations in land usage, and management can profoundly affect soil erosion (Wigley 2005). Axing trees from forestlands removes the protective cover of vegetation and can enhance soil erosion after heavy rainfall. Soil surface residue cover and canopy are necessary to contain both wind and water erosion.

Runoffs and soil erosion increase with loss in vegetative cover. With a higher generation of biomass under climatic changes, more is the vegetative cover and therefore less is the soil erosion. Rise in temperature and scanty rainfall in hot regions can alter the vegetation cover in forests with huge trees to grasslands. In contrast, the situation may be another way around in the subhumid areas (Young et al. 1998). Arid regions with sparse vegetation have a deserted look when exposed to high temperature and low rainfall, which is amenable to more wind erosion.

8.5.3 Cropping Systems

The choice of cropping is influenced by global climate change. Farmers adapt to changing climatic conditions by shifting cropping patterns in the form of crop rotations, varieties, and planting schedules. Farmers may advance their crop planting timings than usual to check soil thawing and warming effects because of higher temperature in silking or having two crops per year. Such planting rescheduling leads to a rise in the cropping season spans. It is observed that the cropping system's effectiveness in the form of a reduction in the wind, as well as water erosion, is influenced by the amount of vegetative cover and the leftover after harvest. Replacement of higher biomass crops by lower biomass ones enhances soil erosion, e.g., in the Midwestern part of the USA, crop rotation shifts from corn, wheat, to soybean in conjunction with a higher degree of precipitation are responsible for increasing soil erosion to the extent of 301% in 2041–2058 as against 1991–1999 (Yuan et al. 2005). Corn to soybean crop rotations aggravate soil erosion because of the reduction in biomass production (Zhang 2006).

8.5.4 Soil Formation

Climate change is predicted as a factor of paramount importance that affects soil formation. Soil property profile is a function of independently variable factors, as shown in Eq. (8.1) (Zhang and Nearing 2005):

$$S = \mathrm{f}\,(\mathrm{Cl}, \mathrm{O}, \mathrm{P}, \mathrm{T}, \mathrm{t}, \ldots) \tag{8.1}$$

where.

S is property of soil.

Cl is climate.

O is biota of soil.

P is principle material.

T is topography.

t is time required for the formation of soil.

Climate is a highly active and productive parameter as it covers vast geographical regions. The significant factors like temperature, precipitation, humidity in relative terms, and radiation affect soil weathering, profile, etc. Temperature apart from rainfall regulates the way soil formation takes place in terms of evaporation velocities, storage of water, leaching, bio-based activities, etc. Soils are vulnerable to more significant erosion in areas having more temperature and precipitation. The local climatic conditions predominantly decide the soil development and degradation rate based on soil management. Climate together with soil development factors like a vegetative cover, organisms in the soil, properties of landscape, a host material, management of soil, and time has a cumulative effect on, e.g., breakdown and deposition of organic contents, retention of water, biomass generation, and the role played by soil organisms, and may further have its own effect on soil development. Climate also checks uptake of nutrients, germination of seeds, and root formation pace, diversity, count, growth, and role of soil organisms. Deserted areas with elevated temperatures and reduced precipitation having a lower vegetative cover, with sparse biomass, are prone to rapid decomposition as against cold and the humid regions possessing more organic contents and microbial activity.

8.5.5 Soil Process Mechanisms

Climatic changes can affect both the kind and magnitude of soil formation processes. The velocity of physicochemical and biochemical degradation may rise with temperature. Weathering due to physical forces dominates chemical one, particularly in dry and hot regions. Precipitation translocates colloids and ions to low-lying areas with the help of eluviation and illuviation processes. The climate which is amenable to an increase in rainfall may be instrumental in bringing about leaching of soluble chemicals. Gradual depletion in inorganic ions (e.g., Na^+, Mg^{2+}, Ca^{2+}) from the topsoil acidifies the soil. Highly acidic soils demand more alkali, such as lime, to restore crop yield. Soil temperature has its effect on chemical processes underlying soil dynamics, dissolution and diffusion of inorganic and organic compounds, and maintenance of equilibria (Aguilera et al. 2003). Rate of reactions, increase in solubility, and diffusion of salts in soil take place with temperature. Small variations to the extent of <5 °C in temperature are responsible for changes in chemical processes that may be detectable after many years. Evaporation exceeds precipitation in semiarid and arid areas. When the evaporation is in excess, dissolved salts in water transport to the top with the help of capillarity action and deposit in the topsoil, causing its salinity. Rise in temperature and decrease in precipitation under changing climatic conditions could facilitate the development of lands in dry areas world over. Climatic changes that are responsible for the rise in sea levels may also leave coastal areas saline.

8.6 Effects of Climatic Alterations on Soil Properties

8.6.1 Temperature

Soil temperature changes are the fallout of air temperature. Soil surface temperature is approximately 1.0 °C higher than the temperature of the air in temperate areas. The rise in soil temperature leads to warming and thawing perpetually of the soils, enhancing available land for cultivation in northern latitude, and can be responsible for croplands to move northward and crop growing seasons to expand.

The rise in soil temperature, in turn, affects soil structure by changing its physical properties such as shrinking to swelling and freezing to thawing and biological properties such as organic matter breakdown and microbial activity. Simulation studies of soil warming and its impacts exhibit a decrease in aggregate stability due to variations in wetting/drying and freezing/thawing reduction with the rise in soil temperature because of fast organic matter degradation and decline in microbial activity. High temperature is also responsible for soil desiccation and crack development and poses a threat to both buildings and crop yields. However, moderate drying will be beneficial to poorly drained, wet, and clayey soils. Higher temperature brings about drying in excess of soils followed by compaction.

8.6.2 Humidity Content

Soil humidity level depends on the amount of precipitation. Thus, soil water regime is influenced by even slight changes in rainfall due to climate change, which would reflect upon variations in biomass generation. Rise in air temperature due to climatic changes is amenable to give an impetus to evapotranspiration and water consumption by plants, cutting down water holding capacity. An Ohio state study observed that soil humidity gets reduced with temperature because of the removal of vegetation cover from no-till farms (Alberts et al. 1995).

8.6.3 Color

Soil weathering and humus decay can aggravate because of an increase in soil temperature. Global warming, floods, and waterlogged conditions may render soils redder (e.g., Oxisols, Ultisols) with a rise in temperature over long periods because of the effect on drainage rates.

8.6.4 Structure-Property Profile

Soil structure, too, is susceptible to alterations in its temperature and degree of precipitation (Altesor et al. 2006). The variation in climatic conditions influences soil structure not directly but as a result of a change in air temperature, microbial activity, plant growth, and precipitation. Alterations in soil structure under the futuristic climatic changes can be understood by:

- Estimating the rainwater infiltration rate with which it gets in the soil or results into a runoff.
- Assessing the soil resilience to forces of erosion by wind and water.
- Regulating water, air, and heat fluxes.
- Degradation of soil organics and soil organism activity.
- Estimates of organic carbon in the soil and the organics fate in the time to come.
- Absorption of buffers and degradation of pollutants (Arnalds 2000).

8.6.5 Precipitation

Intense rainstorms may bring about the damage to the structure of the topsoil and further soil erosion. The rate of soil depletion increases with rainfall enabled erosion. More soil erosion takes place due to more precipitation rates by reducing the thickness of topsoil and exposing soil with poor structure. Rise in rainfall at lower intensity strengthens the structural stability of soil by increasing biomass development as against semiarid and arid regions wherein the situation can be another way around (Blanco-Canqui et al. 2005). Highly clayey soils are vulnerable to greater shrinkage and swelling potential and therefore respond rapidly in tune with the changes in precipitation levels. Rain in excess makes soils moist unstable structurally and amenable to compaction.

8.6.6 Impact of CO_2 Concentration

Higher CO_2 level in the atmosphere has its effect on soil structure by increasing biomass production on the ground both from below and above in both quantitative and qualitative terms. Residues of plants containing lower N contents but with higher carbon/nitrogen ratio are amenable to decomposition at a slower pace leading to maintenance and improvement of soil structure (Briske 1996). The high CO_2 levels although have positive effects on soil structure in the form of increase in biomass generation, the rise in temperature can check it by reducing vegetation cover and soil desiccation.

8.6.7 Soil Biota

Climate change predictions show that it will influence interactions between soil, plants, and animals. Soil microorganisms respond to sudden temperature changes. Variation in biomass generation too affects soil organisms in terms of the number, activity, and diversity (Caldwell and Hodgkinson 1986). Global warming can create conditions conducive to the growth of organisms, including termites and earthworms, which have an important role to play in bringing about the decomposition of organic contents and determining their distribution profile and decomposition rate. Soil organisms like earthworms, termites, and others affect soil structure-property profile and erodibility. Soil structure houses present a suitable environment for the organisms to live, grow, and multiply and further provide all the necessary organic binding agents instrumental in developing soil structure and stability. Soil structure dynamically changes in accordance with changes (temporal and spatial) in climate, management, and biota of soil.

8.6.8 Soil Organic Carbon

Climate change can cut down soil organic carbon content while increasing CO_2 losses. The organic carbon stored in soil controls an increase in temperature and improves the humidity of it. The rise in soil temperature furthers the decay of crop residues. An estimated 3 °C rise in temperature may decrease soil organic carbon content levels by 11% approximately at about 30-cm soil depth while increasing CO_2 emissions by about 8% (Carlassare and Karsten, 2002). The CO_2 emission rates in the initial stages in response to a sudden rise in soil temperature can be of higher order (Chanasyk et al. 2003; Daniel et al. 2002).

8.6.9 Labile and Stable Organic Materials

Soil organic carbon contents may exist as either active or labile and stable or not active, and the former is mainly sensitive to temperature and decomposition behavior than the latter one. The latter fraction is composed of complex aromatic compounds against the former one, which consists more of polysaccharides. The CO_2 emissions predominantly result from the decomposition of the active fraction. Preferentially reduction in dynamic organic carbon contents under the impact of climatic changes lowers down the concentration of carbon contents in soil.

8.6.10 Land Usage Pattern

Depletions and gains in soil organic carbon contents depend on variations in land usage and its management. The shift in land use from carbon sink to carbon source brings about an alteration in the overall terrestrial carbon cycle. The primary causes of the depletion in soil organic carbon and global warming are heavy tillage, deforestation, and removal of crop leftovers. The detrimental effects of global warming on soil organic carbon storage may reduce over time depending on soil type, management, and ecosystem features, e.g., in temperate regions, rise in soil temperature by 5.1 °C by artificial heating enhanced decay in soil organics and raised CO_2 emissions by 27% in the first 6 years. At the same time, the effects declined in the past 4 years of the 10-year investigation, which showed that the carbon loss in the initial stages was transient in nature (De Baets et al. 2006).

8.6.11 Impacts of Climatic Zones

Soil organic content decay rate, however, can be checked with the rise in precipitation, particularly in humid and temperate areas. The decomposition of soil organic contents is found enhanced in cold and moist climatic conditions. The effects of soil heating on variations in organic carbon contents in the soil are region-specific. Rate of organic matter decay and accumulation differs from soil to soil. Variations in soil temperature can affect soil organic carbon dynamics in soils with poor structures. Rise in soil temperature and fall in soil humidity can have effects at a slower pace on clayey soils rather than sandy soils that are vulnerable to erosion, e.g., taking off corn stumps systematically from downslopes and with soils containing silt sensitive to erosion leads to a rise in average soil temperature and fall in soil holding capacity that results into fast depletions in organic carbon contents but affects not much on a clayey but flat soil in a temperate region in particular (Descheemaeker et al. 2006).

8.6.12 Crop Yield

Soil erosion brings about a substantial reduction in the thickness of the topsoil, which in turn cuts down the crop productivity depending on conditions prevailing at the sites (Ruddiman 2003). Soil erosion is responsible for reducing the nutrient-rich soil retention capacity and badly influences crop yields. The soil erosion velocity given climate change in time to come may be too high, which cannot be easily made up by the slow but natural soil renewal rate. The detrimental effects of soil erosion may be weak on sound-structured and fertile soils and high in shallow but stony soils with lower intrinsic fertility and root depth. Extreme runoffs can severely affect the resilience of soil and damage irreversibly soil complexion.

Climatic changes enabled soil erosion to further influence crop productivity in a quite complex manner and, at times, detected after a long span of time. Negative impacts on crop productivity are often overshadowed by the use of more copious amounts of fertilizers and high-yielding seeds. The crop cultivation is undertaken using higher doses of chemical-based fertilizers. However, it may enhance nonpoint source contamination. More soil erosion has detrimental effects at both on-site and off-site levels. Topsoil loss and crop productivity reduction and water pollution are some of the fallouts of it.

Climate change predicted can have a great effect on cultivable lands, forestlands, and pastures. Models of climate and crop predict that crops are, in particular, vulnerable to alterations in CO_2 concentration, precipitation, solar radiation, and temperature (Dormaar and Willms 1998). Change in climate affects planting, harvesting schedule, maturity time, crop productivity, and farm operations and, in turn, an economy based on agricultural activities. Table 8.3 shows the list of agricultural practices that have been employed so far and recommended practices for the future to improve productivity and exacerbate the quantity of carbon in soils.

8.7 Positive Impacts of Climatic Changes

Rise in precipitation, temperature, and CO_2 levels may give an impetus to crop production. It can lengthen the cropping season and permit cropping seasons two in succession to complete in a year. More CO_2 concentration can further enhance efficiency with which water is utilized by plugging stomas while still enhancing photosynthesis. The estimations show that soybean and wheat yield may improve by

Table 8.3 Traditional and recommended agricultural practices for improving yield and carbon in soils

Traditional practice	Recommended practice
Low external inputs	Using fertilizers and integrated nutrient management system judiciously
Constant use of fertilizer	Application of site-centric soil management
Uncontrolled use of water	Use of sound water management/conservation, water table management, irrigation techniques
Monoculture	Enhanced farming systems and numerous crop rotations
Summer fallow	Growing vegetative cover
Plough tillage	No-tillage or conservation tillage
Residue either taken off or burnt	Residue retained in the form of mulch
Draining wetlands	Restoring wetlands

about 40% and 20%, respectively, from 2050 and 2059 (Drewry 2006). A 9% rise in CO_2 concentration boosts wheat production by 4% in the Midwest of the USA (Dunn and Dabney 1996). Higher soil temperature will enhance nitrogen mineralization and microbial activity, thereby making the nutrients of the soil productive. The availability of the nitrogen in the soil would boost the growth of plants and carbon retention in tissues to contain carbon losses. The carbon storage in hardwood trees increases the liberation of nitrogen by heating soil in an artificial manner to increase the temperature by 4.9 °C, equivalent to losses from the soil over a span of 9 years in a forest area (Dyksterhuis 1949). Since the rise in soil temperature enhances organic matter decay, generation of biomass with more carbon/nitrogen ratio, however, in turn, checks decomposition and boosts gains in soil organic carbon.

8.8 Adverse Effects of Climate Change

Global warming causes the growth of the plants in mid- and higher latitude nations such as Alaska, Canada, Russia, and the USA, unlike arid, semiarid, and tropical regions, which may have adverse effects (e.g., Amazon). Plant and soil organisms at higher latitudes are, at times, adaptable to variations in rainfall and temperature, unlike those in hot regions that are more vulnerable to climate changes. A little change in temperature does not have much effect at high latitudes, although it can have a reasonable effect at low latitudes. However, in the USA, crop productivity in northern parts improves against the southern ones in the context of changing climatic conditions.

The high CO_2 concentrations are beneficial to biomass generation; however, they are negated by elevating the temperature of the air and lowering precipitation in dry areas. Higher soil temperature can enhance evaporation and minimize water retention, leading to a decrease in crop yield. Higher temperature too causes prematurity of crops and lower production. Arid and semiarid areas may increase demand and competition for resources of water. Increased vegetation growth due to more CO_2 liberation and soil humidity too dwindles nutrients, which ultimately cuts down plant growth. Soils poorly fertile are affected a lot by enhanced nutrient use. Supplemental use of fertilizers is required to make up for the large nutrient recovery. More mineralization of soil organic contents causes losses of nitrogen. An increase in temperature promotes insect and pest proliferation in comparatively colder areas.

8.8.1 Complex Interactions

Rise or fall in biomass production depends on rising air temperature accompanied by parameters governing plant growth, namely, CO_2 liberation and rainfall. Moreover, the growth of vegetation is affected by the availability of water and essential nutrients to plants apart from air temperature and atmospheric CO_2 concentration levels. Vegetation growth is the cumulative effect of all the parameters mentioned above. The rate at which CO_2 is absorbed depends on age and type of crop. The C_3 category plants such as soybean, rice, and wheat are more vulnerable than C_4 type plants like sugarcane, sorghum, and corn to changes in CO_2 flux.

8.9 Soil Warming Simulation Studies

Simulation studies of global warming effects in the future on soil characteristics, organic carbon, and erosion have been undertaken by artificially heating plots of land. Numerous methods of warming soil artificially are adopted as follows:

8.9.1 Underground Electric Cables

The use of underground electric cables is the most common method to heat small plots of land. The technique involves a burial of electric cables parallel to one another in soil separated 10 or 20 cm from one another and 10 cm deep. A temperature gradient of 5–10 °C in between the control and artificially warmed plot is kept during simulations. An automatic data logger is used for controlling the temperature on land plots. It provides for switching off and switching on when the temperature is above or below the margin, respectively, to keep temperature of the soil within ± 0.1 °C. A network in the form of an array of thermistors is employed to keep a vigil on the temperature of plots of land regularly.

8.9.2 Overhead Heating Facilities

It involves the use of electrically operated heaters, which are kept suspended on soil under experimentation at a given distance, say, 2 m or 3 m (Elliott and Carlson

2004). Heating of the plot uniformly is assured by applying radiometrical and temperature-based measurement techniques, which demand reflectors with sound design to reduce the dissipation of heat outside plots. Light generated artificially by the heater should be made to propagate in such a way that it doesn't interfere with the process of photosynthesis as well as heaters kept suspended above the plots without blocking sunlight.

8.10 Modeling of Effects of Climatic Alterations

The climatic models have been in use for predicting the effects of climatic changes on soil erosion, carbon dynamics, and crop development. A host of scenarios of climatic changes, soil hydrological mechanisms, vegetation, numerous ecosystems, and their management are developed and simulated to study their impact on soil erosion. Various models, too, are in use to extrapolate the data from small plots of land or laboratories, landscapes, and watershed projects. The models on climatic changes used so far are as follows:

- LISEM (The Limburg Soil Erosion).
- USLE (Universal Soil Loss Equation).
- RUSLE (Revised Universal Soil Loss Equation).
- KINEROS (Kinematic Runoff and Erosion).
- SWAT (Soil and Water Assessment Tool).
- WEPP (Water Erosion Prediction Project) (FAO 2000; Florine et al. 2006).

Models are put into use to simulate the below-mentioned variations that straight-away influence soil erosion:

- Variations in atmospheric CO_2 levels.
- Changes in rainfall frequency and intensity and precipitation.
- Spontaneous or progressive change in temperature of soil and air.
- Types of vegetation with shifts in land usage and management.
- Alterations in the evapotranspiration velocities, water holding capacity of soils, and drought spans.
- Changes in soil erosion tendency influenced by soil organic matter and surface characteristics.
- Variations in cropping patterns because of alterations in cropping seasons, schedules, and prices.
- Crop rotations in accordance with soil types, topography, and the effect of fluctuations in water and heat.

8.10.1 Water Erosion Prediction (WEPP)

It is one of the most widely used models for making predictions about the soil erosion behavior with the changes in climate as it provides for simulating:

- Complexity and dynamism in rainfall and temperature patterns.
- Residue decay rates due to effects of integration of variations in soil temperature, water content, and microbial activity.
- Rainwater percolation, compaction of soil, evapotranspiration velocities, vegetation and canopy, surface heterogeneities, and vegetation decay rates are key factors responsible for soil damage and runoff according to WEPP model.
- Model predictions show a rise in soil erosion and runoff velocities to the extent governed by conditions prevailing locally and regionally.

8.11 Adapting to Effects of Global Warming

Climatic variations are predicted to bring about more negative effects on the conservation of water and soil. The rise in wind and water erosion is the fallout which has already come to the fore (Goldsmith 2006). Unless sound and concrete conservation approaches are adopted, climatic changes in foreseeable future will bring about a reduction in soil productivity and promote wind and water erosion risks accompanied by deterioration in quality standards of water. Modern soil conservation approaches are aimed at designing, developing, and managing data pertaining to past climate and not based on projected climate, which therefore demands reengineered and redesigned conservation practices taking into account the climatic changes in the future. Present conservation strategies ought to be useful to sustain futuristic strong and heavy precipitation and more significant amounts of runoff. Soil erosion is caused by infrequent and intense rainstorms (Guretzky et al. 2005). Conservation practices have to be redesigned adapted to the climatic changes so as to contain greenhouse gases and water pollution and enhance crop yield. Management of soil in a prudent manner is of pivotal importance to address global warming problems ecologically. The most preferred way of doing it is to harvest the carbon. All those strategies that invest atmospheric CO_2 and give an impetus to terrestrial carbon sink in soil should be encouraged.

The practices used are:

- Controlled tillage, no-tillage.
- Use of organic manures.
- Postharvest return of vegetative cover.
- Growing trees and grasses.

Soil is a potential source of CO_2 liberation on axing trees, heavily ploughed, and vegetative cover removed postharvest. Reforestation of marginal and eroded soils that are deserted can increase CO_2 uptake. Since the introduction of intensive agricultural activities, there has been a loss of carbon ranging from 40×10^9 to 90×10^9 Mg (Haan et al. 1994). Employing no-tillage farming with crops that protect or mulch of residue on a rotation basis is an alternative to harvesting organic carbon in soil and counterbalancing net emissions of CO_2 (Hall 1998). Models of predictive nature show that no-tillage is the better option to cut down the rise in soil erosion influenced by the new climate change (Herrick and Lal 1995). It reduces wind erosion in the drier region through water conservation and brings down evaporation rates. Till now, at times, no-tillage practice is employed in countries such as Australia, Brazil, Canada, South America, and the USA. The small, marginal, and resource-stricken farmers from the southern part of Asia and Africa where erosion problems are too grave, aren't able to employ much of the tillage practice (Hunt et al. 2003). Therefore, these regions are hit hard under climate change as predicted. Thus, innovative strategies for the adoption of no-till and relevant conservation practices are effective enough to cope up with climate change vulnerable to erosion susceptible environment. Developing nations are resource-deficient wherein farmers use residues of crops to feed animals and burn as a fuel. Hence, no-till practice with no return of vegetative cover is as detrimental as plowing soil subjected to tillage (Johnson 2003).

Crop residue management is vital to combat the influence of climatic changes. The other prominent practices are the use of bioenergy crops as cover, rotating crops with nitrogen-fixing legume forages, and rotating plants for a short period as fuel and food. Strategies that help restore vegetative cover on surface permanently present organic contents to better the soil resilience. Checking overgrazing or the use of wheel traffic can be employed to contain soil erosion and compaction. Grass and vegetative cover strips as buffers for conservation are formed at the croplands that catch hold of and degrade sediment deposits and nutrients at the bottom, thereby stemming down pollution. There is no universal conservation strategy that is applicable, confirming to soil conditions across the board. Therefore site-centric management based on the nature of individual ecosystem and soil has to be engineered and redesigned to nullify the potential adverse effects of climatic changes (Kamm 2004).

8.12 Impact of Soil Erosion on Nutritional Security

Climate change has its own deleterious effects leading to rise in soil erosion, decline in food production levels, and nutrient levels (Kouwen 1992). The following are the areas to be looking into preferentially in this regard:

8.12.1 Availability of Nutrients

Climatic changes affect the availability of food and nutrients, and therefore the following measures can be taken for preserving and bettering the nutrient levels:

- Initiatives to further quantify and check uncertainties in regard to processes finding a place in crop models employed for contemplating effects on food availability keeping intact their nutrient levels. It demands further experimental studies based on variations in temperature, CO_2, soil humidity, and ozone, separately and together, for a host of crops of significance in food and therefore nutrient security. The experimentation will be typically crucial in tropical areas wherein they have been almost not in existence in the past. Another approach that can be adopted involves testing current ecophysiological models given the variations in productivity and weather conditions, at scales varying between independent fields and regions in entirety. Further similar outcomes employed are in time series and panel-based fronts (Kouwen and Li 1981).
- They are furthering the efforts to measure and check uncertainties on the front futuristic climatic changes. Outcomes of a host of climate models are to be used. Upgradation of downscaling techniques will help a lot, but not by overestimating such initiatives, as downscaling is of vital importance for rainfall. In contrast, the implications of rainfall patterns are comparatively low as against tropical areas.
- A lot much of work is required to evaluate the reliable water supply in areas under irrigation, a parameter used in conjunction direct with the effects on productivity as well as the impacts of stratospheric ozone, pathogens, weeds, pests, and sea level rise on crop and nutrient yield that are yet to be ascertained and demand scrutiny in quantitative terms. Incorporating complexity to models may not necessarily be the only goal of futuristic research challenges, as at times it may not substantially change outcomes. We need to make out the domains, their sizes, and the key factors wherein present models are inadequate. At times, a model that works or otherwise in a given situation and set of conditions is either accepted or rejected in other circumstances.
- There are many issues about how rapidly and effectively various measures can be adapted that involve how well farmers understand and take trends in climatic conditions, the resilience of their crops to these changing patterns, how fast they are at learning and implementing novel technologies, and the risk potentials and possibility of success on adaptations. All such issues should be subjected to evaluation, in particular, by using the latest data from tropical regions, which are warming fast. The other vital questions are regarding the potential for the development of technology, and better liaison between the models and scientists of crops will be of great help. However, the effects of frontier technologies are not so easy to predict unless we know what the technologies would be like in the future.
- Not much of work has so far been carried out to assess the true potential of expansion of cropland in cold areas. Further, now how will the soils in these regions be amenable to crop production with the right balance of nutrients, and how rapidly will expansion happen (Tubiello et al. 2007)?

8.12.2 Accessibility of Nutrients

The following are the issues to be taken care of as far as maintenance of food production level and nutrient balance are concerned:

- The majority of poor populations from rural areas have their incomes substantially affected by yield. Many economic assessments have assumed agriculture independent GDP growth, although not so typically in the developing countries. Futuristic works should preferentially consider the effects on income and nutrient security in foods.
- Not relying solely on agriculture shows a viable adaptation approach for populations insecure in terms of nutrients.
- We are bereft of basic understanding of how climatic changes are aimed at bettering agricultural productivity in terms of nutrient contents and in turn incomes by using fertilizer and better agrarian technology as the central strategy to give an impetus to rural livelihoods and for minimizing potential risk in the event of a fluctuation in climatic conditions. Another question is: Will the change in climatic conditions in the future inhibit technology adoption (Wang et al. 2008)?

8.12.3 Utilization of Nutrients

The following questions have to be answered in respect of nutrient security and balance:

- Higher levels of CO_2 in the atmosphere will lower down protein and micronutrient concentration, and people are caught unaware of the health implications of such developments. How to address this issue?
- Will these changes be more or less critical as against corresponding changes in calorie intake, and how do they interact with one another?
- Which management strategies are in existence to prevent a decrease in proteins or micronutrient concentration in crops?
- How will the climatic changes affect the rate of exposure and contraction of numerous diseases in humans?
- How will such changes affect variations in nutrient security?
- How are the hunger and disease correlated with one another, and what should be intervention strategies for getting better outcomes (Xiong et al. 2007)?

8.12.4 Stability of Nutrients

The food is secure for a region, a family, or a person provided:

- It is available with enough nutritious value all along with the life.

- They should have access to food even in the event of a financial or climate-related crisis or recurring events such as a threat to food security in specific seasons.
- The stability of nutrients embodies the accessibility, abundance, and use aspects related to food security with adequacy in nutritional levels.

8.13 Conclusion

The climate change predicted at the global level shows that variations in precipitation and temperature profile may enhance soil erosion risks. Consistently increasing oceanic temperature, sea level, meltdown from glaciers and snow-covered hills, flood recurrence, wind, and cyclonic storms, in tropical regions, and variations in temperature and rain patterns are the fallouts of global climatic changes that affect soil stability and various dynamics of erosion. Runoff velocities are expected to rise by 31–41% at higher latitudes, and wind erosion is expected to rise in semiarid and arid areas. The degree to which water and wind erosion increase is expected to be area-centric and depends on the kind of ecosystem, topographical features, and management practices employed. Impacts of climatic changes on soil erosion are likely to be more severe in developing countries, having eroded soils and constrained access to formidable means to combat soil erosion. Elevated temperature and decreased precipitation velocities enhance wind erosion in semiarid and arid areas. Presently, about 24% of cultivable lands are affected already drastically by wind erosion. The further rise in soil erosion reduces biomass generation. Lowered vegetation and biomass can, in turn, raise erosion and have a cascading effect in furthering soil damage and desertification.

Field experiments and modeling are two strategies employed to get an understanding of the potential effects on soil erosion problem of projected global climate change. Underground electric cables and overhead heaters are put into use to artificially heat small plots of land for simulating global warming effects. The soil responds based on its qualities, and crop yield to artificial heating is then monitored over a given span of time. Models provide for the simulation of a host of patterns of climatic changes such as alterations in both precipitation and temperature, the concentration of CO_2 in the atmosphere, soil erosion behavior, management, vegetation, tillage, and cropping patterns. Approaches used to combat global warming cover the use of soil conservation approaches like zero tillage, rotation of crops, crop cover, bioenergy crops, trees, and trees used as fuel, food, and buffers. A strategy that restores surface vegetation cover permanently reduces variations in temperature and contains the ill effects of precipitation on soil erosion. Conservation practices perform differently to the changes in the projected climate. Therefore, the contemporary conservation practices have to be reengineered and redesigned, and new strategies in tune with the requirements of the projected climate changes have to be developed to check the damages due to soil erosion, and crop yield is improved upon.

We should never lose sight of predicting the futuristic challenges in terms of improving food productivity as well as maintaining the nutritional balance. Various models help in providing invaluable insight by using our knowledge and transforming it into possibilities of outcomes of our interest. In the twenty-first century, if sustenance in agriculture, food, and nutrient security is to be maintained in a consistently heating globe, all-out efforts have to be taken. The extent of climatic stability observed at the advent of agriculture is a past while unpredictable changes that describe the futuristic challenges will characterize it. This may not necessarily have disastrous effects on the Earth as far as food and nutrient security is concerned. Still, it is better not to underestimate the enormous challenges and ecological threats at hand, which we may have to face in the future.

8.14 Policy Recommendations

The following are the state-of-the-art recommendations to maintain food security against the backdrop of climate change and its implications on soil erosion:

- Bettering the information in qualitative terms i.e., sound and timely information pertaining to uncertain situations and risks can certainly make difference in influencing the ecosystem.
- It is reaching susceptible village farming communities with meaningful inputs about climate change and its consequences on soil fertility.
- Constant monitoring of weather and better scientific knowledge about climate change and its fallouts in terms of soil erosion, precipitation, rainfall distribution patterns, and consequently cropping systems and food security.
- Adequately preparing for foreseeable natural disasters by using the predictions made with the help of simulations and modeling.
- Upkeeping agro-meteorological data.
- Designing tools and techniques for evaluating drastic weather effects and therefore providing necessary guidance about adaptation to changing weather conditions.
- Agro-ecological zoning for effect modeling and evaluation of vulnerability.
- Sound mapping of land vegetation cover.
- Sound assessments of crops and forest resources at a local, national, and global level.
- Developing techniques for managing agricultural water more efficiently.
- Investing in resilient agricultural systems.
- Developing stress-tolerant varieties of crops.
- Incorporating insurance schemes for climate change-related risks.
- Use of integrated soil fertility management practice.

References

Aguilera MO, Steinaker DF, Demaria MR (2003) Runoff and soil loss in undisturbed and roller seeded shrublands of semiarid Argentina. J Range Manag 56:227–233

Alberts EE, Nearing MA, Weltz MA et al. (1995) Chapter 7. Soil component. In: Flanagan DC, Nearing MA (eds) USDA-Water Erosion Prediction Project (WEPP). Hillslope profile and watershed model documentation. National Soil Erosion Laboratory (NSERL) Report #10, West Lafayette, Indiana

Altesor A, Pineiro G, Lezama F et al (2006) Ecosystem changes associated with grazing in sub-humid south American grasslands. J Veg Sci 17:323–332

Arnalds O (2000) Desertification: an appeal for a broader perspective. In: Arnalds O, Archer S (eds) Rangeland desertification. Academic Press, London, pp 5–16

Bergh J, Linder S (1999) Effects of soil warming during spring on photosynthetic recovery in boreal Norway spruce stands. Glob Chang Biol 5:245–253

Blanco-Canqui H, Lal R (2007) Soil and crop response to harvesting corn residues for biofuel production. Geoderma 141:355–362

Blanco-Canqui H, Lal R, Owens LB et al (2005) Strength properties and organic carbon of soils in the north Appalachian region. Soil Sci Soc Am J 69:663–673

Blunden J, Arndt DS (2016) State of the climate in 2015. B Am Meteorol Soc 97(8):S1–S275

Boardman J, Favis-Mortlock DT (1993) Climate change and soil erosion in Britain. Geogr J 159:179–183

Briske DD (1996) Strategies of plant survival in grazed systems: a functional interpretation. In: Hogdson J, Illius AW (eds) The ecology and management of grazing systems. CAB, Wallingford UK, pp 37–67

Buol SW, Sanchez PA, Weed SB et al (1990) Predicted impact of climate warming on soil properties and use. In: Kimball BA, Rosenberg NJ, Allen LH (eds) Impact of carbon dioxide, trace gases and climate change on global agriculture, vol 53. ASA Publication, pp 71–82

Caldwell MM, Hodgkinson KC (1986) Ecophysiology of rangeland plants. In: Joss PJ, Lynch PW, Williams OB (eds) Rangelands: a resource under siege proceedings of the second international rangeland congress. Cambridge University Press, Cambridge UK, pp 423–472

Carlassare M, Karsten HD (2002) Species contribution to seasonal productivity of a mixed pasture under two sward grazing height regimes. Agron J 94:840–850

Chanasyk DS, Mapfumo E, Willms W (2003) Quantification and simulation of surface runoff from fescue grassland watersheds. Agric Water Manag 59:137–153

Ci L (1998) Mechanism of desertification and sustainable strategies to combat desertification in China. Quat Sci 5:98–107

Daniel JA, Potter K, Altom W et al (2002) Long-term grazing density impacts on soil compaction. Trans ASAE 45:1911–1915

De Baets S, Poesen J, Gyssels G et al (2006) Effects of grass roots on the erodibility of topsoils during concentrated flow. Geomorphology 76:54–67

Descheemaeker K, Nyssen J, Poesen J et al (2006) Runoff on slopes with restoring vegetation: a case study from the Tigray highlands, Ethiopia. J Hydrol 331:219–241

Dormaar JF, Willms WD (1998) Effect of forty-four years of grazing on fescue grassland soils. J Range Manag 51:122–126

Drewry JJ (2006) Natural recovery of soil physical properties from treading damage of pastoral soils in New Zealand and Australia: a review. Agric Ecosyst Environ 114:159–169

Dunn GH, Dabney SM (1996) Modulus of elasticity and moment of inertia of grass hedge stems. Trans ASAE 39:947–952

Dyksterhuis EJ (1949) Condition and management of rangeland based on quantitative ecology. J Range Manag 2:104–105

Elliott AH, Carlson WT (2004) Effects of sheep grazing episodes on sediment and nutrient loss in overland flow. Aust J Soil Res 42:213–220

FAO (2000) Grassland resource assessment for pastoral systems. In: Plant production and protection paper, vol 162. FAO, Rome, Italy, p 2000

Florine SE, Moore KJ, Fales SL et al (2006) Yield and composition of herbaceous biomass harvested from naturalized grassland in southern Iowa. Biomass Bioenergy 30:522–528

Gao Q, Ci L, Yu M (2002) Modeling wind and water erosion in northern China under climate and land use changes. J Soil Water Conserv 57:46–55

Goldsmith W (2006) Soil strength reinforcement by plants. http://www.ieca.org/membersonly/resources/proceedings/2006/003%20goldsmith.pdf. Accessed 14 Mar 2008

Groisman PY, Knight RW, Thomas T (2001) Heavy precipitation and high streamflow in the contiguous United States: trends in the twentieth century. Bull Am Meteorol Soc 82:219–246

Guretzky JA, Moore KJ, Brummer EC et al (2005) Species diversity and functional composition of pastures that vary in landscape position and grazing management. Crop Sci 45:282–289

Haan CT, Barfield BJ, Hayes JC (1994) Design hydrology and sedimentology for small catchments. Academic Press, Cambridge, MA

Hall MH (1998) Forages. In: Serotkin N, Tibbets S (eds) The Penn State agronomy guide 1999–2000. Penn State University, University Park, PA, pp 169–210

Harte J, Torn MS, Chang FR et al (1995) Global warming and soil microclimate -results from a meadow-warming experiment. Ecol Appl 5:132–150

Herrick JE, Lal R (1995) Soil physical property changes during dung decomposition in a tropical pasture. Soil Sci Soc Am J 59:908–912

Hunt ER, Everitt JH, Ritchie JC et al (2003) Applications and research using remote sensing for rangeland management. Photogrammetric Eng Remote Sensing 69:675–693

IPCC (2001) Intergovernmental panel on climate change, climate change 2001: the scientific basis. Cambridge University Press, Cambridge, UK

IPCC (2007) Climate change 2007: the physical science basis. Summary for Policymakers. Contribution of Working Group I to the Fourth Assessment Report to the Intergovernmental Panel on Climate Change (IPCC). Paris, February, 2007

Johnson K (2003) Terminology for grazing lands and grazing animals. Purdue Univ Agron Extension http://wwwagrypurdueedu/ext/forages/rotational/indexhtml. Accessed 14 Mar 2008

Kamm J (2004) A new class of plants for a biofuel feedstock energy crop. Appl Biochem Biotechnol 113:55–70

Kouwen N (1992) A modern method for the design of grassed channels. J Irrig Drain Eng 118:733–743

Kouwen N, Li RM (1981) Flow resistance in vegetated waterways. Trans ASAE 24:684–690

Lal R (2006) Influence of soil erosion on carbon dynamics in the world. pp 23–35

Lal R (2007) Constraints to adopting no-till farming in developing countries. Soil Tillage Res 94:1–3

Larson WE, Lindstrom MJ, Schumacher TE (1997) The role of severe storms in soil erosion: a problem needing consideration. J Soil Water Conserv 52:90–95

Lavelle P, Bignell D, Lepage M et al (1997) Soil function in a changing world: the role of invertebrate ecosystem engineers. Eur J Soil Biol 33:159–193

Lee JJ, Phillips DL, Dodson RF (1996) Sensitivity of the U.S. corn belt to climate change and elevated CO_2.2. Soil erosion and organic carbon. Agric Syst 52:503–521

Lukewille A, Wright RF (1997) Experimentally increased soil temperature causes release of nitrogen at a boreal forest catchment in southern Norway. Glob Chang Biol 3:13–21

Melillo JM, Steudler PA, Aber JD et al (2002) Soil warming and carbon-cycle feedbacks to the climate system. Sciences 298:2173–2176

Nash LL, Gleick PH (1991) Sensitivity of streamflow in the Colorado basin to climatic changes. J Hydrol 3–4:221–241

Nearing MA, Jetten V, Baffaut C et al (2005) Modeling response of soil erosion and runoff to changes in precipitation and cover. Catena 61:131–154

NOAA (National Oceanic and Atmospheric Administration) (2016) National Centers for environmental information, Accessed February 2016. www.ncei.noaa.gov

O'Neal MR, Nearing MA, Vining RC et al (2005) Climate change impacts on soil erosion in Midwest United States with changes in crop management. Catena 61:165–184

Pruski FF, Nearing MA (2002) Runoff and soil-loss responses to changes in precipitation: a computer simulation study. J Soil Water Conserv 57:7–16

Rastogi M, Singh S, Pathak H (2002) Emission of carbon dioxide from soil. Curr Sci 82:510–517

Reay D (2007) Spring-time for sinks. Nature 446:727–728

Rillig MC, Wright SF, Shaw MR et al (2002) Artificial climate warming positively affects arbuscular mycorrhizae but decreases soil aggregate water stability in an annual grassland. Oikos 97:52–58

Ruddiman WF (2003) The anthropogenic greenhouse era began thousands of yr ago. Clim Chang 61:261–293

Sarah P (2005) Soil aggregation response to long- and short-term differences in rainfall amount under and Mediterranean climate conditions. Geomorphology 70:1–11

Shipitalo MJ, Gibbs F (2000) Potential of earthworm burrows to transmit injected animal wastes to tile drains. Soil Sci Soc Am J 64:2103–2109

Southworth J, Pfeifer RA, Habeck M et al (2002a) Changes in soybean yields in the midwestern United States as a result of future changes in climate, climate variability, and CO_2 fertilization. Clim Chang 53:447–475

Southworth J, Pfeifer RA, Habeck M et al (2002b) Sensitivity of winter wheat yields in the Midwestern United States to future changes in climate, climate variability, and CO2 fertilization. Clim Chang 22:73–86

SWCS (Soil and Water Conservation Society) (2003) Conservation implications of climate change: soil erosion and runoff from cropland. A report from the SWCS. Ankeny, Iowa

Tubiello FN, Amthor JS, Boote KJ, Donatelli M, Easterling W, Fischer G, Gifford RM, Howden M, Reilly J, Rosenzweig C (2007) Crop response to elevated CO_2 and world food supply: a comment on "Food for Thought..." by Long et al., Science 312:1918–1921, 2006. Eur J Agron 26(3):215–223

Wang J, Mendelsohn RO, Dinar A, Huang J, Rozelle S, Zhang L (2008) Can china continue feeding itself? the impact of climate change on agriculture. World Bank Policy Research Working Paper Series, number 4470

Wigley TML (2005) The climate change commitment. Sciences 307:1766–1769

Xiong W, Lin E, Ju H, Xu Y (2007) Climate change and critical thresholds in China's food security. Climate Change 81(2):205–221

Young IM, Blanchart E, Chenu C et al (1998) The interaction of soil biota and soil structure under global change. Glob Chang Biol 4:703–712

Yuan F, Xie ZH, Liu Q et al (2005) Simulating hydrologic changes with climate change scenarios in the Haihe River basin. Pedosphere 15:595–600

Zhang XC (2006) Spatial sensitivity of predicted soil erosion and runoff to climate change at regional scales. J Soil Water Conserv 61:58–64

Zhang XC, Nearing MA (2005) Impact of climate change on soil erosion, runoff and wheat productivity in Central Oklahoma. Catena 61:185–195

Minimizing Weather-Related Risks in Agriculture Through Agromet Advisory Services in India

9

Santanu Kumar Bal and M. A. Sarath Chandran

Abstract

The major factor influencing agriculture, the mainstay of majority of the Indian population, is climate. Both long-term climate change and intra-seasonal climate variability impact the decision-making of farmers. The provision of timely and accurate agromet advisories assumes great importance in this context. This chapter begins with weather-related agricultural risks and climate information needed by farmers in decision-making before and during the crop season. It then discusses the agromet advisory services (AAS)—history, development and present status in India. The dissemination and outreach programmes in the form of farmers' awareness programmes are also included. It further explores the role of AAS in effective operational decision-making, improvement in crop production and economic impact assessment of AAS in India. The chapter ends with constraints, future challenges and opportunities for AAS in India.

Keywords

Agromet advisory · Weather risks · Economic impact · Constraints · Opportunities

9.1 Introduction

Agricultural production is highly influenced by climate. The provision of reliable weather information can be of great help for the decision-making of farmers before and during the crop season for arranging the inputs and their optimum utilization. A

S. K. Bal (✉) · M. A. Sarath Chandran
ICAR-Central Research Institute for Dryland Agriculture, Santoshnagar, Hyderabad, Telangana, India
e-mail: Santanu.bal@icar.gov.in

V. K. Hebsale Mallappa, M. Shirur (eds.), *Climate Change and Resilient Food Systems*, https://doi.org/10.1007/978-981-33-4538-6_9

delay in onset or mid-season drought can hamper the agricultural production in the country, which in turn has manyfold effects on the economy of the country. This points to the necessity for a scientific weather forecasting system and development of agromet advisories, which are crop- and region-specific. A well-timed agromet advisory can save inputs (fertilizers, seeds, plant protection chemicals, etc.), labour and the crop (especially at the harvest time after the crop reaches physiological maturity).

Along with this, climate change has led to increased intensity and frequency of extreme weather events such as heavy precipitation, cloud bursts, hailstorm events, drought, etc. The twenty-first century has witnessed a giant leap in the advancement of science and technology. Our improved understanding of weather forecasting, coupled with the latest information technology tools, has ensured reliable and timely information delivery to the target population, the farmers. Yet, there exists a wide gap between the information provider and the farmer. Hence, the weather- and climatic stress-related challenges to be covered under agromet advisory services are increasing and becoming complex. This chapter includes information on the needs for weather and climate information by farmers and the ways and means to improve future communication to them to facilitate effective operational decision-making. The chapter will also focus on the challenges and opportunities associated with the agromet advisory services in India.

9.2 Weather-Related Agricultural Risks

It is a well-known fact that the variability in the weather elements (particularly rainfall and temperature) is the main factor behind the inter-annual variability of crop production. Sudden onset of weather extremes (dry spells, droughts, floods, heat waves and hailstorm) adversely affects crop yield leading to a low level of productivity. Moreover, weather can affect agricultural production at various levels (vegetative stage, harvest, transportation, storage, etc.). Floods, droughts, heat waves, cold spells, hailstorms and other natural disasters are great sources of risk for farmers (Bal et al. 2014; Bal and Minhas 2017; Sarath Chandran et al. 2017). These disasters can result in lesser yields, leaving uninsured farmers with little income for the season. To cope with erratic weather, farmers often plant low-risk, low-return crops (such as early maturing varieties, drought-resistant) instead of investing in more profitable crops that are more sensitive to weather fluctuations and extremes. Furthermore, farmers wary of bad weather may hesitate to make other investments in their farms, such as increasing fertilizer use. As a result, extreme weather events can trap farmers in a cycle of low productivity.

The climate of a region and seasonal weather affects many risks associated with crop production, which finally decides the farm income (Rao and Bapuji Rao 2013) (Fig. 9.1). This necessitates the vulnerability assessment of any place or region (Singh et al. 2019b).

Risks in production; events such as cold/heat wave, dry spell, flood and hailstorm; the outbreak of pest and diseases; financial risk and market risk; etc., are affected by

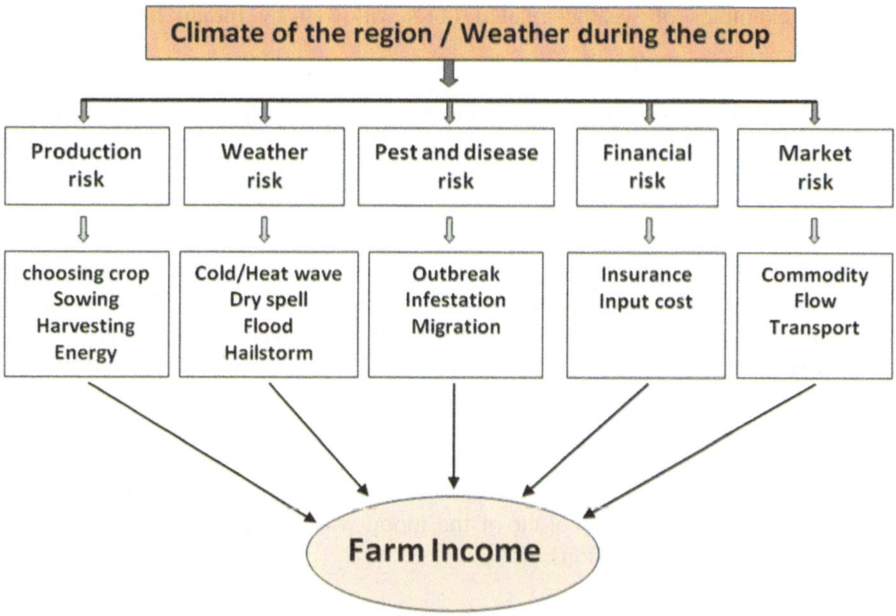

Fig. 9.1 Climate-related risks in agriculture (Rao and Bapuji Rao 2013)

climate/seasonal weather of a region. However, due to technological advancement in agricultural science, adaptation and mitigation options are now available to cope with the adverse impacts of the stresses caused due to these disasters (Bal et al. 2017; Chattopadhyay et al. 2016; Vashisht and Jalota 2018). Ultimately, the decisions taken regarding these risks are reflected in the farm income. It is a well-known fact that Indian agriculture is dependent mainly on southwest monsoon. Any reduction in the total rainfall during this season, monsoon break during the season, delayed onset and early withdrawal can have a significant impact on the crop production of the country. Hence, timely information on anticipated weather helps farmers for optimal utilization of their resources, which will ultimately improve agricultural production, qualitatively and quantitatively.

9.3 Weather-Related Indigenous Technology Knowledge (ITKs) Followed in Agriculture

The accumulated knowledge of a community from a particular region, based on their understanding, observation and experiences with the surrounding within their area, is called traditional knowledge (De Boef et al. 1993). The decision-making of farmers related to farm management is based on their accumulated experience over generations and current information available to them. It provides problem-solving ideas to the farmers in day-to-day life, and it is an underutilized resource.

Some examples of weather-related ITKs followed by Indian farmers are as follows:

- Farmers in the upper northwest Himalayas used a visible spectrum around the sun and moon for weather prediction. An event of rainfall was predicted in a day or two if the spectrum around the sun had more diameter than that around the moon (FAO 2019).
- Sowing of maize and soybean in dry soil after soaking in seeds in the cow urine for 10–12 h to protect seeds from insects/diseases in lower plains of districts of Himachal Pradesh under the late onset of monsoon (Rana et al. 2018).
- A study on colour of the moon and the related weather conditions during post-monsoon (first October to 31st December 2014) was documented by Palampur centre of All India Coordinated Research Project on Agrometeorology (AICRPAM). It was observed that the moon was seen in 92% of cases. In 43.1%, 6.9% and 1.4% of cases, the yellow-coloured moon was found to be associated with cloudy, rainy and windy conditions, respectively. In 18.3% and 5.6% of cases, the white colour of the moon was found to be associated with windy and cloudy conditions.

Farmers do understand that rainfall has different implications for different crop species based on the quantity, intensity and duration. They have developed indigenous ways of weather forecasting based on the experience of many generations, which are passed on mainly verbally. So, proper documentation of such ITKs is necessary. Once it is documented and validated, such useful information can be made part of agromet advisory services, which will add value to the conventional AAS system.

9.4 Climate Information Needed by Farmers

The climate information can impact the following key decisions regarding crop production by farmers (Madhavan and Rengalakshmi 2015) as described in Table 9.1.

Table 9.1 Climate information that can impact key decisions of crop production

Key decisions	Climate information that impacts the decision
Sowing period	Onset of monsoon
Selection of crops/variety	Total season rainfall forecast and its distribution within the season
Irrigation: Timing and quantity	
Resource use allocation: Labour and finance	
Fertilizer application: Amount, type, timing	Rainfall distribution forecast across the growth stages of crop
Pesticide application: Timing	Wind direction, speed, rainfall
Time of harvest	Rainfall forecast during the crop maturity stage

Source: Madhavan and Rengalakshmi (2015)

The major types of climate information needed by farmers before and during the crop production are as follows (FAO 2019):

- Before the commencement of cropping season, agrometeorological crop-risk analysis is carried out to select the most suitable crop(s) based on crop water requirement and forecasted seasonal rainfall.
- Determination of optimum date of planting for a specific region based on the onset of rainfall.
- Seasonal climate outlooks which can be used to adapt farmers to various situations that may arise during the season due to the uncertainty in the long-term weather forecasts. Crop contingency plans can be prepared for multiple possible scenarios.
- Three-day weather forecasts throughout the growing season, with emphasis on probability for extreme weather events; 10-day agromet advisories regarding the best crop management practices for the forecasted weather.

9.5 Introduction to Agromet Advisory Services (AAS) in India

AAS is defined as 'all agrometeorological and agro-climatological information that can be directly applied to improve and/or protect the livelihood of farmers' (Stigter 2011). In India, India Meteorological Department started AAS in 1945 as farmers' weather bulletin (FWB) through All India Radio. IMD started AAS from its state meteorological centres in 1976, in collaboration with state agricultural departments. From 1991 onwards, the National Centre for Medium-Range Weather Forecast (NCMRWF) started to prepare quantitative weather forecasts in the medium range (3 days) for a total of 5 agrometeorological field units (AMFUs), at a spatial resolution of 250 km. The spatial resolution was improved to 150 km in 1993 and to 75 km in 1999. The temporal resolution of the forecast was increased from 3 to 5 days in 2006 at the agro-climatic zone (ACZ) level.

In 2007, these two systems (forecast system of IMD and NCMRWF) merged into one single system, as the AMFU network expanded from 5 units in 1991 to 130 to cover all ACZs. From first June 2008 onwards, IMD has started issuing a multi-model ensemble weather forecast (50 km spatial resolution and 5-day temporal resolution) at the district level. At present, AAS is a multi-institutional and multidisciplinary project. Besides IMD, it involves the Indian Council of Agricultural Research (ICAR), state agricultural universities, state agricultural departments, nongovernmental agencies, media, etc. The flow diagram of AAS and inter-institutional linkages are depicted in Fig. 9.2.

The process of development and dissemination of district-level medium-range weather forecast is depicted as a flow chart in Fig. 9.3.

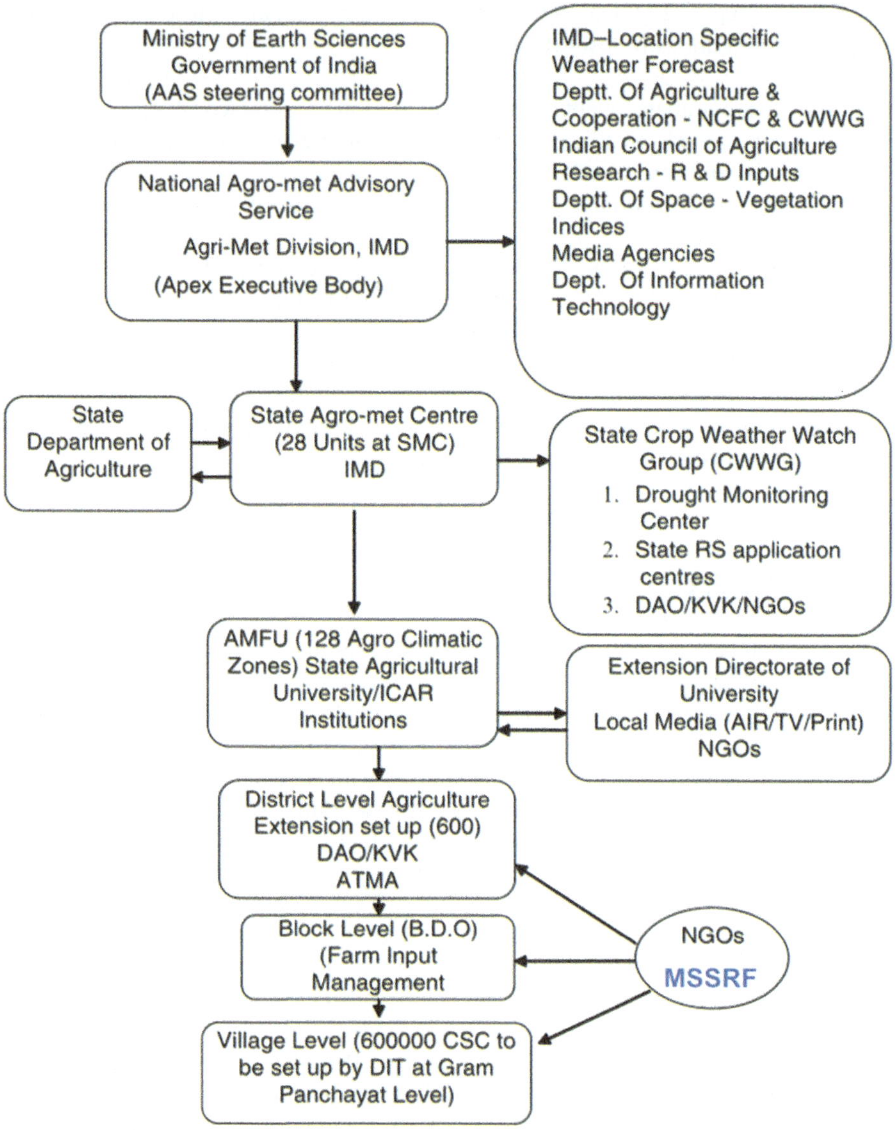

Fig. 9.2 Flow diagram of integrated AAS and inter-institutional linkages (Rathore et al. 2011)

9.5.1 Components of Agromet Advisory

The formulation of crop- and region-specific AAS bulletin requires interdisciplinary skills because of its dynamic nature. A thorough understanding of crop-weather interactions is a critical input for the successful preparation of advisory bulletins. Scientific background of crop-weather interaction includes the development of site-

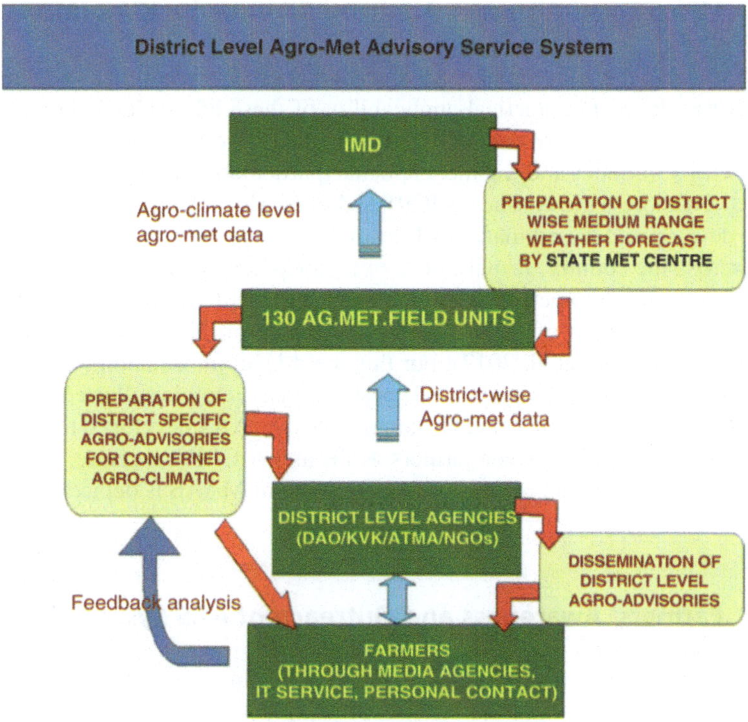

Fig. 9.3 Development and dissemination of district-level medium-range weather forecast by IMD (Rathore et al. 2011)

specific crop-weather models, thumb rule-based pest forewarning models, crop-weather calendars, etc. The use of crop growth simulation models and satellite imagery to monitor the crop health condition is becoming critical input in providing proper advisories to farmers. Vulnerability to weather hazards of a location should be identified and should be taken into consideration while preparing advisory bulletins.

Essential components of agrometeorological advisory are:

- Weather forecast.
- Warning of extreme weather events.
- Information related to location-specific normal sowing practices of crops.
- Farm management information such as irrigation scheduling, spraying of pesticides to control pests and diseases, nutrient (e.g. N, P, K) application, etc.
- Suggested measures for forecasted weather for appropriate action to be taken by the farmers suiting to their condition.
- Special warning for the outbreak of pest and disease and its appropriate control measures.
- Suggestions for livestock management and animal husbandry.

9.5.2 Micro-Level Agromet Advisory Services by ICAR (MAAS)

Though the spatial resolution of medium-range weather forecast has been improved to the district level, the practical applicability of these advisories is grossly inadequate to implement at farmers' level. This is due to the large spatial variability of types of crops, varieties and weather elements (particularly rainfall) that exists within a district. To address these issues, AICRPAM of ICAR took up a pilot project during 2011 to develop and disseminate block-level AAS through its 25 cooperating centres spread across the country. It makes use of a block-level weather forecast issued by IMD. In this, the agrometeorologist of the AICRPAM centre develops the AAS bulletins with the help of subject matter specialists at Krishi Vigyan Kendras—KVKs (Vijaya Kumar et al. 2017). For this, field-level information (types of crops, growth stage of crops, the incidence of pests/diseases, etc.) is collected by the Field Information Facilitator (FIF), and he disseminates the AAS developed to the farmers. FIF acts as an interface between farmers, KVK and AICRPAM centre. The methodology used for the development and dissemination of MAAS is depicted in Fig. 9.4.

9.5.3 Farmers' Awareness and Outreach of AAS System

Awareness generation is an important aspect of technology adoption. Unless the farmers are convinced about the benefits of using agromet advisory services, there will not be any value for the information generated. It is also a mechanism for sensitizing farmers about the scientific crop production coupled with the weather forecast. These programmes also improve the linkages between farmers and providers of AAS, viz. IMD, ICAR and SAUs. The main aim of these programmes is to make farmers self-reliant by providing timely information about the anticipated weather and suitable crop management advisory so that resource utilization is optimized and losses are minimized. In India, such programmes are organized by IMD, ICAR, SAUs, local NGOs and other stakeholders.

The methods to reach out to farmers include personal visits of technical experts, farmers' field schools, etc. In these meetings, the importance and advantages of using AAS in crop production are explained to them. Farmers are provided with leaflets containing general guidelines for weather-based farming, packages of practices for specific crops, details of pests and diseases of crops and their control measures, crop contingency plans for the district, etc. Rain gauges are provided to progressive farmers' groups by IMD during these programmes as a measure to improve the linkage, which encourages the farmers in recording rainfall data and sharing it with the technical experts. IMD has organized 243 farmers' awareness programmes at the district level by AMFUs (Chattopadhyay and Chandras 2018). The feedback of farmers and their perception about impact of climate and climate change on crop production is also documented.

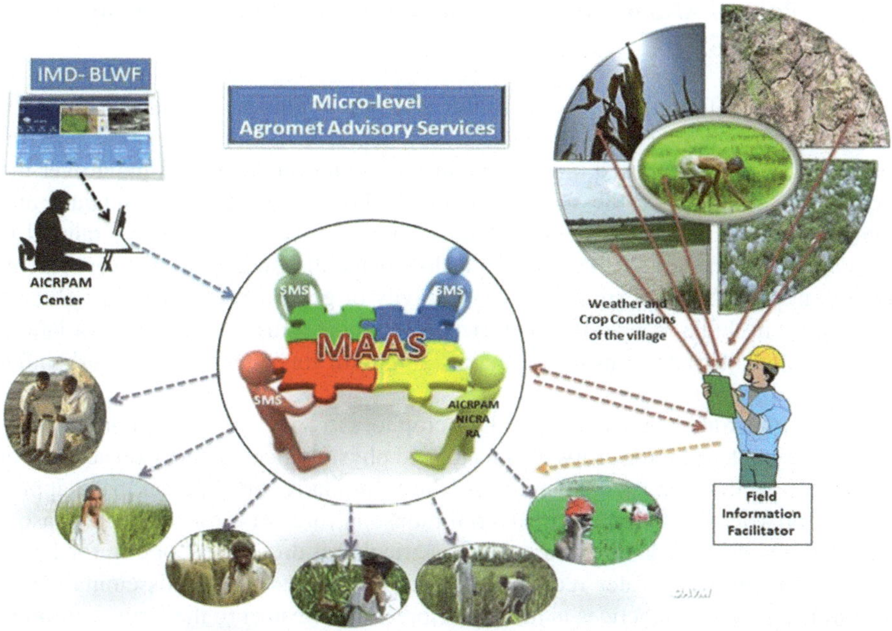

Fig. 9.4 Development and dissemination of MAAS (Vijaya Kumar et al. 2017)

9.5.4 AAS and Effective Operational Decision-Making

The ability to make a correct decision depends on the accuracy of the information provided and interpretation by farmers at the right time (Hansen 2002). But the factors affecting the 'correct decision-making' are so intricate that it makes the whole process difficult (Nesheim et al. 2017). The effectiveness of the AAS system mainly depends on its specificity and reliability of the information (Venkatasubramanian et al. 2014). The specificity of AAS refers to the location specificity, and crop specificity and reliability refers to the confidence with which the farmer accepts it based on previous success rate. A variety of factors influence the decision-making of a farmer, including seasonal rainfall, timely availability of inputs, agricultural subsidies, market demand, cultural traditions and adaptive capacity (Stone and Meinke 2006). Nesheim et al. (2017) had conducted a study in Maharashtra to understand the role of AAS in the decision-making of farmers, and they reported that the relevance of agromet information differs depending on the situation under which decision is made. Some farmers doubted the credibility of the AAS provided, while others positively used the information provided. They mainly valued the advisory provided to undertake precautionary measures. This emphasizes the need to design location-specific weather-based agromet advisories (Balasubramanian et al. 2014; Balasubramanian et al. 2016).

9.5.5 Role of AAS in Improving Crop Production and Ensuring Food Security

The sensitivity of crop yields to seasonal variations in climate is a major reason for inter-year crop production. Drought, flood, unseasonal rain before the harvest, hailstorm, etc., have caused a reduction in food grain production in India over the years. The country has witnessed an increase in climate variability, both spatially and temporally. The number of days with no rainfall (dry spells) and heavy rainfall is increasing (Singh et al. 2014), and a well-distributed monsoon season is very rare in the recent past. A recent study on the sensitivity of grain yield to historical climate variability in India concluded that rice is more sensitive to climate variation, compared to alternate grains, viz. maize, sorghum, finger millet and pearl millet (Davis et al. 2019).

A timely agromet advisory can reduce crop loss to a greater extent. For example, an advisory to harvest the crops which attained physiological maturity on account of forecasted rainfall can save the entire produce. Similarly, farmers can make appropriate crop selection based on the overall monsoon forecast for a particular season. This may avoid crop loss in the event of agricultural drought by opting for hardy crops with low crop water requirements. Another aspect is the dissemination of scientific cultivation practices in tune with forecasted weather to the farmers through AAS. This will ensure that farmers will take up all agronomic, crop protection activities on time, which will boost crop production. Correct AAS can also avoid wastage of input materials (fertilizer, plant protection chemicals, planting material) and save labour cost in the event of heavy rainfall.

9.5.6 Economic Impact Assessment of AAS

The eventual aim of AAS is to increase the profit of farmers by providing timely information about anticipated in-season weather so that optimum utilization of resources is ensured and wastage/losses are reduced. To assess the viability of any project, impact assessment is the most important tool. The economic impact of the medium-range weather forecast issued by NCMRWF was assessed in a pilot study in 15 out of 127 agromet field units (AMFUs) under which AAS was issued (Maini and Rathore 2011). They had selected six cropping seasons (three *kharif* [southwest monsoon] and three *rabi* [winter] seasons) that were chosen for the study during 2003–2007. The crops selected for the study included food grains, cash crops, oilseeds, fruits and vegetables. The sample size of the study consisted of 80 farmers, out of which 40 were AAS following and 40 were AAS non-following. The main objective of the study was to quantify the change in yield and net return due to AAS. The major finding from the study was farmers who followed AAS fetched 10–15% higher yield and 2–5% reduction in the cost of cultivation compared to the AAS non-followers.

Table 9.2 Benefit/cost ratio of micro-level AAS adopted and non-adopted farmers under AICRPAM-NICRA project (Adopted from NICRA 2018)

Region	Village	Crop	B:C ratio	
			AAS farmer	Non-AAS farmer
2016–2017				
Akola, Maharashtra	Yelgaon	Soybean	2.17	1.76
Bangalore, Karnataka	Patrenahalli and Nayanahalli	Grape	9.21	5.45
Palampur, Himachal Pradesh	Dhamrol	Maize	2.74	0.73
Parbhani, Maharashtra	Shekta	Cotton	2.26	1.88
Udaipur, Rajasthan	Nakli	Maize	1.93	1.28
Anantapur, Andhra Pradesh	Yagantipalle	Pigeon pea	3.1	2.48
Kovilpatti, Tamil Nadu	Allikundam	Okra	2.41	1.98
2017–2018				
Ramanathapuram, Tamil Nadu	Malangudi	Chilli	1.99	1.46
Kanpur Dehat, Uttar Pradesh	Baghpur	Rice	1.71	1.50
Sirsa, Haryana	Farwain Kalan	Cotton	1.71	1.33

The economic impact assessment of MAAS by AICRPAM under NICRA (National Innovations on Climate Resilient Agriculture) project was also carried out. The results are presented in Table 9.2 (NICRA 2018).

The higher benefit obtained by farmers who adopted MAAS can be mainly attributed to the implementation of event-wise agromet advisories provided by the AICRPAM-NICRA centres based on the accurate weather forecast. This has helped them to save labour cost, input costs such as fertilizer, irrigation water, plant protection chemicals, and timely harvest (which saved the entire produce from heavy rainfall). Cases of losses were also reported due to erroneous weather forecasts.

Another study is conducted by Vashisth et al. (2013) at villages near national capital region of India (New Delhi) regarding impact of AAS among users and nonusers. The results revealed that users could reduce input cost to the tune of 9.6% in carrot, 7% in rice and 6% in wheat. This resulted in increase of net profit by 4%, 3% and 0.9% in rice, carrot and wheat, respectively, for those who followed AAS when compared to the nonusers of AAS. The authors had attributed the increase in profit of AAS users to the accurate weather forecast, thereby saving inputs like irrigation water, plant protection chemicals, labour cost, etc.

A study conducted by the National Centre for Agricultural Economics and Policy Research (NCAP) concluded that farmers received 10–25% economic benefit due to the adoption of AAS. Another assessment by the National Council of Applied Economic Research (NCAER) revealed that economic benefit of AAS as ₹ 50,000 crores per year in 2009, when extrapolated, increases to ₹ 211,000 crores, if all the farmers of the country are using AAS in their decision-making during crop season (Chattopadhyay and Chandras 2018). NCAER had again conducted another study and reported that only 24% of the total farmers of India had access to AAS from SMS services, which accounted for an economic profit of ₹ 42,000 crores (NCAER 2015). They had also reported that if AAS is utilized by all farming households, it has the potential to create economic benefit up to ₹ 3.3 lakh crores on the 22 principal crops.

9.5.7 Constraints Involved with AAS

The main bottleneck involved in the dissemination of AAS is the linkage problems between institutional actors due to shortcomings in quality weather data availability, limited share of data, coordination and communication. Accuracy of the medium-range forecast (lead time of 5 days) is the primary determinant for the success of AAS as the accuracy of seasonal forecasts (lead time of 1 month) is currently doubtful, meant for policymakers and not for direct use of farmers. Another reason is the availability of dedicated human resource for preparation and dissemination of advisories. This is the reason why advisories always usefully combine weather and agriculture data (Gopalakrishnan and Subramanian 2020). Attempt to involve farmers for improvement of the content in AAS is also lacking. This causes the development of AAS without knowing the need and priorities of the farmers. Other reasons are inappropriate research programmes and inadequate use of information and communication technologies (ICTs), which ultimately results in a low rate of adoption of AAS (Singh et al. 2019a). The contents of AAS should be very location-specific, targeting the specific crops grown, livestock, market facility, etc. Remote areas have a disadvantage in the case of access to timely information due to poor communication infrastructure and services. The twenty-first century has witnessed a giant leap in information and communication technologies (ICTs), and information dissemination is becoming less costly day by day and providing AAS as text messages have become the most effective way of dissemination to a large number of farmers across the country. But a fact that remains is that there are many extremely poor and marginalized farmers in India who cannot afford a mobile phone. So, the AAS providers should devise strategies to include them also in this information dissemination cycle.

9.6 Future Challenges

- The increase in extreme weather events such as heavy precipitation, cloud burst and hailstorms is causing widespread crop damage across many locations of India. Although IMD is issuing 'nowcast bulletins' and 'special weather bulletins', the forecast accuracy needs to be improved.

- The advent of ICTs for the dissemination of AAS has enabled the service to provide agencies to cover masses of farmers. But, still, there are many rural farmers who don't have mobile phones and access to the Internet.
- It has been observed that the same farmer is receiving contrasting AAS for the same time period from different AAS providing agencies, which is creating confusion for them.
- Although IMD is issuing block-level weather forecast, AAS based on the block-level weather forecast is not upscaled to the national level.
- The ultimate aim of AAS will be customized advisories at the farmer level. That will require weather forecast with farm-level spatial resolution, farmer-level crop and soil information and huge computing skills for the automation of AAS.

9.7 Opportunities Ahead and Policy Implications

- Let's hope that the spatial resolution, as well as the accuracy of weather forecasts, will be improved with the advancement in computing skills over the years as microscale advisories are the need of the hour.
- Creating a dedicated platform for quality weather- and climate-related data sourced from both public and private institutions/agencies. The Ministry of Earth Sciences and the Ministry of Agriculture and Farmers' Welfare Government of India need to coordinate for the same.
- Implementation of AAS using block-level weather forecast for regions under varied agro-ecologies will require automation for development and dissemination processes with minimal human interference. Though IMD has already taken initiative in this aspect, it needs to strengthen collaboration with other partners.
- Capacity enhancement in agrometeorology, an interdisciplinary subject, needs to be expanded at all the agricultural universities. The subject, agricultural meteorology, should be compulsorily included under postgraduate programme. There also exists an excellent scope for the development of 'extension agrometeorology' as a full-fledged way for awareness generation, advisory dissemination and feedback collection from farmers regarding AAS.
- The use of ICTs for the dissemination of AAS may further be explored for increasing the number of farmers covered. Exploring the use of ICTs and involvement of other stakeholders (public and private institutions, farmer organizations, farmers) should be taken up for overall improvement of the AAS.
- Farmers' traditional knowledge and practical experiences should be documented and embedded in preparation of AAS.

9.8 Conclusion

The importance of agromet advisory services is increasing in this era of climate change. The increasing trend of extreme weather events is posing a great threat to sustainable crop production in the world. This chapter has discussed in detail about the weather-related risks in agriculture, what kind of information is needed by farmers, the existing system of AAS development and dissemination in India, their economic impact, constraints and future challenges. The twenty-first century has witnessed a significant improvement in computing power and hence the dissemination of AAS using ICTs. Despite these advances, the dissemination of accurate AAS at the right time to the farmers is lacking if we consider the country as a whole. There has been considerable improvement in the spatial resolution of the weather forecast, which is yet to be exploited in the form of block-level AAS. Climate variability and climate change are adding more threats and challenges to the existing system of development and dissemination of AAS. Automation of AAS, further exploration of extension agrometeorology, etc., are suggested as opportunities to be explored in the near future.

References

Bal SK, Minhas PS (2017) Atmospheric stressors: challenges and coping strategies. In: Minhas PS et al (eds) Abiotic stress management for resilient agriculture. Springer, Singapore, pp 9–50. https://doi.org/10.1007/978-981-10-5744-1_2

Bal SK, Saha S, Fand BB, Singh NP, Rane J, Minhas PS (2014) Hailstorms: causes, damage and post-hail management in agriculture. Technical bulletin no. 5, ICAR-National Institute of abiotic stress management, Malegaon, Baramati 413 115, Pune, Maharashtra, India. p. 44

Bal SK, Minhas PS, Singh Y, Kumar M, Patel DP, Rane J, Suresh Kumar P, Ratnakumar P, Choudhury BU, Singh NP (2017) Coping with hailstorm in vulnerable Deccan plateau region of India: technological interventions for crop recovery. Curr Sci 113(10):2021–2027

Balasubramanian TN, Jagannathan R, Maragatham N, Sathyamoorthi K, Nagarajan R (2014) Generation of weather windows to develop agro advisories for Tamil Nadu under automated weather forecast system. J Agrometeorol 16(1):60–68

Balasubramanian TN, Jagannathan R, Maragatham N, Sathyamoorthi K, Nagarajan R, Vanangamudi M, Sathyamoorthy NK, Poonguzhali S, Sakthivel P, Satarji S, Prakash PA, Kumar PR, Hameed JA (2016) Designing agromet advisories for selected weather windows under automated weather based advisory system in Tamil Nadu – a case study. J Agrometeorol 18(1):34–40

Chattopadhyay N, Chandras S (2018) Agrometeorological advisory services for sustainable development in Indian agriculture. Biodivers Int J 2(1):13–18. https://doi.org/10.15406/bij.2018.02.00036

Chattopadhyay N, Vyas SS, Bhattacharya BK, Chandras S (2016) Evaluating the potential of rainfall product from Indian geostationary satellite for operational agromet advisory services in India. J Agrometeorol 18(1):29–33

Davis KF, Chhatre A, Rao ND, Singh D, DeFries R (2019) Sensitivity of grain yields to historical climate variability in India. Environ Res Lett 14:064013. https://doi.org/10.1088/1748-9326/ab22db

De Boef W, Amanor K, Wellard K, Bebbington A (1993) Cultivating knowledge, genetic diversity, farmer experimentation and crop research. Intermediate Technology Publications, London, UK

FAO (2019) Handbook on climate information for farming communities – what farmers need and what is available. FAO, Rome, p 184

Gopalakrishnan T, Subramanian K (2020) Agrometeorological advisory Services in India: an assessment. Centre for Science and Environment, New Delhi, India

Hansen JW (2002) Realizing the potential benefits of climate prediction to agriculture: issues, approaches, challenges. Agric Syst 74:309–330
Madhavan M, Rengalakshmi R (2015) Seasonal climate information for ensuring agricultural sustainability and food security of small holder Rainfed farmers: experience from India. M.S. Swaminathan Research Foundation, Chennai, India. http://docplayer.net/29188604-Seasonal-climate-information-for-ensuring-agricultural-sustainability-and-food-security-of-small-holder-rainfed-farmers-experience-from-india.html. Accessed 20 June 2020
Maini P, Rathore LS (2011) Economic impact assessment of the Agrometeorological advisory Service of India. Curr Sci 101(10):1296–1310
National Innovations on Climate Resilient Agriculture—NICRA (2018) Research highlights 2016-18. ICAR-Central Research Institute for Dryland Agriculture, Hyderabad. p. 128
NCAER (2015) National Council of Applied Economic Research, Report on 'Economic benefits of dynamic weather and ocean information and advisory services in India and cost prizing of customized products and services of ESSO-NCMRWF & ESSO-INCOIS'
Nesheim I, Barkved L, Bharti N (2017) What is the role of agro-met information services in farmer decision-making? Uptake and decision-making context among farmers within three case study villages in Maharashtra, India. Agriculture 7(70):1–16. https://doi.org/10.3390/agriculture7080070
Rana RS, Kaundal M, Kalia V, Pathania R, Katoch A (2018) Indigenous traditional knowledge in agricultural activities Vis-a-Vis climate change in North Western Himalayas. J Agrometeorol 20:37–43
Rao VUM, Bapuji Rao B (2013) Role of agromet advisories in climate risk management. Ann Agric Res New Ser 34(1):15–25
Rathore LS, Roy Bhomik SK, Chattopadhyay N (2011) Integrated agrometeorological advisory services in India. In: Attri SD, Rathore LS, Sivakumar MVK, Dash SK. Challenges and opportunities in agrometeorology. Springer, Cham, pp. 195-205
Sarath Chandran MA, Subba Rao AVM, Sandeep VM, Pramod VP, Pani P, Rao VUM, VishaKumari V, Srinivasa Rao C (2017) Indian summer heat wave of 2015: a biometeorological analysis using half hourly automatic weather station data with special reference to Andhra Pradesh. Int J Biometeorol 61(6):1063–1072. https://doi.org/10.1007/s00484-016-1286-9
Singh D, Tsiang M, Rajaratnam B, Diffenbaugh NS (2014) Observed changes in extreme wet and dry spells during the south Asian summer monsoon season. Nat Clim Chang 4:456–461
Singh M, Ghanghas BS, Sharma V, Sharma BC (2019a) Minimize weather risk in agricultural planning and management through agromet advisory services in rural areas. In: Ahamad S, Sharma JP (eds) Transformation of Indian agriculture through innovative technologies. Daya Publishing House, Delhi, India, pp 11–21
Singh NP, Singh S, Anand B, Bal SK (2019b) Climate vulnerability assessment in semi-arid and arid region of Rajasthan, India: an enquiry into the disadvantaged districts. J Agrometeorol 21 (2):197–202
Stigter CJ (2011) Agrometeorological services: reaching all farmers with operational information products in new educational commitments; CAgM report 104. Geneva, Switzerland, World Meteorological Organization (WMO)
Stone RC, Meinke H (2006) Weather, climate, and farmers: an overview. Meteorol Appl 13:7–20
Vashisht BB, Jalota SK (2018) Impact of temperature variability and management interventions on productivity of wheat. J Agrometeorol 20(1):11–15
Vashisth A, Singh R, Das DK, Baloda R (2013) Weather based Agromet advisories for enhancing the production and income of the farmers under changing climate scenario. Int J Agric Food Sci Tech 4(9):847–850

Venkatasubramanian K, Tall A, Hansen J, Aggarwal P (2014) Assessment of India's Agro-Meteorological Advisory Service from a Farmer Perspective. CGIAR Research Program on Climate Change, Agriculture and Food Security (CCAFS Working Paper No. 54). Copenhagen, Available online: https://cgspace.cgiar.org/handle/10568/43733?show=full. Accessed 10 June 2020

Vijaya Kumar P, Subba Rao AVM, Sarath Chandran MA, Venkatesh H, Rao VUM, Srinivasa Rao C (2017) Micro-level Agromet advisory services using block level weather forecast – a new concept based approach. Curr Sci 112:227–228

Farmers-Led Adaptation Strategies to Climate Change for Sustaining Coastal Agriculture in Sundarbans Region of India

10

Suchandra Dutta, Sanchita Garai, Subhasis Mandal, and Sanjit Maiti

Abstract

Climate change-accelerated sea-level rise is the cause of devastating impact on the Sundarbans, the *World Heritage Site*, which is a deltaic complex and is highly prone to extreme climatic events-led disaster. The erratic rainfall pattern and unpredictable cyclones had resulted in deterioration of soil and water quality due to the inundation of saltwater or occurrence of brackish groundwater level at shallow depth, lack of irrigation water during the dry season, and drainage congestion during monsoon months. Majority of the local population is dependent on agriculture and its allied activities. For the inhabitants of Sundarbans, changing climate has become part of their everyday struggle for survival. They adjusted to the expected climatic stimuli, which moderate the catastrophic effect of changing climate. This chapter focuses on farmers-led adaptation strategies to climate change in the Indian Sundarbans region. Therefore, this chapter discusses the adaptation techniques that farmers follow in the fields of crop farming, livestock rearing, and fisheries. Mixed farming comprised of the crop, livestock, and fish component is the predominant farming system practiced by the farmers of Indian Sundarbans, and about 11 adaptation strategies in mixed farming are followed by them. It turned to be apparent that challenging issues, such as changing climate and its associated uncertainties, need adaptation strategies and an emphasis on resilience at the farm level. This chapter, therefore, concludes that

S. Dutta · S. Garai · S. Maiti (✉)
ICAR-National Dairy Research Institute, Karnal, Haryana, India
e-mail: Sanjit.Maiti@icar.gov.in

S. Mandal
Central Soil Salinity Research Institute (CSSRI), Regional Research Station, Canning, West Bengal, India

V. K. Hebsale Mallappa, M. Shirur (eds.), *Climate Change and Resilient Food Systems*, https://doi.org/10.1007/978-981-33-4538-6_10

to combat changing climatic conditions, the farming community needs to adopt the integrated farming system (IFS) in which by-product of one system is used as an input for another system. Thus, the promotion of integrated farming system (IFS) model should be encouraged to increase the adaptive capacity of the farmers of the Indian Sundarbans region to cope up with the changing climatic scenario.

Keywords

Climate change · Farmers-led adaptation strategies · Indian Sundarbans · Agriculture

10.1 Introduction

Climate change, caused by rising temperatures, thermal expansion of ocean water, and increased precipitation, is likely to trigger the rise in sea level by 50 cm by 2050. This means increased coastal erosion and flooding and greater tidal heights amid cyclonic storms (Hazra 2012). The Sundarbans, *World Heritage Site* as declared by UNESCO in 1984 which falls under the complex-diverse-risk-prone (CDR) agroecosystem, is a deltaic complex (Mahadevia and Vikas 2012). Indian Sundarbans is renowned for its diversity in species of plants and animals, detritus ecosystem, complicated shoreline, countless islands, crossover tributaries, and residents of a quite conventional population and is highly prone to extreme climatic events-led disaster (Singh et al. 2012). The world's largest delta is facing subsidence due to tectonic movements that results in a tiltation toward the east. Coastal regions are vulnerable to seasonal ocean currents, tides, waves, winds, and cyclones. Heavy monsoonal rain and river flow almost partially inundate this region most of the year. The reduction in mangrove area has reduced sedimentation and accelerated erosion along with removing the primary protection against floodwaters and high waves (Roy 2010). Climate change causes inexorably rise of sea level that disappeared many islands with buffeted cyclones and changing rainfall pattern which results in deterioration of soil and water quality due to inundation of saltwater or occurrence of brackish groundwater level at shallow depth, lack of irrigation water during the dry season, and drainage congestion during monsoon months making the lives of farmers more challenging and improvement of the area more extravagant (Mimura 2013). Furthermore, there have been significant disruptions in the hydrological conditions and alterations in fishery practices that bring devastating results for fisherfolk (Mahadevia and Vikas 2012).

Adaptation to climate change is characterized as modification in natural or anthropogenic systems regarding the impacts of real or anticipated climatic stimuli that mitigate possible harm or take full advantage of desirable possibilities (Forino et al. 2014). Adaptation is commonly defined as the reaction of individuals, groups, and governments to changing climate or other intervention that is used to alleviate their susceptibility to detrimental effects (Bradshaw et al. 2004). It denotes any change in procedures, methods, and mechanisms to respond to the real or assumed dangers of climate change, as well as changes in social and institutional frameworks and technical alternatives (Howden et al. 2007). Agrarian adaptation to changing climate is a dynamic, multifaceted, and multiphase procedure that involves a variety of forms. A vast range of agriculture sector adaptations to climate fluctuations and transition are accessible (Bryant et al. 2000; Smit and Skinner 2002; Bryan et al. 2009). Adaptations that are usually applicable for addressing climate-linked hazards are likely to be recognized; few make the argument that it is not so straightforward to recognize adaptations that are appropriate for mitigating innumerable threats, namely, recessions in commodity markets and adjustments in the policy and aid of government (Dolan et al. 2001; Bradshaw et al. 2004). Farmers also made farm-related improvements, such as the introduction of improved crop varieties, expanding the use of fertilizers, investing in better land management practices, and modifying the scheduling of farming activities (Wood et al. 2014). Climate change has a negative impact on the productive and reproductive performance of livestock and milk production lactation length. Occurrence of livestock diseases and the parasitic infestation has been increased with the increased rate of mortality, decreased trend of feed and fodder resources and fresh drinking water, and decreased length and intensity of estrous period as well as conception rate. These lead to the change in livestock and herd composition (large ruminants vs. small ruminants) during adverse climatic conditions (Rötter and Geijn 1999). To reduce the threat of saline water penetration, farmers are taking certain corrective action, such as increasing the height of pond dykes, repairing and strengthening dykes, planting on dykes, removal of saline water and addition of fresh or rainwater, incorporation of chemicals or lime or dung in the pond, and inclusion of tree branches in ponds for hideaways (Chand et al. 2012a).

Sundarbans has a population of humans that is financially, academically, and socially marginalized and occupies a region which is largely unreachable owing to inferior amenities and infrastructure. As per the Government of India's 2011 census, approximately 4.37 million people are residing in and around the Sundarbans delta. The significant majority of the local population in Indian Sundarbans is dependent on agriculture and its allied activities due to the absence of any industrial sector. Consequently, due to inadequate irrigation facilities and fundamental framework, agricultural activities are rudimentary and reliant on periodic rainfall, reducing crop yield that led to an unpredictable means of livelihood. Both tenant farmers and landowners enhance their earnings by harnessing mangrove forests, catching fish, aggregating, and cultivating tiger prawn seeds in saltwater, causing extensive destruction to coastal ecosystems. Hence, Sundarbans, a coastal and poorly developed area, is, therefore, one of the most vulnerable survivors of climate change in India. Changing climate has become a part of people's everyday struggle for survival. While global climate change mitigation agreements have remained ambiguous, sea-level rise inevitably, cyclones have pummeled, and rainfall patterns have continued to worsen, making life harder and developing the region more costly. The most crucial understanding that is lacking is how the farmers cope up with the vulnerable ecosystem that has an impact on their lives and livelihoods. The small and marginal farmers of this region are changing themselves by adopting several mechanisms to cope up with the changing climatic scenario. Hence, it is need of the hour to find out these adaptation strategies followed by the farming community of the Indian Sundarbans to combat with the detrimental effect of changing climate for maintaining sustainable productivity of their farm.

10.2 The Indian Sundarbans: A World Heritage Site

In the map, the Sundarbans delta (Fig. 10.1) is spreading across West Bengal state of India and the nearby country Bangladesh (between $21^0$32'N–$23^0$31'N latitude and $88^0$10′E–$92^0$15′E longitude). The unified Sundarbans is about 26,000 sq. km and one-third of it falls in India. The origin of this unique ecosystem is a fascinating subject of study which baffles scientists, historians, and naturalists. In 1893, a very precise theory stated that the deltas situated between the river Hooghly on the west and Meghna in the east, including the Sundarbans, have been formed by debris deposits brought down by the mighty rivers, Ganges, Brahmaputra, and their tributaries. Silt deposition remains ongoing and is very active as demonstrated by the recent emergence of the new Moore Island. Despite the cause of the origin of this deltaic alluvium, it has been estimated that the present-day Sundarbans has been formed about 4000 years ago. The river system of Sundarbans is quite complex. The origin of the major rivers that traverse through Sundarbans is *Matla, Bidyadhari, Saptamukhi, Thakuran, Gosaba, and Raimangal.* Other mentionable rivers of Sundarbans are *Muri Ganga, Ichamati, Piyali, Kalindi, Hogol, Netidhopani, Mridangabhanga, Gomor, Chulkati, Durgadoani, Duttargang, Baro Sahebkhali, Raidighi, Pitchkhali, Maya, Hatania Doania, Chaltaboni*, and *Dhulibhasani.* The soil of Sundarbans can be classified into three types, such as fine silty clay in the northernmost part, peaty deposits in the middle, and swampy areas close to the sea face along with the sandy clay and dunes. Based on salinity, the soil can be divided into (1) saline soil, where there is intrusion or flushing of brackish water, and (2) saline-alkali soil, when the sodium salts flow over a fresh lowland enhanced with alkali and nitrogen ions (Mandal et al. 2013a).

Located in a special bioclimatic zone within a traditional geographical position in the Bay of Bengal coastal region, it is a symbol of the ancient history of mythological and historical events. Offering spectacular scenic scenery and natural wealth, it is known globally for its mangrove forests, species diversity of plant, and animal both on land and in the oceans. The Sundarbans is of fundamental significance to endangered species worldwide including the Royal Bengal Tiger, Ganges and Irrawaddy dolphins, estuarine crocodiles, and the critically endangered native river terrapin (*Batagur baska*). It is the world's only mangrove habitat for the species *Panthera tigris tigris.* The Sundarbans is the habitats of diversified marine, terrestrial, and amphibian species.

In 1987, Sundarbans was designated as a *World Heritage Site* by UNESCO. The World Heritage property consists of three wildlife sanctuaries which constitute the core breeding area of several endangered wildlife species. The property also contains areas of exceptional natural beauty, ethnobotanical significance, special marine wildlife significance, rivers, creeks, beaches, swamps, estuaries, mudflats, and tidal flats. Property boundaries cover all the main forms of mangrove habitat, areas with

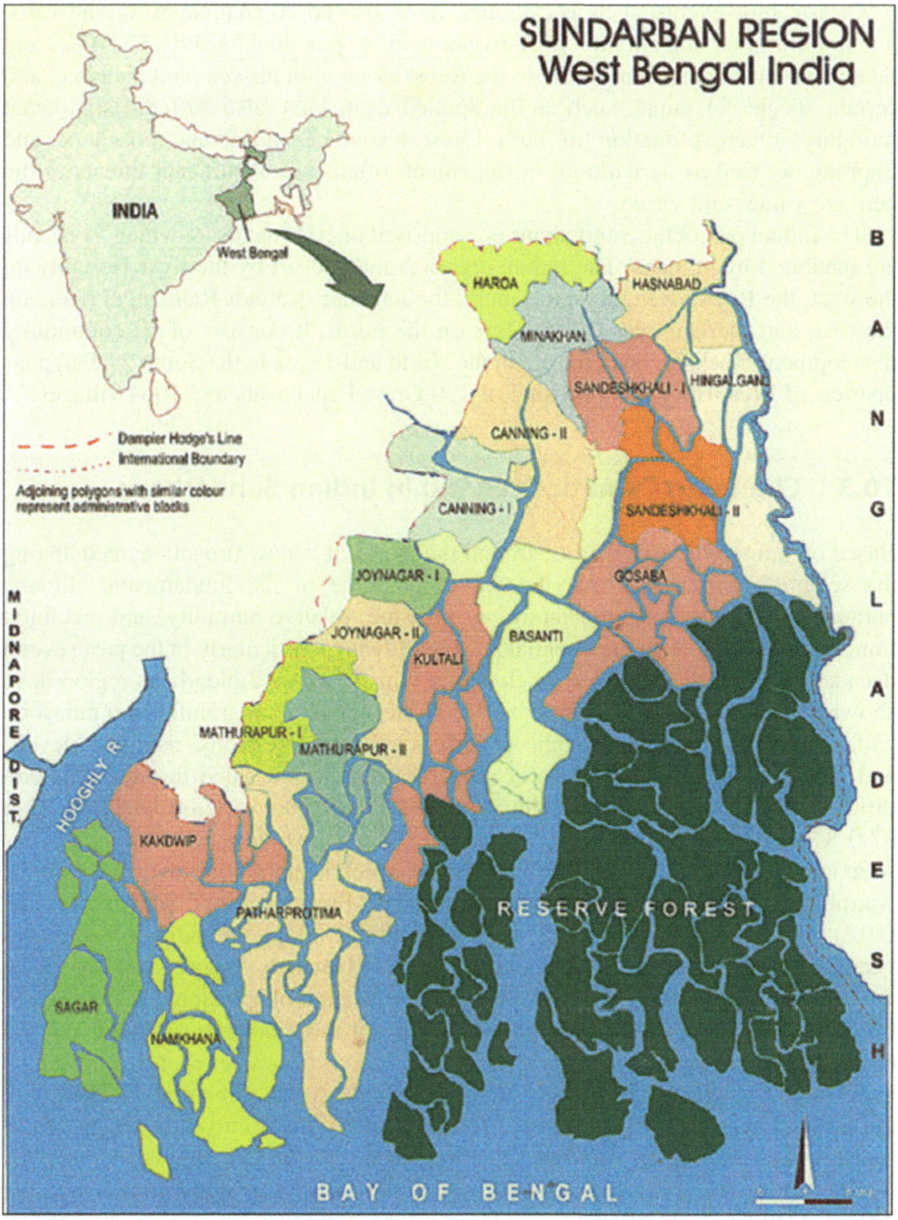

Fig. 10.1 Indian Sundarbans comprising 19 community development blocks (Chowdhury et al. 2016)

high floral and faunal resources, and important bird habitats. The property's integrity is further reinforced by adjacent terrestrial and marine buffer zones, which are not part of the designated property.

Catastrophic events, such as cyclones, have also posed challenges to land value due to infiltration and siltation of saltwater causing potential hazards. Cyclones and tidal waves have done some harm to the forest along with the sea-land interface, and certain species of fauna, such as the spotted deer, have also suffered significant mortality. Overexploitation of both forest resources and fauna, poachers, and trapping, as well as agricultural infringement, often pose significant threats to the heritage values and virtue.

The Indian part of the Sundarbans is comprised of 102 islands, of which 54 islands are inhabited by humans. The Indian region is marked off by the river Hooghly on the west, the Bay of Bengal on the south, the Ichamati-Kalindi-Raimangal rivers on the east, and the Dampier-Hodges line on the north. It consists of 19 community development blocks, of which 6 are in the North and 13 are in the South 24-Parganas districts of West Bengal with a total of 190 Gram Panchayats and 1064 villages.

10.3 Changing Climatic Scenario in Indian Sundarbans

Based on numerous studies from around the world, it is now broadly agreed among the scientific community that the values of some of the fundamental climatic parameters, namely, air temperature, air pressure, relative humidity, and precipitation, are increasing at an exponential pace worldwide, particularly in the past several decades (IPCC 2013). Studies on changing climate in the Sundarbans region have shown variation in air temperature, surface water temperature, rainfall and monsoon pattern, salinity regimen, cyclonic disturbances and depressions, rising sea levels, and erosion and accretion (Raha et al. 2012). Studies revealed that the temporal differences of air temperature in the Sundarbans region are varied from 11.96 °C to 37.0 °C (Pitchaikani et al. 2017; Mitra 2019). The temperature of surface water has been increasing at a rate of 0.5 °C per decade which is eight times the rate of global warming which is at a rate of 0.06 °C per decade (Mitra et al. 2009; Sundaresan et al. 2013). A rise in sea level at a rate of 2.5 mm annually along the Indian coastline has been observed from the 1950s (Udayakumar 2014) which is closely related to the rise in sea surface temperature that favors the formation of a cyclone. Experience of tropical storm and cyclone is an annual feature of the Sundarbans region. Ericson et al. (2006) have approximately calculated that the rise in sea level of Bay of Bengal is the world's highest at more than 10 mm annually. At the same time, UNESCO in 2006 has forecasted 45 cm sea-level rise by the end of the twenty-first century. In its fourth assessment report, the Intergovernmental Panel on Climate Change (IPCC) reported clear observation-based testimony of climatic change in the coastal area and provided evidence of elevated ocean temperature, precipitation changes, subsequent upstream river drainage, and increasing sea level. This will usually cause higher coastal flooding and higher salinity (Parry et al. 2007). According to some estimates, the present-day sea-level rise in coastal areas of Sundarbans is 6–8 mm per year (Khan et al. 2008).

The delta people endure a long, intense, and hot summer, while the winters are getting shorter, warmer, and drier (Mandal et al. 2019). Disturbances such as depressions and cyclonic storms emerged in the Bay of Bengal during 120 years

(1891–2010) at a rate of 10.79 annually. However, the overall number of disturbances in the last 40 years (1971–2010) has decreased, but the frequency of extreme storms and intensity increased noticeably (Chand et al. 2012b). The northern part of Bay of Bengal has recorded four cyclones, namely, *Sidr, Nargis, Bijli,* and *Aila*, during the last part of the last decade (2006–2009) (Roy 2010; Nautiyal et al. 2016). A study of cyclone frequencies on India's east and west coasts between 1891 and 2000 revealed that almost 358 cyclones (of which 103 were severe) impacted the East Coast. Over the same time, the West Coast was struck by 48 tropical cyclones, of which 24 were extreme cyclonic disturbances. More than 58% of the cyclones that form in the Bay of Bengal approach and cross the East Coast in October and November (Burman et al. 2013). Along the East Coast, tidal action during both the southwest and northeast monsoons seems to be noteworthy. Exaggerated wave incidents during strong tropical cyclones are common during northeast monsoon in the Bay of Bengal (Kumar et al. 2006). The entire coastal region of India is climatically vulnerable and exposed to Nor'wester and cyclonic storms because of its proximity to the sea. The East Coast is more prone to the cyclone and other climatic disturbances compared to the West Coast (Patwardhan et al. 2003). It is predicted that the intensity of tropical cyclones would increase by 15%, which would drastically pose a serious threat to people residing in cyclone-prone coastal areas of India, notably the Indian Sundarbans region (Rishi and Mudaliar 2014).

Adaptation includes modifications in the social-ecological structures in response to climate change's real and anticipated impacts. Adaptation plans and activities may vary from short-term mitigation to longer-term, broader changes aimed at achieving more than just climate change priorities, which may or may not be effective in moderating damage or leveraging favorable opportunities (Moser and Ekstrom 2010; Leal et al. 2018). Mixed farming is the predominant farming system in the Indian Sundarbans region, and major components of this farming system are crops, livestock, and fisheries.

10.4 Present Agricultural Scenario of the Indian Sundarbans Region

India's coastal area is traditionally backward and marginalized by farmers with low agricultural productive output and weak livelihood stability. The coastal region's ecosystem is highly precarious and susceptible to destruction by anthropogenic practices. Small, marginal, and landless farmers dominate the farming community in this region, the majority of whom belong to backward communities. The agricultural production in coastal areas is outdated with a cluster of problems like degeneration of soil and water quality due to inundation of saltwater or the presence at a shallow depth of a brackish groundwater table, lack of irrigation water during the dry season, drainage congestion during monsoon months, weather adversities, etc. By adopting various adaptation strategies, the agriculture of this region has experienced certain transformations.

Agriculture in coastal regions was dependent on rains which mostly occur only in monsoon months, and the cropping pattern was completely mono-cropped with conventional rice varieties in *Kharif* that are of long duration and lower yield. After Kharif, the land remains nearly fallow throughout the year, leading to high poverty and unemployment among rural farmers. The coastal areas lagged far behind many inland areas in terms of agricultural production and the farmers' livelihood protection (Danda and Rahman 2019). The vast resource potentials of the coastal region are extracted by the farming community by adopting various adaptation strategies like land shaping, crop diversification, liming, etc., to enhance the agricultural productivity of coastal lands.

In Sundarbans, the essence of farming is different as fragments of farmland are not intertwined, rendering equal water distribution a challenge. The region's lack of sweet water represents a significant irrigation problem. Thus, most land cultivation remains mono-cropped because all farmers can rely solely on the monsoon season. The farming community thus compels to grow more food by adopting various adaptation strategies leading to advancement by fulfilling the needs of expanding population.

Livestock and poultry rearing in Sundarbans were mainly tended by women. Most of the cattle were indigenous nondescript. Black Bengal goat and Garole sheep being the native of West Bengal were found in the region. Problems faced by the farmers on livestock and poultry are reduced productivity of their livestock with an increased rate of mortality, reduced growth rate, and highly susceptible to diseases which are the impact on livestock due to changing climate. Thus, to mitigate this condition, farmers started to adopt various adaptation strategies to combat this situation against climate vagaries.

While fishing continues throughout the season, production begins to increase at the beginning of the monsoon and reaches its peak in winter (November–January). Aquaculture (aquafarming) and marine fishing are the two main types of fishing activities in Sundarbans, which have considerable commercial value and require capital investment. Commercial and subsistence fisheries are now under danger. Overexploitation is one of the main factors behind declining fish populations. There is also a continuing dispute between fishing and forest security as fishing is not allowed in the protected Project Tiger zones in the Sundarbans' eastern region. Nevertheless, illicit fishing persists due to the constraints of livelihoods and poverty. To combat this scenario, brackish water aquaculture is a rapidly expanding farming activity. There are many areas in Indian Sundarbans particularly near the brackish water river and seacoast which are highly saline year-round and not good for growing crops. Considering the presence of clayey soil and tidal water in these areas, the farming community has adopted and modified brackish water aquaculture to enhance their livelihood. The adaptation strategies have paved the way for the development of agriculture and raised the socioeconomic condition of the farmers.

Further, we will discuss the adaptation strategies followed by the farmers of the Indian Sundarbans in their farming enterprises, namely, crop farming, livestock rearing, and fisheries to cope up with the changing climate. All the farmers of this region having high experienced toward climatic disaster follow integrated farming system (IFS) in which the by-product of one system is used as an input for another system, as the main adopting strategies to sustain their livelihood, and as a barrier against changing climate.

10.4.1 Adaptation Strategies Followed in Crop Farming

10.4.1.1 Land Shaping

Most of the coastal land is low-lying with around flat topography, resulting in deep waterlogging (about 50 cm) and drainage congestion following heavy rainfall during monsoon months. Heavy waterlogging and drainage stagnation are the main obstacles to crop cultivation in the coastal area. The increase in sea levels following global warming has a significant effect on coastal agriculture, resulting in land loss due to seawater intrusion into agricultural property. This leads to more land and water salinity, increased freshwater shortage, and increased drainage congestion issues in the coastal area. The current sea-level rise in coastal areas of Sundarbans is expected to be 6–8 mm per year, according to some estimates (Khan et al. 2008). It is also estimated that 10 cm rise in sea level (SLR) will inundate 15% of Sundarbans, 25 cm SLR will inundate 40%, 45 cm SLR will inundate 75%, 60 cm SLR will inundate 100%, and 100 cm SLR will wipe out the entire Sundarbans (Oppenheimer et al. 2019).

Land shaping (Fig. 10.2) is characterized as land surface modification primarily for rainwater harvesting for irrigation, reducing the impact of brackish groundwater

Fig. 10.2 Land shaping

at shallow depth on salinity buildup, reducing drainage congestion, and growing multiple and diversified crops all year round. In promoting the adaptation of agriculture to climate change in coastal areas, land shaping technology is likely to play an important role as it increases freshwater availability through rainwater harvesting in the proper reservoir developed on the field, enhances land drainage, and reduces soil salinity buildup. Land shaping increases carbon sequestration through increased production of biomass on fallow land resulting in better sequestration of carbon for better environmental quality.

Land shaping techniques create different types of land situations like farm pond, deep furrow and high ridge, and shallow furrow and medium ridge for fish cultivation and rainwater harvesting in the farm (Velmurugan et al. 2018). Reduced waterlogging and salinity of soil in the highlands provides scope for the cultivation of diverse multi-crops other than rice during *Kharif* and other seasons. This strategy accommodates cash crops (vegetables), field crop (paddy), and pisciculture in the same piece of land that was only fit for paddy-based mono-cropping earlier. So the farmers are assured against climatic vagaries leading to sudden flooding and continuous water stagnation. Farmers were growing traditional rice varieties like '*Dudheswar*', '*Matla*', '*Nonabokra*', '*Talmugur*', '*Lalgetu*', '*Sadagetu*', '*Hamilton*', '*Taalsaree*', and '*Gobindobhog*' which are saline and flood tolerant. With this method, farmers with limited land holdings will be able to maximize the income from multipurpose land use. Bandyopadhyay et al. (2009) reported that by following this technique, the soil salinity was much less as compared to the normal land. This technique aided in better water management and lesser methane emission. In Indian Sundarbans, land shaping strategies, especially farm pond, high ridge and furrow, and paddy-cum fish models, constitute distinctive technology for resolving primary constraints such as land degradation (salinity), drainage congestion, and freshwater paucity for irrigation and, in effect, have the potential to boost production, productivity, income, and employment (Mandal et al. 2013b).

A study by Burman et al. (2013) stated that major constraints for the adoption of land shaping techniques are "marginal landholdings that too divided into several parcels," "high initial investment," "presence of acid sulfate soil layer after certain depth at places," "distance from residential areas," "scarcity of labor availability in time," and "low marketable surplus" and hence "high marketing cost" or "lack of remunerative price," "high input prices," "poor input supply and output delivery," "difficult to reverse the land shaping to original land," "availability of quality crop and fish seed," and "lack of supervision by family members." They suggested that "community-based rainwater harvesting" and "common pool wasteland" may be motivated in this direction.

Although this adaptation strategy has been well implemented at the farm level, there is a lack of knowledge on wider watersheds or hydrological impacts at basin level, such as downstream rainwater availability, groundwater recharge, etc. Measuring hydrological impact on a larger scale is complex. The different conceptual models may help in quantifying the long-term implication of rainfall variability on the hydrological impact of different land shaping techniques both locally and watershed scale. There is indeed very few research on the continued management of rainwater harvesting structures, the time frame for managing silt buildup, and the reduction of harvesting potential over time. More comprehensive studies should be conducted to tackle these issues to ensure that land formation for sustainable agricultural production in the salt-affected coastal region is implemented on a large scale.

10.4.1.2 Change from Mono-Cropping System to Multi-Cropping System

All plants have similar properties in mono-cropping systems and use the same assets. The only paddy was grown in the *Kharif* season in the coastal region of Sundarbans due to waterlogged conditions. In addition to this, there was a shortage of freshwater for irrigation in winter and summer, along with salinity of the soil, which restricted the growth of vegetables. By the adoption of appropriate land shaping techniques, the problem of salinity, waterlogging, and land degradation were mitigated and also improved the productivity of coastal land. This led to enhanced multiple cropping systems (Fig. 10.3).

Fig. 10.3 Change from mono-cropping system to multi-cropping system

In regions with more salinity to soil, sweet potato, cotton, groundnut, and sunflower were grown which increased the cropping intensity and assured healthy economic return. During Rabi season, tomato, cauliflower, French bean, bottle gourd, bitter gourd, etc., were cultivated on the same land. The multiple and diverse crop cultivation including agriculture-aquaculture-livestock production in the region in place of mono-cropping has led to better nutritional security and food security to poor farming communities as well as better soil health. Multiple cropping systems employ plant interactions to improve the production of crop with lower water and nutrient inputs. The temporal and spatial mix of species selected for an association should, therefore, use separate resources or promote cooperative growth, and/or sowing densities and spatial arrangements could reduce the competition and minimize harmful environmental effects such as greenhouse gas emissions and nitrate leaching.

In the broadest sense, multiple cropping systems can control pests by preventing their growth, reproduction, or dispersal (Gaba et al. 2015). Multiple cropping minimizes soil erosion and the subsequent loss of nutrients (Dabney 1998). The vegetables grown all along the field increased biodiversity in a mono-cropped land. Previous studies stress that crop diversification can boost resilience by improving the potential to eradicate pest occurrences and lessen the transference of pathogen, as well as mitigating crop production from the outcome of increased climate variability and severe incidents (Lin 2011).

10.4.1.3 Ail Cultivation

It is described in terms of land embankment cultivation. Farmers developed and extended their ail (bund/embankment) that was primarily employed to delimit territory, soil and water protection, and effortless mobility to the field and started growing vegetables. Along with the upper portion of the embankment, farmers used both sides of it (Fig. 10.4). *Amaranthus*, ladies' finger, chilli, hybrid tomato, dolichos bean, and cucurbits (ridge gourd, snake gourd, bottle gourd, and pumpkin)

Fig. 10.4 Ail cultivation

are the main crops grown on the bunds. Many a time, two plants of different heights are grown concurrently by the farmers to optimize the nutrients, light, and moisture. Climate change adaptation to integrate the rising magnitude and frequency of heatwaves, floods, and cyclone is a huge challenge. Since the inception of the farming operation, farmers have adapted to their local climate. Studies display that the embankment cultivation helps to mitigate crop loss during continuous water stagnation in the main field during the flood (Basu et al. 2009). These activities are undertaken because they help to implement individuals and improve their farm business efficiency. Changes in crop timing and crop-mix shifts are major adaptation strategies observed that lead to changes in crop yields. According to Aisabokhae et al. (2011), simulations of crops show that changes to the planting date and crop mix will decrease the effects of climate change and suggest that this is the most important adaptation.

10.4.1.4 Management of Acid Sulfate Soil by Liming

Indian Sundarbans lies in the coastal and saline agro-climatic zone of West Bengal. The low-lying area suffers heavy drainage congestion throughout the rainy season, due to its topographical disadvantage, which leads to prolonged water stagnation. Acid sulfate soils are formed under waterlogged conditions. When such soils are washed, excavated, or exposed to the air by water infiltration, the sulfides react with oxygen to form sulfuric acid. Releasing this sulfuric acid again from soil can, in turn, introduce iron, aluminum, and other heavy metals (especially arsenic) into the soil. When activated in this way, acid and metals can produce several adverse effects such as vegetation destruction, contamination, and acidification of groundwater and surface water sources, destroying fish and other aquatic species. The application of lime and a higher dose of phosphatic fertilizers and green manures are extremely effective for the treatment of acid sulfate soil (Bandyopadhyay and Maji 1995, 1999; Bhowmick et al. 2020).

10.4.2 Adaptation Strategies Followed for Livestock Rearing

10.4.2.1 Shifting from Large Ruminants to Small Ruminants

The erratic rainfall pattern and unpredictable cyclones had resulted in a scarcity of feed and fodder due to saline water inundation in the grazing field. It was found that farmers preferred small ruminants over large ruminants. This is evident from their increasing trend of rearing of native Garole sheep and Black Bengal goat. Farmers were of the opinion that small ruminants can sustain more during fodder scarcity situation than large ruminants and have robust survival capabilities and produce in adverse circumstances. The mortality rate of small ruminants is low during climatic disasters and is highly prolific. Small ruminants are usually effective converters of forage feeds, whether they are farmed under temperate, arid, or semitropical conditions. Their greatest advantage compared to large ruminants is perhaps their low cost, small size, suitability for smallholdings, and, in many developing countries, their triple use for meat, milk, and fiber (Timon and Hanrahan 1986).

Temperature humidity index (THI) is always high (above 80) during the summer season due to high temperature and humidity. It was observed that the small ruminants are highly disease resistant and can survive in harsh climatic conditions with very little care. Kurukulasuriya and Rosenthal (2003) presented evidence that livestock diversification has been successful in combating climate change-associated diseases and pest outbreaks such as anthrax. Climate change directly affects the growth of livestock, the occurrence of diseases and mortality, rates of animal reproduction, and the quality of dairy products. Small ruminants were used as working capital by the resource-poor farm families in Indian Sundarbans.

From a study in Africa (Seo et al. 2009), it was clear that farmers switched to both livestock and crop diversification to combat the changing climate. Furthermore, farmers were found to increase dependence on livestock amid dry and hot conditions, shifting to sheep and goats as compared to cattle and chickens when temperature increased, and when precipitation increased, they preferred more goats and chickens than cattle and sheep. Henry et al. (2012) and Rowlinson (2008) showed that changing in breeding strategies would increase the resistance of livestock to heat stress and diseases, thereby enhancing the reproduction of livestock. Farmers perceived that after the destructive *Aila* (cyclone) that attacked West Bengal on the 25th of May 2009, it was found that there was a decreased performance in growth and lactation, a significant increase in mortality rate, and decreased proliferation of the large ruminants (Sejian et al. 2012). Maiti et al. (2014) also observed the shifting trend from large ruminants to small ruminants in coastal West Bengal. Hence, preferring small ruminants over large ruminants (Fig. 10.5) is a sustainable adaptation strategy as it strengthens resilience by reducing vulnerability and risk to a wider set of climatic catastrophe and is supported by diversified livelihoods in Indian Sundarbans.

10.4.2.2 Preferring Garole Sheep

Garole sheep is an indigenous breed of Indian Sundarbans. Garole, a highly fecund breed of sheep, can be characterized as resistant to foot rot disease which is very common due to seasonal fluctuation. The goat has natural physiological adaptability for sustaining in a saline environment. This variety can graze in knee-deep water and can adapt to hot humid weather conditions. Farmers preferred Garole sheep due to their quick multiplication rate, resilience to climate change, and easily convertible to cash (selling them for meat and wool purpose as and when required by the farm family) (Karim and Shinde 2007). Garole sheep are typically known for the standard quality of felt produced from their fleece (Banerjee 2009). Felt is classified in the nonwoven class as no thread enters the fabric composition. To manufacture felt or nonwoven goods, the unique property of wool fibers to create an irreversible structure by rubbing under certain conditions is used (Australian Felt Specialist Pty Ltd. 1999). Thus, farmers preferred Garole sheep cultivation (Fig. 10.6) with less management.

Fig. 10.5 Shifting from large ruminants to small ruminants

10.4.2.3 Providing Clean and Fresh Drinking Water Frequently

Unpredictable increases in temperature during the summer season resulted in increased body temperature and heat stress of the livestock. To overcome this stressed condition, farmers used to provide two to three times more water than normal to their animals for drinking (Fig. 10.7), so that their body temperature comes to a normal level and reduces their irritation. Many researchers are studying the response of rising temperatures to livestock water demand like Upadhyay et al. (2009) and Sirohi and Sirohi (2010) who concluded that a sufficient amount of clean, fresh, and cool water is necessary for reducing livestock heat stress.

Fig. 10.6 Preferring Garole sheep (Banerjee 2009)

Fig. 10.7 Providing clean and fresh drinking water frequently

10.4.2.4 Housing with Cross Ventilation for Keeping Livestock

Farmers perceived the need for cross ventilation of animal housing to eliminate the heat, odors, and moisture produced by livestock and to refill oxygen supply by taking in drier, cleaner air from outside. Sufficient exchange of air often eliminates gases such as ammonia (NH_4), hydrogen sulfide (H_2S), and methane (CH_4) that can affect animal health. Ventilation also provides air circulation that facilitates ventilation for confined animals and increases air quality.

For effective cross ventilation, many livestock rearers used electric fans in the shed of their herds. They also covered the four walls of the shed with a net so that insects' entry in the shed can be prevented and provision of air movement is ensured.

Fig. 10.8 Cattle housing with cross ventilation

The rise in temperatures between 2 and 3 °C across the region, along with the increase in humidity due to changing climate, is expected to exacerbate the heat stress in livestock emanating in decreased growth, reduced breeding rates, and milk production (Das 2017). Cross ventilation in animal shelters is highly beneficial to reduce the heat load.

Livestock is homoeothermic, which means that to remain healthy and efficient, they must sustain their body temperature within a fairly narrow range. The temperature humidity index (THI) is widely used to show the stress level of dairy cattle. THI less than or equal to 74 is normal, THI 75–78 is alert, THI 79–83 is danger, and THI value 84 and above is an emergency condition (Eigenberg et al. 2007). Cross ventilation helps to maintain proper THI of the animal in their shed. Cross ventilation removes excess heat, reduces excess perspiration of the animal due to extreme temperature rise, and removes microbes, dust, and gases with standard air circulation. Skuce et al. (2013) found that the influence of changing climate on livestock farming will be crucial, both directly via impact on animals generally and indirectly by increasing exposure to pests and pathogens. Cross ventilation in livestock housing (Fig. 10.8) plays an important role in keeping the animal environment comfortable. During hot weather, high air exchange levels are required to help extract heat from the cattle's body.

10.4.3 Adaptation Strategies Followed in Fisheries

10.4.3.1 Practicing Brackish Aquaculture

Throughout the year, farmers' lands near the brackish water river or seashore remain extremely saline and are not ideal for crop cultivation. There is the presence of clayey soil and a good supply of tidal water which is suitable for coastal aquaculture producing finfish and shellfish. Several farmers perform brackish water fishing with tiger shrimp (*Penaeus monodon*) along with brackish fishes like mullets, *Liza tade* (local name: golbhangon/bhangon) and *Mugil cephalus* (local name: aansbhangon), in an area of approximately 0.13–0.4 ha. Farmers shift from freshwater aquaculture to brackish water aquaculture because of saline water inundation which results in infringement of fish deposits and mass mortality, escape of existing fish stocks and fish diseases, the introduction of unwanted species, disruption of growth, and degradation of water quality resulting in harm to the ecosystem of the ponds (Chand et al. 2012a). Changes in monsoonal rain patterns (Goswami et al. 2006) and occurrences of extreme climatic events like flood and storminess, in general, forced the farmers to adapt brackish aquaculture.

10.4.3.2 Cultivation of Stress-Tolerant Monosex Tilapia

The area is prone to regular cyclones and storms, making it susceptible to brackish water entering from the nearby river through rupturing the embankment. It triggers mass mortality of current freshwater fish, such as snakehead fish, catfish, carps, and other native species of fish. However, the fish tilapia (*Tilapia nilotica*), with its vast range of salinity tolerance, is found to be uninfluenced by the sudden change in their environment. The fish (Fig. 10.9) is quite hardy and omnivorous in nature and therefore can tolerate fluctuations in water temperature. Farmers have found that tilapia has a good consumer preference as well as is popular with the local farming community. However, owing to its high prolific breeding habit, it is quite easy for the fish to overcrowd the pond, thereby reducing the growth of all the fish present resulting in lowering the production of fish. Studies reported that the production of tilapia (*Oreochromis* sp.) all-male populations is important in aquaculture to prevent energy consumption in reproduction and for the production of sex with higher growth potential (Macintosh and Little 1995; Green et al. 1997; Dan and Little 2000; Sayed et al. 2016).

10.4.3.3 Practicing Composite Fish Culture

Composite fish culture is a mixed culture of a group of compatible, mutually complementary and supplementary freshwater species of fast-growing food fish, collectively grown in ponds across a duration of time. Under this polyculture system, almost all the available ecological niches in the pond ecosystem have been

Fig. 10.9 Cultivation of stress-tolerant monosex tilapia

Table 10.1 Different species with its feeding habit and zone

Species	Feeding habit	Feeding zone
Indian major carp		
Catla	Zoo plankton feeder	Surface feeder
Rohu	Omnivorous	Column feeder
Mrigal	Detritivorous	Bottom feeder
Exotic carps		
Silver carp	Phytoplankton feeder	Surface feeder
Grass carp	Herbivorous	Surface, column, and marginal areas
Common carp	Detritivorous/omnivorous	Bottom feeder

efficaciously exploited by cultivating compatible fish species of distinct feed habits. The practice made use of Indian and exotic carps such as catla (*Catla catla*), rohu (*Labeo rohita*), mrigal (*Cirrhinus mrigala*), silver carp (*Hypophthalmichthys molitrix*), grass carp (*Ctenopharyngodon idella*), and common carp (*Cyprinus carpio*), which graze from distinct zones of the pond. Table 10.1 shows different species with its feeding habit and zone.

These six species of Indian and exotic origin were cultivated under a management system involving the preparation of ponds, elimination of unwanted stocks, fertilization, manuring, feeding of these fish through supplementary feeding, and manipulation of stock through periodic harvesting and stocking. Thus, fish species were not competitors to each other for food and space. The foods of the different layers in pond water are fully utilized by the carps in the composite fish culture or mixed culture. Induced breeding can also be done in the exotic carps and they are fast-growing, and the total production of fishes is very high and the cost of production is less. According to FAO (2009), planktonophagous and herbivorous fish culture, which feeds at a lower level of the food chain and others such as seaweed and shellfish, could contribute to carbon sequestration and thus mitigate the impact of climate change.

10.5 Conclusion

This chapter focuses on farmers-led adaptation strategies to climate change in the Indian Sundarbans region. This chapter concluded that farmers of Indian Sundarbans have experienced heavy rainfall within the short span of time; unpredictable tropical cyclone and tropical storm create waterlogging condition during monsoon months; and the destruction of soil and water quality is due to saltwater inundation or presence of brackish groundwater level at shallow depth and shortage of irrigation water throughout the dry season. Livestock of this region suffered from high-level thermal humid combination due to increasing temperature as well as humidity. An increase in sea surface temperature in the Sundarbans region affects fish stocks. Hence, the farming community in Indian Sundarbans adopted different adaptation strategies, viz. land shaping to harvest and retain rainwater for future use, shifting from mono-cropping to multi-cropping system, cultivating crops on the land embankment (*ail*), shifting from large ruminants to small ruminants, rearing stress-tolerant animals like Garole sheep, housing with cross ventilation for keeping livestock, providing clean and fresh drinking water regularly to diminish heat stress, practicing brackish aquaculture and composite fish culture, and cultivating stress-tolerant fish like tilapia, thereby practicing integrated farming system to cope up with the negative impact of changing climate.

The scientific basis of all these adaptation strategies was spotted and assembled in this chapter. The reinvention of these adaptation strategies was also observed according to the local situation. In this chapter, it was found that farmers of the Indian Sundarbans were highly experienced toward climatic disasters and followed adaptation strategies to cope up with it. Mixed farming comprised of crop, livestock, and fish component is the predominant farming system practiced by the farmers of Indian Sundarbans.

To combat with changing climatic scenario, the small and marginal farmers of Indian Sundarbans had adopted the integrated farming system in which by-product of one system is used as an input for another system. It is a farming system with synchronous activities including both crops and animals. The main aim of integrated farming is that the farming components support one another, thereby reducing the external inputs (Sharmin et al. 2018). This practice reduces stress due to biotic and abiotic factors. Seo (2010) suggests that integrated farming practices are more resilient against climate change. The farming community emphasizes the judicious use of available resources with an appropriate combination of enterprises available with farmers as the main adopting strategies to sustain their livelihood and as a barrier against changing climate. This aimed at effective sustainable management of resources for improved production in the crop system, livestock rearing, and fishing distributed spatially and temporarily around the same unit of land for the best use of available resources (Ravishankar et al. 2007). This is the only need of the hour to cope with the catastrophic consequences of climate change and guarantee their family's sustained income throughout the year. Thus, the promotion of the IFS model should be encouraged to increase the adaptive capacity of the farmers. To mitigate with the climate change impact like a saltwater intrusion, etc., for longer term, promotion of all the discussed adaptation strategies should be done, so that the rate of adoption of all the adaptation strategies increased.

For agriculture, several policy-based choices for adaptation to climate change have been established (Agrawal 2008). It is increasingly recognized that efforts to improve the resilience of agricultural systems need to understand and draw on local coping strategies (Eriksen et al. 2005). Studies have highlighted that the robust adaptive capacity shown by farmers in relation to climate and other stresses is their sophisticated strategies to cope with stress (Newsham and Thomas 2011). The integrated farming system is useful as it has the potential to combat against changing climate in a very cost-effective, participatory, and sustainable manner. All the agricultural adaptation measures should be integrated so that it addresses non-climatic stresses and risks and enhance a great chance of effectiveness. The IFS model utilizes the by-products of one component of the farming system as an input in other for ensuring supplementary and complementary enterprise relationship, hence maximizing return from a unit area. Policymakers should recommend a model of IFS that includes most of the abovementioned adaptation strategies. This chapter, therefore, concludes that to combat changing climatic conditions, the farming community needs to adopt an integrated farming system in which by-product of one system is used as an input for another system. Thus, the promotion of integrated farming system (IFS) model should be encouraged to increase the adaptive capacity of the farmers of the Indian Sundarbans region to cope up with the changing climatic scenario.

10.6 Policy Recommendations

- Through the adoption of an integrated farming system (IFS), the major issues of climate change that induced constraints to farming (such as fluctuating water supplies, deteriorating soil quality, vagaries in farm income, etc.) can be addressed successfully. Thus, the promotion of the IFS model should be encouraged to increase the adaptive capacity of the farmers.
- The farming community of the Indian Sundarbans is recognizing indigenous livestock breed, viz. *Garole* sheep and Black Bengal Goat, than the cross-bred. Indian Sundarbans is also the breeding track of these breeds. Therefore, livestock development department may promote this tendency to popularize climate-resilient livestock farming in the climate-sensitive Indian Sundarbans region.

References

Agrawal A (2008) The role of local institutions in adaptation to climate change. Paper prepared for the social dimensions of climate change. Social Development Department, the World Bank, Washington, DC

Aisabokhae RA, McCarl BA, Zhang YW (2011) Agricultural adaptation: needs, findings and effects. In: Dinar A, Mendelsohn RO (eds) Handbook on climate change and agriculture. Edward Elgar, Cheltenham and Northampton, pp 327–341

Australian Felt Specialist Pty Ltd (1999) From sheep to shop: a story about the process of making of pressed woolen felt. Australian Felt Specialists Pty Ltd, Blackburn. http://www.ausfelt.com/Schools.htm. Accessed 3 Sep 2009

Bandyopadhyay BK, Maji B (1995) Nature of acid soils of Sundarbans delta and suitability of classifying them as acid sulphate or potential acid sulphate soils. J Indian Soc Soil Sci 43:251–255

Bandyopadhyay BK, Maji B (1999) Lime and phosphate requirements of coastal acid saline soils for rice cultivation. J Indian Soc Coastal Agric Res 17:96–100

Bandyopadhyay BK, Burman D, Sarangi SK, Mandal S, Bal AR (2009) Land shaping techniques to alleviate salinity and water logging problems of mono-cropped coastal land for multi-crop cultivation. J Indian Soc Coast Agric Res 27(1):13–17

Banerjee S (2009) Utilisation of Garole sheep wool: a step towards the alleviation of poverty. Anim Genet Resour Inf 45:85–89

Basu D, Banerjee S, Goswami R (2009) Vegetable cultivation on ail (bund): a successful farmer-led technology transfer. Acta Hortic 809:155–160

Bhowmick MK, Srivastava AK, Singh S et al (2020) Realizing the potential of coastal flood-prone areas for rice production in West Bengal: prospects and challenges. In: Rakshit A, Singh H, Singh A et al (eds) New frontiers in stress management for durable agriculture. Springer, Singapore, pp 543–577

Bradshaw B, Dolan H, Smit B (2004) Farm-level adaptation to climatic variability and change: crop diversification in the Canadian prairies. Clim Change 67:119–141

Bryan E, Deressa TT, Gbetibouo GA et al (2009) Adaptation to climate change in Ethiopia and South Africa: options and constraints. Environ Sci Policy 12:413–426

Bryant CR, Smit B, Brklacich M et al (2000) Adaptation in Canadian agriculture to climatic variability and change. Clim Change 45:181–201

Burman D, Bandyopadhyay BK, Mandal S et al (2013) Land shaping—a unique technology for improving productivity of coastal land (Technical Bulletin No. CSSRI/Canning Town/Bulletin/2013/02:38), Central Soil Salinity Research Institute, Karnal, Haryana

Chand BK, Trivedi RK, Biswas A et al (2012a) Study on impact of saline water inundation on freshwater aquaculture in Sundarban using risk analysis tools. Explor Anim Med Res 2:170–178

Chand BK, Trivedi RK, Dubey SK et al (2012b) Aquaculture in of changing climate Sundarban survey report on climate change vulnerabilities, aquaculture practices and coping measures in sagar and basanti blocks of Indian Sundarban. University of Animal & Fishery Sciences, Kolkata

Chowdhury AN, Brahma A, Mondal R, Biswas MK (2016) Stigma of tiger attack: study of tiger-widows from Sundarban Delta, India. Indian J Psychiatry 58:12–19

Dabney SM (1998) Cover crop impacts on watershed hydrology. J Soil Water Conserv 53 (3):207–213

Dan NC, Little DC (2000) The culture performance of monosex and mixed-sex new-season and overwintered fry in three strains of Nile tilapia (*Oreochromis niloticus*) in northern Vietnam. Aquaculture 184(3–4):221–231

Danda AA, Rahman M (2019) Transformation of the Sundarbans eco-region: lessons from past approaches and suggested development options. In: Sen H (ed) The Sundarbans: a disaster-prone eco-region, vol 30. Springer, Cham, pp 61–89

Das S (2017) Impact of climate change on livestock, various adaptive and Mitigative measures for sustainable livestock production. Appro Poult Dairy & Vet Sci 1(4):64–70. https://doi.org/10.31031/APDV.2017.01.000517

Dolan AH, Smit B, Skinner MW et al (2001) Adaptation to climate change in agriculture: evaluation of options. Occasional paper 26, Department of Geography, University of Guelph, Guelph

Eigenberg RA, Brown BTM, Nienaber JA (2007) Development of a livestock weather safety monitor for feedlot cattle. Appl Eng Agric 23(5):657–660

Ericson JP, Vörösmarty CJ, Dingman SL et al (2006) Effective Sea-level rise and deltas: causes of change and human dimension implications. Global Planet Change 50(1–2):63–82

Eriksen SH, Brown K, Kelly PM (2005) The dynamics of vulnerability: locating coping strategies in Kenya and Tanzania. Geogr J 171:287–305

FAO (2009) Yearbook of fisheries statistics-fishery commodities. Food and Agriculture Organisation, Rome, p 187

Forino G, Meding JV, Brewer G et al (2014) Disaster risk reduction and climate change adaptation policy in Australia. Procedia Econ Financ 18:473–482

Gaba S, Lescourret F, Boudsocq S et al (2015) Multiple cropping systems as drivers for providing multiple ecosystem services: from concepts to design. Agron Sustain Dev 35(2):607–623

Goswami BN, Venugopal V, Sengupta D et al (2006) Increasing trend of extreme rain events over India in a warming environment. Science 314(5804):1442–1445

Green BW, Veverica KL, Fitzpatrick MS (1997) Fry and fingerling production. In: Egna HS, Boyd CE (eds) Dynamics of pond aquaculture. CRC Press, Boca Raton, pp 215–243

Hazra S (2012) Climate change adaptation in coastal region of West Bengal. Regional policy action platform on climate change (RPAPCC): policy paper, 2

Henry B, Charmley E, Eckard R et al (2012) Livestock production in a changing climate: adaptation and mitigation research in Australia. Crop Pasture Sci 63(3):191–202
Howden SM, Soussana JF, Tubiello FN et al (2007) Adapting agriculture to climate change. Proc Natl Acad Sci U S A 104(50):19691–19696
IPCC (2013) In: Stocker TF, Qin D, Plattner GK et al (eds) Climate change 2013: the physical science basis. Contribution of working group I to the fifth assessment report of the intergovernmental panel on climate change. Cambridge University Press, Cambridge, p 1535
Karim SA, Shinde AK (2007) Pasture based feeding system for small ruminant production and its relevance in tropics. In: International conference on tropical animal nutrition, NDRI, Karnal, 3–5 Oct 2007
Khan MH, Sarwar GM, Islam J (2008) Climate change and strategic adaptation provisions for coastal Bangladesh. In: Proceedings of national seminar on mangrove for sustainable livelihood and adaptations to climate change, Dhaka, pp 28–61
Kumar VS, Pathak KC, Pednekar P et al (2006) Coastal processes along the Indian coastline. Curr Sci 91(4):530–536
Kurukulasuriya P, Rosenthal S (2003) Climate change and agriculture: a review of impacts and adaptations. Environment department papers no. 91. Climate change series. World Bank, Washington, DC
Leal FW, Modesto F, Nagy GJ et al (2018) Fostering coastal resilience to climate change vulnerability in Bangladesh, Brazil, Cameroon and Uruguay: a cross-country comparison. Mitig Adapt Strat Glob Chang 23(4):579–602
Lin BB (2011) Resilience in agriculture through crop diversification: adaptive management for environmental change. Bioscience 61(3):183–193
Macintosh DJ, Little DC (1995) Nile tilapia (*Oreochromis niloticus*). In: Bromage NR, Roberts RJ (eds) Broodstock management and egg and larval quality. Blackwell, Cambridge, p 424
Mahadevia GK, Vikas M (2012) Climate change—impact on the Sundarbans: a case study. Int Sci J Environ Sci 2(1):7–15
Maiti S, Jha SK, Garai S et al (2014) Adaptation strategies followed by the livestock rearers of coastal Odisha and West Bengal to cope up with climate change. Indian J Anim Sci 84 (6):652–659
Mandal B, Mukherjee A, Banerjee S (2013a) A review on the ichthyofaunal diversity in mangrove based estuary of Sundarbans. Rev Fish Biol Fisher 23(3):365–374
Mandal S, Sarangi SK, Burman D et al (2013b) Land shaping models for enhancing agricultural productivity in salt affected coastal areas of West Bengal—an economic analysis. Indian J Agric Econ 68(3):389–401
Mandal UK, Maji B, Mullick S et al (2019) Global climate change and human interferences as risk factors, and their impacts on geomorphological features as well as on farming practices in Sundarbans eco-region. In: The Sundarbans: a disaster-prone eco-region. Springer, Cham, pp 405–437
Mimura N (2013) Sea-level rise caused by climate change and its implications for society. P Jpn Acad B-Phys 89(7):281–301
Mitra A (2019) Estuarine pollution in the lower Gangetic Delta threats and management. Springer, Cham
Mitra A, Banerjee K, Sengupta K et al (2009) Pulse of climate change in Indian Suindarbans: a myth or reality? Natl Acad Sci Lett 32(1):19
Moser SC, Ekstrom JA (2010) A framework to diagnose barriers to climate change adaptation. Proc Natl Acad Sci 107(51):22026–22031
Nautiyal S, Schaldah R, Raju KV et al (2016) Climate change challenge (3C) and social-economic-ecological Interface-building—exploring potential adaptation strategies for bio-resource conservation and livelihood development: epilogue. In: Nautiyal S, Schaldach R, Raju K et al (eds) Climate change challenge (3C) and social-economic-ecological Interface-building. Environmental Science and Engineering. Springer, Cham, pp 631–639

Newsham AJ, Thomas DS (2011) Knowing, farming and climate change adaptation in north-Central Namibia. Glob Environ Chang 21(2):761–770

Oppenheimer M, Glavovic BC, Hinkel J et al (2019) In: Pörtner HO, Roberts DC, Masson-Delmotte V et al (eds) Sea level rise and implications for low-Lying Islands, coasts and communities: a special report on the ocean and cryosphere in a changing climate. IPCC, Geneva, pp 321–445

Parry ML, Canziani OF, Palutikof JP et al (2007) IPCC, 2007: climate change 2007: impacts, adaptation and vulnerability. Contribution of working group II to the fourth assessment report of the intergovernmental panel on climate change. Cambridge University Press, Cambridge

Patwardhan A, Narayanan K, Parthasarathy D et al (2003) Impacts of climate change on coastal zones. In: Shukla PR, Sharna SK, Ravindranath NH et al (eds) Climate change and India: vulnerability and adaptation. Universities Press, Hyderabad, pp 326–359

Pitchaikani J, Sensarma K, Bhattacharyya S (2017) First time report on the weather patterns over the Sundarban mangrove forest, east coast of India. Indian J Geomarine Sci 46(4):766–770

Raha A, Das S, Banerjee K et al (2012) Climate change impacts on Indian Sundarbans: a time series analysis (1924–2008). Biodivers Conserv 21(5):1289–1307

Ravishankar N, Pramanik SC, Rai RB et al (2007) Study on integrated farming system in hilly upland areas of Bay Islands. Ind J Agronomy 52:7–10

Rishi P, Mudaliar R (2014) Climate stress, behavioral adaptation and subjective well being in coastal cities of India. Am J Appl Psychol 2(1):13–21

Rötter R, van de Geijn S (1999) Climate change effects on plant growth, crop yield and livestock. Clim Change 43(4):651–681

Rowlinson P (2008) Adapting livestock production systems to climate change – temperate zones. In: Rowlinson P, Steele M, Nefzaoui A (eds) Livestock and global climate change. University Press, Cambridge, pp 61–63

Roy A (2010) Vulnerability of the Sundarbans ecosystem. J Coast Environ 1(2):169–181

Sayed AEDH, Mahmoud UM, Mekkawy IA (2016) Erythrocytes alterations of monosex tilapia (*Oreochromis niloticus*, Linnaeus, 1758) produced using methyltestosterone. Egypt J Aquat Res 42(1):83–90

Sejian V, Naqvi SMK, Ezeji T et al (2012) Introduction. In: Sejian V, Naqvi SMK, Ezeji T et al (eds) Environmental stress and amelioration in livestock production. Springer, Berlin Heidelberg, pp 1–12

Seo SN (2010) Is an integrated farm more resilient against climate change? A micro econometric analysis of portfolio diversification in African agriculture. Food Policy 35(1):32–40

Seo SN, Mendelsohn R, Dinar A et al (2009) Adapting to climate change mosaically: an analysis of African livestock management by agro-ecological zones. J Econ Anal Policy 9(2):1–35. https://doi.org/10.2202/1935-1682.1955

Sharmin S, Rashid MHA, Begum R et al (2018) Relative profitability of farming systems research and development (FSRD) project farmers and non-project farmers of integrated farming systems in Tangail district of Bangladesh. J Bangladesh Agric Univ 16(1):117–122

Singh AK, Maitra NJ, Chatterjee P et al (2012) Enhancement of adaptive capacity to climate change in vulnerable regions of Sundarban (WB). Indian Res J Ext Educ 1:31–37

Sirohi S, Sirohi SK (2010) Vulnerability of milk producers to climate change: technological and policy options for livelihood security. In: Upadhyay RC et al (eds) Climate change & livestock productivity in India. ICAR-NDRI, Karnal, pp 143–152

Skuce PJ, Morgan ER, van Dijk J et al (2013) Animal health aspects of adaptation to climate change: beating the heat and parasites in a warming Europe. Animal 7(s2):333–345

Smit B, Skinner MW (2002) Adaptation options in agriculture to climate change: a typology. Mitig Adapt Strat Gl 7(1):85–114

Sundaresan J, Sreekesh S, Ramanathan AL et al (eds) (2013) Climate change and island and coastal vulnerability. Springer, Dordrecht

Timon VM, Hanrahan JP (1986) Small ruminant production in the developing countries. FAO animal production and health paper 58, Food and Agriculture Organization of United Nations, Rome, p. 234

Udayakumar P (2014) Effects of sea level change on vulnerable east coast of India. Res J Mar Sci 2 (1):1–5

Upadhyay RC, Ashutosh RV, Singh SV et al (2009) Impact of climate change on reproductive functions of cattle and buffaloes. In: Aggarwal PK (ed) Global climate change and Indian agriculture. Indian Council of Agricultural Research, New Delhi, pp 107–110

Velmurugan A, Ambast SK, Swarnam TP et al (2018) Land shaping methods for climate change adaptation in coastal and island region. In: Sivaperuman C, Velmurugan A, Singh AK et al (eds) Biodiversity and climate change adaptation in tropical islands. Academic Press, Amsterdam, pp 577–596

Wood SA, Jina AS, Jain M et al (2014) Smallholder farmer cropping decisions related to climate variability across multiple regions. Glob Environ Chang 25:163–172

Using Beneficial Microorganisms to Promote Sustainable Crop Production and Resilience of Smallholder Agroecosystems to Changing Climate

11

Ezekiel Mugendi Njeru and Gilbert Koskey

Abstract

Agriculture today faces a uniquc challcngc of producing sufficient and nutritious food for the rising human population under finite natural resources, shrinking economies, and changing climate. Soil fertility is pertinent to sustainable agricultural production and mitigating the consequences of climate change drivers to crops. To promote healthy crop production practices, the search for alternative soil amelioration and plant disease management strategies is on the rise. Among the most feasible alternatives are beneficial soil microorganisms, which are central to many agroecological cycles and improvement of crop nutrient and water uptake and resistance to biotic and abiotic stresses. Since the majority of smallholder farmers in many parts of the world cannot afford inorganic fertilizers, there is a pressing need to develop sustainable and affordable soil fertility management strategies that focus on low-input cropping systems which is crucial for attaining agricultural sustainability and global food security. Therefore, this review explores the potential of beneficial microorganisms to promote sustainable crop production and resilience of smallholder agroecosystems to global climate change drivers.

Keywords

Beneficial microorganisms · Climate change · Food security · Microbial inoculants · Smallholder farmers

E. M. Njeru (✉)
Department of Biochemistry, Microbiology and Biotechnology, Kenyatta University, Nairobi, Kenya
e-mail: njeru.ezekiel@ku.ac.ke

G. Koskey
Institute of Life Sciences, Scuola Superiore Sant'Anna, Piazza Martiri della Libertà, Pisa, PI, Italy

V. K. Hebsale Mallappa, M. Shirur (eds.), *Climate Change and Resilient Food Systems*, https://doi.org/10.1007/978-981-33-4538-6_11

11.1 Introduction

Meeting the increasing food demands of the rapidly growing population in the face of limited resources and changing climate presents an enormous challenge to several developing nations. Per capita agricultural production and food availability have decreased over time (Funk and Brown 2009), and many people, especially within the developing world, are in danger of widespread food insecurity. Smallholder farmers predominate in an environment of dwindling natural resources, especially arable land, rising population, changing climate, and environmental pollution (Aniah et al. 2019). Low soil fertility presents one among the main challenges to sustainable crop production since most smallholders cannot afford to invest in high-cost chemical fertilizers (Clair and Lynch 2010). Given that the majority of the populace in developing nations depends on smallholder agriculture, there is a pressing need to understand how soil fertility can be sustainably improved on a small scale and contribute to achieving the United Nations Sustainable Development Goals (UN SDG). This includes a strategic combination of the various alternative soil organic amendments and beneficial microorganisms with specifically adapted crop varieties under diverse agroclimatic conditions. Resource-saving strategies and land-use intensification require the development of more viable plant nutrition strategies in crop production as alternatives to the prevailing use of nonrenewable mineral fertilizers.

Beneficial soil biota provides essential agroecosystem services, especially in low-input agroecosystems, where the use of external inputs is usually limited. These include improving the nutritional status of their host plants and protecting them from deleterious effects of drought, high temperature, soilborne plant pathogens, and heavy metals (Goswami and Deka 2020; Jacoby et al. 2017). Beneficial microorganisms that support plant growth and development include plant growth-promoting rhizobacteria (PGPR), symbiotic fungi such as arbuscular mycorrhizal fungi (AMF), *Trichoderma* spp., and endophytic microorganisms. These promote plant growth and impart plant resilience to global climate change through a plethora of mechanisms including enhancing nutritional and water uptake, rhizoremediation, production of phytohormones, siderophores, secondary growth metabolites, and suppression of pathogenic microorganisms (Fig. 11.1).

11.2 Characteristics of Smallholder Agroecosystems, Opportunities, and Challenges

Agricultural ecosystems in the sub-Saharan African region are inherently diverse, indigenous, and resilient multifunctional complex systems that are sustainably managed to meet the farmers' subsistence needs (Mburu et al. 2016). They thrive without relying much on chemical fertilizers, mechanization, or other sophisticated modern technologies (Awazi and Tchamba 2019). The smallholder systems are the key drivers of the rural economies. Agroecologists acknowledge that they can provide resilient solutions to food security amidst several uncertainties challenging

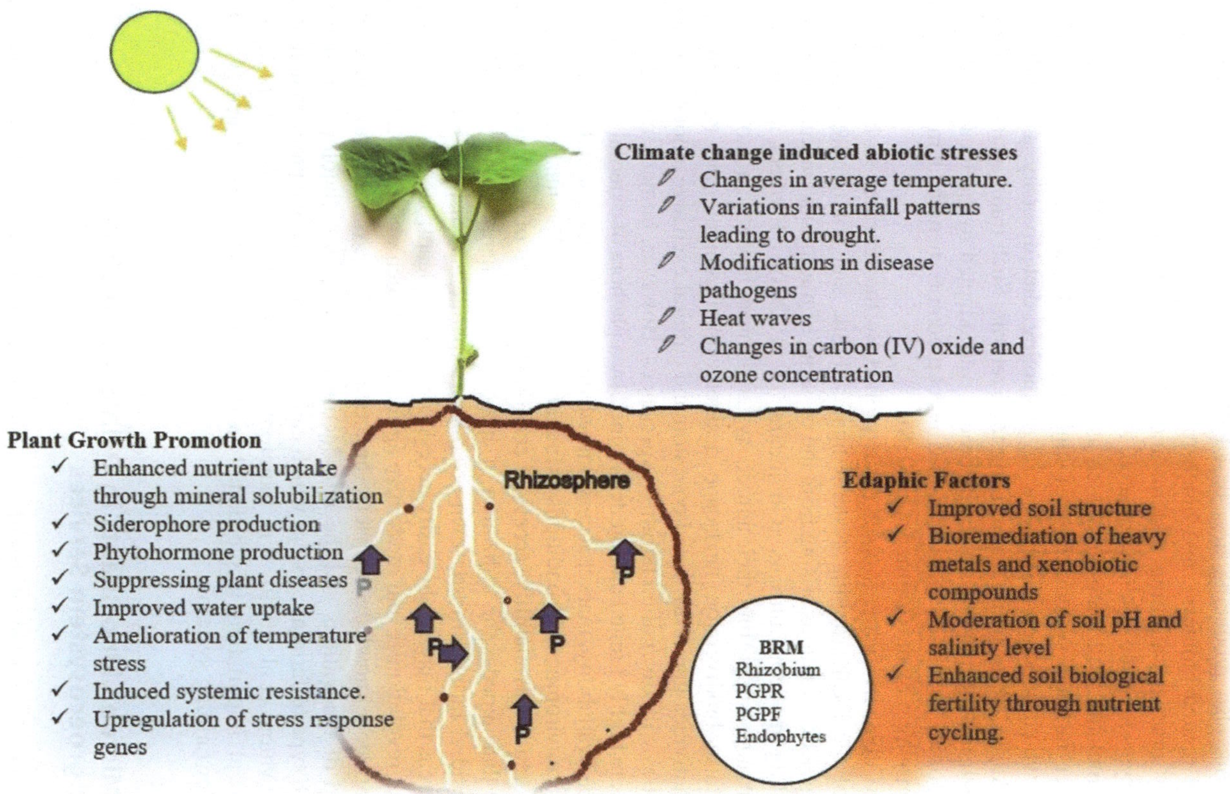

Fig. 11.1 Different ways of plant growth promotion by beneficial microorganisms under changing climate. *BRM* beneficial rhizospheric microorganisms; *PGPR* plant growth-promoting rhizobacteria; *PGPF* plant growth-promoting fungi

human existence such as climate change and economic and energy crisis (Altieri et al. 2012). Despite the characteristically highly fragmented small pieces of land, farmers engage not only in producing food (crops, animals, and their derived products) but also producing fiber-based products (cotton, sisal, and silk), fuel (wood and related biomass), and biochemicals (natural medicinal products). The dominantly practiced farming systems include crop rotation, intercropping, and agroforestry, which are often associated with conferring beneficial effects on the soil by stimulating soil microbiota such as AMF and PGPRs (Raimi et al. 2017). Smallholder farming systems have a high degree of plant and animal species diversity, which is a good strategy in promoting diet diversity and reducing unpredicted risk losses while maximizing farm returns. Such traditional ecosystems that are rich in wild populations of landraces well adapted to the local conditions could harbor a wide variety of pollinators, insect predators, beneficial microbes, and nutrient-enriching plants (Oruru et al. 2018).

Genetic diversity richness could be exploited by the breeders to heighten the stability of the local cropping systems against biotic and abiotic stresses and to promote genetic variations arising from the crossing of traditional landraces with the high-yielding modern cultivars (Govindaraj et al. 2015). Farmers are the key decision-makers supporting their agroecosystems through water and soil conservation, on-farm biodiversity, landscape aesthetic management, and engaging in off-farm activities that bring social cohesion and cultural exchange among different communities with diversified farming experiences. However, farmers' management decisions are highly limited and predetermined by a plethora of external factors such as resource availability, market access, knowledge and exposure, social and cultural needs, and environmental conditions (Mungai et al. 2016). The mismatch between farm size and the availability, access or economies of scale of machines remains a challenge that impedes the utilization of appropriate mechanization in smallholder farms. Besides, farmers mostly own or co-own separate and dispersed agricultural fields in areas that may not be easily accessible, and this leapfrogs any technocratic attempts to revolutionize and modernize agriculture in such setups (Van Loon et al. 2020). Farmer's knowledge capacity and educational needs still undermine the scaling up of innovations and the introduction of any new technology to such a constricted type of farming system which calls for farmers' active participation and approval (Muoni et al. 2019). It is, therefore, imperative to involve farmers in any technological importation if substantial positive changes are to be achieved in smallholder farming systems.

11.3 Microbial Inoculants in Delivery of Multiple Agroecosystem Services

The application of microbial inoculants selected based on functional trait approach and that are adapted to the highly intensified smallholder agroecosystems is highly advocated in agroecology. Besides, the adoption of diversified farm management practices that complement microbial inoculation could optimize the provision and delivery of the essential agroecosystem services. For instance, the cultivation of legumes as relay intercrops with cereals and the inclusion of rhizobia inoculation enhance N acquisition through BNF and N transfer to non-legumes. Other microbial-derived agroecosystem services important in smallholder farming systems include P and K acquisition, secretion of stimulatory phytohormones, and siderophores that help in phytopathogen control, biotic and abiotic stress management.

11.3.1 Biological Nitrogen Fixation

Nitrogen (N) is one of the essential nutritional components of plants that makes up a large portion of plant proteins and nucleic acids regulating the primary productivity of the soil ecosystem (Ahemad and Kibret 2014). Naturally, N exists in various chemical forms and oxidation states, and microorganisms play a key role in catalyzing the different N transformations into forms readily utilizable by plants (Pajares and Bohannan 2016). Biological nitrogen fixation (BNF) is a microbiologically driven process where microorganisms transform atmospheric nitrogen into forms that can be assimilated by the plants. The mechanism of BNF is a complex process that is majorly catalyzed by the nitrogenase enzyme complex and regulated by the *nif* genes produced by diazotrophic N-fixing microorganisms (Wang et al. 2013; Choudhary and Varma 2017).

Major groups of diazotrophs known for N-fixing abilities are the Cyanobacteria, green sulfur bacteria, *Azorhizobium*, *Sinorhizobium*, *Rhizobium*, *Azospirillum*, *Thiobacillus*, *Herbaspirillum*, *Bradyrhizobium*, *Frankia* sp., and *Clostridium pasteurianum* (Yeager et al. 2005; Mus et al. 2018). Legumes are the biggest beneficiaries of the BNF process as they are able to form a symbiotic association with most of the diazotrophs to meet their N demands. However, the efficiency of the symbiotic partnership in delivering the much-needed N to the plants varies highly and depends on the host genotype, soil conditions, microbial strain, and climatic conditions (Maingi et al. 2001; Mabrouk et al. 2018). In smallholder farming systems, the legume cultivation substantially reduces the overall amount of external N inputs required to sustain the present and subsequent crop production (Nyoki and Ndakidemi 2018). This could be ameliorated further by inoculating legumes with effective N-fixing microorganisms.

Inoculating legumes using a combination of rhizospheric nitrogen-fixing bacterial strains improves soil health, quality, and fertility. It enhances plant-microbe interactions through improved root exudation and signaling leading to better root development, nodulation, phytopathogen suppression, and water and nutrient acquisition (Ouma et al. 2016; Koskey et al. 2017; Mabrouk et al. 2018). Various stakeholders have initiated research-based projects aimed at empowering farmers with technologies that would contribute to improvement in their household income through the introduction of N-fixing inoculants that are effective in soil fertility restoration and crop productivity. N2Africa has actively researched on African indigenous rhizobia strains associated with chickpea, soybean, faba bean, common bean, and groundnut legumes since 2009 in more than ten sub-Saharan African

(SSA) countries (Giller et al. 2019). The Microbial Resources Centre Network (MIRCEN), in collaboration with the University of Nairobi in Kenya and other commercial private stakeholders, developed rhizobia inoculants known as Biofix® that is more cost-effective compared to chemical N fertilizers available in the Kenyan market (Odame 1997). In Nigeria, the International Institute of Tropical Agriculture (IITA) introduced promiscuous soybean (*Glycine max* L. Merrill) cultivars that nodulate with a wide diversity of the African *Bradyrhizobium* bacteria (Santos et al. 2019). Other recent researches on N-fixing rhizobia have been carried out across the SSA, and more effective indigenous rhizobia inoculants that could be used by farmers as single or a consortium of different strains are available (Koskey et al. 2018; Grönemeyer and Reinhold-Hurek 2018; Musyoka et al. 2020). However, large-scale production, commercialization, continuous adoption, and accessibility remain the most prominent challenges impeding the use of these elite inoculants by the SSA smallholder farmers (Oruru and Njeru 2016).

11.3.2 Phosphate Solubilization

Phosphorus (P) in the soil exists largely in the form of insoluble compounded deposits. It is one of the main essential macronutrients required by plants for growth and development (Walpola and Yoon 2012). Despite the presence of substantial P deposits in the soil layers, P content available for plant use in many smallholder farms is critically limited. Essential plant physiological, molecular, and biochemical processes such as metabolism, signal transduction, genetic, structural formations, energy storage and transfer, cell and tissue formation require the presence of the P element for optimal functioning (Dissanayaka et al. 2018). The uptake of P from the soil by plants is mainly in the form of orthophosphate anions, which are formed through a bacteria-mediated mechanism of acidic solubilization of inorganic phosphates (Lobo et al. 2019). Alternatively, P mobilization occurs via organic phosphate mineralization, a process carried out by soil bacteria capable of producing phosphatase enzymes such as phosphoesterases, phytases, phosphodiesterases, and phospholipases that catalyze the breakdown of phosphoric esters (Walpola and Yoon 2012; Novo et al. 2018).

Smallholder farmers often rely on phosphatic chemical fertilizers, which are expensive and unavailable and prone to precipitation by metal-cation complexes such as Ca^{2+}, Al^{3+}, and Fe^{3+} found in the soil leading to soil fertility depletion (Dissanayaka et al. 2018). Thus, there is a need for low-priced sustainable techniques that are environmentally friendly and efficient enough to supply adequate P to the plants. Microorganisms such as phosphate-solubilizing microbes (PSMs) form an integral part of the natural P cycle. Research on PGPR and plant growth-promoting fungi (PGPF) with the capability to solubilize and mobilize the insoluble organic and inorganic soil phosphates from the soil rocks to the plants has been done (Sharma et al. 2013; Alori et al. 2017; Selvi et al. 2017; Giovannini et al. 2020). This has led to the upsurge development of various commercial microbial inoculants containing effective PGPRs and PGPFs for use by smallholder farmers to increase their crop production (Tabassum et al. 2017).

The potential PSMs from the bacteria genera include *Pseudomonas putida, Pseudomonas calcis, Pseudomonas fluorescens, Pseudomonas striata* (Mohammadi 2012; Thakur et al. 2014), *Pseudomonas canescens* (Alam and Rashid 2002), *Rhizobium leguminosarum* (Walpola and Yoon 2012; Hajjam and Cherkaoui 2017), *Rhizobium meliloti, Thiobacillus ferrooxidans* (Sharma et al. 2013), *Mesorhizobium mediterraneum* (Peix et al. 2001), *Bacillus subtilis, Bacillus polymyxa, Bacillus megaterium, Bacillus circulans, Bacillus fusiformis, Bacillus coagulans*, and *Bacillus chitinolyticus* (Chen et al. 2006; Thakur et al. 2014; Satyaprakash et al. 2017). The members from the fungal genera include *Aspergillus niger, Aspergillus fumigatus, Aspergillus parasiticus, Aspergillus terreus, Aspergillus candidus, Penicillium simplicissimum, Penicillium rubrum,* nematophagous fungus *Arthrobotrys oligospora, Trichoderma viride,* and *Trichoderma* spp. (Reddy et al. 2002; Aseri and Jain 2009; Selvi et al. 2017). Other important microorganisms include arbuscular mycorrhizal fungi (Giovannini et al. 2020) and actinomycetes such as *Streptomyces albus, Streptomyces cyaneus,* and *Streptoverticillium album* (Kumar et al. 2018). Mixed cultures of bio-inoculants and multiple crop-stage inoculations are known to increase the potential effectiveness of PSMs in enhancing plant growth, shoot development, and yield productivity (Muthukumar and Udaiyan 2018).

AMF have a greater inter- and intraspecific biodiversity due to its ability to colonize the roots of 80–90% of plants (Oruru and Njeru 2016). They thus can be fully exploited as a resource in smallholder farming systems by selecting AMF isolates, species, and strains with the highest colonization efficiency, P solubilization, and siderophore production, among other indirect beneficial traits (Giovannini et al. 2020). Many studies have reported significant contributions of AMF bio-enhancers based on their use either as individual or a consortium of AMF inoculants in improving the yield productivity and nutrient quality of cereals, legumes, vegetables, fruits, and agroforestry trees (Njeru et al. 2017; Avio et al. 2018; Musyoka et al. 2020). Among the AMF communities, the most commonly available commercial inoculants are derived from the species *Funneliformis mosseae* and *Rhizophagus irregularis*. These AMF species coincidentally are broad symbionts widely spread throughout the sub-Saharan African soils and other tropical zones of the world predominated by smallholder farmers (Oruru and Njeru 2016; Giovannini et al. 2020).

11.3.3 Potassium (K)-Solubilizing Microorganism (KSMs)

Potassium (K) is available in the soil in various forms such as exchangeable and non-exchangeable K, mineral, and soluble K. However, depending on the soil type, most of the K^+ ions are bounded and are unavailable for direct uptake by the plants (Etesami et al. 2017). Apart from growth and developmental roles, K promotes plant resistance against pest and disease and takes part in the activation of over

80 physiological processes, including starch synthesis, energy metabolism, sugar degradation, photosynthesis, and nitrate reduction (Gallegos-Cedillo et al. 2016; Hussain et al. 2016). A group of microorganisms, K-solubilizing bacteria (KSBs), have the potential to solubilize the fixed forms of K through various mechanisms such as acidolysis, complexolysis, chelation, exchange reactions, and production of low-molecular-weight organic and inorganic acids (Meena et al. 2014; Etesami et al. 2017). Inoculation of KSBs on to K-deficient soil has been shown to boost seed germination, growth vigor, and yield effect of various crops (Bakhshandeh et al. 2017; Meena et al. 2014; Xiao et al. 2017). Some of the potential KSBs reported solubilizing K include *Enterobacter hormaechei, Burkholderia sp., Aminobacter sp., Acidithiobacillus ferrooxidans, Pseudomonas sp., Bacillus circulans, Bacillus mucilaginosus*, and *Paenibacillus glucanolyticus* (Singh et al. 2010; Sangeeth et al. 2012; Liu et al. 2012).

Arbuscular mycorrhizal fungi (AMF) have also been reported to indirectly play a key role in K solubilization by releasing H^+ protons and organic acid anions that increased K absorption and accumulation on the plant shoot and fruits (Dominguez-Nuñez et al. 2016). In maize (*Zea mays* L.), *Glomus mosseae* and *Rhizophagus intraradices* inoculation increased K uptake significantly compared to non-inoculated plants (Wu et al. 2005). Other fungi with K solubilization traits include *Aspergillus niger, Aspergillus terreus,* and *Penicillium sp.* (Sangeeth et al. 2012). In addition to K solubilization, they play a key role also in the secretion of low-molecular-weight organic acids that dissolves other nutrients bounded to the soil cations such as phosphates (Meena et al. 2014). The application of microbial K solubilization in smallholder settings should, however, be taken keenly as the process is profoundly affected by soil factors such as pH, clay percentage, oxygen concentration, the type of K-bearing minerals, and the microbial strains used (Sheng 2005; Etesami et al. 2017). Optimal conditions for KSB should be determined before inoculation for farmers to achieve good yields.

11.3.4 Siderophore Production

Siderophores are relatively low-molecular-weight complexes, iron-specific chelating molecules whose presence is highly dependent on the amount of iron (Fe) concentration in the soil (Mohammadi 2012). Siderophores are produced by specific strategy II plants and soil microorganisms, which play a key role in transporting iron molecules (Novo et al. 2018). Rhizospheric iron deficiency in smallholder production systems presents a great challenge, especially in parts of SSA with the calcareous type of soils. Its limitation causes leaf chlorosis, reduced photosynthesis, and stunted plant growth leading to reduced crop yields (Lewis et al. 2019). In most of the cases, under aerobic soil conditions, iron is compounded in the form of Fe^{3+} ions and oxy-hydroxides, making them unavailable for plants and microbes that readily utilize the Fe^{2+} forms (Pahari and Mishra 2017). Iron starvation in the soil triggers a specific group of bacteria, fungi, actinomycetes, and algae to synthesize and secrete ferric ion-specific chelating biomolecules. The secreted biomolecules not only improve the colonization of the microbes on the plant rhizosphere but also stimulate antagonistic reactions against phytopathogens and the acquisition of iron nutrients by the plant (Novo et al. 2018). Siderophores such as salicylate, hydroxamate, carboxylate, and catecholate are produced by bacteria, including *Salmonella* sp., *Vibrio anguillarum, Aerobacter aerogenes, Yersinia sp., Aeromonas sp., Enterobacter sp.,* and *Escherichia coli.* Fungal species with siderophore-producing traits include *Penicillium citrinum, Penicillium chrysogenum, Ustilago sphaerogina, Ustilago maydis, Rhizopus sp., Rhodotorula minuta, Mucor sp., Trametes versicolor, Aspergillus versicolor, Aspergillus fumigatus, and Aspergillus nidulans,* while actinomycetes include *Streptomyces griseus, Nocardia asteroides,* and *Actinomadura madurae* (Ahmed and Holmström 2014; Kannahi and Senbagam 2014). Precaution has to be taken while selecting siderophore-based microbial inoculants as some of the human pathogenic microbes have this capacity, which may lead to hazardous effects.

11.3.5 Microbial Secretion of Stimulatory Phytohormones

Crops are constantly exposed to environmental stresses such as salinity, drought, heavy metal contamination, floods, extreme temperature, and radiations that are detrimental to their growth, development, and productivity (Goswami and Deka 2020). Biotic stress caused by pathogenic soil microorganisms, toxic root exudates secreted by higher plants, and toxins released from animals will reduce plant productivity (Gouda et al. 2018). Smallholder farming systems, often faced with a wide range of challenges which include but not limited to financial constraints, are highly exposed to the adverse effects brought by these abiotic and biotic stresses. Thus, there is a need for self-sustaining exogenous biochemical techniques that not only mitigate stress responses but also regulate plant hormonal and nutritional balance and induce systemic tolerance (Egamberdieva et al. 2017). Soil harbors a diverse pool of beneficial hormone-producing microorganisms that colonizes the plant roots where signaling and exchange of nutrients occur. Microorganisms in the nutrient-rich rhizosphere synthesize active biomolecules such as auxins, gibberellins, cytokinin, abscisic acid, antifungal compounds, degrading enzymes, and other beneficial solute metabolites that promotes plant growth, nutrition, stress tolerance, and resistance (Ahemad and Kibret 2014).

Many free-living, endophytic, and symbiotic root-associated microorganisms have been identified or engineered to produce phytohormones that alleviate plants against specific stresses. For instance, *Acinetobacter, Marinobacterium, Pseudomonas, Bacillus, Sinorhizobium, Pantoea,* and *Rhizobium* isolated from halophytic weed *Psoralea corylifolia* L. were shown to produce IAA (indoleacetic acid) auxins that enhanced wheat seed germination (Sorty et al. 2016). A review by Gouda et al. (2018) noted that *Pseudomonas aeruginosa, Pseudomonas putida, Stenotrophomonas maltophilia, Mesorhizobium cicero, Azotobacter chroococcum, Klebsiella oxytoca, Enterobacter asburiae,* and *Rhizobium leguminosarum secrete gibberellin,* kinetin, auxin, and ethylene phytohormones linked to plant root invigoration. Likewise, phytohormones promoting shoot invigoration have been associated with PGPRs, including *Pantoea agglomerans, Bacillus subtilis, Pseudomonas fluorescens, Rhizobium leguminosarum, Paenibacillus polymyxa,* and *Rhodospirillum rubrum* (Prathap and Ranjitha 2015). Actinomycetes and fungi such as *Streptomyces, Nocardia, Spirillospora, Micromonospora, Microbispora, Nocardiopsis,* and *Aspergillus* have been isolated in mandarin and other medicinal plants and produce IAA phytohormones (Ruanpanun et al. 2010; Shutsrirung et al. 2013; Lin and Xu 2013). The introduction of microbial inoculants with the capability to regulate plant hormone production could be a critical step to revolutionize smallholder crop production systems and improve crop qualities. However, caution has to be taken when applying the phytohormone-inducing microbes as studies have shown that the biosynthesis of phytohormones differs depending on the microbial strain, environmental stress levels, and host genotype (Egamberdieva et al. 2017).

11.3.6 Use of Plant Growth-Promoting Rhizobacteria (PGPRs) in Stress Management

The most common stressful conditions affecting crops include drought (water stress), heat, salinity, floods, and metal toxicity (Egamberdieva et al. 2017). Under stress conditions, plants increase the production of reactive oxygen (ROs) species and OH^- radicals causing an oxidative stress condition that damages membrane lipids, proteins, nucleic acids, and photosynthetic pigments, leading to a progressive plant physiological system shutdown (Foyer et al. 2016). Beneficial microorganisms play an active role in stress management against biotic and abiotic agents to induce stress tolerance or resistance (Table 11.1). Extreme temperature, drought, floods, salinity, and wind can result in up to 70% yield losses and, therefore, could threaten smallholder farmers' food security status. Tolerance to these stresses could be induced by exogenous application of PGPRs that stimulates the accumulation of osmolytes critical in maintaining the plant cellular integrity such as proline, trehalose, glycine betaine, enzymatic and nonenzymatic antioxidants, abscisic acid, superoxide dismutase, glutathione, ascorbate peroxidase, and ascorbic acid (Agami et al. 2016; Gouda et al. 2018).

Heavy metal pollution is a common phenomenon in agricultural fields located close to the mining zones. In phytoremediation studies by Baharlouei et al. (2011) on canola and barley plants and Dourado et al. (2013) on tomatoes, it was reported that Cadmium (Cd) soil contamination can be alleviated by PGPRs such as *Pseudomonas fluorescens*, *Pseudomonas putida,* and *Burkholderia* sp. SCMS54 due to their ability to scavenge and translocate Cd^{2+} from the soil. Islam et al. (2016) similarly showed that chromium (Cr) toxicity that could significantly affect maize production could be stabilized using PGPR *Proteus mirabilis* isolates T2Cr and CrP450. There has also been a strong link reported between the presence of PGPRs and drought tolerance (Ngumbi and Kloepper 2016), PGPRs, and salinity stress (Cardinale et al. 2015; Habib et al. 2016). Some of the actively used PGPRs against salinity stress in wheat, maize, and barley include *Ensifer garamanticus* E110, *Curtobacterium*

Table 11.1 Overview of ways through which beneficial microorganisms promote plant growth of various plants under stress conditions

Microorganisms	Plant	Stress type	Plant changes	Reference
Bacillus methylotrophicus SMT38, *Bacillus aryabhattai* SMT48, *Bacillus aryabhattai* SMT50, and *Bacillus licheniformis* SMT51	*Spartina maritima*	Heavy metal pollution	Reduced respiration of the roots and oxidative stress	Mesa-Marín et al. (2018)
Bacillus cereus and *Pseudomonas moraviensis*	Wheat	Heavy metal pollution	Decreased biological accumulation coefficient and translocation factor	Hassan et al. (2017)
Funneliformis mosseae (*Fm*) and *F. caledonium*	Sunflower (*Helianthus annuus* L.)	Heavy metal pollution	Reduced heavy metal concentration in the shoots	Zhang et al. (2018)
Mixed culture of AMF was used which mainly comprised of *Funneliformis* species	Maize (*Zea mays* L.)	High temperature	Regulation of photosystem II heterogeneity	Mathur and Jajoo (2020)
Bacillus safensis and *Ochrobactrum pseudogrignonense*	Wheat (*Triticum aestivum* L.)	High temperature	Antioxidant signaling and reducing chloroplast and membrane injury	Sarkar et al. (2018)
Funneliformis mosseae and *Paraburkholderia graminis* C4D1M	Tomato (*Solanum lycopersicum* L.)	Chilling stress	Increasing the efficiency of photosystem II, reduced cell membrane injuries	Caradonia et al. (2019)
Funneliformis mosseae and *Rhizophagus intraradices*	Tomato (*Solanum lycopersicum*)	Water stress	Enhanced water use efficiency, net photosynthetic rate	Chitarra et al. (2016)
Bacillus megaterium and *Enterobacter* sp.	Okra (*Abelmoschus esculentus* L.)	Salinity	Reactive oxygen species scavenging enzymes	Habib et al. (2016)
Rhizophagus irregularis and *Funneliformis mosseae*	Durum wheat (*Triticum durum* Desf.)	Salinity	Greater stability of plasma membranes	Fileccia et al. (2017)
Pseudomonas fluorescens, Enterobacter hormaechei, and *pseudomonas migulae*	Foxtail millet (*Setaria italica* L.)	Drought	Stimulated seed germination and seedling growth	Niu et al. (2018)
Indigenous arbuscular mycorrhizal fungi	*Leymus chinensis* and *Hemarthria altissima* grasses	Drought	Altering antioxidant enzyme activities and photosynthesis	Li et al. (2019)

flaccumfaciens E108 (Cardinale et al. 2015), *Bacillus licheniformis, Bacillus subtilis, Arthrobacter* sp. (Upadhyay et al. 2012), *Enterobacter* sp. (Sorty et al. 2016), and *Pseudomonas* sp. (Mishra et al. 2017).

11.3.7 Biological Control

The use of microorganisms antagonistic to plant pathogens has revolutionized modern agriculture and could be considered as a substitute for the environment-insensitive chemical pesticides (Raimi et al. 2017). This remains one of the innovative ways of the twenty-first century in which smallholder farmers can sustainably control plant pathogens. For example, inoculation of crops with formulations containing PGPRs that colonize the rhizosphere and act against phytopathogens through competition for resources, niche exclusion, and induction of plant systemic resistance has been demonstrated (Fukami et al. 2018). Among the PGPRs, *Bacillus* sp. and *Pseudomonas* sp. are the most studied and mainly effective biocontrol agents that have been commercially produced over the past decades. They produce antagonistic secondary metabolites (antibiotics) that act against a wide range of pathogens and, most importantly, the biomolecules which are biodegradable, unlike the case of many agrochemicals (Wang et al. 2015).

Pseudomonas sp., for instance, produces antimycotics (pyrrolnitrin, 2,4-diacetylphloroglucinol, ecomycins B and C, oomycin A, visconamide, phenazine-1-carboxamide, phenazines, rhamnolipids, sulfonamides, pyocyanin, and butyrolactones), antitumor agents such as cepafungins, and antiviral agents such as Karalicine (Ramadan et al. 2016). Similarly, *Bacillus* sp. produces antimycotics and active antibacterial molecules that are mainly derived from non-ribosomal and ribosomal sources. These include subtilosin A, mycobacillin, difficidin, bacillaene, chlorotetain, and rhizocticin, surfactin, bacillomycin, and iturin molecules (Wang et al. 2015; Gouda et al. 2018). *Trichoderma* sp., a fungal biocontrol agent, has been of great importance in the crop protection industry. *Trichoderma viride* and *Trichoderma harzianum* species have been demonstrated to effectively control *Fusarium graminearum* pathogen that causes *Fusarium* head blight in wheat (*Triticum aestivum* L.) (Panwar et al. 2014). Spraying of non-aflatoxigenic *Aspergillus flavus* on the reproductive structures of *Zea mays* L. cobs could reduce the levels of aflatoxin contamination (Lyn et al. 2009). Entomopathogenic fungi, *Beauveria bassiana* GHA, and *Metarhizium brunneum* strains have been effectively used to control soybean aphids (Clifton et al. 2018). Despite the success, up to date, a broad-spectrum biocontrol agent that is more efficient and persistent remains a challenge. In addition, the efficacy of antifungal agents greatly varies and is determined mainly by the moisture content, number of applications, and target stage of the pest.

Rhizobium etli, a nitrogen-fixing symbiotic bacteria, have been shown to have antagonistic properties against soil nematodes by activating ethylene and jasmonic acid biochemical pathways that induce resistance against root-knot nematode in tomatoes (Martinuz et al. 2012). Enzyme-secreting bacteria that hydrolyze proteins, cellulose, chitin, and hemicellulose of phytopathogens could be used as biocontrol agents to suppress plant diseases. Bacteria bearing such important traits should be able to colonize the target niche of the plant for an effective pathogen suppression to be achieved. A lot of bacteria, including *Lysobacter* sp. and *Myxobacteria* sp., produce chitinases, glucanases, proteases, cellulases, and lytic enzymes that effectively hydrolyze the fungal cell wall components. These bacteria can be utilized by farmers to suppress the incidence of diseases caused by *Rhizoctonia*, *Pythium*, and *Sclerotium* species (Saraf et al. 2014).

11.3.8 Microbial Inoculants, Yield Quality, and Human Health

In the developed world, microbes have been used extensively by farmers to boost soil health, plant growth, and productivity and to strengthen the plant's resilience and adaptability to the constantly changing climatic conditions. Indeed, especially in organic and low-input agricultural setups, microbial inoculants have positively impacted on agriculture and well-being of the farmers (Alori and Babalola 2018). The African continent is known for its biodiversity, and her soil harbors a wide range of soil microorganisms (bacteria and fungi) that are beneficial. Proper utilization of its rich biodiversity could revolutionize the deteriorating African agriculture, which is mainly characterized by resource-constrained smallholder farming systems (Grönemeyer and Reinhold-Hurek 2018).

Unlike in other regions with developing economies such as Asia, the issue of adopting green energy sources to address low soil fertility, malnutrition, and food insecurity remains underutilized in sub-Saharan Africa. Various initiatives, which have been undertaken to address this challenge, including the establishment of projects like N2Africa, AgBiome, and UNESCO-MIRCEN, among others, that incorporate different groups of stakeholders. To ensure the success in adopting technologies developed from soil microorganisms, participatory field-based experiments for demonstrations should be conducted in strategic locations where farmers can access as this will benefit and empower the local communities. In most cases, experienced researchers, universities, funding agencies, private commercial companies, and African farmers collectively steer the exploration, identification, field testing, evaluation, and adoption of microbial inoculants and ensure their sustainability for farmers' use.

Previous studies have also shown that adaptation of the microbes to the local soil and environmental conditions is a huge factor to consider when exploring beneficial microbial inoculants because they are better adapted to the ecological conditions (Ouma et al. 2016; Koskey et al. 2017). Grönemeyer and Reinhold-Hurek (2018) reported a hidden high diversity of *Bradyrhizobium* species, with exceptionally heat-tolerant traits, that form a symbiotic partnership with legume pulses such as soybean (*Glycine max* L. Merrill), peanut (*Arachis hypogaea* L.), cowpea (*Vigna unguiculata* L.), and bambara groundnut (*Vigna subterranean* L.) commonly grown by smallholder farmers in SSA. These types of microbes can be utilized in most parts of the SSA, where heat is the primary ecological stress for the crops. Modern breeding methods have greatly affected scaling up food production in the SSA

Fig. 11.2 A smallholder farm in Embu County, Kenya, showing healthy and high-yielding cowpea crop following inoculation with effective indigenous rhizobia isolates. Indigenous rhizobia isolates are cheaper and more adapted to the local agroclimatic conditions and develop positive microbial interactions with existing soil microflora compared to exotic commercial isolates

region. In Nigeria, the development and introduction of promiscuous soybeans, by the IITA, that form symbiotic association with diverse *Bradyrhizobium* strains has been an eye-opener to many smallholder farmers (Gabasawa 2020). In addition, Oruru et al. (2018) have shown the benefits of using modern cowpea (*Vigna unguiculata* L.) cultivars in Kenya in enhancing AMF root colonization, NPK uptake, and growth compared to the wild-type cultivars (Fig. 11.2).

Other co-inoculation studies have also shown the importance of soil biodiversity in managing the depleted soil nutrients. N'cho et al. (2013) demonstrated that co-inoculation of *Bradyrhizobium* spp. (RACA6), *Trichoderma harzianum* (Eco-T), and AMF (Rhizatech) with the commercial Agrolyser and Agroleaf foliar fertilizers increased soybean nodulation, shoot P, and grain yields significantly. Under intense water stress, Musyoka et al. (2020) demonstrated through a greenhouse experiment that a consortium of AMF isolates *Glomus etunicatum, Rhizophagus irregularis*, *Glomus aggregatum,* and *Funneliformis mosseae* could increase green grams' (*Vigna radiata* L. Wildzek) shoot and root biomass and P uptake compared to the *Bradyrhizobium* inoculants and uninoculated controls. However, some studies suggest that functional identity could be more important under specific situations, mainly when using selective microbial inoculants that colonize specific crop genotypes. For instance, a study by Njeru et al. (2017) clearly demonstrated that functional identity of individual AMF isolates *Funneliformis mosseae* IMA1 and *Rhizoglomus intraradices* IMA6 could play a more significant role than their compounded diversity in enhancing AMF root colonization and fresh weight of marketable tomato fruits in specific tomato genotypes.

It wasn't until the mid-1990s that AMF was discovered to harbor mycorrhizospheric helper (MH) endobacteria that synergistically interact with the AMF and are reportedly known to associate with the spores, plant roots, and hyphae, thus extending the hyphae-absorbing network (Bianciotto et al. 1996). The MH bacteria are affiliated with Pseudomonales, Burkholderiales, Bacillales, Rhizobiales, and Actinomycetales, which are known for their beneficial plant growth-promoting traits. Further, these MH bacteria were discovered to take part in the establishment of plant-AMF symbiosis and promote spore germination, hyphal growth, and root colonization (Agnolucci et al. 2015; Giovannini et al. 2020). The multiple beneficial traits of AMF and their associated endo-bacteria could be efficiently exploited in smallholder food production through further research on the best effective combinations that can work well in the context of ecological conditions of the SSA. However, their cost of production has to be drastically reduced for the farmers to access cheap, high-quality, and well-packaged microbial products for their farm use.

11.4 Enhancing Healthy Plant-Microbe Interactions in Smallholder Agroecosystems

To achieve a sustainable food production for the growing population, amid the rising cases of climatic instability in SSA, more attention should be given to the innovations that promote self-sustainability of the natural ecosystems and those that advocate for agricultural biodiversity at genetic, species, and habitat management levels (Costanzo and Bàrberi 2014). Soil fertility restoration through mycorrhiza, BNF, and PGPRs aided symbiosis processes that is a complex phenomenon and does not always result in significant improvement in soil quality. This is true in the context of SSA where the soils are exposed continuously to various mechanisms that minimize the benefits that could be gained via the interactions (Raimi et al. 2017). Indigenous AMF, which are beneficial to plants and soil in increasing P solubilization, nutrient availability, carbon sequestration, soil aggregation, and plant stress resistance, do require appropriate agronomic management practices for their maximum contribution to be realized (Giovannetti et al. 2004; Avio et al. 2006). Some of the commonly used agricultural practices that support plant-soil-microbial tripartite interactions include reduced physical soil disturbance (minimum or no-tillage), continuous and rotational cropping, organic amendments, intercropping, cover cropping, use of microbial inoculants, and balanced nutrient management (Njeru 2013). The idea of promoting the presence of high microbial diversity in agricultural soil is to ensure that critical soil functions are carried out by different groups of soil microorganisms at a particular time when other redundant groups are unable or unavailable due to the drastic change in climatic or physical conditions (Mburu et al. 2016). In this way, a highly diverse microbial community offers the much-needed insurance to the farmers that the soil processes that maintain and support plant growth under changing environmental conditions are performed. Long-term experiments on frequent tillage against minimum or no-tillage have been carried out, and results indicated that reduced tillage leads to increased bacterial and fungal biomass, which are the main drivers of natural ecosystems (Frey et al. 1999; Marzaioli et al. 2010). Intensive cultivation, which is often practiced by smallholder farmers in the SSA region, has been shown to lead to soil organic matter depletion progressively, higher CN ratio, reduced biodiversity, and microbial functionality, which consequently lead to reduced crop performance (Ventorino et al. 2012).

Continuous cropping has been shown to add more organic carbon to the soil and maintains obligate beneficial soil microorganisms such as AMF that would not survive without a living host (Hontoria et al. 2019). Crop rotation, on the other hand, has been studied, and results indicate that it plays a critical role in maintaining higher biodiversity. However, it does not always hold that the higher the above-ground plant diversity, the better the soil microbial diversity. It depends on other factors like the plant genotype used during the rotation. The inclusion of legume cover crops, cultivar mixtures, and other mycotrophic crops highly encourages the colonization of symbiotic N-fixing bacteria and P-solubilizing AMF fungi (Njeru et al. 2015; Lazzaro et al. 2018). Therefore, farmers should have some knowledge of the crop types they choose to have a better structured and functional microbial diversity that will aid crop production. The addition of organic amendments helps to sustain high energy demanding soil processes like microbial degradation and nutrient recycling, maintains nutrient and water retention, stabilizes soil structure, and creates a favorable resource-rich microhabitat for plants and other microbial dependents (García-Orenes et al. 2013; Nyamwange et al. 2018). None of the opportunities, as mentioned above, would succeed without the farmers' knowledge. The local knowledge of the farmer is essential for the adoption of modern innovations and agricultural practices. Therefore, unlocking the power of the smallholder farmers' understanding of modern agronomic management practices that promote microbial functions, improves soil fertility and crop productivity and could revolutionize the agricultural sector in many parts of the world.

11.5 Towards the Development of Effective Microbial Inocula

Microbial inoculants carrying beneficial microorganisms have been widely adopted in organic and low-input agricultural systems because of their ability to deliver target microorganisms into the root rhizosphere. For instance, after AMF inoculation, AMF interacts with the plant host-symbiont and develops extra-radical mycelia that colonizes the rhizosphere, solubilizes nutrients, and interacts with other rhizospheric microorganisms and plants (Avio et al. 2006). Ideally, the generation of microbial inoculants starts with the identification of microorganisms with the target trait-effect, which are then grown as lab cultures, followed by lab or greenhouse testing on target plants for their efficacy. The promising microbial candidates are then tried under natural field conditions (Ouma et al. 2016). For effective inoculation and plant growth to be achieved, several factors need to be considered, including mode of delivery, colonization ability, and efficacy. Colonization ability refers to the strength of the microorganism to rapidly and extensively colonize the root and its surroundings. At the same time, efficacy is determined by the symbiotic performance of the microbe on the plant host in enhancing plant growth, development, and nutrition (Giovannini et al. 2020).

A right microbial culture for use in agricultural inoculation should depict a high colonization ability and should compete with other existing native microbial populations in the soil. There are external factors that could affect the colonization ability of the inoculant, which include soil pH, salinity, and environmental conditions such as water stress, heat, and radiations. Inoculants should, therefore, be prepared from the native microbial strains that have adapted to the local climatic and soil conditions over a long time. It is also vital for commercial companies to prepare formulations with known shelf lives as this could critically determine the number of viable microbial cells after inoculation. Some microbial products are incompatible with other commonly used agricultural inputs, hence the need for proper labeling and declaration by the manufacturers.

For maximum efficacy to be achieved, microorganisms must overcome the soil barriers and competition from the resident microbiota and establish large, active, and functional populations that would confer an observable effect on soil health and crop productivity (Lewis et al. 2019). Thus, understanding the physiology and growth requirements of a specific microbial inoculant strain is essential to enhance the growth efficiency, functionality, and stability of the inoculant strain. Nowadays, formulations are available either in liquid or solid form depending on the manufacturer's choice, market demands, and storage. Overcoming desiccation and temperature stresses is mostly considered while choosing the type of microbial formulations to be used in preparing inoculants targeting seed dressing.

Seed companies have introduced "custom inoculation" where seeds are inoculated with specific microbial inoculant strains only on farmers' demands after sale and "pre-inoculation" where seeds are inoculated prior to sale (Deaker et al. 2012). These two approaches relieve the farmer from the hustle of inoculation of seeds on-farm however; there is some extra cost that the farmer needs to incur. Remarkably, the need to bioprospect for better and effective microbial inoculants has led to the production of broad-spectrum combinations of elite strains unlike in the past where first companies produced inoculants with only one or two specific microbial strains (Santos et al. 2019). The idea is supported by targeting a combination of different strains that can carry out different microbial processes efficiently and ultimately produce higher yields. Co-inoculation of seeds with BNF-associated microbes (*Rhizobium* sp.), phytohormone producers (*Pseudomonas* sp., *Azospirillum* sp.), P solubilizers (*Bacillus* sp.), and biocontrol agents (*Trichoderma* sp., *Bacillus* sp.) is commonly used by the SSA farmers (Trabelsi and Mhamdi 2013). With the increasing concerns about the changing climates evident by prolonged droughts, frequent heat waves, flooding, and extreme temperatures, the performance of the currently available microbial inoculants may not be guaranteed. It is, therefore, mandatory to do further research on inoculants that are well adapted to the current and incoming stressful conditions.

11.6 The Role of Microorganisms in Supporting the Resilience of Smallholder Agriculture Systems

The agricultural productivity of the African smallholder systems is gradually declining owing to soil impoverishment instigated by the changing climatic conditions and insufficient adoption of modern sustainable farming technologies (Ngetich et al. 2012). The decreasing soil fertility is further exacerbated by the high decomposition rates of organic matter, rapid soil weathering conditions, high soil acidity, excessive leaching, and intensive cultivation practices (Mukhongo et al. 2016). The appropriate and consistent use of bio-inoculants could offer farmers various biological and agronomic benefits, including nutrient solubilization and uptake, growth stimulation, yield increase, cost reduction, soil health, and fertility restoration (Masso et al. 2015). Therefore, the role played by microbial inoculants is critical in revolutionizing smallholder agricultural systems, maintaining the soil nutrient balance, and crop productivity.

Recently, the use and adoption of effective microbial bio-inoculants such as AMF, rhizobial, and phosphatic inoculants in SSA have slightly increased among smallholder farmers, although the increase is much lower compared to that of other regions of the world (Raimi et al. 2017). Commercial companies such as Dudutech Ltd. in Kenya and Mycoroot Pty Ltd. in South Africa have increased their AMF production due to the increasing demand by the smallholder farmers in their respective regions (Mukhongo et al. 2016). It is estimated that the global market for bio-inoculants could increase progressively at a rate of 12.5% per annum, and by the year 2025, its global value would have reached the US $4092 million, up from the US $1254 million in 2016 (Transparency Market Research 2017). The cost of peat-based rhizobia fertilizers for white clover and faba beans ranges from US $0.25 to 6.5 ha^{-1}, and this could fit the constrained budget of the smallholder farmers. Comparably, these prices are far much below the cost of mineral N fertilizers that could be needed to supply the same quantity of nutrients (Raimi et al. 2017).

The profitability benefits of using bio-inoculants are primarily based on the amount of nutrient fixed/solubilized and/or yields. For instance, in Ghana, Masso et al. (2016) used Legumefix, a rhizobia-based bio-inoculant, to grow soybean and common beans and found that farmers could profit from the inoculation with a net value-cost ratio of >3 when compared to uninoculated control. Nitrogen-fixing rhizobia bio-inoculants increase legume yield, a factor considered by the majority of smallholder farmers. According to a review by Ngetich et al. (2012), legumes through the BNF process can naturally contribute about 48–300 kg of N/ha in a season, and this amount could significantly increase if legumes are inoculated with effective bio-inoculants. Soybean *Bradyrhizobium* inoculants increased soil organic matter and yield and fixed about 80% of the total soybean N requirements in smallholder farm settings in Southern Africa (Kasasa et al. 1999). In West Africa, Osunde et al. (2003) demonstrated that about 54% (the equivalent of 78 kg N/ha) of the total N requirement is fixed by inoculated soybean, and farmers could minimize the supply of external inputs required for subsequent cereal cropping systems. Similarly, in the rice experiment conducted by Rose et al. (2014) using a farmer participatory approach, bio-inoculants were reported to ease the chemical N fertilizer supply by about 52% without significant yield loss. As an alternative source of inorganic N fertilizer, Gebre and Lelago (2017) showed that cyanobacteria bio-fertilizer could be used to reclaim the nutrient-poor alkaline soils, improve soil resilience, and increase yields of kales in Eastern Africa. Bio-inoculants with specific strains of *Bacillus* and *Pseudomonas*, which produce hydrolytic phosphatase enzymes that mineralize organic P, can save African smallholder farmers by adding up to 30–50 kg/ha of P_2O_5 fertilizers (Richardson and Simpson 2011).

Most parts of the SSA are arid and semiarid areas, characterized by long drought seasons that expose plants to frequent water and salinity stresses (Falkenmark and Rockström 2008). Smallholder farmers would benefit from using bio-fertilizers that contain microorganisms that would increase plant tolerance to salinity and water stress. Inoculation of plants with auxin-, cytokinin-, and gibberellin-producing microbes has been reported to improve plant tolerance to water stress and reduce the risk of yield losses significantly (Goswami and Deka 2020). Gururani et al. (2013) observed an increased tolerance to salt and water stress in potatoes inoculated with *Bacillus* spp. That stimulates the production of 1-aminocyclopropane-1-carboxylate (ACC) deaminase. Likewise, plant-mycorrhizal symbiotic relationships can be exploited to enhance root development, moisture, and P uptake, thus ensuring agricultural resilience and productivity under water stress and low soil P conditions (Oruru and Njeru 2016; Musyoka et al. 2020). According to Masso et al. (2016) and Raimi et al. (2017), smallholder farmers in drought-prone areas of the SSA could get more yield benefits by inoculating sweet potato, maize, and rice with effective AMF strains. The extensive hyphal network of AMF reduces localized competition for limited water and nutrients, thus supporting plant biodiversity and maintaining the sustainability of the agricultural ecosystems (Mukhongo et al. 2016).

Bacterial and fungal diseases often cause substantial yield losses in African smallholder production systems, and using low-cost bio-inoculants that produce antimycotic and antibacterial substances could assist in suppressing the associated crop losses (Strange and Scott 2005). In Kenya, Masso et al. (2016) demonstrated that *Trichoderma* inoculants could control late blight disease in tomatoes much better ($p < 0.05$) than the commercial pesticide Ridomil that is commonly used by the farmers. Similarly, other cost-effective inoculants containing *Bacillus* spp., *Pseudomonas fluorescens,* and *Sinorhizobium* spp. can be used in controlling *Fusarium* wilt in pigeon pea, bacterial soft rot in potato, and *Rhizoctonia solani* infections in pepper (Kumar et al. 2010). Co-inoculation of AMF and rhizobia not only increases NPK availability and uptake (Tairo and Ndakidemi 2014) but also enriches the soil with trace minerals such as calcium (Ca), iron (Fe), sulfur (S), zinc (Zn), copper (Cu), and manganese (Mn) that are not often externally supplied by the smallholder farmers (Bambara and Ndakidemi 2010). Furthermore, AMF are known to suppress soilborne pathogens, and this offers crop protection services to the farmers and reduces the overdependence of nonselective pesticides that are harmful to other beneficial soil microbiota (Mukhongo et al. 2016).

11.7 Conclusion

Notwithstanding the plausible contribution of beneficial microbial communities in smallholder agriculture, their adoption and conservation have remained significantly low across many developing nations. Among the various reasons for this is inadequate training and research on beneficial microorganisms in smallholder farms. Therefore, there is a need for the formulation of policies and subsidy programs on bio-fertilizers and biopesticides, besides investment in research and development. Furthermore, a more holistic approach including laboratory, greenhouse, and field experiments established through participatory research and used for on-farm assessment of low-cost microbial inocula is imperative, since some promising microorganisms may not necessarily harbor favorable survival characteristics in the field (Parnell et al. 2016). It is envisaged that the effective, low-cost inocula would be widely adopted by farmers and used to promote sustainable food production, cash generation, and resilience of smallholder agroecosystems to changing climate.

Acknowledgments This work was supported by The Future Leaders – African Independent Researchers (FLAIR) Fellowship Programme, which is a partnership between the African Academy of Sciences and the Royal Society funded by the UK Government's Global Challenges Research Fund, Research (Grant number FLR\R1\190944).

References

Agami R, El Khalek A, Abd El Mola IA, Medani R, Abd El-Mola I, Taha R (2016) Exogenous application with plant growth promoting rhizobacteria (PGPR) or proline induces stress tolerance in basil plants (Ocimum basilicum L.) exposed to water stress. Int J Environ Agric Res 2:5

Agnolucci M, Battini F, Cristani C, Giovannetti M (2015) Diverse bacterial communities are recruited on spores of different arbuscular mycorrhizal fungal isolates. Biol Fertil Soils 51:379–389. https://doi.org/10.1007/s00374-014-0989-5

Ahemad M, Kibret M (2014) Mechanisms and applications of plant growth promoting rhizobacteria: current perspective. J King Saud Univ Sci 26:1–20

Ahmed E, Holmström SJM (2014) Minireview siderophores in environmental research: roles and applications. J Microbial Biotechnol 7(3):196–208. https://doi.org/10.1111/1751-7915.12117

Alam S, Rashid M (2002) In vitro solubilization of inorganic phosphate by phosphate solubilizing microorganisms (PSM) from maize rhizosphere. Int J Agric Biol 4:454–458

Alori ET, Babalola OO (2018) Microbial inoculants for improving crop quality and human health in Africa. Front Microbiol 9:2213. https://doi.org/10.3389/fmicb.2018.02213

Alori ET, Glick BR, Babalola OO (2017) Microbial phosphorus solubilization and its potential for use in sustainable agriculture. Front Microbiol 8:971. https://doi.org/10.3389/fmicb.2017.00971

Altieri MA, Funes-Monzote FR, Petersen P (2012) Agroecologically efficient agricultural systems for smallholder farmers: contributions to food sovereignty. Agron Sustain Dev 32:1–13. https://doi.org/10.1007/s13593-011-0065-6

Aniah P, Kaunza-Nu-Dem MK, Ayembilla JA (2019) Smallholder farmers' livelihood adaptation to climate variability and ecological changes in the savanna agro ecological zone of Ghana. Heliyon 5:e01492–e01492. https://doi.org/10.1016/j.heliyon.2019.e01492

Aseri G, Jain N (2009) Hydrolysis of organic phosphate forms by phosphatases and phytase producing fungi of arid and semi-arid soils of India. Am J Agric Environ Sci 5:564–570

Avio L, Pellegrino E, Bonari E, Giovannetti M (2006) Functional diversity of arbuscular mycorrhizal fungal isolates in relation to extraradical mycelial networks. New Phytol 172:347–357. https://doi.org/10.1111/j.1469-8137.2006.01839.x

Avio L, Turrini A, Giovannetti M, Sbrana C (2018) Designing the ideotype mycorrhizal symbionts for the production of healthy food. Front Plant Sci 9:1089. https://doi.org/10.3389/fpls.2018.01089

Awazi NP, Tchamba MN (2019) Enhancing agricultural sustainability and productivity under changing climate conditions through improved agroforestry practices in smallholder farming systems in sub-Saharan Africa. African J Agric Res 14:379–388. https://doi.org/10.5897/AJAR2018.12972

Baharlouei J, Khavazi K, Pazira E, Solhi M (2011) Evaluation of inoculation of plant growth-promoting rhizobacteria on cadmium uptake by canola and barley. In: International conference on environmental science and technology. IACSIT Press, Singapore, pp 128–132

Bakhshandeh E, Pirdashti H, Lendeh KS (2017) Phosphate and potassium-solubilizing bacteria effect on the growth of rice. Ecol Eng 103:164–169. https://doi.org/10.1016/j.ecoleng.2017.03.008

Bambara S, Ndakidemi PA (2010) Changes in selected soil chemical properties in the rhizosphere of Phaseolus vulgaris L. supplied with Rhizobium inoculants, molybdenum and lime. Sci Res Essays 5:679–684

Bianciotto V, Bondi C, Minerdi D, Sironi M, Tichy HV, Bonfante P (1996) An obligately endosymbiotic mycorrhizal fungus itself harbors obligately intracellular bacteria. Appl Environ Microbiol 62:3005–3010. https://doi.org/10.1128/aem.62.8.3005-3010.1996

Caradonia F, Francia E, Morcia C, Ghizzoni R, Moulin L, Terzi V, Ronga D (2019) Arbuscular mycorrhizal fungi and plant growth promoting rhizobacteria avoid processing tomato leaf damage during chilling stress. J Agron 9:299

Cardinale M, Ratering S, Suarez C, Montoya AM, Geissler-Plaum R, Schnell S (2015) Paradox of plant growth promotion potential of rhizobacteria and their actual promotion effect on growth of barley (Hordeum vulgare L.) under salt stress. Microbiol Res 181:22–32. https://doi.org/10.1016/j.micres.2015.08.002

Chen YP, Rekha PD, Arun AB, Shen FT, Lai WA, Young CC (2006) Phosphate solubilizing bacteria from subtropical soil and their tricalcium phosphate solubilizing abilities. Appl Soil Ecol 34:33–41. https://doi.org/10.1016/j.apsoil.2005.12.002

Chitarra W, Pagliarani C, Maserti B, Lumini E, Siciliano I, Cascone P, Schubert A, Gambino G, Balestrini R, Guerrieri E (2016) Insights on the impact of arbuscular mycorrhizal symbiosis on tomato tolerance to water stress. Plant Physiol 171:1009–1023. https://doi.org/10.1104/pp.16.00307

Choudhary DK, Varma A (2017) Nitrogenase (a key enzyme): structure and function. In: Rhizobium biology and biotechnology. Springer, Cham, pp 293–307

Clair SB, Lynch JP (2010) The opening of Pandora's box: climate change impacts on soil fertility and crop nutrition in developing countries. Plant and Soil 335:101–115

Clifton EH, Jaronski ST, Coates BS, Hodgson EW, Gassmann AJ (2018) Effects of endophytic entomopathogenic fungi on soybean aphid and identification of Metarhizium isolates from agricultural fields. PLoS One 13:e.0194815. https://doi.org/10.1371/journal.pone.0194815

Costanzo A, Bàrberi P (2014) Functional agrobiodiversity and agroecosystem services in sustainable wheat production. A Review Agron Sustain Dev 34:327–348

Deaker R, Hartley E, Gemell G (2012) Conditions affecting shelf-life of inoculated legume seed. Agri 2:38–51. https://doi.org/10.3390/agriculture2010038

Dissanayaka DMSB, Plaxton WC, Lambers H, Siebers M, Marambe B, Wasaki J (2018) Molecular mechanisms underpinning phosphorus-use efficiency in rice. Plant Cell Environ 41:1483–1496. https://doi.org/10.1111/pce.13191

Dominguez-Nuñez JA, Benito B, Berrocal-Lobo M, Albanesi A (2016) Mycorrhizal fungi: role in the solubilization of potassium. In: Meena V, Maurya B, Verma J, Meena R (eds) Potassium solubilizing microorganisms for sustainable agriculture. Springer, New Delhi, pp 77–98

Dourado MN, Martins PF, Quecine MC, Piotto FA, Souza LA, Franco MR, Tezotto T, Azevedo RA (2013) Burkholderia sp. SCMS54 reduces cadmium toxicity and promotes growth in tomato. Ann Appl Biol 163:494–507. https://doi.org/10.1111/aab.12066

Egamberdieva D, Wirth SJ, Alqarawi AA, Abd-Allah EF, Hashem A (2017) Phytohormones and beneficial microbes: essential components for plants to balance stress and fitness. Front Microbiol 8:2104. https://doi.org/10.3389/fmicb.2017.02104

Etesami H, Emami S, Alikhani HA (2017) Potassium solubilizing bacteria (KSB): mechanisms, promotion of plant growth, and future prospects - a review. J Soil Sci Plant Nutr 17:897–911

Falkenmark M, Rockström J (2008) Building resilience to drought in desertification-prone savannas in sub-Saharan Africa: the water perspective. Nat Resour Forum 32:93–102. https://doi.org/10.1111/j.1477-8947.2008.00177.x

Fileccia V, Ruisi P, Ingraffia R, Giambalvo D, Frenda AS, Martinelli F (2017) Arbuscular mycorrhizal symbiosis mitigates the negative effects of salinity on durum wheat. PLoS One 12:9. https://doi.org/10.1371/journal.pone.0184158

Foyer CH, Rasool B, Davey JW, Hancock RD (2016) Cross-tolerance to biotic and abiotic stresses in plants: a focus on resistance to aphid infestation. J Exp Bot 67:2025–2037

Frey S, Elliott ET, Paustian K (1999) Bacterial and fungal abundance and biomass in conventional and no-tillage agroecosystems along two climatic gradients. Soil Biol Biochem 31:573–585. https://doi.org/10.1016/S0038-0717(98)00161-8

Fukami J, Cerezini P, Hungria M (2018) Azospirillum: benefits that go far beyond biological nitrogen fixation. AMB Express 8:73

Funk CC, Brown ME (2009) Declining global per capita agricultural production and warming oceans threaten food security. Food Secur 1:271–289

Gabasawa AI (2020) Prospects for developing effective and competitive native strains of rhizobium inoculants in Nigeria. In: Current microbiological research in Africa. Springer, Cham, pp 223–256

Gallegos-Cedillo VM, Urrestarazu M, Álvaro JE (2016) Influence of salinity on transport of nitrates and potassium by means of the xylem sap content between roots and shoots in young tomato plants. J Soil Sci Plant Nutr 16:991–998

García-Orenes F, Morugán-Coronado A, Zornoza R, Scow K (2013) Changes in soil microbial community structure influenced by agricultural management practices in a Mediterranean agro-ecosystem. PLoS One 8:e80522. https://doi.org/10.1371/journal.pone.0080522

Gebre E, Lelago A (2017) Effect of cyanobacterial biofertilizer on soil quality in kale (Brassica Oleracea L) crop growing field at Ziway area, Ethiopia. Food Sci Qual Manage 61:1–6

Giller KE, Kanampiu F, Hungria M, Vanlauwe B (2019) The role of nitrogen fixation in African smallholder agriculture. Agric Ecosyst Environ 285:106601. https://doi.org/10.1016/j.agee.2019.106601

Giovannetti M, Sbrana C, Avio L, Strani P (2004) Patterns of below-ground plant interconnections established by means of arbuscular mycorrhizal networks. New Phytol 164:175–181. https://doi.org/10.1111/j.1469-8137.2004.01145.x

Giovannini L, Palla M, Agnolucci M, Avio L, Sbrana C, Turrini A, Giovannetti M (2020) Arbuscular mycorrhizal fungi and associated microbiota as plant biostimulants: research strategies for the selection of the best performing inocula. Agronomy 10:106. https://doi.org/10.3390/agronomy10010106

Goswami M, Deka S (2020) Plant growth-promoting rhizobacteria—alleviators of abiotic stresses in soil: a review. Pedosphere 30:40–61

Gouda S, Kerry RG, Das G, Paramithiotis S, Shin HS, Patra JK (2018) Revitalization of plant growth promoting rhizobacteria for sustainable development in agriculture. Microbiol Res 206:131–140

Govindaraj M, Vetriventhan M, Srinivasan M (2015) Importance of genetic diversity assessment in crop plants and its recent advances: an overview of its analytical perspectives. Genet Res Int 2015:431487–431487. https://doi.org/10.1155/2015/431487

Grönemeyer JL, Reinhold-Hurek B (2018) Diversity of bradyrhizobia in sub-Sahara Africa: a rich resource. Front Microbiol 9:2194. https://doi.org/10.3389/fmicb.2018.02194

Gururani MA, Upadhyaya CP, Baskar V, Venkatesh J, Nookaraju A, Park SW (2013) Plant growth-promoting rhizobacteria enhance abiotic stress tolerance in Solanum tuberosum through inducing changes in the expression of ROS-scavenging enzymes and improved photosynthetic performance. J Plant Growth Regul 32:245–258. https://doi.org/10.1007/s00344-012-9292-6

Habib H, Kausar H, Saud HM (2016) Plant growth-promoting rhizobacteria enhance salinity stress tolerance in okra through ROS-scavenging enzymes. Biomed Res Int 2016:6284547. https://doi.org/10.1155/2016/6284547

Hajjam Y, Cherkaoui S (2017) The influence of phosphate solubilizing microorganisms on symbiotic nitrogen fixation: perspectives for sustainable agriculture. J Mater Environ Sci 8:801–808

Hassan TU, Bano A, Naz I (2017) Alleviation of heavy metals toxicity by the application of plant growth promoting rhizobacteria and effects on wheat grown in saline sodic field. Int J Phytoremediation 19:522–529. https://doi.org/10.1080/15226514.2016.1267696

Hontoria C, García-González I, Quemada M, Roldán A, Alguacil MM (2019) The cover crop determines the AMF community composition in soil and in roots of maize after a ten-year continuous crop rotation. Sci Total Environ 660:913–922. https://doi.org/10.1016/j.scitotenv.2019.01.095

Hussain Z, Khattak RA, Irshad M, Mahmood Q, An P (2016) Effect of saline irrigation water on the leachability of salts, growth and chemical composition of wheat (Triticum aestivum L.) in saline-sodic soil supplemented with phosphorus and potassium. J Soil Sci Plant Nutr 16:604–620. https://doi.org/10.4067/s0718-95162016005000031

Islam F, Yasmeen T, Arif MS, Riaz M, Shahzad SM, Imran Q, Ali I (2016) Combined ability of chromium (Cr) tolerant plant growth promoting bacteria (PGPB) and salicylic acid (SA) in attenuation of chromium stress in maize plants. Plant Physiol Biochem 108:456–467. https://doi.org/10.1016/j.plaphy.2016.08.014

Jacoby R, Peukert M, Succurro A, Koprivova A, Kopriva S (2017) The role of soil microorganisms in plant mineral nutrition—current knowledge and future directions. Front Plant Sci 8:1617. https://doi.org/10.3389/fpls.2017.01617

Kannahi M, Senbagam N (2014) Studies on siderophore production by microbial isolates obtained from rhizosphere soil and its antibacterial activity. J Chem Pharm Res 6:1142–1145

Kasasa P, Mpepereki S, Musiyiwa K, Makonese F, Giller KE (1999) Residual nitrogen benefits of promiscuous soybeans to maize under field conditions. African Crop Sci J 7:375–382

Koskey G, Mburu SW, Njeru EM, Kimiti JM, Ombori O, Maingi JM (2017) Potential of native rhizobia in enhancing nitrogen fixation and yields of climbing beans (Phaseolus vulgaris L.) in contrasting environments of eastern Kenya. Front Plant Sci 8:443. https://doi.org/10.3389/fpls.2017.00443

Koskey G, Mburu SW, Kimiti JM, Ombori O, Maingi JM, Njeru EM (2018) Genetic characterization and diversity of rhizobium isolated from root nodules of mid-altitude climbing bean (Phaseolus vulgaris L.) varieties. Front Microbiol 9:968. https://doi.org/10.3389/fmicb.2018.00968

Kumar H, Bajpai VK, Dubey RC, Maheshwari DK, Kang SC (2010) Wilt disease management and enhancement of growth and yield of Cajanus cajan (L) var. Manak by bacterial combinations amended with chemical fertilizer. Crop Prot 29:591–598. https://doi.org/10.1016/j.cropro.2010.01.002

Kumar A, Kumar A, Patel H (2018) Role of microbes in phosphorus availability and acquisition by plants. Int J Curr Microbiol App Sci 7:1344–1347. https://doi.org/10.20546/ijcmas.2018.705.161

Lazzaro M, Costanzo A, Bàrberi P (2018) Single vs multiple agroecosystem services provided by common wheat cultivar mixtures: weed suppression, grain yield and quality. F Crop Res 221:277–297. https://doi.org/10.1016/j.fcr.2017.10.006

Lewis RW, Islam A, Opdahl L, Davenport JR, Sullivan TS (2019) Comparative genomics, siderophore production, and iron scavenging potential of root zone soil bacteria isolated from 'Concord' grape vineyards. Microb Ecol 78:699–713. https://doi.org/10.1007/s00248-019-01324-8

Li J, Meng B, Chai H, Yang X, Song W, Li S, Lu A, Zhang T, Sun W (2019) Arbuscular mycorrhizal fungi alleviate drought stress in C(3) (Leymus chinensis) and C(4) (Hemarthria altissima) grasses via altering antioxidant enzyme activities and photosynthesis. Front Plant Sci 10:499–499. https://doi.org/10.3389/fpls.2019.00499

Lin L, Xu X (2013) Indole-3-acetic acid production by endophytic streptomyces sp. En-1 isolated from medicinal plants. Curr Microbiol 67:209–217. https://doi.org/10.1007/s00284-013-0348-z

Liu D, Lian B, Dong H (2012) Isolation of Paenibacillus sp. and assessment of its potential for enhancing mineral weathering. Geomicrobiol J 29:413–421. https://doi.org/10.1080/01490451.2011.576602

Lobo CB, Juárez Tomás MS, Viruel E, Ferrero MA, Lucca ME (2019) Development of low-cost formulations of plant growth-promoting bacteria to be used as inoculants in beneficial agricultural technologies. Microbiol Res 219:12–25

Lyn ME, Abbas HK, Zablotowicz RM, Johnson BJ (2009) Delivery systems for biological control agents to manage aflatoxin contamination of pre-harvest maize. Food Addit Contam 26:381–387. https://doi.org/10.1080/02652030802441521

Mabrouk Y, Hemissi I, Salem IB, Mejri S, Saidi M, Belhadj O (2018) Potential of rhizobia in improving nitrogen fixation and yields of legumes. In: Rigobelo EC (ed) Symbiosis. InTech, London, pp 107–119

Maingi JM, Shisanya CA, Gitonga NM, Hornetz B (2001) Nitrogen fixation by common bean (Phaseolus vulgaris L.) in pure and mixed stands in semi-arid south-East Kenya. Eur J Agron 14:1–12. https://doi.org/10.1016/S1161-0301(00)00080-0

Martinuz A, Schouten A, Sikora RA (2012) Systemically induced resistance and microbial competitive exclusion: implications on biological control. Phytopathology 102:260–266. https://doi.org/10.1094/PHYTO-04-11-0120

Marzaioli R, D'Ascoli R, De Pascale RA, Rutigliano FA (2010) Soil quality in a Mediterranean area of southern Italy as related to different land use types. Appl Soil Ecol 44:205–212. https://doi.org/10.1016/j.apsoil.2009.12.007

Masso C, Ochieng JRA, Vanlauwe B (2015) Worldwide contrast in application of bio-fertilizers for sustainable agriculture: lessons for sub-Saharan Africa. J Bio Agricul Healthcare 5:34–50

Masso C, Mukhongo RW, Thuita M, Abaidoo R, Ulzen J, Kariuki G, Kalumuna M (2016) Biological inoculants for sustainable intensification of agriculture in sub-Saharan Africa smallholder farming systems. In: Climate change and multi-dimensional sustainability in African agriculture: climate change and sustainability in agriculture. Springer, Cham, pp 639–658

Mathur S, Jajoo A (2020) Arbuscular mycorrhizal fungi protects maize plants from high temperature stress by regulating photosystem II heterogeneity. Ind Crop Prod 143:111934. https://doi.org/10.1016/j.indcrop.2019.111934

Mburu SW, Koskey G, Kimiti JM, Ombori O, Maingi JM, Njeru EM (2016) Agrobiodiversity conservation enhances food security in subsistence-based farming systems of Eastern Kenya. Agric Food Secur 5:19. https://doi.org/10.1186/s40066-016-0068-2

Meena VS, Maurya BR, Verma JP (2014) Does a rhizospheric microorganism enhance K+ availability in agricultural soils? Microbiol Res 169:337–347. https://doi.org/10.1016/j.micres.2013.09.003

Mesa-Marín J, Del-Saz NF, Rodríguez-Liorente ID, Redondo-Gómez S, Pajuelo E, Ribas-Carbó M, Mateos-Naranjo E (2018) PGPR reduce root respiration and oxidative stress enhancing Spartina maritima root growth and heavy metal rhizoaccumulation. Front Plant Sci 9:1500. https://doi.org/10.3389/fpls.2018.01500

Mishra SK, Khan MH, Misra S, Dixit VK, Khare P, Srivastava S, Chauhan PS (2017) Characterisation of Pseudomonas spp. and Ochrobactrum sp. isolated from volcanic soil. Antonie van Leeuwenhoek, Int J Gen Mol Microbiol 110:253–270. https://doi.org/10.1007/s10482-016-0796-0

Mohammadi K (2012) Phosphorus solubilizing bacteria: occurrence, mechanisms and their role in crop production. Resour Environ 2012:80–85. https://doi.org/10.5923/j.re.20120201.10

Mukhongo RW, Tumuhairwe JB, Ebanyat P, AbdelGadir AH, Thuita M, Masso C (2016) Production and use of arbuscular mycorrhizal fungi inoculum in sub-Saharan Africa: challenges and ways of improving. Int J Soil Sci 11:108–122

Mungai LM, Snapp S, Messina JP, Chikowo R, Smith A, Anders E, Richardson RB, Li G (2016) Smallholder farms and the potential for sustainable intensification. Front Plant Sci 7:1720. https://doi.org/10.3389/fpls.2016.01720

Muoni T, Barnes AP, Öborn I, Watson CA, Bergkvist G, Shiluli M, Duncan AJ (2019) Farmer perceptions of legumes and their functions in smallholder farming systems in East Africa. Int J Agric Sustain 17:205–218. https://doi.org/10.1080/14735903.2019.1609166

Mus F, Alleman AB, Pence N, Seefeldt LC, Peters JW (2018) Exploring the alternatives of biological nitrogen fixation. Metallomics 10:523–538

Musyoka DM, Njeru EM, Nyamwange MM, Maingi JM (2020) Arbuscular mycorrhizal fungi and Bradyrhizobium co-inoculation enhances nitrogen fixation and growth of green grams (Vigna radiata L.) under water stress. J Plant Nutr 43:1036–1047. https://doi.org/10.1080/01904167.2020.1711940

Muthukumar T, Udaiyan K (2018) Coinoculation of bioinoculants improve Acacia auriculiformis seedling growth and quality in a tropical Alfisol soil. J For Res 29:663–673. https://doi.org/10.1007/s11676-017-0497-1

N'cho CO, Yusuf AA, Ama-Abina JT, Jemo M, Abaidoo RC, Savane I (2013) Effects of commercial microbial inoculants and foliar fertilizers on soybean nodulation and yield in northern Guinea savannah of Nigeria. Int J Adv Agric Res 3:73–82

Ngetich FK, Shisanya CA, Mugwe J, Mucheru-Muna M, Mugendi D (2012) The potential of organic and inorganic nutrient sources in sub-Saharan African crop farming systems. In: Whalen JK (ed) Soil fertility improvement and integrated nutrient management—a global perspective. IntechOpen, London, pp 135–159

Ngumbi E, Kloepper J (2016) Bacterial-mediated drought tolerance: current and future prospects. Appl Soil Ecol 105:109–125

Niu X, Song L, Xiao Y, Ge W (2018) Drought-tolerant plant growth-promoting rhizobacteria associated with foxtail millet in a semi-arid agroecosystem and their potential in alleviating drought stress. Front Microbiol 8:2580. https://doi.org/10.3389/fmicb.2017.02580

Njeru EM (2013) Crop diversification: a potential strategy to mitigate food insecurity by smallholders in sub-Saharan Africa Institute of Life Sciences. J Agric Food Syst Community Dev 3:63–69

Njeru EM, Avio L, Bocci G, Sbrana C, Turrini A, Bàrberi P, Giovannetti M, Oehl F (2015) Contrasting effects of cover crops on 'hot spot' arbuscular mycorrhizal fungal communities in organic tomato. Biol Fertil Soils 51:151–166. https://doi.org/10.1007/s00374-014-0958-z

Njeru EM, Bocci G, Avio L, Sbrana C, Turrini A, Giovannetti M, Bàrberi P (2017) Functional identity has a stronger effect than diversity on mycorrhizal symbiosis and productivity of field grown organic tomato. Eur J Agron 86:1–11. https://doi.org/10.1016/j.eja.2017.02.007

Novo LAB, Castro PML, Alvarenga P, da Silva EF (2018) Plant growth-promoting rhizobacteria-assisted phytoremediation of mine soils. In: Bio-geotechnologies for mine site rehabilitation. Elsevier, Amsterdam, pp 281–295

Nyamwange MM, Njeru ME, Mucheru-Muna M, Ngetich F (2018) Soil management practices affect arbuscular mycorrhizal fungi propagules, root colonization and growth of rainfed maize. AIMS Agric Food 3:120–134. https://doi.org/10.3934/agrfood.2018.2.120

Nyoki D, Ndakidemi PA (2018) Root length, nodulation and biological nitrogen fixation of rhizobium inoculated soybean (Glycine max [L.] Merr.) grown under maize (Zea mays L.) intercropping systems and P and K fertilization. Adv Biores 9:173–180

Odame H (1997) Biofertilizer in Kenya: research, production and extension dilemmas. Biotechnol Dev Monit 30:20–23

Oruru MB, Njeru EM (2016) Upscaling arbuscular mycorrhizal symbiosis and related agroecosystems services in smallholder farming systems. Biomed Res Int 2016:1–12. https://doi.org/10.1155/2016/4376240

Oruru MB, Njeru EM, Pasquet R, Runo S (2018) Response of a wild-type and modern cowpea cultivars to arbuscular mycorrhizal inoculation in sterilized and non-sterilized soil. J Plant Nutr 41:90–101. https://doi.org/10.1080/01904167.2017.1381728

Osunde AO, Bala A, Gwam MS, Tsado PA, Sanginga N, Okogun JA (2003) Residual benefits of promiscuous soybean to maize (Zea mays L.) grown on farmers' fields around Minna in the southern Guinea savanna zone of Nigeria. In: Agriculture, ecosystems and environment. Elsevier, Amsterdam, pp 209–220

Ouma EW, Asango AM, Maingi J, Njeru EM (2016) Elucidating the potential of native rhizobial isolates to improve biological nitrogen fixation and growth of common bean and soybean in smallholder farming systems of Kenya. Int J Agron 2016:4569241. https://doi.org/10.1155/2016/4569241

Pahari A, Mishra BB (2017) Characterization of siderophore producing rhizobacteria and its effect on growth performance of different vegetables. Int J Curr Microbiol App Sci 6:1398–1405. https://doi.org/10.20546/ijcmas.2017.605.152

Pajares S, Bohannan BJM (2016) Ecology of nitrogen fixing, nitrifying, and denitrifying microorganisms in tropical forest soils. Front Microbiol 7:1045

Panwar V, Aggarwal A, Singh G, Verma A, Sharma I, Saharan MS (2014) Efficacy of foliar spray of Trichoderma isolates against Fusarium graminearum causing head blight of wheat. J Wheat Res 6:59–63

Parnell JJ, Berka R, Young HA, Sturino JM, Kang Y, Barnhart DM, DiLeo MV (2016) From the lab to the farm: An industrial perspective of plant beneficial microorganisms. Front Plant Sci 7:1110–1110. https://doi.org/10.3389/fpls.2016.01110

Peix A, Rivas-Boyero AA, Mateos PF, Rodriguez-Barrueco C, Martınez-Molina E, Velazquez E (2001) Growth promotion of chickpea and barley by a phosphate solubilizing strain of Mesorhizobium mediterraneum under growth chamber conditions. Soil Biol Biochem 33:103–110. https://doi.org/10.1016/S0038-0717(00)00120-6

Prathap M, Ranjitha BD (2015) A critical review on plant growth promoting rhizobacteria. J Plant Pathol Microbiol 6:1–4. https://doi.org/10.4172/2157-7471.1000266

Raimi A, Adeleke R, Roopnarain A (2017) Soil fertility challenges and biofertiliser as a viable alternative for increasing smallholder farmer crop productivity in sub-Saharan Africa. Cogent Food Agric 3:1400933. https://doi.org/10.1080/23311932.2017.1400933

Ramadan EM, Abdelhafez AA, Hassan EA, Saber FM (2016) Plant growth promoting rhizobacteria and their potential for biocontrol of phytopathogens. Afr J Microbiol Res 10:486–504. https://doi.org/10.5897/AJMR2015.7714

Reddy M, Kumar S, Babita K (2002) Biosolubilization of poorly soluble rock phosphates by Aspergillus tubingensis and Aspergillus niger. Bioresour Technol 84:187–189

Richardson AE, Simpson RJ (2011) Soil microorganisms mediating phosphorus availability. Plant Physiol 156:989–996

Rose MT, Phuong TL, Nhan DK, Cong PT, Hien NT, Kennedy IR (2014) Up to 52% N fertilizer replaced by biofertilizer in lowland rice via farmer participatory research. Agron Sustain Dev 34:857–868. https://doi.org/10.1007/s13593-014-0210-0

Ruanpanun P, Tangchitsomkid N, Hyde KD, Lumyong S (2010) Actinomycetes and fungi isolated from plant-parasitic nematode infested soils: screening of the effective biocontrol potential, indole-3-acetic acid and siderophore production. World J Microbiol Biotechnol 26:1569–1578. https://doi.org/10.1007/s11274-010-0332-8
Sangeeth KP, Bhai RS, Srinivasan V (2012) Paenibacillus glucanolyticus, a promising potassium solubilizing bacterium isolated from black pepper (Piper nigrum L.) rhizosphere. J Spices Aromat Crop 21:118–124
Santos MS, Nogueira MA, Hungria M (2019) Microbial inoculants: reviewing the past, discussing the present and previewing an outstanding future for the use of beneficial bacteria in agriculture. AMB Express 9:205. https://doi.org/10.1186/s13568-019-0932-0
Saraf M, Pandya U, Thakkar A (2014) Role of allelochemicals in plant growth promoting rhizobacteria for biocontrol of phytopathogens. Microbiol Res 169:18–29. https://doi.org/10.1016/j.micres.2013.08.009
Sarkar J, Chakraborty B, Chakraborty U (2018) Plant growth promoting rhizobacteria protect wheat plants against temperature stress through antioxidant signalling and reducing chloroplast and membrane injury. J Plant Growth Regul 37:1396–1412. https://doi.org/10.1007/s00344-018-9789-8
Satyaprakash M, Nikitha T, Reddi EUB, Sadhana B, Vani SS (2017) Phosphorous and phosphate solubilising bacteria and their role in plant nutrition. Int J Curr Microbiol App Sci 6:2133–2144. https://doi.org/10.20546/ijcmas.2017.604.251
Selvi KB, Paul JJA, Vijaya V, Saraswathi K (2017) Analyzing the efficacy of phosphate solubilizing microorganisms by enrichment culture techniques. Biochem Mol Biol J 3:1–15. https://doi.org/10.21767/2471-8084.100029
Sharma SB, Sayyed RZ, Trivedi MH, Gobi TA (2013) Phosphate solubilizing microbes: sustainable approach for managing phosphorus deficiency in agricultural soils. Springerplus 2:587
Sheng XF (2005) Growth promotion and increased potassium uptake of cotton and rape by a potassium releasing strain of Bacillus edaphicus. Soil Biol Biochem 37:1918–1922. https://doi.org/10.1016/j.soilbio.2005.02.026
Shutsrirung A, Chromkaew Y, Pathom-Aree W, Choonluchanon S, Boonkerd N (2013) Diversity of endophytic actinomycetes in mandarin grown in northern Thailand, their phytohormone production potential and plant growth promoting activity. Soil Sci Plant Nutr 59:322–330. https://doi.org/10.1080/00380768.2013.776935
Singh G, Biswas DR, Marwaha TS (2010) Mobilization of potassium from waste mica by plant growth promoting rhizobacteria and its assimilation by maize (Zea mays) and wheat (Triticum aestivum L.): a hydroponics study under phytotron growth chamber. J Plant Nutr 33:1236–1251. https://doi.org/10.1080/01904161003765760
Sorty AM, Meena KK, Choudhary K, Bitla UM, Minhas PS, Krishnani KK (2016) Effect of plant growth promoting bacteria associated with halophytic weed (Psoralea corylifolia L) on germination and seedling growth of wheat under saline conditions. Appl Biochem Biotechnol 180:872–882. https://doi.org/10.1007/s12010-016-2139-z
Strange RN, Scott PR (2005) Plant disease: a threat to global food security. Annu Rev Phytopathol 43:83–116. https://doi.org/10.1146/annurev.phyto.43.113004.133839
Tabassum B, Khan A, Tariq M, Ramzan M, Khan MS, Shahid N, Aaliya K (2017) Bottlenecks in commercialisation and future prospects of PGPR. Appl Soil Ecol 121:102–117
Tairo EV, Ndakidemi PA (2014) Micronutrients uptake in soybean (Glycine max L.) as affected by Bradyrhizobium japonicum inoculation and phosphorus (p) supplements. World J Soil Crop Sci Res 1:1–9
Thakur D, Kaushal R, Shyam V (2014) Phosphate solubilising microorganisms: role in phosphorus nutrition of crop plants—a review. Agric Rev 35:159. https://doi.org/10.5958/0976-0741.2014.00903.9
Trabelsi D, Mhamdi R (2013) Microbial inoculants and their impact on soil microbial communities: a review. Biomed Res Int 2013:11. https://doi.org/10.1155/2013/863240

Transparency Market Research (2017) Biofertilizers (biological nitrogen fixation, phosphate solubilisers and others) market for seed treatment and soil treatment applications. Global industry analysis, size, share, growth, trends and forecast, 2017-2025. In: Transparency market research. Allbany, New York. https://www.transparencymarketresearch.com/biofertilizers-market.html. Accessed 23 Apr 2020

Upadhyay SK, Singh JS, Saxena AK, Singh DP (2012) Impact of PGPR inoculation on growth and antioxidant status of wheat under saline conditions. Plant Biol 14:605–611. https://doi.org/10.1111/j.1438-8677.2011.00533.x

Van Loon J, Woltering L, Krupnik TJ, Baudron F, Boa M, Govaerts B (2020) Scaling agricultural mechanization services in smallholder farming systems: case studies from sub-Saharan Africa, South Asia, and Latin America. Agr Syst 180:102792. https://doi.org/10.1016/j.agsy.2020.102792

Ventorino V, De Marco A, Pepe O, De Santo AV, Moschetti G (2012) Impact of innovative agricultural practices of carbon sequestration on soil microbial community. In: Piccolo A (ed) Carbon sequestration in agricultural soils. Springer, Berlin Heidelberg

Walpola CB, Yoon MH (2012) Prospectus of phosphate solubilizing microorganisms and phosphorus availability in agricultural soils: a review. Afr J Microbiol Res 6:6600–6605. https://doi.org/10.5897/AJMR12.889

Wang Q, Quensen JF, Fish JA, Lee TK, Sun Y, Tiedje JM, Cole JR (2013) Ecological patterns of nifH genes in four terrestrial climatic zones explored with targeted metagenomics using FrameBot, a new informatics tool. MBio 4(5):e00592. https://doi.org/10.1128/mBio.00592-13

Wang X, Mavrodi DV, Ke L, Mavrodi OV, Yang M, Thomashow LS, Zheng N, Weller DM, Zhang J (2015) Biocontrol and plant growth-promoting activity of rhizobacteria from Chinese fields with contaminated soils. J Microbial Biotechnol 8:404–418. https://doi.org/10.1111/1751-7915.12158

Wu SC, Cao ZH, Li ZG, Cheung KC, Wong MH (2005) Effects of biofertilizer containing N-fixer, P and K solubilizers and AM fungi on maize growth: a greenhouse trial. Geoderma 125:155–166. https://doi.org/10.1016/j.geoderma.2004.07.003

Xiao Y, Wang X, Chen W, Huang Q (2017) Isolation and identification of three potassium-solubilizing bacteria from rape rhizospheric soil and their effects on ryegrass. Geomicrobiol J 34:873–880. https://doi.org/10.1080/01490451.2017.1286416

Yeager CM, Northup DE, Grow CC, Barns SM, Kuske CR (2005) Changes in nitrogen-fixing and ammonia-oxidizing bacterial communities in soil of a mixed conifer forest after wildfire. Appl Environ Microbiol 71:2713–2722. https://doi.org/10.1128/AEM.71.5.2713-2722.2005

Zhang Y, Hu J, Bai J, Wang J, Yin R, Wang J, Lin X (2018) Arbuscular mycorrhizal fungi alleviate the heavy metal toxicity on sunflower (Helianthus annuus L.) plants cultivated on a heavily contaminated field soil at a WEEE-recycling site. Sci Total Environ 629:282–290. https://doi.org/10.1016/j.scitotenv.2018.01.331

Sustainable Production of Edible and Medicinal Mushrooms: Implications on Mushroom Consumption

12

Mahantesh Shirur, Anupam Barh, and Sudheer Kumar Annepu

Abstract

Mushrooms are one of the oldest human foods and sources of medicine. Both the naturally grown and artificially cultivated mushrooms are used for human consumption. Mushrooms are known to be a healthy source of food, having both nutritive and medicinal significance. Mushrooms have a high amount of edible and digestible protein. Though this is the most important criterion for mushrooms to be considered for human diet, their therapeutic effect on several human diseases and illness is equally worthwhile. Diverse edible mushroom production systems are seen the world over. In several countries, different varieties of mushrooms are attempted for commercial production with varying degrees of success. In India, the mushroom varieties under commercial cultivation are very few. Many of the mushroom production models and systems are unsustainable either on economic or ecological criteria. This leads to a higher rate of attrition among the adopters of mushroom cultivation. For the stable food production system, mushroom units have to be ecologically less polluting and economically valuable to the producer. In the immediate run, if the mushroom production is not commercially viable, there is no question of mushroom production continuing sustaining for a long time. This chapter attempts to explore the ways to balance the profitability of commercial mushroom production without affecting the environment and quality of human food system.

M. Shirur (✉)
National Institute of Agricultural Extension Management (MANAGE), Hyderabad, India

A. Barh · S. K. Annepu
ICAR-Directorate of Mushroom Research, Solan, Himachal Pradesh, India

V. K. Hebsale Mallappa, M. Shirur (eds.), *Climate Change and Resilient Food Systems*, https://doi.org/10.1007/978-981-33-4538-6_12

Keywords

Mushroom · Spawn · Oyster · *Agaricus* · Sustainability · Profitability

12.1 Introduction

Vision 2030 document of the Indian Council of Agricultural Research (ICAR) envisages that the demand for high-value commodities is increasing faster than food grains (Vision-2030 2011). It also foresees a rise in the demand of vegetables from 93 million tonnes in 2000 to 180 million tonnes in 2030. A constant supply of cultivable land will certainly make the challenge tougher to meet growing vegetable demand. Mushrooms being an indoor crop can offer solutions to the growing demand of vegetables. Mushrooms are a class of macrofungi known as a health food by virtue of their nutritional and medicinal properties.

The forest dwellers and aborigines have consumed mushrooms in several parts of the world since ancient times. They collected the naturally occurring wild mushroom species such as *Agaricus* sp., *Morchella* sp., *Auricularia* sp., *Grifola* sp., *Pleurotus* spp., etc., from the forests. The indigenous knowledge was developed in many nations about the edibility and nutritional and medicinal value of mushrooms. Mushrooms were also valued for their flavour in the food preparation for several years. Even today, many such mushrooms are collected excessively by the locals leading to their overexploitation, causing loss of microflora and ecological disturbances. Realising their significance, especially for food value, modern man started attempting the artificial mushroom cultivation. Man must have surely tried his luck based on his long observation of nature and resorted to spreading the vegetative parts and mycelium of mushrooms simulating the natural conditions for their vegetative and reproductive growth phases.

12.1.1 Mushroom Production: Global Scenario

Edible mushrooms are cultivated commercially in more than 100 countries on different scales and by different systems. The production of mushrooms across the world is increasing at an annual growth rate of 6–7% (Singh 2011). In some developed countries of Europe and America, mushroom farming has assumed the form of a high-tech industry with mechanisation and automation in all operations. At present, China and the United States are the largest consumers of this protein-rich delicacy (Bose 2016). Present world consumption of mushrooms is around 12.74 million tonnes and is growing at a compounded annual growth rate of 6.4% (Report of Fortune Business Insights entitled "Mushroom market size, share & industry analysis, By type (Button mushroom, Shiitake mushroom, oyster mushroom, & others), form (Fresh mushroom, frozen mushroom, dried mushroom and canned mushroom), and regional forecast, 2019-2026"). The same report predicts that by 2026, the world mushroom production will scale to 20.84 million tonnes.

12.1.2 Mushroom Production in India

In India, commercial mushroom production was started very late in the 1970s, but growth rate, both in terms of productivity and production, has been noteworthy (Shirur et al. 2014). In the 1970s and 1980s, button mushroom was grown as a seasonal crop in the hills and in some part of plains during the winter. But with the development of the technologies for environmental controls and increased understanding of the cropping systems, mushroom production shot up from mere 5000 ton in 1990 to 100,000 ton in 2006 (Wakchaure 2011). Presently, the production of mushroom is estimated to be close to 201088 ton (ICAR-DMR Annual report 2019) al.). Haryana, Maharashtra, Odisha, Punjab and Himachal Pradesh are emerging as the leading states in mushroom production in India (Fig. 12.1).

Though the total mushroom production in India has registered significant growth, the per capita consumption of mushrooms in India is still <100 gm/annum which is significantly low as compared to the requirement (Singh and Shirur 2016; Sharma et al. 2017; Shirur et al. 2018). The low consumption of mushrooms in India is mainly attributed to the nonavailability of mushrooms in all the seasons. Moreover, the diversity of mushroom varieties is also underexploited in India (Shirur et al. 2014). India's varied agro-climatic situation offers tremendous scope to cultivate

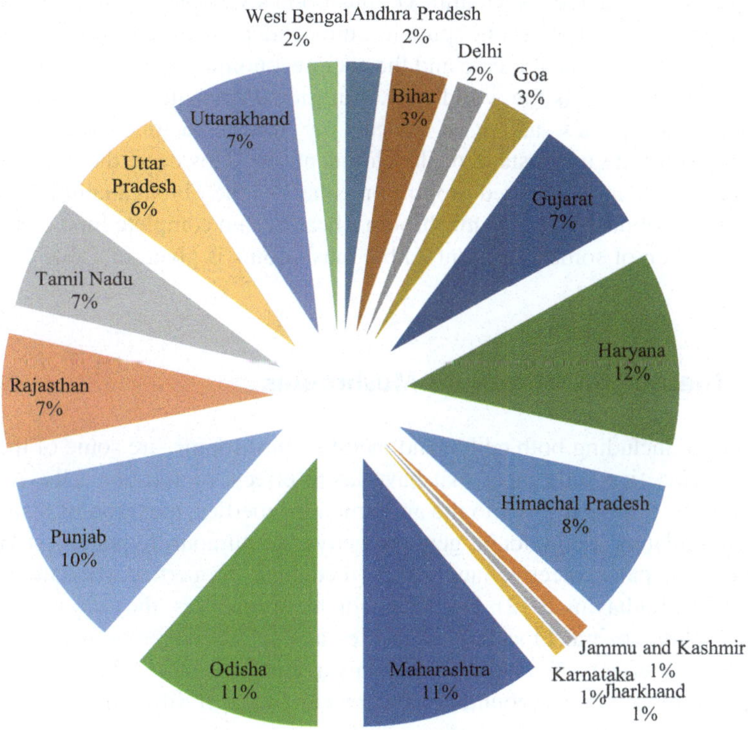

Fig. 12.1 State-wise (major) contribution to total mushroom production

different mushrooms depending on the location and season. Since mushroom cultivation is an indoor activity, it has potential to increase farm productivity without much pressure on the land (Singh 2011). Part of India's annual 800 million tonnes of farm waste can be used as a substrate to cultivate the mushrooms, which is otherwise being allowed to get waste. The labour-intensive mushroom cultivation generates enough employment opportunities, especially to the women and youth. Also, it has the potential to be taken up as an agribusiness activity with a capital investment ranging between Rs 3-50 million.

12.2 Nutritional Attributes and Therapeutic Properties of Mushrooms

Mushrooms are a significant form of new-generation food as it is rich in protein, vitamin D and antioxidants and has been recorded for their therapeutic effects across the globe. The vulnerability of population to the viral pandemics due to the vitamin D deficiency is the much-discussed issue in recent times. Recent studies found a correlation between vitamin D levels and hyperactive immune systems. Vitamin D strengthens immunity and prevents overactive immune responses. Being the only vegetarian dietary source for vitamin D, mushrooms will play a vital role in future dietary supplements. The recent spurt in nutraceutical significance of mushrooms and their usefulness in health care and the socio-economic status of mushrooms has scaled significant heights the world over. Nutritional attributes of mushrooms vary with species to species, and it is affected by several factors such as growing conditions, substrate used, stage of development, etc. However, the availability of essential nutrients in the desired proportion enhances their bioavailability in human beings. This attribute makes the mushroom a healthy and complete food component. The composition of some important edible mushrooms is given in Table 12.1.

12.2.1 Therapeutic Properties Mushrooms

Higher fungi, including both edible and inedible mushrooms, are some of the major sources of bioactive substances that have latent effects on tumour cells. Although much attention has been given to the antitumour properties, mushrooms also exhibit immunomodulating, antioxidant, genoprotective, antitumour, hypocholesterinemic, antidiabetic, hepatoprotective and other medicinal properties (Badalyan 2000; Wasser 2010; Badalyan 2012). Besides medicinal properties, they are a rich source of dietary fibre, several bioactive molecules and prized enzymes with more than 120 therapeutic effects (Wasser 2010; Badalyan 2012). Table 12.2 gives an account of important bioactive compounds and therapeutic uses of different mushrooms.

Table 12.1 Nutritive values of some important edible mushrooms (dry weight basis g/100 g)

Mushroom variety	Carbohydrate	Fibre	Protein	Fat	Ash	Energy (K cal)	Reference
Agaricus bisporus	46.17	20.90	33.48	3.10	5.70	499	Stamets (2005)
Pleurotus pulmonarius	63.40	48.60	19.23	2.70	6.32	412	Stamets (2005)
Lentinula edodes	47.60	28.80	32.93	3.73	5.20	387	Stamets (2005)
Pleurotus ostreatus	37–48	24–31	17–42	0.5–5.0	–	350[a]	Khan (2010)
Volvariella volvacea	12–48	4.0–11.9	20.2–34.1	1–6	0.8–13.0	300[a]	Cheung (1997), Ul Haq et al. (2011)
Calocybe indica	49.2	13.2	21.60	4.96	12.8	350[a]	Pushpa and Purushothoma (2010)
Flammulina velutipes	–	3.30	31.2	5.8	5.60	378	Sharma et al. (2008)

[a]Authors' approximation

12.3 Major Cultivated Mushroom in India

Commercially grown species of mushrooms are *Agaricus bisporus* (white button mushroom), *Lentinula edodes* (shiitake mushroom), *Volvariella volvacea* (paddy straw mushroom), *Pleurotus* spp. (oyster mushrooms), enokitake (*Flammulina velutipes*), etc., followed by other tropical mushrooms such as *Calocybe indica* (milky mushroom), *Auricularia*, *Ganoderma*, etc. The concentrated areas of production in India are the temperate regions for the button mushroom and tropical and subtropical regions for oyster, milky, paddy straw and other tropical mushrooms. Two to three crops of button mushroom are grown seasonally in temperate regions with minor adjustments of temperature in the growing rooms, while one crop of button mushroom is raised in northwestern plains of India seasonally. Oyster, paddy straw and milky mushrooms are grown seasonally in the tropical/subtropical areas from April to October. Figure 12.2 shows the important regions in India, which are known to produce different edible mushrooms.

Tropical/subtropical mushrooms requiring a temperature of about 25–35 °C like oyster mushroom (*Pleurotus* sp.), paddy straw mushroom (*Volvariella* sp.) and milky mushroom (*Calocybe indica*) can be cultivated in most parts of India. The medium- and high-elevated regions are suitable to grow temperate mushrooms like white button mushroom (*Agaricus bisporus*) and shiitake (*Lentinula edodes*) that requires temperatures between 14 °C and 20 °C. Other than these edible mushrooms, many medicinal mushrooms like red reishi (*Ganoderma* sp.) and *Grifola*/maitake (*Grifola frondosa*) can also be cultivated in India (Shirur 2011). It is germane to mention that being a subtropical country, India has the paradox of contributing more

Table 12.2 Bioactive compounds and associated therapeutic effects of different edible and medicinal mushrooms

S. no	Mushroom	Common name	Bioactive compounds	Therapeutic effects
1	*Agaricus bisporus*	White button mushroom	Agariciten	Antitumour property
2	*Lentinula edodes*	Shiitake mushroom	Lentinan	Antitumour and antiageing property
3	*Pleurotus* spp.	Oyster mushroom	Lovastatin	Antiobesity property
4	*Hericium erinaceus*	Monkey head mushroom	Hericenone	Found to induce synthesis of nerve growth factor, which is associated with an ameliorative effect in Alzheimer's dementias
5	*Ganoderma lucidum*	Reishi mushroom	β-Glucans, ganoderic acid, polysaccharides and triterpenes	Hepatopathy, chronic hepatitis, nephritis, hypertension, arthritis, neurasthenia, insomnia, bronchitis, asthma and gastric ulcers
6	Schizophyllum *commune*	Schizophyllum mushroom	Schizophyllan or sonifilan	Antitumour properties
7	*Grifola frondosa*	Maitake	β-Glucan (β-1,6-glucan branched with a β-1,3-linkage)	Antioxidant, anti-inflammatory, Free radical scavenging activities and the antiageing process
8	*Trametes versicolor*	Turkey tail mushroom	Proteoglycan fractions, PSP and PSK	Antitumour properties
9	*Cordyceps militaris*	Cordyceps mushroom	Cordycepin	Biological response modifier
10	*Auricularia polytricha*	Wood ear mushroom	Polysaccharides	Anti-inflammatory effect

Source: Authors' compilation

than 80% of temperate mushrooms among its total mushroom production (Shirur et al. 2018a).

12.4 Constraints in Mushroom Production Activity

The mushroom production world over is rising. However, this rise does not necessarily imply that the commercial mushroom production is devoid of any problems. Mushroom as an industry faces impending challenges like a competitive market, costly raw materials, technology and labour-intensive activity, etc. Mushroom as a protein-rich food product has several substitute protein-rich products such as non-vegetarian food and eggs in most of the countries. Mushroom cultivation in the

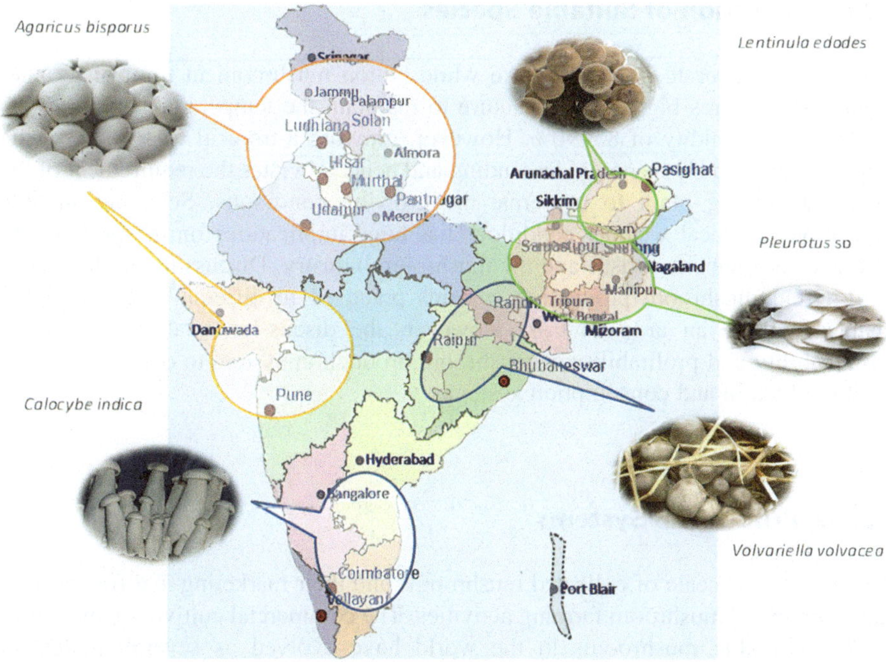

Fig. 12.2 Major areas producing different mushrooms in India

backyard livelihood activity or small-scale farm enterprise mainly suffers due to the scale of economy and poor market opportunities. Due to their importance as an edible food and their quality protein, mushroom production must sustain. For sustainability, it must be done to ensure profitability to its producers, feasibility and affordability to its consumers and not causing serious ecological and environmental disturbances. Farmers and big entrepreneurs adopting mushroom cultivation find it challenging to sustain this agribusiness activity on above accounts, and hence the rate of attrition is very high in mushroom farming (Shirur et al. 2017).

12.5 Sustainable Mushroom Production Practices

In India, the mushroom production systems are mixed type, i.e. both seasonal farming and high-tech industry. Though this system of mushroom production is common for white button mushroom, other mushrooms are also cultivated in seasonal and environment-controlled units. But the type of infrastructure to create the environment-controlled units for different mushrooms differs depending on the type of mushrooms and the ambient temperature of the growing region.

12.5.1 Selection of Suitable Species

Growing a temperate mushroom-like white button mushroom in tropical weather conditions requires bigger infrastructure to maintain the temperature of 16–18 °C and relative humidity of 85–90%. However, growing a tropical mushroom in the tropical weather is less energy demanding and hence obviates the requirement of the high-end chilling plant to moderate the growing conditions. Selecting suitable mushrooms to local weather conditions has huge implications on the profitability of the mushroom enterprises and the mushroom industry. Discussing all the details of different mushroom production systems practised in different regions in this chapter will be an arduous task. However, the discussion shall focus on the sustainability and profitability of mushroom production system to contribute to the food production and consumption system.

12.5.2 Production Systems

The history and scale of collected mushrooms and their marketing is different from the commercial mushroom farming activities. The commercial cultivation models of all major edible mushrooms in the world have evolved as several models of production. There are units with very high investment targeting huge mushroom production intended for cosmopolitan markets and exports on one extreme and units with very modest investment targeting small production to market in the nearby consumption places at other extremes. In the former system of mushroom production, lots of resources are spent to create artificial growing conditions to maintain a particular temperature, relative humidity, light (both duration and quality) and air composition (O_2/CO_2) in the cropping rooms and compost chambers as per the crop and compost requirement. The resources are spent on thermodynamic and aerodynamic equipment to create and maintain an exact set of cropping and compost conditions. Though the cost of raw materials, labour wages and the market price of mushroom decide the profitability, the role of scale of economy is also a very important factor for mushroom production profitability.

12.5.2.1 Mushroom Production Units: Facilities and Infrastructure

The success in mushroom cultivation irrespective of the mushroom variety depends on three main components: quality spawn, quality substrate or compost and suitable growing conditions required by the mushroom variety. Therefore, the commercial mushroom production facility will have a spawn production facility, substrate preparation or compost production facility (e.g. phase 1 and or phase 2 bunkers for white button mushroom production) and mushroom growing rooms with temperature- and humidity-controlled facilities. The growing rooms (Fig. 12.3) these days are constructed with very good insulating materials to cut the cost on electricity by conserving the temperature in the growing rooms.

Fig. 12.3 Modern mushroom production under insulated growing rooms

Table 12.3 Raw materials and treatment of raw materials to grow different mushrooms

Raw material	Treatment of raw material	Facilities required	Mushroom varieties are grown
Dry agriculture residue or substrate (wheat straw and paddy straw are most commonly used)	Composting	Compost yard, phase 1 and phase 2 bunkers	White button mushroom, *Macrocybe*
	Pasteurised	Boilers, sieves, gunny bags, etc.	Milky mushroom, oyster mushroom
	Sterilised	Autoclave, polypropylene bags.	Shiitake, *Flammulina*, *Auricularia*, *Ganoderma*, etc.

12.5.2.2 Raw Materials and Substrate Combinations

Mushrooms are a saprophytic fungus and depend on a variety of decomposing residue or substrate for their growth. In artificial cultivation, the fresh dry residue which is free from dust and any other disease/infection is chosen. Depending on the variety of mushroom to be grown, the substrate is either pasteurised, sterilised or partially decomposed. Table 12.3 gives the idea about the raw materials used, the treatment for raw material, facilities required and the mushroom varieties grown.

Mushroom growers give due attention to compost making as it plays major role in mushroom production and productivity. With the help of raw materials and agriculture residue, different mushrooms can be grown depending on the treatment of the raw materials and using other raw materials. Under the Indian context, the raw materials are mostly the wheat straw and the paddy straw or the combination of both in different proportions. In recent years, these raw materials are mixed with sugarcane bagasse, cotton stalks, maize cobs, sunflower stalks, etc. Different outputs from these raw materials are briefly described here.

Composting

In the composting, the dry residues and raw materials which are ideally rich in cellulose and hemicellulose are decomposed by mixing with a different combination of raw materials. Farmers follow different methods and use a different proportion of raw materials to prepare compost for growing white button mushroom. The compost so prepared is very selective for white button mushroom only by not supporting the growth of other competitor fungi. The most commonly used combination of raw materials is listed in Table 12.4.

Most farmers in North India prepare the compost by following a long method of compost preparation (Vijay 2011). This method of composting takes 28–30 days period over which the mixed materials are watered and frequently turned and made into fresh piles seven to eight times at an interval of 3–4 days so that the compost becomes uniformly pasteurised, homogeneous and free from ammonia.

Since the long method of compost making is time-consuming and causes air pollution and short method is cost-intensive, an alternate method to reduce the time required for compost in a climate-friendly way of zero energy polytunnel (ZEPT) using high-density polyethylene (HDPE) is designed at ICAR-DMR, Solan (Shirur and Sharma 2016). The technique of ZEPT uses the perforated polyvinyl chloride (PVC) pipes to facilitate proper air circulation in the core of the compost pile and moderate the temperature in a pile (Fig. 12.4). The adoption of this technology still needs adoption among the majority of the mushroom growers. Further, the commercialisation and spread of zero energy polytunnel by the industry should be emphasised as this will boost mushroom production and, in turn, mushroom consumption in the society. More of such innovations will make the compost production less labour intensive and environment friendly.

Pasteurisation

Pasteurisation is the process of killing the microorganisms primarily through the exposure to heat/high temperature. Through pasteurisation of a large number of agro-wastes such as cereal straw, sugarcane bagasse, sawdust, jute and cotton waste, dehulled corncobs, peanut shells, dried grasses, etc., two important varieties of tropical mushrooms can be grown; they are oyster mushroom and milky mushroom. Dry agricultural waste or residue is selected that can support the cultivation of oyster mushroom and milky mushroom. While selecting the agriculture residue, ensure that the straw is not too old and not exposed to rains or dust as it may be susceptible to attract the contamination. It should also be completely dry and not have any green leafy portion. Pasteurisation of such agricultural residue helps in achieving quick colonisation of mushroom mycelium in the substrate by killing off harmful competitor moulds and fungi. The pasteurisation process of the growing medium gives the mushroom mycelium an advantage over other harmful fungi or competitor moulds as the mushroom mycelium gets established in the pasteurised substrate while other fungi gets eliminated. Pasteurisation occurs between temperatures of 58 °C and 62 °C. The substrate can also be pasteurised by hot water or steam at this temperature for 4–5 h in a steam chamber or room. Though the chemical pasteurisation using the

Table 12.4 Formulations to prepare compost to grow white button mushroom

Raw materials	DMR Solan 1 (long method)	DMR Solan 2 (short method)	IARI, New Delhi	PAU, Ludhiana	IIHR, Bangalore	Mushroom lab, Solan (short method)	Farmers' practice in North India
Wheat straw (kg)	300	300	350	300	300	1000	1000
Poultry manure (kg)	125	–	300	60	–	400	300
Horse manure (kg)	–	–	1000	–	–		
Cotton seed cake (kg)	–	12	–	–	–		
Wheat bran (kg)	15	21	–	15	30		40
CAN (kg)	–	–	–	6	9		15
Urea (kg)	–	7	–		4	14.5	20
Gypsum (kg)	20	15	30	30	12	30	150
Superphosphate (kg)	–	–	–	3	9		10
Calcium carbonate (kg)	–	–	–	–	10		
BHC (g)	125		–	250	–		
Brewers manure (kg)						72	

DMR: Directorate of Mushroom Research, Solan, India
IARI: Indian Agricultural Research Institute, New Delhi, India
PAU: Punjab Agricultural University, Punjab, India
IIHR: Indian Institute of Horticultural Research, Bengaluru, India

Fig. 12.4 Composting by ZEPT method with HDPE pipes in the compost pile

carbendazim and formaldehyde is also in vogue, it is discouraged strongly as the injudicious use may leave traces of chemical in the pasteurised straw and later detected in the fresh mushrooms.

Sterilisation

Through the sterilisation of substrate, many specialty mushrooms such as shiitake, *Auricularia* and *Flammulina* are grown. Usually, the substrate to grow these mushrooms is sawdust of broad-leaved trees (tuni, mango, safeda, oak, maple and poplar). The substrate preparation stage to produce the synthetic logs is made using 80% sawdust, 19% cereal bran like wheat bran or rice bran and 1% calcium carbonate on weight basis. Ingredients are properly mixed in a mixer and moistened to hold a moisture level of 60–65%. 1.5–2.0 kg of this substrate is filled in double polypropylene bags and kept for sterilisation for 2 h at 121 °C. After sterilisation, they are removed from the autoclave and allowed to cool down at room temperature. Subsequently, they are spawned and kept for incubation (Annepu et al. 2019).

Mushroom Growing Rooms/Structures

Mushrooms require maintenance of proper air, temperature and humidity levels as per the stage of the crop. All varieties of mushrooms have specific temperature and humidity requirement for their vegetative and reproductive phases (details in Table 12.5). Hence, the mushroom growing structures, whether the temporary huts or the permanent rooms, must be designed/planned to achieve these growing ambiences. The task is easy for environmentally controlled growing rooms where the temperature and humidity are controlled through chilling plant and humidifiers. In the seasonal type of huts, the farmers depend on day and night temperatures to maintain the temperature and humidity inside the huts by opening and closing the doors during day or night to allow the movement of cool/hot and dry/humid air. In addition, they resort to frequent spraying of water on the floor to moderate the humidity and temperature to some extent.

Table 12.5 Temperature requirement of different cultivable mushrooms

Type of variety	Name of variety	Spawn-run temperature (°C)	Fruiting temperature (°C)
Tropical mushrooms	*Pleurotus sapidus*	25–30	22–26
	P. flabellatus	25–30	20–26
	P. sajor-caju	25–30	18–26
	P. membranaceus	25–30	22–30
	P. djamor	26–32	28–30
	Calocybe indica	30–35	30–32
	Volvariella volvacea	30–35	30–32
	Ganoderma lucidum	23–25	30–32
	Auricularia polytricha	23–25	30–32
Subtropical mushroom species	*P. ostreatus*	22–26	14–20
	P. ostreatus var. *florida*	22–28	14–22
	P. citrinopileatus	24–28	18–22
	Cordyceps militaris	18–22	18–22
	Hericium erinaceus	23–25	18–20
Temperate mushrooms	*Agaricus bisporus*	23–25	16–18
	Agaricus bitorquis	23–25	18–20
	Lentinula edodes	23–25	18–20
	P. eryngii	20–24	10–15
	Flammulina velutipes	18–24	5–10

This economic activity uses all the agriculture residues. It is in line with the climate-friendly practices of agriculture. However, in environment-controlled units, the growers must be judicious in saving the resources and adopt the seasonal cultivation of mushrooms based on the prevailing temperature in the season. For this, the consumers must also be educated to eat the mushrooms as per the season. In India, the skewed demand for white button mushroom has caused less demand for other tropical mushrooms, and also efforts on their popularisation are lacking.

Structure and design of mushroom huts in North India: In Haryana and Punjab, temporary thatched houses are erected using mainly bamboo, paddy straw and polythene sheets. Varying numbers of huts are juxtaposed (Fig. 12.5) in the space available in their field. Only a single hut is delineated for better understanding the structure of these temporary mushroom houses.

Usually, a single hut is laid out over an area of $22.5' \times 60'$ (three racks arrangement) or $28.5' \times 60'$ (four racks arrangement). Each separate hut comprises of three or four racks, and three to four vertical shelves are created in each of these

Fig. 12.5 Series of mushroom growing huts laid one after the other in the expansive field

Fig. 12.6 Steps in preparation of mushroom growing shed in North India

racks (Shirur et al. 2018a). Generally, 16–18 ton of prepared compost is spread in one such shed. About 2–2.5 ton of fresh button mushroom is obtained in a month's time depending on the compost quality, spawn quality and management practices of the farmers. Figures 12.6 and 12.7 depict the structures of mushroom huts and the steps followed in their construction.

In Himachal Pradesh and Jammu and Kashmir, the growing houses vary from temporary to semipermanent to permanent structures. Some farmers use concrete

Fig. 12.7 Arrangement of racks and shelves in a single mushroom shed

rooms with cemented floor and roof. Most commonly observed mushroom growing rooms are simple unplastered brick walls raised to 12–16′ height and covered with either tin or asbestos sheets. They are covering such sheets with locally available (*Sarkhanda*) grass as an insulating material to avoid excessive heating of the growing rooms. This helps in moderating the temperature of growing rooms, especially during the beginning of season when outside temperature is very high. Some poor farmers erect small huts using locally available wooden poles and straw. Even the very resource-poor farmers in the region were seen making the mushroom sheds using the decrepit materials available with them.

In contrast, the oyster and milky mushrooms are grown in different and diverse structures. The structures range from huts to mud houses to polysheds and concrete buildings. The oyster mushroom bags with compost and spawn are either hanged from the ceiling in three to five tiers (Fig. 12.8) or placed over the shelves. The room temperature during the spawn run is maintained to suit the type of mushroom variety selected. For most of the tropical oyster mushroom varieties, 25–30 °C is maintained for spawn run and fruiting. In case of milky mushrooms, the bags are kept in the incubation room or cropping rooms, which are mostly dug 3–4 ft. below the ground level (Fig. 12.9). The milky mushrooms can also be grown in environment-controlled units with the facility of maintaining temperature and humidity through humidifiers (Fig. 12.10).

Fig. 12.8 Substrate mixed with mushroom spawn in polythene bags kept for spawn run

Fig. 12.9 Commercial milky mushroom unit

12.5.2.3 Implications for Mushroom Consumption

Small and marginal mushroom growers in Himachal Pradesh do not prepare compost on their own but purchase from private commercial mushroom production units. Farmers in North Indian states (Himachal Pradesh, Haryana, Jammu and Kashmir and Punjab) have the opportunity to purchase compost from the state department of horticulture in their states as these governments have set up facilities to prepare compost and sell to encourage mushroom farming as livelihood activity even among the small farmers. Besides, farmers getting benefitted, this measure will improve the availability of mushroom in the remote places which are far off from large

Fig. 12.10 Individual milky mushroom hut

mushroom production units. When there is not much transaction cost, the availability of fresh mushrooms at affordable prices for majority population in hinterland will have access to quality protein food and improve their protein intake.

Many spawn-producing firms are selling ready-to-fruit (RTF) bags by mixing the spawn with the ready compost or pasteurised substrate. This is the easiest way for new entrepreneurs to gain confidence in mushroom cultivation. Like the compost sold by large compost production units, the sale of pasteurised substrate and mixing with spawn has been successful in making the mushrooms available for rural areas and small towns. This will increase the consumption of mushrooms among the masses, thereby increasing their protein intake and achieving food security. This kind of arrangement will give impetus to the mushroom entrepreneurship and livelihood security of mushroom growers.

12.5.3 Spawn

Quality spawn is one of the critical inputs to realise the economic returns in mushroompreneurship. But many small growers do not produce the spawn required in their mushroom unit due to lack of technical knowledge involved in spawn production. The nonavailability of quality spawn is a major constraint for both new and established mushroom units and limits the expansion of mushroom cultivation in new areas.

Spawn is the mushroom seed which is prepared from the pure culture of mushrooms. The culture is allowed to grow on starchy substrates like cereal grains. This mycelium and grain mixture is used to seed the pasteurised substrate or compost for growing the mushrooms. Though the whole description of spawn production process is beyond the scope of this chapter, brief information given below will help to understand its implications for mushroom consumption system.

Regardless of the mushroom variety, techniques for growing mushrooms follow the same pattern, which directly reflects the life cycle of the mushrooms. The mushroom life cycle under artificial conditions begins with the isolation of fungal mass and implanting it in an environment that gives an advantage over the other competitors. The mycelial mass collected from the healthiest fruit body is initially grown on enriched culture media for expansion of inoculum. This mycelium is then transferred on a cereal-based substrate to make the mother spawn. Once the mushroom mycelium has completely colonised the grain, it can be used to inoculate the grain substrate for large-scale multiplication in the form of commercial spawn.

A small-scale mushroom grower can produce a quantity of 10–12 kg spawn per day by spending of Rs 2.5–3.0 lakh on purchase of mini autoclave, laminar flow chamber, BOD incubator and other necessary consumables. For establishing the medium- to large-scale spawn laboratory with a capacity of 20 ton per annum, an expenditure of Rs 15–18 lakh is required for purchasing machinery and for creating infrastructure facilities. By producing the good-quality spawn, this activity may fetch the net profit of Rs 6–7 lakhs per annum to the spawn producers.

The quality of the spawn is usually ascertained by physical examination. The following criteria should be ensured to be considered as quality spawn and avoid economic losses in mushroom cultivation:

1. There should be a proper coating of mycelium around grains used as a substrate for spawn production. No loose grain should be visible in bottles/bags.
2. The growth of the mycelium in the spawn bags should be silky/strandy type. It should not be cottony or slimy type of growth.
3. The growth of fresh spawn is more or less white. As the spawn gets older, it turns gradually into brown colour.
4. There should not be any greenish or blackish spot in the spawn bags. Such spots indicate that the spawn is contaminated with moulds.
5. When the spawn bags are opened for spawning, it should emit a typical mushroom smell.

12.5.4 Implications on Mushroom Consumption

Nonavailability of quality spawn is an important reason for non-adoption or discontinuance of mushroom cultivation in our country. All the sources of spawn fail to meet the demands of the growers during the winter months in North India. Another bottleneck with respect to spawn is the sale of infected or spurious spawn. The growers who put one-month effort to prepare their growing sheds and compost get deceived when they realise that the spawn they bought was either contaminated or immature. They will lose the season and also suffer financial losses because of the expenditure they incurred for growing mushrooms. Proper regulation of spawn standards and quality traits of spawn in India needs immediate attention by the policymakers. Stringent measures must be ensured to regulate the quality of the spawn through an institutional setup. Similar is the case of availability of quality gypsum in sufficient quantity during the peak cultivation season in North India. The sale of poor-quality gypsum lets the mushroom growers suffer loss. The quality of the gypsum mainly determines the final compost quality with respect to its pH. By the time the growers realise that there is improper spawn run due to poor-quality gypsum, they will suffer losses because of investment and also the season advances, giving limited time for seasonal growers to prepare fresh compost to cultivate button mushroom.

12.6 Crop Management and Sustainability

Farmers practising seasonal button mushroom cultivation mostly depend on purchased spawn as it is seasonal agriculture activity for them. Generally, 0.5% seed rate (5 g of spawn/kg of compost) on a wet compost weight basis is adopted among the growers. In Haryana and Punjab, prepared compost is mixed with mushroom spawn and spread over the racks for about 4–6″ height and allowed for the fungal mycelia to spread (spawn run) for about 7–12 days. After the spawn run, a layer of casing soil of about 1″ is spread over compost uniformly. Mushroom pinheads start appearing 8–12 days after spreading the casing soil. Such pinheads develop into full-grown mushroom fruiting bodies after 3–4 days. In case of oyster mushroom, the seed rate is around 2.5–3.0%, while for milky mushroom, it is 5.0% on a wet weight basis. The spawn is mixed alternatively with the layers of straw and placed in a room with suitable temperature and humidity for spawn run. After the incubation period, the spawn run completes and the temperature and humidity are altered to facilitate the fruiting.

Since mushroom cultivation is a very sensitive enterprise for several competitor microorganisms, hygiene is most critical factor in all three facilities. Any chance of infection and contamination either in spawn laboratory, production rooms or composting facility is sure to affect the mushroom quality and production. Usually, the spawn production facility is an integral part of the mushroom production units with 250 ton per annum (TPA) of fresh mushrooms. For successful commercial white button mushroom production, the minimum ideal production capacity is 250 ton per annum. Though many entrepreneurs are trying to set up 100 TPA units with indigenous cooling systems, their B:C (benefit-cost) ratio may be low due to higher operational expenses per unit mushroom production. It is always advisable to consult the experts to decide about the size and scale of mushroom units and accordingly design the type of growing rooms, compost yard and composting bunkers if required.

The reason for failure of mushroom enterprises to achieve the profitability is to start with uneconomical production and not following the cropping cycle to keep the production constant. This is essential to be a credible supplier of fresh mushrooms in the market and use the farm resources appropriately. Since the mushroom crop gives no yield in the initial spawn running period and gives a higher quantity of mushroom in the initial period of cropping, one- or two-room production facility and even four- to five-room structure create problem to achieve uniform supply of mushrooms continuously. Minimum of eight rooms can mitigate such a constraint, and having more cropping rooms will help to offset the uneven distribution of mushroom production in a single room. In the following model (Fig. 12.11), a 12-room

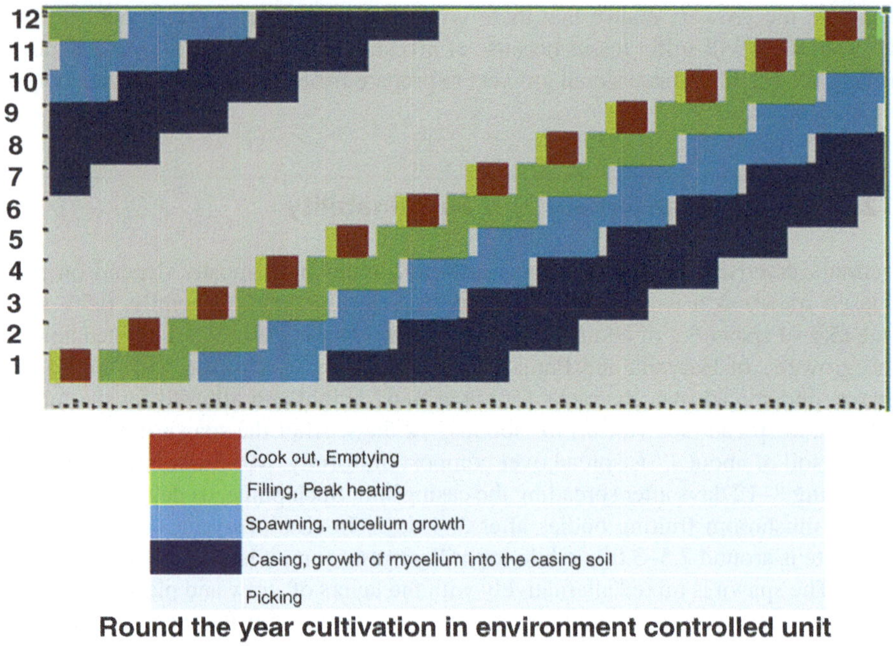

Fig. 12.11 Schematic diagram to fit the round the year cultivation in a 12-room model of white button mushroom unit

mushroom unit for button mushroom is given to demonstrate how the uniformity in production of mushrooms can be achieved over a period of 12 weeks. The cycle will again continue with room number one after it is emptied after taking the crop. In this model, it is essential that the timing is not disturbed because of failure of the compost lot or failure to delay the spawn run and cropping in the cropping room due to improper temperature and humidity or any pest/disease incidence. With the same idea, the other mushrooms can be set to maintain the required quantity of mushrooms to meet the demand in the market.

Unlike the white button mushrooms, the milky mushroom and oyster mushroom can be grown in four to five separate rooms with one room exclusively meant for spawn run to maintain the constant supply of mushrooms for the market. This practice is very important as it helps to maintain the rooms with a defined set of temperatures and humidity as per the crop stages. Since the spawn-run room does not require exclusive aeration facility, spawn run may ideally take 12–15 days, and the more number of bags can be put compared to the growing rooms. Hence, this kind of arrangement will be able to give sufficient bags to be placed for three to five growing rooms. However, this will require labour during the shifting of the spawn-run bags to growing rooms for fruiting.

12.6.1 Round the Year Cultivation

Due to the varied agro-climatic conditions prevailing in the country, all mushrooms can't be grown in every region. For this purpose, the country has been broadly classified into five major mushroom-climatic regions, and five different types of mushrooms are proposed for each region as given in Table 12.6.

12.6.2 Novel Approaches in Cultivation of Speciality Mushrooms

Bottle Cultivation It is a modern method and well suited for mechanisation. The technology includes the polypropylene bottle and requires less space and vertical space. The bottles are sterilised. The bottles are filled with wet substrate with sawdust+ wheat bran in ratio of 80:20 and are sterilised at 121 °C at 22 psi pressure for 1.30 h. The spawning is done in laminar airflow. The method is used for *Flammulina* (Fig. 12.12) and *Pleurotus eryngii.*

Ready-to-fruit (RTF) Technology It is important for getting growing experience of mushrooms. Many countries and institutes develop it. The bags come with handy options of hanging and can be hanged in rooms and kitchen. The objective of the RTF is to provide the growing experience to the beginner growers and to promote the mushrooms in urban and periurban areas. The technology now empowers the mushroom growers to grow mushroom in their moist backyard or kitchen and ready healthy vegetable for the family. The RTF technology can be conveniently used to grow milky mushrooms (Fig. 12.13) and oyster mushrooms.

Short-duration cultivation Technology of Shiitake Mushroom It is possible to get the fruiting in shiitake within 45–60 days compared to the earlier synthetic log technologies where it took 80–110 days. Sawdust is the main ingredient for the preparation of synthetic logs supplemented with some starchy substance like cereal bran as a source of nitrogen and calcium carbonate and calcium sulphate to balance the pH. Ingredients are mixed in a mixer and moistened to hold a moisture level of 60–65%. The prepared substrate is filled in polypropylene bags (1.5 kg) and then sterilised for 2 h at 121 °C. Then the bags will be seeded with grain spawn at 3% on a dry weight basis of the substrate under aseptic conditions. The bags after heat sealing will be shaken to evenly distribute the spawn; sawdust spawn or cereal grain spawn is used in this system. Spawn run at 25 °C with 4 h of light per day takes 18–23 days for optimum growth. Fully colonised blocks are taken out by slicing and peeling off the poly cover and kept for 4 weeks in the environment conducive for browning of the exterior surface, i.e. temperature of about 19–20 °C and 2000–3000 ppm CO_2, and are watered once daily to maintain the humidity levels more than 85%. As the browning process nears completion, pinheads start to form about 1–2 mm beneath the surface. Primordia development is stimulated by soaking the blocks in ice-cold water for 15–20 min. Mushrooms are ready for harvesting approximately after 7–9 days of soaking (Fig. 12.14).

Table 12.6 Region-wise possibilities of growing different mushroom varieties in India

Region	States covered	Average temperatures of the region[a]	Suitable mushroom species	Temperature requirement for fruiting	No. of crops to be grown in a year
Northwestern India	Punjab, Haryana, Himachal Pradesh, Uttarakhand, J&K	15–25 °C	White button mushroom, oyster	16–20 °C	One crop of button mushroom (Nov–Jan) Three crops of oyster mushroom (Feb–April and July–Oct)
Northeast India	Assam, Manipur, Sikkim, Meghalaya, Nagaland, Tripura, Mizoram	16–26 °C	Shiitake, oyster	18–20 °C	One crop of shiitake (Sep–Feb) Two crops of oyster (Feb–April and July–Sep)
Central India	Madhya Pradesh, Maharashtra, Uttar Pradesh	25–35 °C	Oyster, milky	32–35 °C	Two crops of oyster (Sep–Feb) Two crops of milky (Mar–August)
South India	Kerala, Tamil Nadu, Karnataka, Andhra Pradesh, Telangana	25–35 °C	Oyster, milky, paddy straw mushroom	32–35 °C	One crop of oyster (Sep–Nov) Two crops of milky (Jan–April) One crop of paddy straw mushroom (July–August)
East coastal area	Odisha, West Bengal, Chhattisgarh	30–35 °C	Paddy straw mushroom, milky, oyster	30–35 °C	Three crops of paddy straw mushroom (March–August) One crop of oyster (Oct–Nov) One crop of

(continued)

Table 12.6 (continued)

Region	States covered	Average temperatures of the region[a]	Suitable mushroom species	Temperature requirement for fruiting	No. of crops to be grown in a year
					milky (Feb–March)

[a]Average temperatures are excluding the hottest summer months

Fig. 12.12 Bottle cultivation of *Flammulina velutipes*

Bag Cultivation of King Oyster Mushroom For cultivation of *Pleurotus eryngii*, saw dust/wheat straw/paddy straw can be chosen as the basic ingredient. The substrate should be supplemented with organic nitrogen materials such as wheat bran or cereal bran. The substrate is prepared by wetting the straw or sawdust thoroughly in water for 16–18 h. After wetting, the excess water should be drained out, and 20% wheat bran is added to the substrate on a dry weight basis and mixed thoroughly. 1.5–2.0 kg of wet substrate is filled in each heat-resistant polypropylene bag. The bags are then plugged with nonabsorbent cotton by inserting polypropylene ring at the opening of the bag. The filled bags are sterilised in an autoclave for 90–120 min at 22 pounds per square inch of pressure. After sterilisation, the bags are cooled down to room temperature and then inoculated with grain spawn at 10–15% on a wet weight basis and then transferred to the incubation room at 22–25 °C. Spawn run will be completed within 15–20 days. After the completion of spawn run, bags are removed and the blocks shifted to the cropping room. The temperature and relative humidity of the cropping should be maintained at 12–14 °C and 80–85%, respectively. Pinheads start developing after 5–7 days after removing the bags. Matured fruit bodies (Fig. 12.15) are harvested 3–4 days after pinning. The yield may range from 300 to 400 g of fresh mushrooms per kg of dry substrate.

Fig. 12.13 Ready-to-fruit (RTF) bags for milky mushroom cultivation

Fig. 12.14 Fruiting of shiitake under short-duration cultivation technology

12.7 Future Thrust

Mushrooms being a nontraditional vegetable, the practice of cultivating different species of mushrooms in rotation is not yet popular among the growers. Farmers in the plains of Punjab and Haryana take up white button mushroom cultivation only during winter by erecting the temporary bamboo houses in the fields. The growers in coastal Orissa utilise the paddy straw to cultivate the paddy straw mushroom only. The tribals in the northeast harvest the shiitake by adopting the wood log cultivation technology. In Himachal Pradesh also, the mushroom production is largely limited

Fig. 12.15 Growing *Pleurotus eryngii* under controlled environmental conditions

to winter season to produce white button mushroom. Though all these farmers show enough acumen in exploiting the prevailing climatic conditions to grow mushrooms by low-cost technology, they fail to realise the possibility of growing different mushrooms in different seasons. Hence, notwithstanding the high demand for white button mushroom, there is an urgent need in the paradigm shift in the adoption of different mushrooms to achieve the diversification in mushroom cultivation to suit the local conditions and demand. This is an important step to ensure the economic production of mushrooms in climate-friendly manner.

The majority of the seasonal growers in North India prepares the compost by following a long method to cultivate white button mushroom. Long method of composting was replaced by a short method in European countries long back due to its low productivity, disease incidence, long duration and the pollution effect on the environment. However, special structures like pasteurisation tunnel and bunkers requiring heavy investment hinder many Indian farmers from switching to this improved composting technology. Wheat straw, the main raw material used in the preparation of selective compost, is costing Rs 7–8 or more, adding to the high cost of mushroom production. The future research must focus to develop state-of-the-art low-cost alternative structures and try other farm wastes and residues in the composting technology. The idea of cooperatives and community pasteurisation tunnels to prepare the compost is to be strongly encouraged. The United Nations Development Programme (UNDP)- and District Rural Development Agency (DRDA)-sponsored bulk-composting schemes to encourage small mushroom growers need to be replicated in other potential areas of the country. Places are to be identified, offering scope to cultivate button mushrooms under natural conditions during a particular season.

Challenges to Mushroom Industry

Mushroom cultivation is a capital intensive and also increases with the increase in farm size; the financial assistance through institutional agencies at a cheaper interest rate would be the desirable entity. Farmers' cooperative marketing societies must be promoted to take care of surplus quantity of mushroom producer. Mushroom is a highly perishable crop and prone to high temperature; marketing infrastructures such as cold storage facilities are of immense importance (Ram Singh and Subhash Chandra 2008).

Research and Policy Issues

The policy must aim to remove the anomalies in giving subsidy to mushroom production activity. There is no uniform policy among different state governments to give subsidy to seasonal growers by putting temporary growing structures. Also, it is not clear about the subsidy on inputs like chicken manure, gypsum, etc., that are excessively used in mushroom production activity. The electricity tariff for controlled units of mushroom production is also a case of another anomaly. Many state governments are considering mushroom houses as industrial units. However, since mushroom is a vegetable, it must be considered as an agriculture activity for electricity tariff.

The research has to focus on developing the cultivation technology of many wild mushrooms, which are consumed in different parts of the country. Biochemical studies on medicinal and pharmaceutical compounds present in such mushrooms need to be encouraged. Since the morphological features of mushrooms cannot be relied upon for their edibility, possibility of an efficient bioassay to differentiate the poisonous and edible mushrooms if explored can be of great value. Such tools will greatly help to educate the tribals and common people to avoid the tragic deaths occurring due to consumption of poisonous mushrooms.

Mushrooms are promoted as an important agribusiness activity to address the issue of environmental and ecological degradation. Mushroom cultivation though uses the farm wastes and recycles them; the disposal of spent mushroom substrate (substrate remaining after taking of crop) sometimes emanates foul smell causing irritation among the inhabitants in the area. Proper disposal of SMS, their use in vermicomposting, conversion into biofuel, etc., need further attention from researchers. The use of polythene for filling the compost and spawn also contradicts mushroom business as environment-friendly activity. The substitution of polythene with biodegradable materials for the same purpose needs similar kinds of efforts.

Besides unavailability, low per capita consumption of mushrooms among Indians is mainly attributed to unawareness about the mushroom cooking. Mushroom cooking is not much different from other cooking many other vegetables. Deliberate efforts must be made to popularise common mushroom recipes through print and electronic media. Postharvest processing of mushrooms to prepare the value-added products like pickle, biscuits, soup powder, papad, jam, murabba, candy, etc., can augment the consumption of mushrooms along with generating enough employment opportunities for women entrepreneurs and self-help groups in the rural areas. Mushrooms, if served in the ICDS (Integrated Child Development Services) project, would serve the objective of the scheme. ICDS offers supplementary nutrition to nearly ten million beneficiaries through more than 10 lakhs Anganwadi Centers (AWCs) in India. All the Krishi Vigyan Kendras (KVKs) across India must assume a greater role in this endeavour.

Training and Education

Much needs to be done on extension and education on mushrooms. Children and students can be best educated by including the mushrooms in their curriculum. The state agriculture universities must include a comprehensive practical course on mushroom cultivation technology. This would enable the students to foresee and pursue an agribusiness activity involving the production, processing and marketing of mushrooms.

The Directorate of Mushroom Research (DMR), Solan, its more than 30 centres of AICMIPs (All India Coordinated Mushroom Improvement Projects) situated across India, state departments of horticulture and many other NGOs provide short-term training on mushroom cultivation technology. All these trainings do help the people in getting exposed to different mushrooms and impart the knowledge about their cultivation technology. However, it is felt that the long-duration training aiming at imparting skills required for various aspects of mushroom cultivation like the spawn production, substrate preparation, maintenance of growing conditions, systematic and safe disposal of SMS, etc., is the need of the hour. The DMR is offering a long-term (three-month) comprehensive training module on this line. Mushroom cultivation training courses merit similar modules because of the need to understand the scientific and technical aspects involved in its cropping cycle and its activities.

Unlike many other crops and commodities, not much is done in developing a reliable decision support system (DSS) or an expert system on mushroom cultivation technology. In an era of information technology, such a system will help the farmers in making intelligent decisions in selecting a particular mushroom variety to suit to the local climatic conditions prevailing in their locality. The DSS must also address the needs of the existing mushroom growers with respect to the incidence of pest and disease and their control measures for enhanced economic returns to the farmers and entrepreneurs. Similar content can be made available in CD-ROM at an affordable price to cater to the information needs of many farmers. Recent experiences suggest that such content gives better results when used in conjunction with other modes of instruction. Since there is a reluctance to accept computer-assisted instruction (CAI) as a sole teaching method (Herriot et al. 2004), combining the digital content with traditional teaching methods will improve the learning effectiveness. Some of the efforts of the Directorate of Mushroom Research, Solan, to popularise mushroom farming and mushroom consumption are given in Table 12.7.

At a time when agriculture is in distress, farmers look up to activities, which can improve their livelihood security. Mushroom cultivation is one such agribusiness which can help farmers realise better returns compared to many other crops or enterprises. If all the research, extension and policy issues concerning mushroom growers are addressed with empathy, there will be certainly a sea change in the growth prospects of mushroom cultivation in India. Mushrooms have the potential to play a strategic role in maintaining ecological health, reducing rural poverty and unemployment, and thus would make a world of difference to our agrarian community.

Table 12.7 Innovative extension methods tried at ICAR-Directorate of Mushroom Research, Solan (HP)

Title of the innovative extension method	Process (how the method is developed and how it is being used in the field)	Impact on stakeholders
Regional mushroom mela and interaction with farmers (*Pradeshik khumb mela va kisan goshthi*)	Having seen the demand and response of national mushroom mela which is held annually at the ICAR-DMR, Solan, we started organising the regional mushroom mela among the cluster of mushroom growers. Such melas were held in Sonepat, Panipat, Hoshiarpur, Bilaspur (HP) and Jammu	The regional melas and kisan goshthis were received with a lot of expectations by the seasonal mushroom growers. Prior to each such mela, the multidisciplinary team of scientists undertook the survey to understand the situations in the area. Such melas were instrumental in educating the farmers about good mushroom cultivation practices to overcome the incidence of pest and diseases These melas went beyond giving advisory and extension services. They were also used as a tool to popularise mushrooms as a health food among the common public
Web-based decision support system/mushroom information system (Agri Daksh)	The platform gave the opportunities for the farmers and entrepreneurs with net accessibility to interact with experts on the problems faced by them	Farmers and entrepreneurs are logging in to get the information on need basis. All the static information on mushroom cultivation, compost and substrate preparation, value addition and postharvest practices are given in the system
Strategy for need-based training aspects of different duration	Rather than having a uniform or blanket training programme on mushroom cultivation aspects, we have formulated trainings of different duration covering different aspects which are need based. A ten-day training programme highlighting large-scale environment-controlled units, seven-day training for livelihood-based mushroom farming and three-day trainings on spawn production and processing have been formulated	All training programmes are in huge demand with seats getting filled 3–4 months ahead of scheduled date

(continued)

Table 12.7 (continued)

Title of the innovative extension method	Process (how the method is developed and how it is being used in the field)	Impact on stakeholders
Bulk SMS service for publicity	Using the bulk SMS facility, we disseminated information about the extension events and crop advisory services	Aimed to reach the stakeholders in time with less resource wastage and larger outreach compared to traditional means of communication of writing letters
Farmer-scientist-industry interface	This initiative was taken to bring the stakeholders in mushroom cultivation on platform to discuss the road ahead to improve the scope and opportunities of mushroom cultivation in the country	Well appreciated by the industrialists, entrepreneurs and mushroom growers. The scientists were made aware about the farmers' practical problems and find solutions on the problem faced by them
Digital content delivery of mushroom cultivation to new mushroompreneurs	The increasing demand for the training on mushroom cultivation was not met adequately by the ICAR-DMR and state department personnel. Hence, this idea was taken up to cater to the needs of the entrepreneurs to give them the necessary basic information on mushroom cultivation cropping management practices on different varieties of mushrooms and spawn production	The digital content documentaries on mushroom cultivation have been widely received by the prospective mushroom growers. The documentaries have been used to complement the lectures and discussion in the training programmes to show the procedure from the beginning to the end

By preferring mushroom production in place of *Rabi* crops on fertile fields, farmers of Haryana and Punjab take a measured risk. However, our interaction with farmers revealed that climatic conditions in North Indian plains during the winter season support good mushroom production without costly environmental manipulation required for their cultivation. Erection of temporary structures during the winter season also ensures the availability of land for growing field crops during the rest of the year. Materials used for structures such as bamboos are often reused. Hence, ideally, the profit from the mushroom production is much higher in subsequent years as the cost of the sheds is not incurred.

Farmers of Haryana and Punjab shift the site of mushroom production from the previous years to reduce the incidence of soilborne diseases. This practice by many farmers is praiseworthy as widespread occurrence of wet bubble, green mould, brown plaster mould, etc., can be reduced else. It results in a heavy loss of income to farmers (Sharma and Kumar 2000). Few of the farmers also used yellow sticky traps to counter the menace of small mushroom flies in the growing rooms. This cost-effective and efficient practice results in a significant reduction of flies in the mushroom sheds (Kumar and Sharma 1999).

The compost, on which mushrooms are grown, is a selective medium for the growth of white button mushroom mycelium (Sinden and Hauser 1950). Production and quality of mushrooms is decided by compost quality. Compost is prepared by two different methods, the short method and the long method. Short method composting takes less time but is cost-intensive on account of machinery and pasteurisation tunnels (Sinden and Hauser 1950; Vijay 2011). Hence, farmers adopt a long method of composting to reduce the investment. By lab analysis, the best compost for white button mushroom will have 68–72% moisture and pH of 7.2–7.8. Many progressive farmers use pH paper to analyse compost pH and few even get the compost analysed in research laboratories of ICAR-DMR, Solan, ICAR-IARI, New Delhi, and Haryana Agro-Industrial Corporation (HAIC), Murthal, Haryana.

Seasonal mushroom cultivation during winter brings a copious supply of fresh button mushrooms in vegetable markets of North India during winter season. Special mushroom *mandis* are organised by farmer groups and vegetable vendors in several cities. Farmers continue to get good prices for mushroom throughout the season. However, on several occasions, sharp decline in mushroom prices is also seen. In such cases, some farmers decide to sell the produce to nearby canning units or to pickling units. This will at least save their cost of transportation to vegetable markets.

12.8 Conclusion

Innovation and adoption of various technologies by seasonal mushroom growers of four North Indian states have resulted in considerable economic profit to them. It also provides job opportunities for a huge number of labourers. These labourers from adjoining states usually migrate during mushroom growing season to Haryana and Punjab to cater the need. In that way, the core labour group refreshes their skill of mushroom growing year after year, thereby becoming more experienced, and the employers get dependable persons to take care of whole activity. The labourers usually work under a leader on contractual basis, and relationship between employers and labourers becomes enduring. This ensures smooth running of whole process. Economic returns claimed by farmers are huge; many growers with 60–80 huts or above claim to have crossed one crore turn over in a season. Hence, we may opine that the seasonal mushroom growing activity by North Indian farmers has reached a level akin to small-scale industry and efforts are needed to replicate this success in other parts of the country with similar climatic conditions and resources. Interpolation of many such indigenous practices has inherent potential to bring desirable changes in the lives of rural folk. Detailed economic and social impact analysis of such activity needs further investigation and interventions by scientists to further refine the location-specific technologies.

The economic success in mushroom farming has huge implications on the survival of mushroom cultivation as an agri-enterprise and for livelihood security of small and marginal farmers. Therefore, the round the year cultivation and targeting regular supply of fresh mushrooms must be ensured among the farmers. This will also help to improve the consumption of mushrooms over a period of time.

References

Annepu SK, Sharma VP, Barh A, Kumar S, Shirur M, Kamal S (2019) Effects of genotype and growing substrate on bio-efficiency of gourmet and medicinal mushroom, lentinula edodes (berk.) Pegler. Bangladesh J Bot 48(1):1–10

Badalyan SM (2000) Antitumor and immune-modulating activities of compounds from several basidiomycetes. Probl Med Mycol 2:23–28

Badalyan SM (2012) Edible ectomycorrhizal mushrooms. In: Zambonelliand A, Bonito G (eds) Edible ectomycorrhizal mushrooms. Soil biology series, vol 34. Springer, Berlin, pp 317–334

Bose S (2016) Mushroom cultivation and marketing strategies: an untapped source of sustainable development and livelihood in North Bengal. Sumedha J Manag 5(2):121

Cheung PCK (1997) Chemical evaluation of some lesser known edible mushroom mycelia produced in submerged culture from soy milk waste. Food Chem 60:61–65

Herriot AM, Bishop JA, Truby H (2004) The development and evaluation of student training, education and practice for dietetics CD-ROM: a computer-assisted instruction programme for dietetic students. J Hum Nutr Diet 17(1):35–41

ICAR-DMR Annual Report (2019) ICAR Directorate of Mushroom Research, Chambaghat, Solan, Himachal Pradesh, India

Khan MA (2010) Nutritional composition and hypocholesterolemic effect of mushroom: Pleurotus sajor-caju and Pleurotus florida. LAP Lambert Academic Publishing Gmbh, Saarbrucken, pp 1–11

Kumar S, Sharma SR (1999) Yellow coloured light trap- a cheap method for management of mushroom flies. Mushroom Res 8:57

Pushpa H, Purushothoma KB (2010) Nutritional analysis of wild and cultivated edible medicinal mushrooms. World J Dairy Food Sci 5(2):140–144

Ram Singh G, Subhash Chandra AS (2008) Economics of production and disposal pattern of mushroom. Environ Ecol 26(4C):2273–2279

Sharma SR, Kumar S (2000) Studies on wet bubble disease of white button mushroom (Agaricus bisporus) caused by Mycogone perniciosa. In: Grievisven LJLD (ed) Science and cultivation of edible fungi. Balkema, Rotterdam, pp 569–575

Sharma VP, Kumar S, Tewari RP (2008) *Flammulina velutipes*, the culinary medicinal winter mushroom. National Research Centre on Mushrooms, Solan

Sharma VP, Annepu SK, Gautam Y, Singh M, Kamal S (2017) Status of Mushroom Production in India. Mushroom Res 26(2):111–120

Shirur M (2011) Round the year cultivation of mushrooms. In: Singh M, Vijay B, Kamal S, Wakchaure GC (eds) Mushrooms: cultivation, marketing and consumption. ICAR-Directorate of Mushroom Research, Solan

Shirur M, Sharma VP (2016) Doubling farmers' income through technological interventions in mushroom. Indian Hortic 61(2):57–60

Shirur M, Ahlawat OP, Manikandan K (2014) Mushroom consumption and purchasing behavior in India: A study among selected respondents. Mushroom Res 23(2):225–231

Shirur M, Shivalingegowda NS, Chandregowda MJ, Rana Rajesh K (2017) Socio economic analysis of entrepreneurial behaviour of mushroom growers in Karnataka. Indian J Agric Sci 87(6):840–845

Shirur M, Shivalingegowda NS, Chandregowda MJ, Rana Rajesh K (2018) Performance analysis of South-Indian mushroom units: Imperative policy implications for their preparedness for global competitiveness. Curr Sci 115(11):2141–2146

Shirur M, Manikandan K, Wakchaure GC, Mandal K (2018a) Indigenous system of white-button mushroom production for augmenting farm income in northern India. Indian Farm 68 (02):11–16

Sinden JS, Hauser E (1950) The short method of mushroom composting. Mushroom Sci 1:52–59

Singh M (2011) Mushroom production: an agribusiness activity. In: Singh M, Vijay B, Kamal S, Wakchaure GC (eds) Mushrooms: cultivation, marketing and consumption. ICAR-Directorate of Mushroom Research, Solan

Singh M, Shirur M (2016) Mushroom production: a viable and profitable avocation for educated youth. In: Chadha KL, Singh SK, Kalia P, Dhillon WS, Behera TK, Jai P (eds) Doubling the farmers income through horticulture. Daya Publishing House, New Delhi, pp 847–852

Stamets P (2005) Notes on nutritional properties of culinary-medicinal mushrooms. Int J Med Mushrooms 7(1–2):103–110

Ul Haq I, Khan MA, Khan A, Ahmad M (2011) Biochemical analysis of fruiting bodies of Volvariella volvacea strain Vv pk, grown on six different substrates. Soil Environ 30 (2):146–150

Vijay B (2011) Methods of compost production for white button mushroom (*Agaricus bisporus*). In: Singh M, Vijay B, Kamal S, Wakchaure GC (eds) Mushrooms: cultivation, marketing and consumption. ICAR-Directorate of Mushroom Research, Solan, pp 49–68

Vision-2030 (2011) Indian Council of Agricultural Research, New Delhi

Wakchaure GC (2011) Production and marketing of mushrooms: global and national scenario. In: Singh M, Vijay B, Kamal S, Wakchaure GC (eds) Mushrooms: cultivation, marketing and consumption. ICAR-Directorate of Mushroom Research, Solan

Wasser SP (2010) Medicinal mushroom science: history, current status, future trends and unsolved problems. Int J Med Mushroom 12:1–16

Climate-Resilient Vegetable Farming: Approaches for Sustainable Development 13

Sudheer Kumar Annepu, Sunil A. Nair, Shivender Thakur, and Vinay Verma

Abstract

Climate change is an alarming phenomenon with potential impacts on agriculture production systems across the globe. Uneven weather patterns as a result of changing climate traces its origin both from natural and anthropogenic factors. It has been forecasted by several global agencies that, until 2050, climate change-related adversaries will lead to a 2% decline in total agricultural production in every decade. The developing countries with high population density are more vulnerable to this phenomenon. Besides recurring crop failures due to climate change, the immediate consequences may be observed as altered soil fertility status, a resurgence in insect-pest and pathogen levels, and abnormal behavioral changes in the pollinators, which ultimately lead to a reduction in productivity of crops including the vegetables. Vegetable crops are vital sources of nutrition by supplementing the minerals, vitamins, crude fibers, and many other vital dietary elements. In cultivation of vegetable crops, several improved agronomic practices have been evolved over time for different regions. With prudent application, these practices have potential to mitigate the effects of climate change.

Keywords

Climate change · Nutrition · Adaptation · Mitigation · Indigenous vegetable crops

S. K. Annepu (✉) · S. A. Nair · S. Thakur · V. Verma
Department of Vegetable Science, Dr YSPUH&F, Solan, Himachal Pradesh, India

V. K. Hebsale Mallappa, M. Shirur (eds.), *Climate Change and Resilient Food Systems*, https://doi.org/10.1007/978-981-33-4538-6_13

13.1 Introduction

Horticulture, which includes cultivation of fruit crops; vegetable crops; flower crops; medicinal, aromatic, and plantation crops; mushroom cultivation; etc., is an essential sector of agriculture and contributes immensely to the Indian economy. These crops impart the diversification in the agriculture sector and improve farmers' livelihoods by generating additional employment and enhanced farm income. India is bestowed with a wide range of climatic conditions, and many of the horticultural crops can be grown naturally in one or other parts of the country. In the last three decades, the horticulture sector has recorded exceptional growth in India owing to the technological innovations, improved varieties, and policy interventions. Vegetable crops due to their short growing period, high productivity per unit area, and the presence of high nutritional attributes assume special significance among the horticulture crops. Fresh vegetables are considered an effective dietary source of nutrition, thus occupying a prominent role in the regular diet of all categories of people, irrespective of the social and economic status.

13.1.1 Vegetable Crops for Improved Nutrition

Vegetables, being rich in vitamins, minerals, dietary fibers, roughages, and other essential amino acids, play an essential role in ensuring the nutritional security of the burgeoning population. In several tropical and subtropical regions of the world, root and tuber vegetable crops such as potato, sweet potato, cassava, yams, etc., constitute an important source of carbohydrates. Leguminous vegetables, viz., peas, beans, and vegetable soya bean, offer good-quality protein with high digestibility. They also contribute trace minerals, vitamins, roughages, and dietary fibers. Many of the minor and underexploited vegetable crops are rich sources of phytochemicals, antioxidants, and several other secondary metabolites, which help in boosting the human immunity (Table 13.1). Epidemiological studies show that a higher intake of fruits and vegetables reduces the incidence of lifestyle-related diseases, including cancer, diabetes, and cardiovascular diseases.

Living organisms cannot synthesize the minerals biochemically. They need to be supplemented through drinking water and the external diet. Among the minerals, calcium, phosphorus, potassium, sodium, and magnesium are the five major minerals that humans need along with the other trace minerals with specific biological activity. Leafy vegetables such as spinach, amaranth, drumstick, and palak are particularly rich in these mineral elements (Natesh et al. 2017). Intake of vegetables, 300 g/day, is almost sufficient to meet the recommended dietary allowance of most of the nutrients (NIN 2011).

Though traditional cultivation of vegetables attained much progress in China, India, and several other South Asian countries, the per capita consumption and availability of vegetables in Southeast Asia, sub-Saharan Africa, and some of the Latin American countries is far below than the requirement to provide adequate nutrition. Realizing the importance of vegetable crops in nutritional security, the

Table 13.1 Vegetable crops rich in bioactive compounds/phytochemicals

Nutraceuticals/ bioactive compounds	Vegetables	Properties
Allylsulfides, fructan	Onion, garlic	Antibacterial; antifungal
Anthocyanin	Black carrot, beetroot, red amaranth, red lettuce, red cabbage, brinjal	Anti-inflammatory; antitumor
Butylphthalide	Celery	Anti-hypertension
Capsaicin	Chilli	Anti-inflammatory
Carotenoids	Leafy vegetables, carrot, orange cauliflower, orange flesh sweet potato and cassava, muskmelon, pumpkin	Antioxidant; antitumor
Ferulic acid	Turnip	Antioxidant; antiaging
Flavonoids (isoflavones)	Beans	Reported to treat menopausal symptoms
Folate	Chenopodiaceae (spinach, Swiss chard, beet greens)	Prevents DNA damage
Glucosinolates	Cole crops (cabbage, cauliflower, sprouting broccoli, Brussels sprout)	Antibacterial; antifungal; antiviral; antitumor
Hesperitin	Green vegetables	Anti-inflammatory
Luteoline	Cauliflower, celery	Reduces eye disorders
Polyphenols	Potato, brinjal, okra, leafy vegetables, onion	Antioxidant; fights against cardiovascular disease
Resveratrol	Red onion	Anti-inflammatory; antitumor
Silymarin	Compositae plants (artichoke)	Hepatoprotective

Source: Rai et al. (2012), Singh et al. (2020)

majority of the countries fall under these regions, given utmost importance to boost vegetable crop production. Past achievements are significant in many countries. However, the challenges remain much greater than past achievements. For instance, India is feeding 17% of the global population with a mere 2.3% of land area and 4.5% of hydrological resources. In the coming years, the challenges to agriculture, particularly vegetable crops, will continue to increase due to the accumulating problems such as higher population growth, increasing pressure on cultivable land, and shrinking water reserves within the potentially hostile scenario of climate change. The uncertain weather patterns created by the changing climate further aggravate these problems.

13.2 Impact of Climate Change on Vegetable Cultivation

Climate change, a much-discussed phenomenon in recent time, is posing a major challenge in the twenty-first century for ensuring food security and ecological balance. Climate change is defining in many ways, but its consequences ultimately lead to a significant rise in the mean average global temperature.

Being succulent in nature, vegetable crops are highly vulnerable to extreme changes in the growing environment. The environmental factors, viz., temperature, moisture stress, and CO_2 levels, have been reported to affect crop yield adversely. In addition to a reduction in yield levels, frequent crop failures, poor quality, higher incidence of pest, and disease incidence are the common problems arising due to these environmental stresses (Koundinya et al. 2017). Though some of the vegetable crops exhibit positive responses to the elevated atmospheric CO_2 levels, the end results are reported to be undoubtedly negative (Bisbis et al. 2018). In certain crops, higher level of atmospheric CO_2 reported to led to inreased yields. This phenomenon has been conceptualized as carbon fertilization. However, higher levels of warming have been perceived to influence physiology and metabolism in the majority of vegetable crops.

Further, prolonged exposure of plants to the high radiation levels results in accumulation of more antioxidants in the plant parts. Similarly, when plants are exposed to persistent moisture stress conditions, the plants will improve its efficiency to utilize the available water resources. Further, extension in the cropping periods has been observed due to the increased mean temperatures mainly in the northern hemisphere. This phenomenon helps in ensuring the market supply for longer durations.

Despite these beneficial effects, the impact of climate change is found to be significantly detrimental in several areas. It has been widely reported in the literature regarding the loss of nutrient contents and deterioration of morphological quality in the majority of the vegetables due to high-temperature stress. Raise in temperature leads to a higher respiration rate and hastens the postharvest spoilage of perishable vegetables. Continuous rise in mean temperatures shifts the cropping period by accelerating the phonology in many vegetable crops, whereas, in temperate vegetable crops, a higher temperature is a major limiting factor for fulfilling the vernalization requirement and possesses a great challenge to its seed production process. Erratic rainfalls are likely to reduce seed yields. New pests, diseases, and weeds thrive well under the global warming conditions and therefore compete with the main crop (Fig. 13.1).

13.3 Strategies for Climate-Resilient Vegetable Farming

UNFCCC has envisaged two main strategies to counteract climate change, viz., mitigation and adaptation. Mitigation relates to reducing the severity of climatic change by minimizing the greenhouse gas (GHG) emissions employing carbon sink or by imbibing them through plant biomass. The adaptation measures minimize the

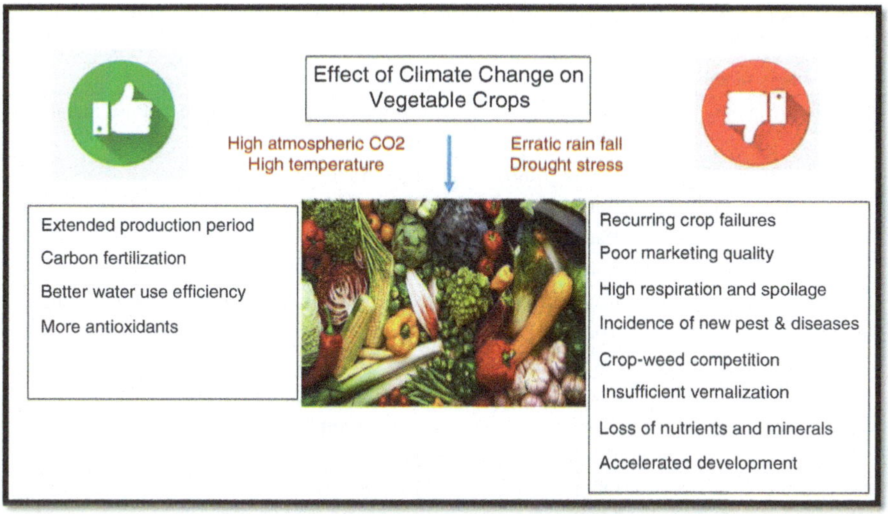

Fig. 13.1 Beneficial and adverse effects of climate change on vegetable crops (Source: Bisbis et al. 2018)

negative aspects of climatic changes or taking measures to reduce the negative aspects and seeking alternatives through suitable agronomic measures. Though both are complementary in nature, few points clearly differentiate adaptation from mitigation. Unlike mitigation, adaptation doesn't consider the causes of climate change. Adaptation takes advantage of positive impacts, while the mitigation reduces both negative impacts. The immediate vulnerability of vegetable crops to the climate extremities can be minimized by the adaptation strategy with direct visible benefits. A mitigation strategy provides a long-term benefit with wider scale visibility. An adaptation measure alone doesn't eliminate the negative impacts. Hence, both mitigation and adaptation have to be followed holistically for optimal results (IPCC 2007; Muller 2009). Breeding techniques and resource conservation technologies are some of the examples of adaptation measures. Minimizing the carbon footprint in agriculture by reducing the use of fossil fuel-based agrochemicals and carbon sequestration are few of the mitigation strategies.

The following are some of the widely recommended strategies for climate-resilient vegetable farming.

13.3.1 Genetic Improvement

Genetic improvement is an adaption strategy that prepares the crop plants to adapt to the expected climatic extremities. The fundamental requisite for the genetic improvement in this direction is the presence of genetic variability in the existing plant population. The diversity can be explored with various selection procedures. Screening techniques such as diffused porometry measure the leaf water

Table 13.2 Varieties with abiotic stress tolerance in different vegetable crops released for cultivation in India

Abiotic stress factor	Crop	Variety
Temperature stress	Tomato	Pusa Sheetal; Pusa Hybrid 1; Pusa Sadabahar
	Cucumber	Pusa Barkha
	Bottle gourd	Pusa Santusthi
	Carrot	Pusa Kesar
	Radish	Pusa Himani
	Potato	Kufri Surya
	Cauliflower	Arka Vimal; Arka Spoorthi, Pusa Meghna
Drought stress	Tomato	Arka Vikas
	Sweet potato	Sree Nandini
	Cassava	Sree Sahya
Salinity stress	Tomato	Sabout Suphala
	Brinjal	Pragati and Pusa Bindu
	Okra	Pusa Sawani
	Musk melon	Jobner 96–2
	Spinach beet	Jobner Green
	Onion	Hissar-2

Source: Koundinya et al. (2017)

conductance; the mini-rhizotron technique can be used for measuring the root penetration, distribution, and its density. Carbon isotope discrimination, drought index measurement, infrared thermometry, and visual scoring are some other techniques helpful to identify the stress-responsive genetic stock. The selection of the resistant plants from the existing populations and subsequent development of new varieties from their progeny is the next step in this process. Further, using this genetic stock in hybridization process helps in the transfer of desirable genes into the agronomically superior cultivars (Table 13.2).

The major result of climate change is a shift in the cropping season, which reduces the duration of favorable environment needed for crop growth. Because of this shift, breeding for short duration varieties is of prime importance to fit the crop into the stress avoidance category. Wild relatives are the reservoirs of the desired gene combinations that confer the resistance to many abiotic stresses. Such genes can be identified and introgressed into the cultivated varieties either by conventional breeding techniques or with the aid of modern biotechnological tools. India is endowed with a large germplasm pool of 58,250 accessions being scientifically conserved in different research organizations. Recent advances in the marker-assisted selection provide an opportunity for effective screening of the germplasm. Phenomics-based studies help in characterization of individual germplasm under controlled growing conditions. High-throughput phenomics facilitates the ultramicroscopic-level observations to improve the precision in selection.

13.3.2 Protected Cultivation

Growing vegetable crops under protected structures is gaining popularity from the last decade intending to meet the off-season market demand. Protected cultivation, either in fully controlled or partially controlled growing structures, is highly beneficial to mitigate the unfavorable environmental conditions that hinder the cultivation of vegetable crops. Besides allowing the crops to survive under varying growing conditions, protected cultivation allows them to extend the harvest period. Better control is possible for pest and disease management with minimal pesticide application. Input use efficiency is relatively higher under protected growing conditions. Hi-tech greenhouses enable to regulate the key growth parameters such as temperature, CO_2 requirement, and humidity through automation. Aeroponics and hydroponics systems are completely free from the soil-related stress factors and also prevent the incidence of soilborne pathogens. At present, protected structures are primarily confined to the production of high-value crops, and it needs to be upscaled to ensure the year-round supply of vegetables.

13.3.3 Agronomic Practices

Agronomic practices such as mulching, zero tillage, crop rotation, and organic soil cover help to mitigate the atmospheric GHGs by reducing the net emissions in the cultivation process. Resource conservation practices minimize the disturbance in soil ecosystem and hasten the process of carbon sequestration. Reduced or zero tillage minimizes the oxidation of organic carbon and its subsequent escape in the form of CO_2 into the atmosphere. Do not till or little till is the present-day slogan to manage the consumption of fossil fuels and climate change-associated problems. It has been estimated that globally conservative agronomic practices can mitigate up to 0.39 t CO_2 equivalent/ha/year under a dry climate and 0.98t CO_2 equivalent/ha/year under a moist climate (Smith et al. 2007). It was also estimated that conservation tillage and effective management of crop residues could reduce the emissions as much as 0.35t CO_2 equivalent/ha/year under warm, dry climate and 0.72 CO_2 equivalent/ha/year under warm, moist climate (Smith et al. 2007; Milder et al. 2011). Baker et al. (2007) reported that 25 Gt of carbon could be sequestered by adapting the conservation tillage in total croplands at the global level. Moreover, slower decomposition of organic matter is associated with the conservation tillage (Drury et al. 2006), thus reducing the CO_2 emission than the conventional tillage.

13.3.4 Grafting

Grafting is a vegetative reproductive technique that joins the tissues of two different plants and allows it to grow into a single plant. Grafting is widely practiced in fruit crops and plantation crops. However, it is not popularized in vegetable cultivation. However, recent studies indicated that its potential could be well exploited to make the plants adapt to climate adversities. Grafting of a susceptible scion onto a rootstock possessing the resistance to specific abiotic stress is the best way to utilize the available diversity (Koundinya et al. 2014). In the recent times, several attempts were made to achieve the high temperature tolerance in tomato and achieved significant level of success. *Solanum melongena* EG 203 (Burleigh et al. 2005) and *S. habrochaites* LA1777 (Venema et al. 2008) were used as rootstocks, to achieve high- and low-temperature tolerance, respectively. Sakata et al. (2007) reported that drought tolerance in watermelon could be achieved by grafting onto ash gourd plants. Grafting onto *Solanum melongena* rootstock helped in bacterial wilt and flooding tolerance in tomato (Palada and Wu 2007).

13.3.5 Integrated Cropping Systems

Integrated cropping systems such as mixed cropping, intercropping, strip cropping, and relay cropping have showed the ability in carbon sequestration, thereby acting as a potential mitigation choice (Wang et al. 2010). Crop failures, reduction in crop yields, and higher pest and disease incidence are the major consequences of climate change. Under such circumstances, integrated cropping systems play a more beneficial role than mono-cropping. The loss due to climate change in one crop can be compensated by another crop. An integrated cropping system mainly aims to increase farm income by means of diversification. Among the vegetables, legumes offer a better opportunity to use them in intercropping and mixed cropping systems. Research work in rainfed areas has shown that intercropping with specific planting geometry and selection of compatible crops is a cost-effective practice to increase the input use efficiency (Goswami et al. 2002). Tree-based intercropping (TBI) systems and agro-forestry are also supposed to be effective in mitigating GHGs.

13.3.6 Farming with Perennials

Perennial vegetable crops such as *Moringa oleifera*, *Parkia roxburghii, Sesbania grandiflora,* and *Piper mullesua* maintain the ground cover for prolonged periods and, in turn, improve the soil health and soil structure. With their deep root system, these crops bind the soil particles together and support the microbial processes that increase the water-stable aggregates and soil organic matter. Growing of perennial crops also reduces soil erosion. Drumstick is well adaptable to dry and hot climates of peninsular India and is considered as drought-resistant vegetable crop. Moreover, farming with perennials acts as a mitigation strategy by means of aerial carbon sequestration.

13.3.7 Organic Farming (OF)

Agricultural systems are not only affected by climate extremities but also contribute to it. About 10–12% of the global greenhouse emissions are generating from the food production systems. In addition, modern agricultural practices led to deforestation and soil degradation. Hence, OF acts both as a mitigation strategy that will address the emissions problem and as an adaptation technique by reducing the pressure on conventional practices (Table 13.3). Organic farming increases the resilience within the agroecosystem. It mimics the natural ecological processes and thus suffers less damage compared to conventional agriculture. It is widely perceived that OF is an alternative way to overcome the problems of global warming and the challenge of sustainability. It has its roots in various terms, biodynamic, regenerative agriculture, nature farming, and permaculture movements which developed in different countries.

In OF, emissions can be reduced through lower inorganic nitrogen input. It results in a reduction of N_2O emission and also eliminates the energy requirement in manufacturing the inorganic fertilizers. OF stores the carbon in soil, and by maintaining the high vegetation and intact soil structure, the CO_2 emission as a result of erosion can be minimized. OF is also found to be beneficial by minimizing the soil erosion than the conventionally managed soils. Thus, it helps in carbon sequestration by taking out the atmospheric CO_2 and stores as soil organic matter. The soils under OF are also more resilient to the floods and droughts by virtue of higher soil organic matter.

Table 13.3 Mitigation and adaptation strategies offered by organic farming

Mitigation	Adaptation
Reducing emissions of GHGs • Self-sufficient in nitrogen • Reduced emissions of nitrous oxide • Nitrogen efficiency	Application of traditional skills and farmers' knowledge • Breeding of indigenous crop varieties • On-farm production of manures and bio-fertilizers • Preparation of natural organic plant protection chemicals
Carbon sequestration • Arresting soil erosion • Carbon gain by organic management • No-tillage and minimum-tillage cropping	Soil fertility-building techniques • In situ nitrogen enrichment by introducing the leguminous plants
Other than cultivation aspects • Changes in consumer behavior and diet patterns • Stopping deforestation	A high degree of diversity • Diversification with a mix of farm enterprises reduces vulnerability

13.4 Diversification of Vegetable Farming with Indigenous and Aquatic Vegetables

Indigenous and minor vegetables are the traditional crop species that are native to that particular region (Fig. 13.2). They are important in view of nutrition, health, and sustainability of the social systems in the region where they have been evolved over a period of time. Traditional vegetables enhance the multiplicity in the regular diet with a balanced source of micronutrients. Unlike annual and biennial vegetable crops, which has a major share in Indian vegetable production, perennials such as drumstick, ivy gourd, pointed gourd, spine gourd, sweet gourd, breadfruit, chow-chow, chekurmanis, etc., are grown and consumed in relatively small scale. These perennial vegetables have a handful of vital nutrients, trace minerals, antioxidants, and medicinally important bioactive compounds. Lack of knowledge in consumers' and farmers' tendency to grow annual vegetables renders these vegetables of minor importance in the human diet. Further, exotic perennial vegetables such as asparagus, rhubarb, artichokes, etc., are not part of the average Indian human diet even today. Indigenous vegetables show substantiate biodiversity and are adapted to specific marginal growing conditions with minimal inputs.

The indigenous perennial vegetables in which leaves are used as a vegetable are *Bacopa monnieri* (Indian *brahmi*), *Basella* spp. (Indian spinach), *Clerodendrum*

a. *Basellarubra* b. *Sauropusandrogynus* c. *Murrayakoenigii*

d. *Moringaoleifera* e. *Trichosanthesdioica* f. *Sesbaniagrandiflora*

Fig. 13.2 Perennial indigenous vegetables

Table 13.4 List of indigenous perennial vegetables and their ethnobotanical use (leafy vegetables)

Scientific name	Local name/s	Ethnobotanical use
Bacopa monnieri	Indian *brahmi*	Acts as a blood purifier and useful in treating diarrhea, epilepsy, ulcers, and other digestion-related disorders
Basella spp.	Indian spinach	Tender leaves, shoots, and leaf stalks are used in the preparation of soups and also used as a stew. It is used to cure digestive disorders
Clerodendrum colebrookianum	East Indian glory	A decoction prepared by boiling the leaves is used to reduce hypertension and rheumatic pains. Roots, along with the barks, are used in treating asthma and bronchitis
Diplazium esculentum	Vegetable fern	The leaves are rich in minerals such as Zn, Mn, and Fe. Leaves are used in culinary preparations and making the pickles
Gmelina arborea	Malay bush beech	Its consumption is found to improve the appetite. The root and bark are useful for treating the stomach pains and also act as antihelmintic
Nymphaea spp.	Water lily	The rhizomes are sweet in taste, and its consumption is found to be useful for treating diarrhea and dysentery
Paederia foetida	Stinkvine	Found to be useful in treating rheumatism, urinary bladder stones, and flatulence
Pandanus amaryllifolius	Indian pandan	The fresh leaves are used as flavoring agent in cooking the sweet dishes and rice preparations
Pisonia grandis	Lettuce tree	Leaves are used as diuretic and antidiabetic
Polygonum spp.	Knotweed	Leaves are consumed as a vegetable after mixing with the crushed ginger
Crambe cordifolia	Greater Sea kale	The tender leaves have a pleasant cabbage-like flavor and are useful as a cure to itching
Zanthoxylum hamiltonianum	Tejamoo	Its consumption is found to improve the appetite. The tender stem is used to brush teeth when there is a toothache

colebrookianum (East Indian glory), *Diplazium esculentum* (vegetable fern), *Gmelina arborea* (Malay bush beech), *Nymphaea* spp. (water lily), *Paederia foetida* (stinkvine), *Pandanus amaryllifolius* (Indian pandan), *Pisonia grandis* (lettuce tree), *Polygonum* spp. (knotweed), and *Urtica spp.* (nettles). The indigenous perennial vegetables in which fruit is used as vegetable are *Coccinia grandis* (ivy gourd), *Moringa oleifera* (drumstick), *Parkia roxburghii* (tree bean), *Piper mullesua* (hill pepper), *Sesbania grandiflora* (agathi), *Solanum indicum* (bush tomato), *S. spirale* (titakuchi), and *Trichosanthes dioica* (pointed gourd) (Tables 13.4 and 13.5). Including these traditional vegetables diversifies the existing production systems and enhances the resilience to climate change.

Table 13.5 List of indigenous perennial vegetables in which fruits and leaves are used as vegetable

Scientific name	Local name/s	Ethnobotanical use
Coccinia grandis	Ivy gourd	The tender fruits are used as a vegetable. Apart from the culinary usage, the juice extracted from the roots and leaves is used as antidiabetic tonic
Moringa oleifera	Drumstick	The pods are rich in vitamins A and C. drumstick leaves and pods are considered as mineral-packed vegetable with high amounts of Fe, P, and Ca
Parkia roxburghii	Tree bean	Tender shoots and flowers are used in culinary preparations and as salads. Tender pods are rich in dietary fibers, protein, and also minerals such as P and Fe
Piper mullesua	Hill pepper	The dried plant parts are found to cure malaria. Roots and fruits are used in Ayurvedic medicines
Sesbania grandiflora	Agathi	It is a folk remedy for many disorders such as a diuretic, dysentery, sore throat, cataract disorders, and night blindness
Solanum indicum	Bush tomato	Fruits are consumed to cure digestive disorders, gastritis, dysentery, etc.
S. spirale	Titakuchi	Green fruits possess anti-malaria properties. The dried fruits are useful to cure gastric problems
Trichosanthes dioica	Pointed gourd	Tender fruits are consumed to overcome the problems of satiation and constipation. It also improves appetite and digestion

Source: Chadha and Patel (2007), Chadha (2009)

13.5 Aquatic Vegetables: An Underexploited Source to Combat the Malnutrition

Southeast Asian countries have access to large water bodies, viz., fresh, saline, and brackish. As this region is characterized by a high population rate, cultivation of vegetables under aquatic conditions may meet the challenges of undernourishment and chronic energy deficiency. Aquatic vegetables thrive in a natural habitat that compensates space and nutrient requirements for completing its life cycle. Edible aquatic plants are commonly grown in Southeast Asia countries such as China, Cambodia, Indonesia, Vietnam, India, etc.

The common aquatic vegetables are water chestnut (*Eleocharis tuberosa*), lotus root (*Nelumbo nucifera*), cattail (*Typha latifolia*), common water plantain mad-dog weed (*Alisma plantago-aquatica*), water bamboo (*Zizania latifolia*), arrowhead (*Sagittaria sagittifolia*), water dropwort (*Oenanthe stolonifera*), water caltrop (*Trapa spinosa*), duck-lettuce (*Ottelia alismoides*), watershield (*Brasenia schreberi*), water cress (*Nasturtium officinale*), water taro (*Colocasia esculenta*), water spinach (*Ipomoea aquatica*), cordon euryale (*Euryale ferox*), and fragrant water lily (*Nymphaea odorata*). Aquatic vegetables, viz., *Trapa bispinosa*, *Ipomoea aquatic*, and *Nelumbo nucifera*, occupy lion's share in food palate with daily consumption exceeding 50 g in many Asian countries.

13.5.1 Importance of Aquatic Vegetables

A poor person in Southeast Asia spends almost three-quarters of their earnings on food alone. It indicates their vulnerability to fluctuations in supply and cost of vegetables and other food materials. Hence, cultivation of aquatic vegetables may play a major role by reducing the spending on vegetables. Moreover, aquatic vegetables are eco-friendly and free from pesticide residues. Thus, consumption of these vegetables by the low-income group of Southeast Asia could reduce protein shortages among the local population. These vegetables have the potential to substitute the traditional vegetables which are widely under cultivation in several Asian countries. The innovations in food processing technologies could improve the palatability of these aquatic vegetables suitable for the culinary needs of the local communities.

In India, several areas receive rainfall more than 110 cm and contain unique wetland habitats ranging from high-altitude Himalayan lakes to the flood plains, brackish cultures, and coastal wetlands especially in the west coast. These unique backwaters and estuaries, areas with below sea level paddy cultivation especially the lower Kuttanad region of Kerala, mangrove swamps, and coral reefs, offer abundant scope for cultivation and expansion of such vegetables. Wetlands are the lifeline of resources encompassed by features such as wildlife habitat, control the runoff and minimizes the soil erosion, nutrient recycling besides maintaining stream-flow and provision of fish, fodder and fuel. Besides the wetlands, the arid region of India, mainly Rajasthan comprising of desert areas, is also host to a diversity of aquatic species with an estimated one-fifth of the 1500 plant species reported (Razvy et al. 2011).

13.5.2 Nutritional Properties of Aquatic Vegetables

These vegetables are basically alkaline in nature, are rich in fibers associated with a lesser occurrence of cardiovascular diseases (Peter 2002), and have low calorific value. Water hyacinth was found to be rich in amino acids such as cysteine, phenylalanine, and lysine. The lotus root, *Cordon euryale*, contains several alkaloids useful for treatment of cardiovascular diseases (Meng et al. 2003). Aquatic vegetables with nutraceutical properties, viz., brahmi, are administered to children for combating nutritional deficiencies. Young shoots of *Hydrolea zeylanica* are reported to possess antiseptic properties. *Vallisneria spiralis* is a rich source of P, Ca, and Fe and can be used as a stomachic and for leukorrhea treatment. *Ottelia alismoides* as astringent have also been reported (Pareek and Kumar 2014).

13.6 Vegetable Seed Production Under Climate Change Scenario

Vegetable seed production is highly subjected to fluctuations in the mean temperature, which controls the physiology and metabolism and, in turn, decides the quality of the seed to be produced (Ayyogari et al. 2014). Erratic precipitation during critical stages of vegetative and reproductive growth has been reported to cause losses in vegetable seed yield (Afroza et al. 2010). The Indian seed industry is on a positive growth phase with a growth rate of 12% compared to <5% growth at the global market. The very own "farmers saved seed" contributes 75% of the total requirement of total seed in the country: the rest a mere 25% from the organized seed sector. There is a large scope for India as a key player in the world seed industry if its domestic seed production by the organized sector is doubled to 50%. The targeted crop production could be achieved without tilling any excess landmass but through evolution and breeding of varieties with the ability to cope with climate change-induced changes (Ceccarelli et al. 2010). The following are the approaches recommended to mitigate the effect of climate change on seed production activity.

13.6.1 Shifting Areas for Seed Production

The measures include shifting areas for seed production as reported during cabbage seed production during 1971–1977 by Arya et al. (1979) wherein seed production was advocated to be implemented in areas with narrow temperature fluctuations. Kumar et al. (2009) observed 40% of reduction in seed yield in cabbage var. Golden Acre (during 1981 to 2004). Seed production in lower altitudes (1200–1450 MSL) was suggested as adaptive measures to ensure better seed yield. Another measure to adapt is by adjusting the sowing/planting dates as reported by Olesen et al. (2011). Singh et al. (2013) suggested changes in the date of sowing/planting dates so as to minimize the effect of elevated temperatures during flowering and seed set. Hu et al. (2017) upheld a similar opinion in case of potato seed production in semiarid growing conditions. Optimum sowing during 10–27th of May was suggested so as to target anticipated precipitation for higher yield.

13.6.2 Plant Breeding

Plant breeding measures include developing hybrids/varieties resistant to various abiotic stresses, viz., drought, salinity, waterlogging heat stress resistance, and biotic stress resistance through waterlogging; identification of suitable landraces for securing food security; and the use of conventional and molecular technologies, viz., marker-assisted breeding, transgenic breeding, and genome-wide selection. Employing stable alleles in self-incompatibility and male sterility mechanisms with the potential to perform well under alleviated temperature offers opportunity to breeders in ensuring seed production in both open-pollinated varieties and F_1

hybrids. The use of specific resistance source of grafted plants, viz., solanaceous vegetables and cucurbits have increased in recent years as an adaptive measure to counter saline, drought tolerance conditions, etc. Farmer's participatory breeding program by identifying farmers as "citizen scientist" facilitates the development of varieties for local needs (Kidane et al. 2017) by taking consideration of the data generated by the farmers. This helps to minimize costs, maximize resources, and engage diverging technical skills across various organizations.

13.6.3 Establishment of Community Seed Banks

Community seed banks with repositories of local genetic diversity serve as a measure to adapt to prevailing abiotic and biotic stresses. "Seed for needs" is an ambitious program by *Bioversity International* to combat climate change. Harnessing existing crop diversity and making it available to farmers is one of the solutions offered by the agency as mostly crop diversity ends up in hands of vested interests for commercial gains. The Navdanya Seed Bank in Uttarakhand, India, with more than 54 community seed banks and 5000 crop varieties; the Svalbard Global Seed Vault in Norway; the National Centre for Genetic Resources (NGCR), Colorado; and the Vavilov Research Institute (VRI), Russia, are some of the promising seed banks operating across the globe as adaptive measure to tackle climate change.

13.7 Challenges and Way Forward

The development of new cultivars with improved resistance to both biotic and abiotic stress factors induced by climate change is the immediate research priority. The germplasm should be evaluated systematically to develop the varieties that can sustain climate extremities by suitably adjusting its physiology. OMICS-based approaches (genomics, proteomics, metabolomics, transcriptomics) aid to target the potential genes that confer resistance to various stress factors. Such genes can be identified and introgressed into the cultivated varieties either by conventional breeding techniques or with the aid of modern biotechnological tools. The major change induced by global warming is the gradual shift in the cropping season. Hence, rescheduling of crop calendars and the development of standard package of practices in accordance with modified seasonal calendars is necessary to minimize the negative effects of climate change. The emerging technologies such as LiDAR (light detection and ranging) and GIS (geographic information system) can be utilized effectively to monitor the insect-pest outbreaks to biographic and physiographic changes of the landscape. Integrated nutrient and pest management strategies should be adopted to reduce the dependence on inorganic agrochemicals. The development of early forecasting and disease monitoring models should be developed for major pathogens that have a direct bearing on the productivity of the vegetable crops and in turn the food security.

The National Action Plan on Climate Change (NAPCC) in India has been announced in 2008 with the National Mission for Sustainable Agriculture (NMSA) among the other schemes with an aim to develop the climate-proof agriculture practices. The knowledge and resources generated from these programs could be effectively integrated with the knowledge of farm sciences to combat climate change. Foreseeing the vast potential for cultivation of aquatic vegetables and underexploited and indigenous perennial vegetables, there is an urgent need of holistic approach to popularization of these vegetables on a large scale to achieve the sustainability in vegetable farming.

References

Afroza B, Wani KP, Khan SH, Jabeen N, Hussain K (2010) Various technological interventions to meet vegetable production challenges in view of climate change. Asian J Hortic 5:523–529

Arya PS, Korla BN, Sharma PP (1979) Effect of would be more informative for any conclusion to be drawn environmental factors on cabbage seed production in Kalpa Valley. S Indian Hortic 27:297–299

Ayyogari K, Sidhya P, Pandit MK (2014) Impact of climate change on vegetable cultivation -a review. Int J Agric Environ Biotechnol 7:145

Baker JM, Ochsner TE, Venterea RT, Griffis TJ (2007) Tillage and soil carbon sequestration – what do we really know? Agric Ecosyst Environ 118:1–5

Bisbis MB, Gruda N, Blanke M (2018) Potential impacts of climate change on vegetable production and product quality- a review. J Clean Prod 170:1602–1620

Burleigh JR, Black LL, Mateo LG, Cacho D, Aganon CP (2005) Performance of grafted tomato in Central Luzon, Philippines: a case study on the introduction of new technology among resource-limited farmers. Crop Manag 4(1):1–10. https://doi.org/10.1094/CM-2005-0701-01-MG

Ceccarelli S, Grando S, Maatougui M, Michael M, Slash M, Haghparast R, Rahmanian M, Taheri A, Al-Yassin A, Benbelkacem A (2010) Plant breeding and climate changes. J Agric Sci 148:627–637

Chadha ML (2009) Indigenous vegetables of India with potentials for improving livelihood Proc. IS on underutilized plants (Eds.: Jaenicke et al.). Acta Hortic 806:579–586

Chadha KL, Patel BV (2007) Prospect of indigenous perennial plants as source of vegetable. Proc. Ist IC on indig. veg. and legumes (Eds. M.L. Chadha et al.). Acta Hortic 752:49–54

Drury CF, Reynolds WD, Tan CS, Welacky TW, McLaughlin NB (2006) Emissions of nitrous oxide and carbon dioxide: influence of tillage type and nitrogen placement depth. Soil Sci Soc Am J 70:570–581

Goswami VK, Khansi SK, Gautam RC (2002) Effect of intercropping and weed control on nutrient uptake and water use efficiency of pearl millet (*Pennisetum glaucum*) under rainfed condition. Indian J Agron 47:504–508

Hu Q, Yang N, Pan F, Pan X, Wang X, Yang P (2017) Adjusting sowing dates improved potato adaptation to climate change in semiarid region, China. Sustainability 9(4):615

IPCC (2007) Climate change 2007: impacts, adaptation, and vulnerability. In: Contribution of the working group II to the fourth assessment report of the IPCC. Intergovernmental Panel on Climate Change, Cambridge University Press, Cambridge

Kidane YG, Mancini C, Mengistu DK, Frascaroli E, Fadda C, Pè ME, Dell'Acqua M (2017) Genome wide association study to identify the genetic base of smallholder farmer preferences of durum wheat traits. Front Plant Sci 8:1230

Koundinya A, Sidhya P, Pandit MK (2014) Impact of climate change on vegetable cultivation–a review. Int J Agric Environ Biotechnol 7(1):145–155

Koundinya AVV, Kumar PP, Ashadevi RK, Hegde V, Kumar PA (2017) Adaptation and mitigation of climate change in vegetable cultivation: a review. J Water Clim Change 9(1):17–36

Kumar PR, Yadav SK, Sharma SR, Lal SK, Jha DN (2009) Impact of climate change on seed production of cabbage in North Western Himalayas. World J Agric Sci 5(1):18–26

Meng QM, Liang J, Wu GF (2003) Advances in research on pharmacological effect of alkaloidal compounds. Lishizhen Med Mater Med Res 14(11):700–702

Milder JC, Majanen T, Scherr SJ (2011) Performance and potential of conservation agriculture for climate change adaptation and mitigation in Sub-Saharan Africa: final report on an assessment of WWF and CARE projects in support of the WWF–CARE Alliance's Rural Futures Initiative. EcoAgriculture Partners, Washington DC

Muller A (2009) Benefits of Organic agriculture as a climate change adaptation and mitigation strategy for developing countries. Discussion paper series–EfD DP 09–09. Environment for development. http://www.ifr.ac.uk/waste/Reports/BenefitsOfOrganicAgriculture.pdf. Accessed 8 Nov 2013

Natesh HN, Abbey L, Asiedu SK (2017) An overview of nutritional and anti-nutritional factors in green leafy vegetables. Hortic Int J 1:00011

National Institute of Nutrition (NIN) (2011) Dietary guidelines for Indians: a manual

Olesen JE, Trnka M, Kersebaum KC, Skjelvag AO, Seguin B, Peltonen-Saino P, Rossi F, Kozyra J, Micale F (2011) Impacts and adaptation of European crop production systems to climate change. Eur J Agron 34:96–112

Palada MC, Wu DL (2007) Increasing off-season tomato production using grafting technology for peri-urban agriculture in Southeast Asia. Acta Hortic 742:125–131

Pareek A, Kumar A (2014) Nutraceutical value of aquatic plants. J Acad 4(1):10–21

Peter JJ (2002) Clinical nutrition: 7. Functional foods-more than just nutrition. J Can Med Assoc 166:1555–1563

Rai SK, Arora N, Pandey N, Meena RP, Shah K, Rai SP (2012) Nutraceutical enriched vegetables: molecular approaches for crop improvement. Int J Pharm Bio Sci 3:363–379

Razvy MA, Faruk MO, Hoque MA (2011) Environment friendly antibacterial activity of water chestnut fruits. J Biodivers Environ Sci 1:26–34

Sakata Y, Takayoshi O, Mitsuhiro S (2007) The history and present state of the grafting of cucurbitaceous vegetables in Japan. Acta Hortic 731:159–170

Singh RP, Prasad PVV, Reddy KR (2013) Impacts of changing climate and climate variability on seed production and seed industry. Adv Agron 118:49–110

Singh S, Selvakumar R, Mangal M, Pritam K (2020) Breeding and genomic investigations for quality and nutraceutical traits in vegetable crops-a review. Indian J Hortic 77(1):1–40

Smith P, Martino D, Cai Z, Gwary D, Janzen H, Kumar P, McCarl B, Ogle S, O'Mara F, Rice C, Scholes B, Sirotenko O (2007) Agriculture. In: Metz O, Davidson R, Bosch PR, Dave R, Meyer LA (eds) Climate change 2007: mitigation. Contribution of working group III to the fourth assessment report of the intergovernmental panel on climate change. Cambridge University Press, Cambridge, pp 492–540

Venema JH, Dijk BE, Bax JM, Van HPR, Elzenga JTM (2008) Grafting tomato (*Solanum lycopersicum*) onto the rootstock of a high-altitude accession of *Solanum habrochaites* improves suboptimal temperature tolerance. Environ Exp Bot 63:359–367

Wang Q, Li Y, Alva A (2010) Cropping systems to improve carbon sequestration for mitigation of climate change. J Environ Prot 1:207–215

Integrating Local Knowledge in the Climate Services for Resilience: A Case of "Haiyan" Fishers

14

Ladylyn Lim Mangada

Abstract

Local knowledge has been recognized by scholars to have valuable contributions to the larger issues of resilience, particularly in the provision of climate services. Data for this study was drawn from 24 key informants and three focus group discussions in six fishing communities in the Haiyan-affected provinces of Leyte and Samar. Due to the lack of climate services, fishers still use the weather patterns transmitted to them by their grandparents by observing the clouds and behavior of plants or animals. By rethinking local knowledge beyond its perceived weaknesses, integrating it in climate services and other government programs and the inclusion of the voices of the fishing communities may result in healthier, safe, and enhanced adaptive capacities even in places where marginality abounds.

Keywords

Local knowledge · Resilience · Climate services · Haiyan fishers

14.1 Introduction

According to the World Risk Report (2017), after Vanuatu and Tonga, the Philippines is the third country with the highest disaster risk worldwide due to its comparatively high exposure to a number of hazards, such as typhoons, floods, earthquakes, volcanic eruptions, and sea-level rise. For these reasons, the Center for Research on the Epidemiology of Disasters has "identified the Philippines as one of

L. L. Mangada (✉)
UP Visayas Tacloban College, Tacloban, Philippines
e-mail: llmangada1@up.edu.ph

V. K. Hebsale Mallappa, M. Shirur (eds.), *Climate Change and Resilient Food Systems*, https://doi.org/10.1007/978-981-33-4538-6_14

the most disaster-prone countries in the world" (Bankoff 2003). In fact, the "cumulative hazard experience has led the people in the Philippines to develop a "culture of disaster" in which risks and hazards become "normalized" as a part of everyday life" (Bankoff 2003).

Before super typhoon (ST) Haiyan, fishing communities in the Leyte Gulf would rely on fishers' indigenous/local knowledge for preparedness and resilience to climate change. However, after ST Haiyan, typhoons became more frequent and severe, affecting the fishing community's livelihoods. Fishers noticed sea-level rise, the unpredictable onset of the dry and wet season, and the change in the distribution and movement of fish species.

Some of the fishers admitted that indigenous knowledge on predicting weather conditions has failed them in the past but remains helpful due to lack of timely, relevant, and accessible climate services. Hence, here are few questions which need to be answered: What are available climate services for the fishers? Do fishers make use of these services? What are the advantages and weaknesses of the climate services and indigenous/local knowledge in preparing for and responding to climate change? What are the barriers to the lack of indigenous knowledge integration in climate services?

This paper argues on government programs for adapting to the consequences of climate change that do not give enough consideration to what really matters to fishers and communities. Adaptation will be more readily and rationally adopted with the incorporation of indigenous/local knowledge in the climate services. This is because indigenous/local knowledge better represents local needs and desires. Fishers use their perceptions to make decisions on coping and adapting to climate change.

Through indigenous/local knowledge, policymakers can understand how fishers perceive and adapt to subtle climate changes before they design climate services for the fishermen to use on climate change preparation and resilience.

The Intergovernmental Panel on Climate Change (IPCC) released the most scientifically valid report to date on climate change, which states that the impacts of climate change will be "severe, pervasive and irreversible" (UNEP 2014). The IPCC (2014) warns that time is running out swiftly to act against climate change if the world wants to avoid catastrophic impacts.

Other leading scientific bodies, for example, the National Aeronautics and Space Administration (NASA), have reported that some of the (e.g., warming of oceans and rising sea levels) adverse effects of climate change are already observable throughout the world. Former US President Barack Obama has termed the problem as urgent, growing, and immediate. A "once distant problem has come into the present" (Obama 2014).

Mary Robinson, the UN special envoy on climate change, emphasized the urgency in tackling the severity of climate as our generation comes to understand the impacts of climate change (Lynch 2015). Scholars and scientists alike have posited that the Pacific Islands will experience seasonal variations causing physical changes in the environment such as temperature, sea level, precipitation, and winds which will affect plants and animals, including fisheries (Kelman 2019).

The effect would last several months and would deviate from the expected behavior of the seasons (Kelman 2019; Power 2010; Sterrett 2011; Tripathi et al. 2016). In fact, the World Risk Report (2017) mentioned that the Philippines is currently the third country with the highest risk worldwide after Vanuatu and Tonga, due to its comparatively high exposure to a number of hazards such as typhoons, floods, earthquakes, volcanic eruptions, and sea-level rise. For these reasons, the Philippines has been assessed by the Center for Research on the Epidemiology of Disasters as a highly disaster-prone country so much so that disasters have become a fact of life that Filipinos have had to learn to live with (Bankoff 2003).

In the coastal areas of the Leyte Gulf, which is part of the Pacific Ocean, and where fishing is the primary source of livelihood, unpredictable weather conditions have posed severe threats to the livelihood sustenance of households. Natural and human-induced climate changes are currently being blamed as the most significant driver of the ongoing trend.

After ST Haiyan devastated the area, several efforts were launched to assist coastal communities and households in mitigating threats and coping with the multiple effects of unpredictable climatic conditions and other hazards. These efforts have also begun catching the attention of some donors and those in public office. However, a major drawback of the present approach to climate change adaptation in the context of the studied area has been the lack of climate services for marginal fishers. Moreover, the centuries-old knowledge/local practices in climate adaptation are barely integrated into the said approach.

Due to the focus on "physical, engineering and technical aspects of the adaptation planning process," there is a lack of substantive studies on the climate adaptation methods of fishers. As such, the perceptions and attitudes of fishers on climate adaptation actions are understudied, partly due to that bias in adaptation planning.

14.2 Methodology

Leyte Gulf, located in the central-eastern part of the Philippines, covers the Islands of Leyte and Samar, including San Pedro Bay, which has a shelf area of 13,147 km^2. The average depth of the Gulf is 38 fathoms at the center and 8 fathoms in the bay. It is considered as one of the richest fishing grounds in the Visayas.

There is an indication of corral growth on the 15-fathom contour of the Gulf along the Northwest Coast, while the bottom is generally muddy. In terms of management, the difficulty is encountered because of its multi-gear, multispecies fishery characteristics. The Leyte Gulf is made up of 13 municipalities.

Six local government units can be found in the Leyte side, and eight municipalities are located in the island of Samar. The fishers residing in the small-scale fishing communities in the Leyte Gulf tend to be poor and uneducated. Most of them rely on local ecosystems for their livelihoods. Rain-fed agriculture augments

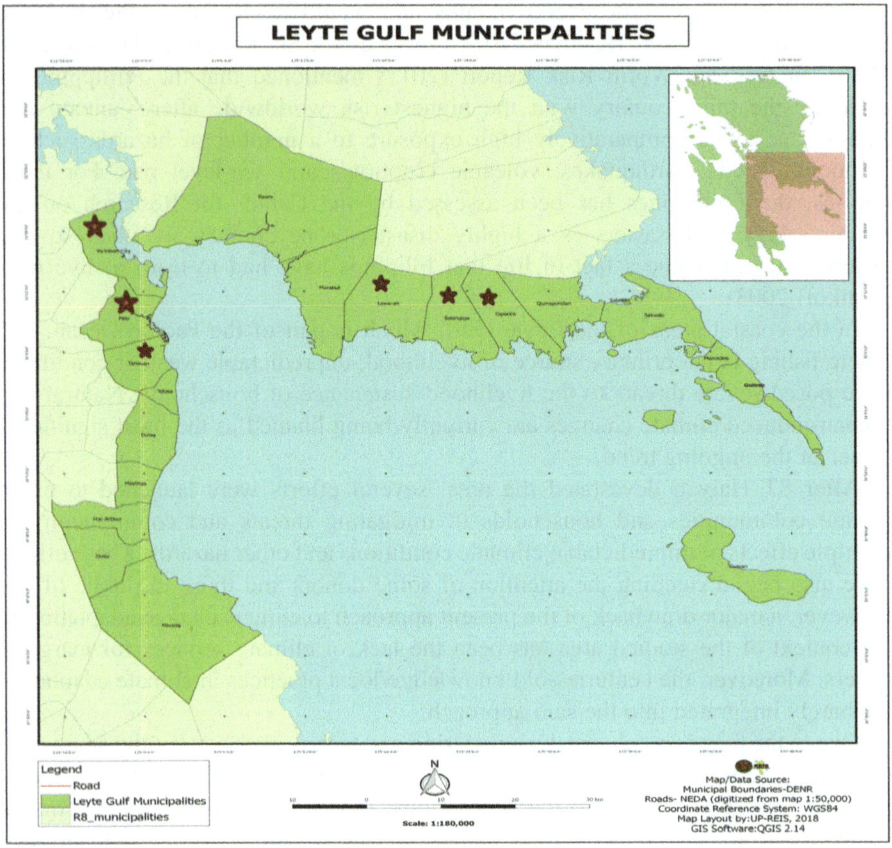

Fig. 14.1 Study sites

fishing during the lean season. The fishers are more likely to face further economic hardship brought about by climate change.

In this study, the municipalities of Giporlos, Balangiga, Lawaan, Palo, Tanauan, and the city of Tacloban were selected principally for the scale of Haiyan devastation (Fig. 14.1). These areas were greatly impacted by ST Haiyan. Methods: For 4 weeks in December 2019, 24 key informant interviews were conducted in the six fishing communities to look into the available climate services for the fishers. Key informants were identified through the author's combination of personal and professional networks.

Three focus group discussions (FGDs) were held for a broader corroboration of information obtained from key informants. Discussions revolved around local knowledge, practices, and climate services and their perceptions of climate change. Most of the participants of the FGDs were referred by the key informants.

They were orally requested to give their consent in the conduct of the FGDs and interviews to give their understanding of the research and their permission to participate and be quoted. There were seven to ten participants per FGD. The average length of time of fishing for the fishers was 25 years. These fishers do not carry mobile phones and, instead, rely on traditional knowledge.

14.2.1 Local Knowledge: What we Know

According to the WIPO (2017), "Traditional knowledge" refers to "knowledge, skills, and practices that are developed, sustained, and passed on from generation to generation within a community, often forming part of its cultural or spiritual identity."

Traditional knowledge "been conceptualized as a cumulative body of knowledge, practice, and belief, evolving by adaptive processes and handed down through generations by cultural transmission, about the relationship of living beings (including humans) with one another and with their environment." Local knowledge, on the other hand, refers to knowledge that people in each community have developed and continue to develop over time.

It is based on experience, often tested over centuries, adapted to the local culture and environment, and embedded in community practices, institutions, relationships, and rituals (UNEP 2014). Local knowledge is commonly understood as the long-standing traditions and practices of certain regional, indigenous, or local communities encompassing the wisdom, knowledge, and teachings of these communities (Wan Talaat et al. 2012). Traditional and local knowledge are ways of understanding culture and, in so doing, in understanding the world. Transmitting a body of knowledge, beliefs, and practices through the use of locally available resources to improve human health and well-being is an important function of culture in traditional societies (Gomez-Baggethun et al. 2012).

For instance, in the absence of basic navigation tools such as maps and compass, the head fisher had to rely on his traditional/local knowledge for navigation and location of fishing points. His tools are his senses and mental images; hence, the entire boat depends on his skills. Hasan (2015) described it as if the head fisher has the map of the sea instead of a physical compass, and his concept is guided by the rising and setting of the key star and constellations, as well as wind direction.

Several studies have been published in recent decades emphasizing on the substantial contribution of traditional knowledge to matters as important as resource management, socio-ecological resilience, and biodiversity conservation (Saylor et al. 2017; Gomez-Baggethun et al. 2012; Berkes et al. 1995, 2000). In terms of disaster risk reduction strategies, Mercer et al. (2012) highlighted the importance of incorporating culture on adaptation and disaster risk reduction strategies and strongly proposed that disaster risk reduction policies should be culturally responsive (Berkes 1993).

Despite not being a focus or institutionally acknowledged by the policymakers in Bangladesh, Anik and Khan (2012) stressed the importance of local knowledge for climate change adaptation. In addition, marginalization of the at-risk population, which are communities lacking in political representation and with poor socioeconomic status, leaves them vulnerable to disaster (Phillips and Morrow 2007).

They were also less likely to receive, analyze, and act appropriately to hazard forecasts and warnings. Indigenous/local knowledge can enrich scientific data as information on change of climate over time is critical for verifying climate models and climate change scenarios. A solid understanding and respect for age-old knowledge system can provide an important foundation for resilience.

14.3 Results and Discussion

14.3.1 Fishers' Local Knowledge and Practices for Climate-Related Events Prior to ST Haiyan

The fishers in the Leyte Gulf would read signs or take a cue from nature—the wind, animal behavior, and the appearance of indicator plants' leaves and flowers—to predict weather conditions. According to respondents, "The color of the sky would alert us of good or bad weather. If it turns red, it means tomorrow, and the succeeding days would have bad weather. If the clouds turn dark, it means stormy and strong winds are approaching."

Based on the weather conditions (waves and sea currents), if it's dangerous to continue fishing, then we secure our boats to higher and safer ground and go home.

There are months that we observe and follow where the catch is very good. For example, the northeastern wind (Amihan) signals plenty of fish, while the northwestern wind (Habagat) tells difficult fish catch.

When we experienced low fish catch for several weeks, we resort to "pagdididwata" to give us good luck. We thoroughly clean our boats and perfume it with herbal plants, and we would look for an object with very unique/strange features resembling a fish, which we would place/hide carefully in our net not to be seen by others. It works. My friend practiced orasyon. He would also offer food in the night or during a full moon.

Others augment it with a tambalan to say a prayer to ward off bad luck. Fishers voluntarily disclosed: Some of our friends practice orasyon and offer food during a full moon. Others would get the services of a faith healer. Five days before ST Yolanda, birds flew in batches/groups; in relation to this, scholars have long recognized the important contribution of local knowledge in understanding and managing climate change.

The different winds have distinct and unique characteristics and weather. For instance, Kanaway is the wind that brings forth stormy weather.

When it clashes against Kabunghan, the turbulent weather intensifies and can cause severe flooding, landslides, or destruction of property. On the other hand, Salatan is seen as the good and pleasant wind that brings with it sunny weather with fairly blowing wind. This is typically experienced from March to April. Many of the fishers claimed that the aforementioned weather patterns showed in Table 14.1 are increasingly becoming challenging to identify accurately.

Table 14.1 Common indigenous/local knowledge possessed by the respondents

Type of local/ indigenous knowledge	Practiced?	Purposes/signifies
Pagdidiwata/pagbulong	Yes	1. Good luck to the fishing vessel 2. Good catch
Kanlaon (south wind)	Sometimes	The calm sea during January to April
Amihan (NE wind)	Sometimes	An abundant supply of fish during October to December/good catch
Habagat (NW wind)	Sometimes	The decrease in the supply of fish during June to august/dire season for fishing
Salatan (SE wind)	Sometimes	A decrease in the supply/catch of fish during June to October
Kanaway (north wind)	Sometimes	Abundant pelagic fish
Formation of clouds	Sometimes	Tells good or bad weather
Position of stars	Yes	Provides approximate time at sea
Strong current	Yes	A small supply of fish

Informants disclosed that traditional indicators to forecast weather and seasons were no longer reliable due to changes in climate condition. They pointed to climate change as the main culprit. As such, people depending most on natural resources for economic survival actively try coping and adapting to changing conditions instead (Salick and Byg 2007; Macchi et al. 2008), especially if these changes critically affect the dwindling resources they rely on (Reedy et al. 2014).

Most fishers also believe that animals are powerful and smart beings that give them signs whenever natural hazards are about to come. As observed by the fishers in the Leyte Gulf, when certain species of birds such as (a) kalaw or rufous hornbill/ Philippine hornbill (*Aceros waldeni*), (b) talusi or Palawan hornbill (*Anthracoceros marchei*), and (c) limukon/kuro-kuro or white-eared brown dove (*Phapitreron leucotis*) suddenly stop making a sound during the day, a typhoon or heaven rainfall is about to take place (Cuaton and Su 2020).

Meanwhile, when insects like kuliglig (Tagalog)/gangis (Waray-Waray) or crickets (Gryllidae) intermittently stop making noises for 2–3 min during the night, an earthquake is believed to come. Furthermore, when wild pigs pile up leaves and twigs to create some sort of shelter, a typhoon or heavy rainfall is expected, which may lead to occurrences of flooding and landslide.

Some of the endemic Philippine wild pig (baboy damu) species are the Visayan warty pig (*Sus cebifrons*), Philippine warty pig (*Sus philippensis*), Oliver's warty pig (*Sus oliveri*), Palawan bearded pig (*Sus ahoenobarbus*), and bearded pig (*Sus barbatus*) (Cuaton and Su 2020). Some fishers acknowledged the following signs as their guide (Table 14.2).

The fishers of the study communities still use the traditional and local knowledge and practices orally transmitted to them over generations to forecast possible climatic occurrences and minimize its impact.

The color, direction, and speed of the clouds and the speed and direction of the wind, as well as the behavior of the sea, provide the fishers signals for good or bad

Table 14.2 Natural guides for fishers

Sign	Meaning/interpretation
Full moon	Fewer to difficult fish catch
Dap-dap flower in full bloom	Fat crabs will be caught
Bees in the boat	Bountiful fish catch
Cockroaches crawling at night	Plentiful fish catch
Seagulls moving inland	Rough seas/turbulence
First to fourth quarter of the moon	Safe to dive for bow fishers
Moths encircling the lamps	Abundant fish in the ocean
Seagull and other seabirds frequenting a particular location in the sea	Guides the fishers the location of the points to the abundant presence of fish in that particular location
Moving seawater similar to boiling water during daytime under calm seas	Indicates the presence of fish in the area
The sudden appearance of too many millipedes	Earthquake is coming

weather. Despite this, fishers notice the shift of the onset of the dry and wet season. All of the informants, including those who participated in the FGDs, indicated that they had in some way experienced the negative impact of climate change as evidenced in their low fish catch and more extended periods in the sea as well as its cost. However, most of them refer to or associate climate change with ST Haiyan.

Climate change is super typhoon Haiyan. It was strangely different from all typhoons we had. It's warmer now. I am confused with the summer season. There are summers when it is very dry and hot, and then the next summer it's suddenly wet. Typhoons have become frequent, almost monthly, and are very destructive.

Before Haiyan, our boats were safe when we tie them to a tree and higher ground. With Haiyan, that practice failed us, and our boats and nets were not spared. The months for the northwest wind have become longer.

It makes fishing more difficult. The shoreline used to be 15 m; these days it's 10 m. Seawater has risen. At present, the fish are fewer and located very far. In the past, around 50 km, we could get fish, but today we reach as far as 200 km, and one is not assured of a good catch.

It is more expensive and more time in the sea. There are fishes that we do not catch anymore; they disappeared. Also, there are fewer sea grasses now. 2–3 months after Haiyan, few fishers went to the sea because boats were destroyed. When we went on fishing, I enjoyed a plentiful catch. In the past 10 years, the dry season was observed from March until the middle of May, while the wet season started during July to early February of the following year. However, 3 years ago, there was a sudden shift in the duration of the dry and wet season.

The frequent occurrence of low-pressure areas prolonged either the dry season or the wet months. Among climate extremes, typhoons, unpredictable stormy weather, high winds, and heavy rainfall were the most dreaded in the four study sites. In fact, fishers are experiencing low yield in fishing not just because of overfishing but mainly due to typhoons and stormy weather occurring at least ten times in a year.

What is clear to the fisher is that ST Haiyan and the succeeding storms led to eroding household incomes and food security challenges. However, fishers indicated that they had not undertaken various adaptation measures to cope with climatic variability. For example, most of the fishers' schedule to go to the sea has remained unaltered. They leave at around three in the morning and return around noon.

They have not adjusted their schedule in spite of their recognition of warmer temperature occurring. Another factor is the place to hide and secure their boats during bad seasons. The fishers have not demanded government assistance regarding the site or place to keep their boats.

Fishers take the appropriate steps to protect their properties, regardless of the level of resistance or resilience. In addition, some of the fishers are not registered. They have not learned their lesson from ST Haiyan that the provision of assistance is usually based on the official list of registered fishers. This indicates the presence of important constraints to fishers' ability to implement more "climate-smart" fishing practices.

Some potential limitations faced by fishers may include a lack of sufficient climate information and insufficient resources or a lack of relevant skills and affordability. Di Falco et al. (2011) previously found that the lack of information is likely the major barrier to understanding and acceptance of climate change adaptation strategies in very remote areas.

14.3.2 Local Knowledge Integration in Climate Services: Challenges and Barriers

As the severe effects of climate change are strongly felt around the globe through changes in weather patterns and the intensification of weather-related disasters, the focus of climate services should be to provide accessible information at the time of need and for the appropriate people or sector. For instance, identifying the climate services of relevant government agencies prepares and assists the fishers in coping with and recovering from climate and socioeconomic challenges of insecurity.

It requires a whole-systems perspective to develop innovative solutions and predict catastrophes. Local fishers revealed that aside from the weather bulletin from the Philippine Atmospheric, Geophysical and Astronomical Services Administration (PAGASA), which is regularly disseminated in TV and radio, no forms of science-based measures exist for the fishers in order to predict the weather (changes in seasonality/unpredictable onset of dry and wet season), sea-level rise, acidification of the sea, alteration in sea temperature, species distribution, and movement of the fish. Consequently, they relied on local knowledge and applied.

Since it has proven to work, it has been difficult for any organization (public or private) to alter or replace fishers' thinking and practices. Local knowledge and experience matter in building resilience. How come public decision-makers continuously ignore fishers' generations-old thinking and practices? The paper presumes that this is because local knowledge and practices are inferior to scientific knowledge.

After ST Haiyan, there is a growing awareness from the ranks of fishers of the limitations of local knowledge, but a strong belief in their potential continues to persist. For government personnel, this is what they have to say: "Local knowledge and practices need validation and should consider the changes happening."

The world has changed significantly. Making use only of local knowledge or practices is problematic. In one area, fishers believe this object to represent something, and then in another place that's not even distant, the said object is replaced by another.

On the other hand, the fishers disclosed: "We have not been asked regarding our local knowledge and practices. But they would introduce something new." Indeed, there is no quick fix to build a sound interface between local knowledge and science and technology unless integrated research is carried out.

In spite of the lack of scientific validation to local knowledge, local knowledge offers valuable insights on hazard evaluation or elements at risks, which are crucial in the reduction of losses: centralized government program menu and mono-disciplinary perspective. Key informant interviews at the regional office of the Bureau of Fisheries and Aquatic Resources (BFAR) disclosed that the office has only recently started to develop a warning system for the onset of the red tide.

What is practiced is a weekly collection of water samples, and the results are transmitted to the village officials for public dissemination. There is no way of checking if fishers received these results. A middle supervisor of BFAR claimed that undertaking research on climate services for fishers had not been one of their mandates, hence their inability to respond to local needs.

They pointed out that the identification and prioritization of fishery agenda are lodged in the central office, which is based in Quezon City, the former capital of the Philippines. Another BFAR personnel stressed that after Haiyan, they were directed to increase fish production through aqua-based livelihood. As such, after Haiyan, government and development interventions discouraged these municipal fishers from returning to capture fisheries.

Small fishers were strongly advised to avail of aqua-based income-generating opportunities. One significant change after the post-Haiyan intervention is teaching fishers not to use bamboo but a plastic material known as HDPE in constructing their fish cages. HDPE is not locally available and is costly.

The office believes that the use of HDPE, though not locally available compared to the bamboo, is advantageous and cost-efficient in the long run. It is because the HDPE fish cages can be submerged during bad weather and, therefore, not prone to destruction. So after a typhoon or strong winds, fishers can continue to fish and earn their livelihood.

These are some of the views expressed by government personnel: "Fishery programs come from the top; people with strong science education determine these. We only implement our role is to follow. Otherwise, it would mean unsatisfactory performance. We submit our accomplishment reports, and we try to indicate the sentiments of the fishers regarding our programs. For every problem presented, it is up to them to provide the corresponding solution. They will laugh at us if we invoke local knowledge."

On the other hand, local fishers commented: "Why can't government officials see fishery needs and interest from the lens of marginal fishers? Even before ST Yolanda, the government's main activity is either instructing us to use their technology or asking questions, but our answers are not considered. There is one-way information flow from the government to the local fishers." Other fishers maintained: "The government does not see value in local knowledge and practices. If one invokes local knowledge, government personnel will turn their back, and others would secretly laugh."

To incorporate local knowledge in the climate, services for resilience require an inclusive approach. This means embracing bottom-up community participation and proactive as well as engaging multiple stakeholders with different needs and abilities to take action. Appropriating and implementing fishery solutions that are suited and adapted to the local context is essential.

Lack of locally translated technical materials, poor sharing of information and appropriate personnel warrant serious attention. Key informants pointed out that communicating with the public is one of the biggest problems they face. The BFAR employees, as well as their local government counterpart, added that they lack technical and other information, education, and communication (IEC) materials.

They even volunteered that some of the IEC materials are difficult to be translated in the dialect due to their technical nature. These activities require a social science person, which the office has difficulty in employing. An employee commented: "We do not have social science people, we hired more criminology graduate. The office sees law enforcement as the main problem in fisheries."

A few local fishers asked: "We understood sea-level rise, but what is the ocean acidification that some officials are talking about? What are the signs of ocean acidification? Who will teach or warn us?"

Planning and decision-making without appropriately drawing upon the knowledge and preferences of fishers can lead to measures that fail to support resilience or have unintended negative consequences on development. A lack of fishers' involvement in livelihood decisions can lead to policies inimical to the growth of their income. And in some point, this can leave the fishing community to choose between using the traditional practices and struggling with the new, expensive fishing technology introduced by the government.

14.3.3 The Lapu-Lapu Fish Kill: A Different Story

On December 16, 2018, less than 2 weeks before Christmas, a lapu-lapu fish kill occurred in Tacloban City. Marginal fishers were extremely frustrated as they hoped to sell the lapu-lapu at an excellent price. The lapu-lapu fish would usually fetch at PHP300.00 during ordinary days, but it costs PHP600.00 during the holiday season. Fishers narrated that it was the first time that fish kill happened in Tacloban City. The night before, the color of sea water turned blue, and the following day, our lapu-lapu fish were dead.

BFAR personnel came, and they got a live lapu-lapu as a sample to test. The result of the test revealed that the sea water lacks oxygen, the sea water was polluted due to the several rivers and drainages nearby, and we did not follow the distance required for fish cages. We do not agree because it is the first time that we encountered our fish dying.

Instead, we believe that it was caused by a supernatural event, known as *buyag*. Some of the fishers narrated this experience: The night before, the color of the seawater turned blue, and the following day, our lapu-lapu fish were dead. BFAR personnel came, and they got a live lapu-lapu to test. The result of the test revealed that the seawater lacks oxygen, the seawater was polluted due to several rivers and drainages nearby, and we did not follow the distance required for fish cages. We do not agree because it is the first time that we encountered our fish dying. Instead, we believe that it is caused by a supernatural event, known as buyag. The seawater does not change. It's still the same from the beginning. We just continue the practices we were used to in fishing.

The BFAR personnel who attended to resolve this problem shared that the size of the lapu-lapu made them ready for harvest. But the fishers wanted them to grow more to command a good price. Big-sized lapu-lapu needs more oxygen, and the seawater in the area could not provide that due to pollution and close distances of fish cages. They claimed to have "overstocking."

Indeed, the fishers of the study community do not have scientific resources to predict seawater changes and understand environmental phenomena. They need to be coached on the biophysical-chemical composition of seawater, so they do not end up relying on traditional and local knowledge system to interpret climate-related conditions. The local knowledge that they have cannot show or explain complex climate and environmental processes.

Due to the challenges in obtaining scientific methods and tools for reliable climate forecasts and predictions, fishing communities continually depended on these knowledge systems and practices. In climate services, integration of local and traditional climate change adaptation efforts into the scientific system should be mandatory in order to decrease the vulnerability of nonurban fishing communities to disasters. This will avoid future risks to mitigate future harm.

14.3.4 The Future of Science-Based Information/Technology Usage by the Fishers

Since fishers predominantly use local knowledge, experience, and perception in interpreting climate-related events and decision-making, it is essential to be familiar with fishers' beliefs, customs, and insights on climate change. While gaps and disparities between fishers and government implementers exist in understanding climate change and its impact on their livelihood, fishers are open to listening to experts' views or knowledge. They also welcome advice in a manner that they can comprehend.

According to some fishers, "We believe in the old way of predicting and mitigating bad weather, but you cannot completely depend on it. That is why a TV or radio set is important. However, the idea and practices passed on to us should not be discarded because it is true, and it happens." "BFAR personnel come here when there is a problem. Some of them keep on interviewing us. We do not know if what we give them or our answers to their questions have value because when they return, it's different." The people from BFAR pointed out: "We teach and help the fishers to improve their livelihood income, but in the end, when we leave the area, they go back to their old ways of fishing."

From this remark, it is clear that government efforts to deliver sustainable post-disaster livelihood packages to fishers often fall short because they fail to account for the complex interactions between sociocultural, economic, and environment. Key informants commented that their voices tend to be neglected or ignored by the government and other providers.

They feel that public institutions and the policymakers have either overlooked or neglected them and that they have received minimal support and services from the government. Could this mean that their fishing practices are viewed as being backward or irrelevant in the modern world compared to external (usually technology-based) knowledge, which is seen as modern and representing the future? Most of them are hungry for experts' views on climate change, particularly on securing climate-resilient income-generating activities. Meeting this need might facilitate better local adaptation results.

The government provides the least attention to small fishers. They prefer big-time fishers such as those we hear about the "tuna highway," fishes for export, which can bring in revenue. Presently, the BFAR, through its FishCorral project, is disseminating the adoption of aquaculture to solve the problem of scarce fish supply in the region.

While it is presenting the socioeconomic benefits of aqua-based livelihood, fishers are not aware of the detrimental effects of these livelihood interventions, especially the feeds, to the health of the seawater and the eventual sustainable adaptation to climate events. Effective adaptation strategies should be aimed at securing communities' well-being in the face of climatic changes (Somah 2013).

14.4 Conclusion

It is important not to under-acknowledge the role of traditional and local knowledge in enhancing sustainable adaptation to climate change. Selected lessons can be derived from the traditional and local practices that have been applied by communities over the centuries to predict the changing climate events. However, in order to integrate these knowledge systems into climate services, they must be understood with adequate and continuous community education and evaluation to minimize maladaptation practices.

Typhoons, irregular stormy weather/damaging strong winds, intense rainfall, and big waves were the climate events with the most significant adverse effects on fishers. However, as experienced by the local fishers, traditional and local knowledge poorly predict the onset, length, and intensity of the dry and wet season, seawater acidification, and sea-level rise. Even though they do not provide reliable forecasts at all times, these traditional and local knowledge systems for climate-related events are rooted in the Leyte Gulf.

Luckily, the fishers of the study communities are open to affordable science-based information and technology to help them achieve climate change-resilient fishing livelihoods. The major challenge is to equip the fishers and their communities to adapt and raise their incomes without losing the sound social and cultural values and practices that underpin their traditional way of life.

Disasters of various types and intensities have become more frequent. By focusing on fishing communities and climate services, this paper may have succeeded in highlighting the challenges of marginal and vulnerable people. They believe in and practice local knowledge in responding to natural hazards. This study also highlighted public institutions, from national to local, and lack of openness to the use/application of local knowledge in climate service delivery. In short, findings of this study are applicable to all archipelagic countries that frequently experience disasters where community fishers are most susceptible and suffer the most.

14.5 Policy Implications

There is an emerging common ground from which scientific and local knowledge can dialogue and complement to better assist fishers in regaining their ability to live in harmony with nature lowering risks and decreasing losses and damages through local and scientific knowledge-based actions by way of community vulnerability assessments.

An initial step would be for government offices and development organizations to appreciate the potential value of local knowledge in achieving sustainable climate adaptation. Including the views of the fishers in the adaptation, plans will be a good appreciation. Secondly, the integration of local practices with modern scientific knowledge on climate change may become easier when there is a list or inventory of local practices to provide as a baseline.

Also, inclusionary policies and practices are crucial for post-disaster adaptation. These ensure the participation of fishers' groups and networks, development actors, and government offices is instituted and installed to promote sustainable livelihood practices. Inclusion is more than a representation of the marginal fishers in the planning and decision-making structures.

Instead, it means that local knowledge and capacities of the fishers should be respected and factored in. Local knowledge and practices should be part of development work. Operationally, this means understanding the local contextual factors (e.g., environmental, social, and cultural), proper community identification and assessment of local risks to their livelihood, prioritizing adaptation policies according to local needs, and regular community information regarding policies for adaptation.

Overall, this means that livelihood initiatives should not occur without the active collaboration of the people concerned. Donors and various government offices should avoid imposing the nature of development and how it should take place. There is so much to learn from the fishers. Understanding the local worldview of the fishers, that is, seeing things from their eyes and hearing from their ears, will significantly help to moderate and mitigate climate change impacts.

References

Anik SI, Khan MASA (2012) Climate change adaptation through local knowledge in the north eastern region of Bangladesh. Mitig Adapt Strateg Glob Chang 17(8):879–896. https://doi.org/10.1007/s11027-011-9350-6

Bankoff G (2003) Cultures of disaster:society and natural hazards in the Philippines. Routledge, Abingdon, UK

Berkes F (1993) Traditional ecological knowledge in perspective. In: Inglis J (ed) Traditional ecological knowledge: concepts and cases. Canadian Museum of Nature/International Development Research Centre, Ottawa, pp 1–9

Berkes F, Folke C, Gadgil M (1995) Traditional ecological knowledge, biodiversity, resilience and sustainability. In: Perrings C, Mäler KG, Holling CS, Jansson BO (eds) Biodiversity conservation. Kluwer Academic Publishers, Dordrecht, The Netherlands, pp 281–299

Berkes F, Colding J, Folke C (2000) Rediscovery of traditional ecological knowledge as adaptive management. Ecol Appl 10(5):1251–1262. https://doi.org/10.1890/1051-0761(2000)010

Cuaton GP, Su Y (2020) Local indigenous knowledge on disaster risk reduction:insights from Mamanwa indigenous peoples in Basey. Samar after typhoon Haiyan in the Philippines 48:1–12. doi.https://doi.org/10.1016/j.jdrr.2020.101596

Di Falco S, Verones M, Yesuf M (2011) Does adaptation to climate change provide food security? A micro-perspective from Ethiopia. Am J Agric Econ 93(3):829–846. https://doi.org/10.1093/ajae/aar006

Gomez-Baggethun E, Reyes-Garcia V, Olsson P, Montes C (2012) Traditional ecological knowledge and community resilience to environmental extremes: a case study in Donana SW Spain. Glob Envi Change 22:40–650. https://doi.org/10.1016/jgloenvcha.2012.0.05

Hasan Z (2015) Artisan fishers' perceptions of, and adaptation to, climate change in the southeast coast of Bangladesh. Unpublished Dissertation, The University of Adelaide

IPCC (2014) Food security and food production systems. In: Field CB, Barros VR, Dokken DJ (eds) Impacts, adaptation, and vulnerability. Cambridge University Press, Cambridge, UK

Kelman I (2019) Pacific island regional preparedness for El Niño. Environ Dev Sustain 21:405–428. https://doi.org/10.1007/s10668-017-0045-3

Lynch S (2015) Mary Robinson says 2015 will be crucial year for climate change. The Irish times. http://www.irishtimes.com/news/world/mary-robinson-says-2015-will-be-crucial-year-for-climate-change-1.2078812. Accessed Feb 3, 2015

Macchi M, Oviedo G, Gotheil S, Cross K, Boedhihartono A, Wolfangel C, Howell M (2008). Indigenous and traditional peoples and climate change. https://cmsdata.iucn.org/downloads/indigenous_peoples_climate_change.pdf

Mercer J, Gaillard JC, Crowley K, Shannon R, Alexander B, Day S, Becker J (2012) Culture and disaster risk reduction: lessons and opportunities. Environ Hazards 11(2):74–95. https://doi.org/10.1080/17477891.2011.609876

Obama B (2014) Remarks by the President at United Nations (UN) Climate Change Summit. https://www.whitehouse.gov/the-press-office/2014/09/23/remarks-president-un-climate-change-summit

Phillips BD, Morrow BH (2007) Social science research needs: focus on vulnerable populations, forecasting, and warnings. Nat Hazards Rev 8(3):61–68

Power AG (2010) Ecosystem services and agriculture: tradeoffs and synergies. Philos Trans R Soc B: Biol Sci 365(1554):2959–2971. https://doi.org/10.1098/rstb.2010.0143
Reedy D, Savo V, McClatchey W (2014) Traditional climatic knowledge: orchardists' perceptions of and adaptation to climate change in the Campania region (southern Italy). Plant Biosyst 148 (4):699–712. https://doi.org/10.1080/11263504.2013.793753
World Risk Report (2017). World risk Report analysis and prospects 2017. Bündnis Entwicklung Hilft. pp 15–30 https://reliefweb.int/sites/reliefweb.int/files/resources/WRR_2017_E2.pdf
Salick J, Byg A (2007) Indigenous peoples and climate change. Tyndall Centre, Oxford, UK
Saylor CR, Alsharif DKA, Torres H (2017) The importance of traditional ecological knowledge in agroecological systems in Peru. Int J Biodivers Sci Ecosyst Serv Manag 13(1):150–161. https://doi.org/10.1080/21513732.2017.1285814
Somah TP (2013) Climatic change impacts on subsistence agriculture in the Sudano- Sahel zone of Cameroon – constraints and opportunities for adaptation. Ph.D. Thesis, Brandenburgische Technical University
Sterrett C (2011) Review of climate change adaptation practices in South Asia Research Report No. 7: 65–164. Oxfam
Tripathi A, Tripathi DK, Chauhan DK, Kumar N, Singh GS (2016) Paradigms of climate change impacts on some major food sources of the world: a review on current knowledge and future prospects. Agric Ecosyst Environ 216:356–373. https://doi.org/10.1016/j.agee.2015.09.034
UNEP (2014) Climate change threatens irreversible and dangerous impacts, but options exist to limit its effects. https://www.unenvironment.org/news-and-stories/press-release/climate-change-threatens-irreversible-and-dangerous-impacts-options. Accessed 14 June 2018
Wan Talaat WIA, Mohd Tahir N, Husain ML (2012) Traditional knowledge on genetic resources: safeguarding the cultural sustenance of indigenous communities. Asian Soc Sci 8(7):184–191. https://doi.org/10.5539/ass.v8n7
World Intellectual Property Organization (WIPO) (2017) Traditional knowledge. http://www.wipoint/tk/en/tk/. Accessed Apr 10, 2017

Climate-Resilient Livestock Farming to Ensure Food and Nutritional Security 15

Shahaji Phand and Prabhat Kumar Pankaj

Abstract

Like agriculture, livestock is affected by global phenomena of climate change. It also, in turn, contributes to climate change. The global greenhouse gas (GHG) emissions contributed by the livestock sector are major challenges. The sources of GHGs such as feed production, enteric fermentation, and manure management are further influenced by several other factors. The adverse effects of climate change can be effectively mitigated through livestock management, breeding program, animal nutrition and health management, developing farmers' capabilities, manure management, carbon sequestration, fertilizer management, changing the human diet pattern, etc. These issues are discussed in the chapter along with relevant facts and countering strategies to address global climate change.

Keywords

Livestock feed · Methane emission · Greenhouse gas · Livestock management

15.1 Introduction

Global warming and climate change have posed several challenges in the form of erratic rainfall pattern, frequent droughts, long and extreme heat waves, long dry spells, flood, weather extremes, hailstorms, etc., which has been adversely affecting

S. Phand (✉)
Centre for Agri-Allied Extension, National Institute of Agricultural Extension Management (MANAGE), Hyderabad, Telangana, India

P. K. Pankaj
ICAR-Central Research Institute for Dryland Agriculture, Hyderabad, Telangana, India

V. K. Hebsale Mallappa, M. Shirur (eds.), *Climate Change and Resilient Food Systems*, https://doi.org/10.1007/978-981-33-4538-6_15

livestock productivity and production directly or indirectly. Due to declining natural resources such as water scarcity, fodder availability, and pastureland degradation, livestock food production system has been severely impacted, and in addition to it, emerging livestock diseases have raised serious concern over human food and nutritional security worldwide. The global human population is estimated to reach 8.5 billion by 2030 from the present level of 7.7 billion. The livestock products are critical for human food and nutritional security since they supply 17% of energy and 33% of protein for human diet (Rosegrant et al. 2009). It is forecasted that the demand for livestock products will further increase by 80–100% in 2030, wherein nearly 60% increase in demand will be attributed to change in consumption pattern and frequency of intake and remaining 40% due to change in population (Robinson and Pozzi 2011). This chapter aims to describe the contribution of livestock toward climate change and the effect of climate change on livestock population and mitigation and adaptation strategies to ensure food and nutritional security.

15.2 Contribution of Livestock Sector in Climate Change

The livestock sector is not only affected by climate change but also contributes to climate change effects. Greenhouse gas (GHG) emissions are observed throughout the livestock production activities. The contribution of livestock sector in climate change is mainly contributed by:

1. Livestock feed production.
2. Enteric fermentation.
3. Manure and animal waste management practices.

A large portion of natural resources is utilized in livestock production system which is contributing to 14.5% of total anthropogenic GHG emissions (Fig. 15.1). Among these, methane and nitrous oxide are the major culprits. Methane is produced by enteric fermentation in ruminants as well as manure storage and is reported to have grown 28 times higher global warming potential (GWP) in a reference of 100 years. Nitrous oxide is reported to have a GWP of 265–268 times higher than carbon dioxide in the reference 100 years and is mostly produced due to excreta/by-product storage and fertilizer usage in feed/fodder production.

Globally, feed production and its processing are contributing around 45% of GHG emissions out of the total GHG emissions from livestock sector. It is closely matched by the enteric fermentation (39%) to the GHGs. Among the other contributors, the manure storage accounts for about 10%, and the remaining 6% of GHGs is attributed to processing and transportation of animal products (Gerber et al. 2013).

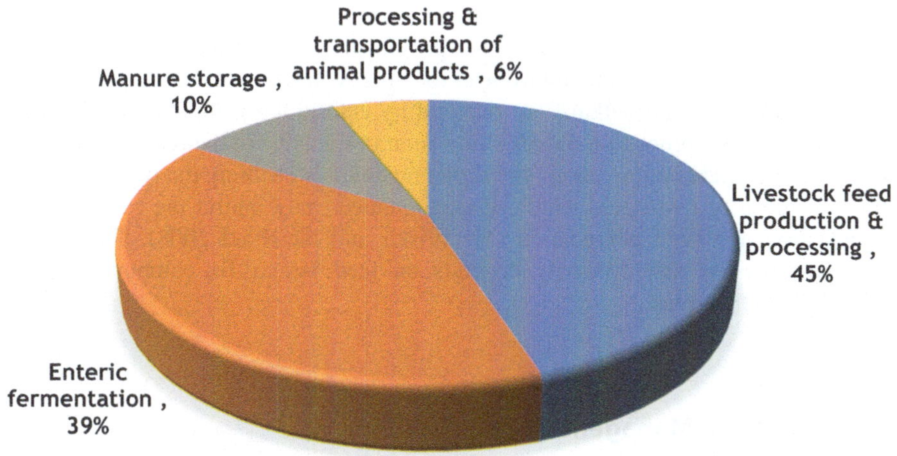

Fig. 15.1 Sources of GHG in livestock production

15.2.1 Feed Production

The GHG emission emanating from livestock feed production is mainly attributed to:

(a) Change in land-use pattern.
(b) Fertilizer and pesticide usage.
(c) Storage of manure and its application to crop fields.
(d) Agricultural operations.
(e) Feed production and processing.
(f) Transportation of feed.

Global livestock production system utilizes nearly 60% of the biomass harvested as feed (Krausmann et al. 2008), accounting approximately 30% of the water used for agriculture (Peden et al. 2007). The manure and animal waste application to crop fields generates enormous nitrous oxide, which represents almost half of these emissions (Gerber et al. 2013). As reported by Sonesson et al. (2009), about 60–80% of GHG emissions come from feed production required for poultry egg, chicken, and pork production alone and the rest 35–45% from milk and beef subsectors.

15.2.2 Enteric Fermentation

In the enteric fermentation, the microbes (bacteria, protozoa, and fungi) present in rumen of ruminants will facilitate the digestion of the feed and fodder, and during this digestive process, the volatile fatty acids (VFA), CO_2, and methane are produced. This process is known as enteric fermentation. After ruminal digestion, broken feed converted into VFA, which go into the liver through the circulatory system bypassing the rumen wall to meet the energy needs of the animal.

Carbon dioxide and methane are released by enteric fermentation and removed from the rumen by eructation process. Emission of methane in the reticulo-rumen is an evolutionary adaptation among the ruminant animals enabling the rumen ecosystem to dispose of hydrogen, which, if accumulated, may inhibit the carbohydrate fermentation and fiber degradation (McAllister and Newbold 2008). The enteric methane emission rate primarily depends on feed intake, its composition, and digestibility of the feed.

15.2.3 Manure Management

Animal waste, mainly manure, is a source for methane and nitrous oxide whose emission rate mainly depends on the environmental conditions, composition, and management of manure. The warm and wet conditions of environment increase the emissions from manure. The methane emission is linked to the organic matter content of manure, whereas the nitrogen content is linked to the nitrous oxide emission. The manure management such as long storage in tank can accelerate anaerobic decomposition leading to an increase in methane and carbon dioxide production. In contrary, when manure is applied as solid (dung) to the fields, the nitrous oxide emission increases, while the methane emission is not there, or even if it is there, it is in negligible quantity. Nitrous oxide is also generated through nitrification and denitrification processes of the nitrogen present in manure, which is mainly in the organic form (e.g., proteins) and in inorganic form as ammonium and ammonia.

15.3 Impact of Climate Change on Livestock

The impact of climate change on livestock can be direct or indirect which is summarized as follows:

- Effect on feed intake.
- Effect on livestock production.
- Effect on livestock reproduction.
- Effect on disease occurrence in livestock.
- Effect on biodiversity.
- Effect on feed resources.

15.3.1 Effect on Feed Intake

Extreme warming employs adverse impacts on livestock pasture grazing comforts. Increased atmospheric temperatures as a consequence of climate change lead to heat stress (HS) which causes reduced feed intake resulting in poor growth rate and production performance in bovines. The researchers have shown that heat stress in high-yielding lactating cows especially crossbreeds results in a dramatic reduction in feed intake and rumination (Collier et al. 1982). The reduced appetite due to heat stress is a result of increased body temperature aggravated by the specific dynamic action of feed and may be caused by gut fill. Reduced roughage consumption causes decreased VFA production which alters the ratio of acetate and propionate in the animal body. Further, the pH of rumen turns into acidic during heat stress (Collier et al. 1982) that further hinders gut and ruminal motility. Heat stress in the animal may induce sodium and potassium deficiencies resulting in metabolic alkalosis (Chase and Cherney 2012). The effect of HS on biomolecular expressions in a lactating animal is shown in Fig. 15.2.

15.3.2 Effect on Livestock Production

The anticipated rise in temperature between 2.3 and 4.8 °C together with increased precipitation resulting from climate change is likely to aggravate the heat stress in livestock, adversely affecting their productive and reproductive performance and

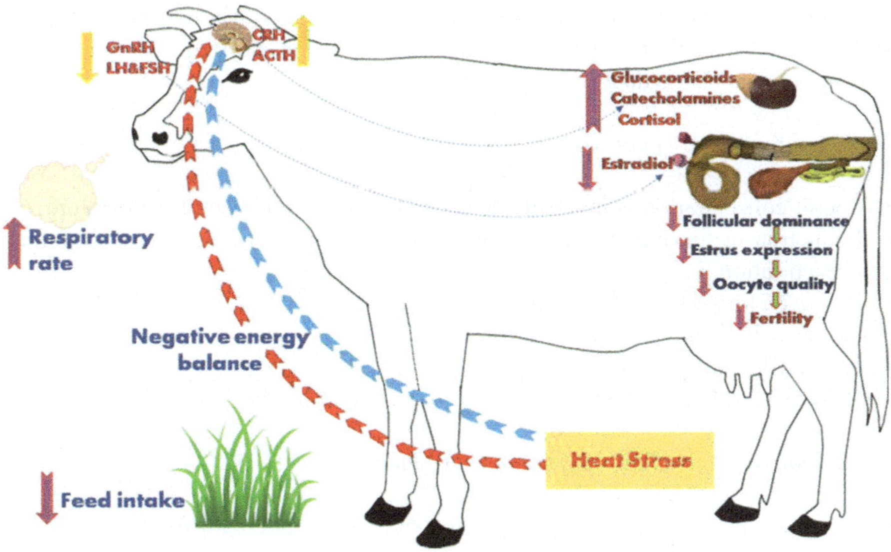

Fig. 15.2 Effect of heat stress on biomolecular expressions in lactating animal

hence reducing the total area, where high-yielding dairy cattle could be economically reared. The heat stress causes a chain reaction of physiological, behavioral, and anatomical changes leading to a reduction in growth and productive and reproductive functions (Pankaj et al. 2013). In addition, there is a decrease in activity, increase in respiration and body temperature, increased peripheral blood flow, and alterations in endocrine functions. Heat stress has an undesired impact on animal welfare, body conditions, body growth and development, animal body resistance, and milk yield. Sudden changes in environmental temperature may cause a decline in milk yield as well as a change in its composition. It is reported that due to climate change, presently, India is losing 1.8 MT (2%) of milk in terms of production decline annually. Further, it is forecasted that global losses in milk production due to climate change by 2050 will be to the tune of 15 million tons. McDowell et al. (1976) observed reduced milk production up to 15% along with 35% decrease in the efficiency of energy utilization in lactation Holstein cows when they are transferred from 18 °C to 30 °C ambient temperature. Response to high ambient temperature is also affected by the stage of lactation in dairy cows, where lactating cows in mid-stage (100–200 days) are the most heat-sensitive as compared to other stages of lactation.

15.3.3 Effect on Livestock Reproduction

Reproductive efficiency of both male and female livestock is vulnerable to climate change; however, the impacts are observed in terms of delay in puberty by a few weeks more in crossbreds and buffaloes. These effects are observed more prominently in buffaloes compared to cattle which may be due to the high thermal sensitivity of buffaloes. During heat stress, manifestations of estrus symptoms are compromised (Nobel et al. 1997), and anestrus incidence and frequent silent ovulation are noticed (Gwazdauskas et al. 1981).

(a) *Climate change effects on male reproduction.*

- Sperm concentration is inversely correlated with environmental temperature.
- Reduction in survivability of sperm due to decreased total protein and albumin concentration.
- Increased abnormality of sperm.
- Decreased sperm motility.
- Increased dead sperms.
- Reduced level of testosterone leading to decreased libido.

(b) *Climate change effects on female reproduction.*

- Alteration in gonadal activity, reproductive rhythm, and maturity.
- Increased cortisol level blocking estradiol secretion resulting into reduction in sexual behavior especially silent heat.

- Increased requirement of number of artificial inseminations per conception.
- Reduced diameter and volume of developing follicles.
- Increased early embryonic losses.
- Decreased fetal growth and calf birth weight.
- Decreased intensity and duration of estrous affecting conception rate.

(Madan and Prakash 2007)

15.3.4 Effect on Disease Occurrence in Livestock

Temperature and humidity change are the key factors affecting the incidence of animal diseases. The frequency and incidence of mastitis and foot and mouth disease (FMD) has been increased in crossbred cows. Climate change can have direct and/or indirect effects on animal health (Nardone et al. 2010). The morbidity and death are the potential direct impact, while the indirect effect is the proliferation of vector-borne diseases, foodborne diseases, and change in host resistance (Tubiello et al. 2008; Thornton et al. 2009; Nardone et al. 2010).

(a) ***Effects on Pathogens.***

Climate change effects may increase the growth rate of development of certain pathogens or parasites that have some part of life cycle stage outside their host. Apart from this, there may be emergence of new diseases due to impact of climate change on genetic mutations of pathogens.

(b) ***Effects on Vectors.***

Vectors causing diseases are impacted more by climate change, especially these conditions are more aggravated during hot–humid weather conditions.

(c) **Effects on Hosts.**

Climate change may bring a new set of livestock to be exposed to outbreaks of some diseases due to substantial shifts in disease distribution.

15.3.5 Effect on Biodiversity

The animal genetic resources are at risk of being lost even as the impacts of climate change are through weather extremes and emergence of diseases directly, whereas the indirect impacts are through changing the adaptation capability of animals to extreme climatic conditions (Banik et al. 2015). Climate change is estimated to have extinction potential of 15% to 37% for all species in the world (Thomas et al. 2004). The higher temperature has an obvious impact on reproduction, mobility, mortality, and distribution of different species (Steinfeld et al. 2006).

15.3.6 Effect on Feed Resources

India, with only 2.29% of land area of the world, is maintaining nearly 17% of world human population and 10.70% of livestock (more than 535.82 million heads) exerting huge pressure on land, water, and other resources. The country is having just about 5% of its cultivable land under fodder production which is already being impacted by rapid urbanization and industrialization. Among different feed resources, crop residues are major which are meeting more than 50% of the livestock sector feed demand in the country. At prevailing livestock productivity and production levels, livestock sector is facing severe feed and fodder shortage. In a report released by the ICAR-Indian Grassland and Fodder Research Institute (IGFRI), India has suggested that there is a deficit of 23.40% in the availability of dry fodder, 11.24% in green fodder, and 28.90% for concentrates required by livestock in India (Roy et al. 2019). Greater incidences of drought (once in 2 or 3 years) will further widen this gap of feed and fodder availability. Apart from this, the quality of fodder will also be affected by the variation in concentrations of water-soluble carbohydrates and nitrogen. Changes in meteorological variables will impact on differential growth rate of pastures altering the pasture composition and grazing preferences (Thornton et al. 2009). Plant competition dynamics is also influenced by seasonal shifts in water availability (Polley et al. 2013).

15.4 Climate Change Adaptation Measures

Adaptation activities should be designed to preserve ecosystems and the biological diversity on which we rely. They can be:

- Livestock management systems.
- Breeding strategies.
- Animal nutrition.
- Building farmers' capabilities.

15.4.1 Livestock Management Systems

(a) *Diversification.*

Area-specific suitable diversification farming system can increase drought tolerance through better resource use efficiency and might enhance livestock production when animals are exposed to extreme weather stress (Kurukulasuriya and Rosenthal 2003; Batima et al. 2005).

(b) *Integration.*

Suitable integration of systems (agriculture, horticulture, forestry, animal husbandry) as a land-use management approach can aid in achieving the sustainability and neutralize emissions from this subsector. This approach has the potential to improve ecosystem through better cycling of nutrients (Jose 2009; Smith et al. 2013).

(c) *Modifications of Farm Operations.*

Environmentally sustainable and economic viable modification in livestock and crop production location could bring dynamism in the farm operations and reduce the soil erosion and improve soil moisture content and nutrient retention capacity (Kurukulasuriya and Rosenthal 2003). Apart from this, adjustments in crop rotation and timing of operations could be adapted to match with growing seasons and environmental variables (Batima et al. 2005; Kurukulasuriya and Rosenthal 2003).

(d) *Mixed Livestock Farming Systems.*

Suitable change in mixed crop–livestock systems as per farming situations can enable the farmer to produce more outputs per unit of land using superior nutrient use efficiency imparted by the farming system.

15.4.2 Breeding Strategies

Identification of native breeds, which are suitable for adoption to local climatic conditions and their gradual upgradation to improve production performance, is necessary to sustain animal production. Selection of heat-tolerant animals within the breed for future breeding program would be highly useful. Changing breeding time and use of modern technologies such as sex selection, frozen storage of oocytes and embryos, in vitro fertilization, and transfer of gene are some of the molecular breeding strategies which can help the animals in terms of better heat tolerance and sustained production performance under changing climate.

15.4.3 Animal Nutrition

The availability of feed and fodder to sustain livestock production is one of the critical issues in view of climate change. There are many crops whose potential needs to be explored as such as alternative fodder production systems. Some of well-proven methods are as follows:

(a) *Intensive Forage Production Systems.*

Intensive forage production systems aim at efficient utilization of land and other inputs for maximum fodder production per unit area per unit time. In this production system, three to four fodder crops are cultivated on same piece of land in a calendar year for a continuous supply of fodder to livestock throughout the year.

(b) *Hydroponics and Azolla Production.*

Hydroponics is a soilless method of growing crops using mineral nutrient solutions in a water solvent with high water productivity. The fodder grown through hydroponics techniques is having 20–25% more protein, 10–15% higher mineral content, and 7–10% more essential fatty acids and can save 25–40% cattle feed. It is also observed that milk production can be increased by 10–15% with added effect of enhanced body resistance to animal diseases. Similarly, *Azolla* is highly productive aquatic ferns. It doubles its biomass in 1.9 days or more depending on conditions and can be utilized as animal feed efficiently.

(c) *Efficient Use of Crop Residue.*

According to the report of the Ministry of New and Renewable Energy (MNRE), India generates almost 500 million tons of crop residue from different crops per year. Majority of these crop residues are used as fodder and fuel for other domestic and industrial purposes. However, there is still a surplus of 140 million tons, out of which 92 million tons is burned each year, which can be potentially used for animal feed after processing and fortifying (Bhuvaneshwari et al. 2019). Through the National Innovations on Climate-Resilient Agriculture (NICRA) project, there are ample of proven high-yielding varieties of fodder and technologies such as silage making, hay-making, and urea-molasses treatment available for crop residue. However, the adoption of such climate-resilient technologies needs to be promoted all over the country.

(d) *Fodder Production: Tank Beds, Farm Ponds, Horti-pasture, Silvi-pasture.*

Alternate land-use system utilizing wasteland and fallow land such as horti-pasture, silvi-pasture, and agro-horti-silvi-pasture should be promoted, adopted, and upscaled in different agroclimatic zones. Opportunities in joint forest management (JFM), watershed, and rural development programs in a participatory mode need to be harnessed. Degraded forestland and cultivable wasteland should be restored to sustain fodder production.

(e) *Perennial Nonconventional Fodder Production Systems.*

Development of improved varieties of locally suited perennial grasses, legumes, and fodder trees as per the water availability for grasslands and silvi-pasture is need

Fig. 15.3 Nonconventional fodder production through thornless cactus cultivation

of the hour; however, presently, the seed availability of forage crops is just 15–20% of national requirement which is quite worrisome (Fig. 15.3).

15.4.4 Farmers' Perception and Adaptive Capacity

Most of the farmers are realizing that climate change is a bitter fact and their own capability to recognize it and adopt suitable climate change strategies will help them in the long run. The appropriate adaptation and mitigation measures are the biggest challenge they are facing (Jones et al. 2013). It is more relevant to collate information about farmers' perceptions to make the research more qualitative and reliable. After understanding such perceptions and including them in rural development policy framing, there will be a greater chance of accomplishing food and nutritional security with environmental conservation for sustainability (Barnes 2013; Oliver et al. 2012). Farmers' decision-making ability and means of redressal can be improved through increased risk perception.

15.5 Climate Change Mitigation Measures

Mitigation measures involve reducing the release of greenhouse gas emissions by harnessing mitigation options and integrating climate resilience measures through:

- Increasing carbon sequestration.
- Enteric fermentation.
- Manure management.
- Fertilizer management.
- Reduction in livestock population.

- Lowering livestock product consumption.

By implementing the abovementioned technologies and practices, there is huge potential to reduce GHG emissions from livestock sector. Some of the technical options are as follows:

15.5.1 Carbon Sequestration

The process of removing carbon from the atmosphere and depositing it in the suitable reservoirs is known as carbon sequestration. The following options may be utilized for carbon sequestration:

(a) *Forests as a sink*: Afforestation/reforestation/plantation/agroforestry are some of the activities helpful for carbon sequestration as the plant and trees are natural sequesters of carbon which is utilized either for photosynthesis process or stored as biomass or wood.
(b) *Wetland restoration*: Wetlands are the natural carbon pool or sink which has potential to absorb 14.5% of the soil carbon found in the world. Thus, restoration of such wetlands is needed to be promoted as one of the mitigation measures.
(c) *Oceans as a sink*: Oceans absorb CO_2 from the atmosphere because the concentration of CO_2 in the atmosphere is greater than that in the oceans which can be sequestered in the oceanic biomass.
(d) *Geological sequestration*: The concentration of carbon dioxide can be reduced in the atmosphere by storing them in oil and gas reservoirs and other geological materials.
(e) Improving land and water management efficiency can facilitate sequestration.
(f) Carbon sequestration potential of the land could be improved by scientific pasture management or management of common property resources (CPR).

15.5.2 Enteric Fermentation

Emission per unit of animal product (milk/meat) varies greatly by adopting different practices and technologies which impose the scope of significant reduction in emission using them. Increasing the productivity of animal can be a very ideal and effective strategy for reducing GHG emissions per every unit of livestock product produced. The researchers have shown that:

(a) Increase in quality green fodder into the diet decreased methane production by 5.7% (Singhal and Madhu 2002).
(b) Enhancing the forage digestibility and higher intake of digestible forage will generally reduce GHG emissions from rumen fermentation (and stored manure), per unit of animal product.

(c) Legume silages do have an advantage over grass silage due to their lower fiber content and the additional benefit of replacing inorganic nitrogen fertilizer. Effective silage preservation will improve forage quality on the farm and reduce the intensity of GHG emission.
(d) Providing higher quality forage increases digestibility resulting in lower methane production.
(e) Inclusion of concentrate feeds in the ration of ruminants will decrease enteric methane emissions per unit of animal product, particularly when there is above 40% of dry matter intake.
(f) Increasing 1% dietary fat content can decrease emission of methane up to 5%.
(g) Increasing the concentrate in the diet can reduce methane emission by 15–32% (Singh and Madhu 1999).
(h) Feed supplements, feed antibiotics, growth hormones, probiotics, etc., which tend to improve digestibility, production, and weight gain.
(i) Methane inhibitors, such as bromo-chloromethane (BCM), 2-bromo-ethenesulfonate (BES), chloroform, etc., have the potential to reduce CH_4 production up to 50% in ruminants (Knight et al. 2011).
(j) Feeding with a nitrate supplement can reduce enteric methane production up to 50%.
(k) Proper grazing management practices can improve the quality of pastures, thereby improving the digestibility of fodder resulting in reduction in methane production.

15.5.3 Manure Management

The bulk of the methane emitted from manure and animal wastes are resulting from its storage and anaerobic treatment. Generally, animal manure is stored in open space in the field conditions which accounts for more than 25 million tons of methane emission globally every year. It can also cause nitrous oxide emissions. The GHG emission from manure is dependent on the ambient temperature, timing of application, and duration of the storage. Therefore, most mitigation practices involve reducing the duration of the storage, application of additives, suitable timing and application of manure, use of anaerobic digesters, covering of storage of manures, use of solid separator, proper change in the animal diets, etc. (ICF International 2013). Diet can have a significant impact on chemical composition of manure (feces and urine) and therefore on the subsequent GHG emissions during storage and following land application. Promotion of community biogas plant as biodigester can achieve 50–75% reduction in methane emissions from manure. GHG emissions can also be reduced by proper blend of dietary proteins and feed supplements. If protein intake is reduced, the nitrogen excreted by animals is also reduced. Supplements such as tannins have the potential to reduce emissions (Hess et al. 2006; Dickie et al. 2014).

15.5.4 Fertilizer Management

Soil health analysis-based suitable dose of fertilizer application on feed and fodder crops can decrease the quantity of nitrous oxide emissions and hence can act as an effective mitigation strategy. Nitrogen use efficiency can be enhanced through:

(a) Plant breeding techniques and genetic modifications with better nutrient uptake and utilization efficiency.
(b) Proper use of fertilizers in terms of dose and application timings.
(c) Proper management of soil health and use of soil health cards.
(d) Use of quality organic fertilizers.
(e) Combining legumes with grasses in pasture for reducing the requirement of supplementary nitrogen.
(f) Regular soil testing by scientific method of sampling and analysis (at least once in 3 years).
(g) Using technologically advanced fertilizers like bio-fertilizers with a suitable consortium of microbes, which regulate the release of nutrients from the fertilizer as per the demand of the crop.

15.5.5 Reduction in Livestock Population

There is greater debate on the issue of reducing animal numbers in most parts of the world. The discussion is also around the fact that animal rearing and livestock management is the single most influential climate change mitigation strategy, which can significantly reduce the GHG emissions. The reduction of livestock population without compromising energy and protein needs of human nutrition, increase in animal productivity through improving animal genetics, supply of balance and adequate nutrition, efficient reproduction, health, and overall management of the animal operations can be the best strategy to implement it. In the Netherlands, milk production increased from 6270 kg/cow to 8350 kg/cow in a span of 18 years which resulted into decrease in methane production from 17.6 to 15.4 g/kg milk (Bannink et al. 2011). Thus, increasing milk yield from 3.6 l/day/cattle up to 9.0 l/day/cattle can potentially reduce the methane production from 2.29 to 1.38 Tg/year. under temperate climatic conditions (Blummel et al. 2009). Though this will help, the absolute quantity of methane produced due to livestock production needs to be managed on a scientific basis.

15.5.6 Lowering Livestock Product Consumption

Most often, the discussion regarding the reduction of livestock product consumption is on moral values. The choice of consumption of food is individualistic. However, the ecological balance and sustainability of animal population and their production need special attention. Lowering the consumption of meat, milk, and milk products in areas having a high standard of living (people relying more on non-vegan diets) will be an effective short-term response to the GHG mitigation and nutrition security of the people who have less purchasing power. Europe, North America, and Soviet Union countries produced 46.3% of ruminant meat and milk energy and only 25.5% of the enteric CH_4 emissions in 2005 (O'Mara 2011). In contrast, Asia, Africa, and Latin America produced a similar amount (47.1%) of ruminant meat and milk energy but a large proportion (almost 69%) of enteric CH_4 emissions. Though there are multiple and complex issues behind these emissions, further research will help to find out lowering livestock product consumption to reduce GHG emissions. An improvement in productivity level and farming as per the carrying capacity of land may certainly help to optimize the mitigation strategy.

15.6 Conclusion and Implications

As the growing human population will create a demand for meat and dairy products, the livestock sector has to grow to meet such rising demand. This will certainly lead to endangering several natural resources to produce the livestock and the feed for them. Such a scenario will have an impact on the vulnerable climate. The wise thing is to balance the production without overexploiting the resources and ensuring sustainability. Broadly, the following research and policy interventions should be pursued religiously if the livestock sector has to be sustainable and at the same if it has to contribute to the well-being of the world through effective climate adaptation and mitigation strategies.

The adverse effects of climate change can be effectively mitigated through modern scientific livestock management that should be the first line of action to mitigate global climate change.

The animal breeding program must be driven by the needs of the people and the existing natural resources and stress for the animal from biotic and abiotic factors.

Developing farmers' capabilities must be holistic and should not be driven by short-term goals. The educated farmers can play a pivotal role in good practices in manure management, crop production, understanding the market dynamics, human diet pattern, etc.

The recent advances such as information and communication technology for weather forecasting, crop pest and pathogen surveillance, animal disease diagnostics, market intelligence, data management, etc., should be emphasized sufficiently. All the ICT tools and applications must help for effective decision-making by the farmers so that climate resilience is a new norm in livestock management.

References

Banik S, Pankaj PK, Naskar S (2015) Climate change: impacts on livestock diversity in tropical countries. In: Pradeep KM, Bhatta R, Takahashi J, Richard AK, Cadaba SP (eds) Livestock production and climate change. CAB International, Boston, MA, pp 162–182

Bannink A, Schijndel V, Jan MW, Jan D (2011) A model of enteric fermentation in dairy cows to estimate methane emission for the Dutch National Inventory Report using the IPCC tier 3 approach. Anim Feed Sci Technol 166–167:603–618

Barnes AP (2013) Heterogeneity in climate change risk perception amongst dairy farmers: a latent class clustering analysis. Appl Geogr 41:105–115

Batima P, Natsagdorj L, Gombluudev P, Erdenetseteg B (2005) Observed climate change in Mongolia. AIACC working paper no.12. Available via DIALOG. http://www.start.org/Projects/AIACC_Project/working_papers/Working%20Papers/AIACC_WP_No013.pdf. Accessed 28 July 2020

Bhuvaneshwari S, Hiroshan H, Meegoda JN (2019) Crop residue burning in India: policy challenges and potential solutions. Int J Environ Res Public Health 16(5):832

Blummel M, Anandan S, Prasad CS (2009) Potential and limitations of by-product based feeding systems to mitigate greenhouse gases for improved livestock productivity. Research gate. Available via DIALOG. https://www.researchgate.net/publication/283738415_Potential_and_Limitations_of_byProduct_Based_Feeding_Systems_to_Mitigate_Greenhouse_Gases_for_Improved_Livestock_Productivity. Accessed 28 July 2020

Chase LE, Cherney DJ (2012) Using grass forages in dairy cattle rations. In Proceedings of the Tri-State Dairy Nutrition Conference, pp. 75–85

Collier RJ, Doelger SG, Head HH, Thatcher WW, Wilcox CJ (1982) Effects of heat stress during pregnancy on maternal hormone concentrations, calf birth weight and postpartum milk yield of Holstein cows. J Anim Sci 54:309–319

Dickie A, Streck C, Roe S, Zurek M, Haupt F, Dolginow A (2014) Strategies for mitigating climate change in agriculture: Abridged report. Climate Focus and California Environmental Associates, prepared with the support of the Climate and Land Use Alliance. Available via DIALOG. https://climatefocus.com/sites/default/files/strategies_for_mitigating_climate_change_in_agriculture.pdf. Accessed 28 July 2020

Gerber PJ, Steinfeld H, Henderson B, Mottet A, Opio C, Dijkman J, Falcucci A, Tempio G. (2013) Tackling climate change through livestock: a global assessment of emissions and mitigation opportunities. Rome: FAO Available via DIALOG. http://wwwfaoorg/3/a-i3437epdf. Accessed 28 July 2020

Gwazdauskas FC, Thatcher WW, Kiddy CA, Paape CJ, Wilcox CJ (1981) Hormonal patterns during heat stress following PGF2α induced luteal regression in heifers. Theriogenology 16:271–277

Hess H et al (2006) Strategic use of tannins as means to limit methane emission from ruminant livestock. In. International conference on greenhouse gasses and animal agriculture, Elsevier international congress series, 2006. vol 1293. Zurich, Switzerland, pp 164–167

ICF International (2013) Greenhouse gas mitigation options and costs for agricultural land and animal production within the United States. Available via DIALOG. http://www.usda.gov/oce/climate_change/mitigation_technologies/GHG_Mitigation_Options.pdf. Accessed 28 July 2020

Jones AK, Jones DL, Edwards-Jones G, Cross P (2013) Informing decision making in agricultural greenhouse gas mitigation policy: a best-worst scaling survey of expert and farmer opinion in the sheep industry. Environ Sci Pol 29:46–56

Jose S (2009) Agroforestry for ecosystem services and environmental benefits: an overview. Agrofor Syst 76(1):1–10

Knight T, Ronimus RS, Dey D, Tootill C, Naylor G, Evans P (2011) Chloroform decreases rumen methanogenesis and methanogen populations without altering rumen function in cattle. Anim Feed Sci Technol 16:101–112

Krausmann F, Erb KH, Gingrich S, Lauk C, Haberl H (2008) Global patterns of socioeconomic biomass flows in the year 2000: a comprehensive assessment of supply, consumption and constraints. Ecol Econ 65:471–487

Kurukulasuriya P, Rosenthal S (2003) Climate change and agriculture: A review of impacts and adaptations. Paper No. 91 in Climate Change Series, Agriculture and Rural Development Department and Environment Department, World Bank, Washington, DC

Madan ML, Prakash BS (2007) Reproductive endocrinology and biotechnology applications among buffaloes. Soc Reprod Fertil Suppl 64:261–281

McAllister TA, Newbold CJ (2008) Redirecting rumen methane to reduce methanogenesis. Aust J Exp Agric 48:7–13

McDowell RE, Hooven NW, Camoens JK (1976) Effect of climate on performance of Holsteins in first lactation. J Dairy Sci 59:965–971

Nardone A, Ronchi B, Lacetera N, Ranieri MS, Bernabucci U (2010) Effects of climate changes on animal production and sustainability of livestock systems. Livest Sci 130(1):57–69

Nobel RL, Jobst SM, Dransfield MBG, Pandolfi SM, Balley TL (1997) Use of radio frequency data communication system, heat watch, to describe behavioral estrus in dairy cattle. J Dairy Sci 179

O'Mara FP (2011) The significance of livestock as a contributor to global greenhouse gas emissions today and in the near future. Energy Fuels 166(1):7–15

Oliver DM, Fish RD, Winter M, Hodgson CJ, Heathwaite AL, Chadwick DR (2012) Valuing local knowledge as a source of expert data: farmer engagement and the design of decision support systems. Environ Model Softw 36:76–85

Pankaj PK, Ramana DBV, Rani R, Nikhila M, Sudheer D, Chowdary KP (2013) Thermotolerance of sheep in relation to coat color. J Agrometeorol 15:163–169

Peden D, Tadesse G, Misra AK (2007) Water and livestock for human development. In: Molden D (ed) Water for food, water for life: a comprehensive assessment of water management in agriculture. Earthscan, London, pp 485–514

Polley HW, Briske DD, Morgan JA, Wolter K, Bailey DW, Brown JR (2013) Climate change and north American rangelands: trends, projections, and implications. Rangel Ecol Manag 66 (5):493–511

Robinson TP, Pozzi F (2011) Mapping supply and demand for animal-source foods to 2030. Animal production and health working paper. No. 2. Rome. Available via DIALOG. http://www.fao.org/3/i2425e/i2425e00.pdf. Accessed 28 July 2020

Roy AK, Agrawal RK, Bhardwaj NR, Mishra AK, Mahanta SK (2019) Revisiting National Forage Demand and availability scenario. Available via DIALOG https://wwwresearchgatenet/publication/336230799_Revisiting_National_Forage_Demand_and_Availability_Scenario. Accessed 28 July 2020

Rosegrant MW, Fernandez M, Sinha A (2009) Looking into the future for agriculture and AKST. In: McIntyre BD, Herren HR, Wakhungu J, Watson RT (eds) International assessment of agricultural knowledge. Science and Technology for Development (IAASTD). Agriculture at a crossroads. Island Press, Washington, DC, pp 307–376

Singh GP, Madhu M (1999) Effect of different levels of monensin in diet on rumen fermentation, nutrient digestibility and methane production in cattle. Asian Aust J Anim Sci 12(8):1215–1221

Singhal KK, Madhu M (2002) Uncertainty reduction in methane and nitrous oxide gases emission from livestock in India. Project report, Dairy Cattle Nutrition Division, National Dairy Research Institute, Karnal, India, p 62

Smith P, Ashmore M, Black H, Burgess PJ, Evans C, Quine T, Thomson AM, Hicks K, Orr H (2013) The role of ecosystems and their management in regulating climate, and soil, water and air quality. J Appl Ecol 50:812–829

Sonesson U, Cederberg C, Berglund M (2009) Greenhouse gas emissions in milk production. Decision support for climate certification Report 2009:3 Available via DIALOG http://wwwklimatmarkningense/wp-content/uploads/2009/12/2009-2-feedpdf. Accessed 28 July 2020

Steinfeld H, Gerber P, Wassenaar T, Castel V, Rosales M, Haan C (2006) Livestock's long shadow: environmental issues and options. FAO, Rome

Thomas C, Alison C, Rhys G, Michel B (2004) Extinction risk from climate change. Nature 427:145–148

Thornton P, Van de Steeg J, Notenbaert MH, Herrero M (2009) The impacts of climate change on livestock and livestock systems in developing countries: a review of what we know and what we need to know. Agric Syst 101:113–127

Tubiello F, Schmidhuber J, Howden M, Neofotis PG, Park S, Fernandes E, Thapa D (2008) Climate change response strategies for agriculture: challenges and opportunities for the 21st century. Agriculture and rural development discussion paper, p. 42

Climate Change and Food Security: Two Parallel Concerns

16

Philip Kuriachen, S. Aiswarya, and K. S. Aditya

Abstract

Agriculture and climate change effects exhibit two-way causation. Agriculture contributes to climate change and is also affected by climate change. Agricultural production systems need to be continuously adjusted, so as to mitigate the climate change as well as to adapt to it to minimise the negative effects of climate change on agricultural production and food and nutritional security. In this chapter, we review the potential impact of climate change on the three dimensions of food security, namely, food availability, food accessibility and food utilisation. We also examine some adaptation measures in ameliorating the negative impacts of climate change on food security.

Keywords

Climate change · Food security · Food availability · Food accessibility · Food utilisation

16.1 Introduction

Climate change is one of the major concerns of the twenty-first century. Global mean surface temperatures are predicted to increase by 1.8 °C to 4.4 °C (IPCC 2007), and India's surface temperature is predicted to be 4.4 °C warmer by 2100 (Krishnan et al. 2020). Correspondingly, sea levels are also predicted to increase by 0.26–0.77 mm by 2100. The frequency of extreme weather events like droughts, floods and heatwaves is expected to increase due to changes in the hydrological cycle induced

P. Kuriachen (✉) · S. Aiswarya · K. S. Aditya
Division of Agricultural Economics, ICAR-Indian Agricultural Research Institute, New Delhi, India

V. K. Hebsale Mallappa, M. Shirur (eds.), *Climate Change and Resilient Food Systems*, https://doi.org/10.1007/978-981-33-4538-6_16

by global warming (Childers 2015). The onset of seasons and the duration will also change, because of which the suitability of the regions for agriculture and crop suitability for the region will also change. In India, the surface temperatures are predicted to increase by 2.0 °C over the next 50 years under high emission scenarios. In the absence of suitable adaptation strategies, the changes in climate can cause a loss in income and increase the variability in incomes and thus exacerbate poverty. Hence, battling climate change is recognised as one of the major goals by United Nations Development Programme under Sustainable Development Goals (SDGs[1]).

Eradication of hunger and ensuring food security is a major target under Sustainable Development Goals. Although India has witnessed substantial growth in its economy in the post-liberalisation era, undernutrition and hunger still remain key priorities. India couldn't achieve the target of zero hunger proposed by the Millennium Development Goals (MDGs[2]), and also India's 12 states come under the 'alarming category' in the global food security report (Chakrabarty 2016). For achieving SDGs, India's performance is crucial in terms of not only food security and zero hunger but in all the 15 SDGs. Further, the demand for food grains is predicted to increase substantially in the immediate future driven by rising population, rapid urbanisation and changing food habits (Mittal 2008). Meeting the increased demand for food grains through domestic production alone is challenging due to low growth in yield of major crops (Kumar et al. 2009). The challenge increases manifold when we account for the possible adverse effects of climate change on agriculture.

Agriculture sector performance is crucial not only in ensuring food security but also in poverty alleviation as agriculture is the source of livelihood for the majority of the rural population. Climate change has been recognised as a principal factor contributing to the vulnerability of food systems in general and agriculture in particular. It could serve as an impediment in eradicating hunger and ensuring food security (Gregory et al. 2005). Food security is a complex phenomenon which comprises four major components food availability, food accessibility, food affordability and food stability, all of which are intrinsically related to agriculture and are affected by climate change. In this paper, we explore the implications of climate change on the different dimensions of food security with a special focus on the Indian context. We also discuss some adaptation strategies that need to be implemented to reduce the risks to food security due to climate change.

[1]Refer Annexure I

[2]Refer Annexure II

16.2 Climate Change and Implications for Food Security

16.2.1 Climate Change Impact on Food Availability

Food availability is defined as the total quantity of food available in a region comprising of domestic production, food imports and food stocks after accounting for food exports. This section explores the impact of climate change on various aspects of food availability, which includes food production, processing, storage and distribution.

16.2.1.1 Climate Change Impact on Food Production

Climate change is expected to influence food production and agricultural activities through several pathways. It is well established that climate change will affect crop yields which will have ramifications for food chains both at regional and macro levels. Herein, we review the scientific evidence on the impact of climate change on crop yields. Further, we also explore the effects of climate change on food supplies and agricultural supply chains. The yield and, consequently, food production are affected by climate change, mainly through an increase in CO_2 concentration, ozone pollution and a rise in surface temperatures (Yadav et al. 2018). Few studies have documented the positive effect of the increase in CO_2 concentration on crop yields. An increase in CO_2 fertilisation due to increase in CO_2 is expected to have a positive correlation with both plant growth and yield (Food and Agricultural Organization of the United Nations 2008a). Experimental evidence reveals that under optimal growth, increase in CO_2 concentration to 550 ppm raises the yield of C_3 plants like rice and wheat by 10–30%, while yields of C_4 plants like maize and sorghum increase by 0–10% (IPCC 2007). Global crop yields are projected to increase by 1.8% every decade, with an increase in CO_2 (IPCC 2001).

However, recent studies have argued that potential benefits of CO_2 fertilisation on food crops are roughly half of that of previous studies (Long et al. 2005) and elevated CO_2 concentrations are unlikely to affect world food supply (Schmidhuber and Tubiello 2007). Moreover, the beneficial effects of CO_2 increase will be offset by the detrimental impact of an increase in O_3 concentration and a rise in temperature and other changes in climate. Increase in O_3 concentration is likely to affect crop yields detrimentally. It is predicted that by 2050, ozone pollution would negate yield gains through CO_2 fertilisation in C_3 plants and reduce the yield of C_4 plants by 5% (Nelson et al. 2009). Global mean surface temperatures are also predicted to increase by 1.8–4.4 °C by 2100 (IPCC 2007). The increase in temperature is likely to have a beneficial impact on agriculture in temperate regions. The rise in temperature would increase the growing period in temperate regions and therefore increase biomass accumulation leading to higher yields (Fischer et al. 2005). On the other hand, in lower latitudes, crop yields are likely to decrease due to shortened grain filling period and increased thermal stress during flowering and seed setting stages. Subsequently, in temperate regions, area under cultivation is predicted to expand, while in the tropics and subtropics, increase in evapotranspiration and a decrease in soil moisture

Table 16.1 Impact of climate change on major food crops in India

Study	Crop	Findings
Kumar et al. (2014a)	Wheat	Wheat yields will reduce by 6–23% by 2050 and by 15–25% by 2080. Wheat growing regions in Central India will be affected more adversely by warming. Regions with mean maximum and minimum temperatures over 27 °C and 13 °C are likely to experience yield decline
Soora et al. (2013)	Rice	Yields of irrigated rice could decrease by 7% in the 2050s and by 10% in the 2080s; for rainfed rice, projected yield reduction is marginal amounting to 2.5%
Byjesh et al. (2010)	Maize	Yields in mid-indo-Gangetic plains are predicted to decrease by 5%, 7% and 13% in 2020, 2050 and 2080, respectively, for monsoon crop. Yields in upper indo-Gangetic plains during Rabi season are predicted to increase by 5% till 2050. However, by 2080, yields are projected to decrease by 25%
Srivastava et al. (2010)	Sorghum	Climate change is projected to reduce sorghum yields by 7% in 2020 and 11% in 2050, in Rabi season. The southwest zone is predicted to be affected adversely by climate change in comparison with central and southern zones

levels would render currently cultivated regions unsuitable for cropping activities (IPCC 2007).

Further, we review the potential impact of climate change on major food crops in India. The findings have been enlisted in Table 16.1. A projected increase in surface temperature is the main contributor for crop yield reductions in India, compensating the potential yield increase due to CO_2 fertilisation. The review revealed that yields of rice and wheat, the two major food crops of India, are likely to decline by 2.5–7% and 6–23% by 2050. As noted earlier, in both the cases, the increase in mean temperatures is likely to offset the beneficial effects of CO_2 fertilisation in the medium- and long-term under moderate emission scenarios (Kumar et al. 2014a; Soora et al. 2013).

The impact of climate change on crop yields directly affects food availability. Herein we review evidence of the impact of climate change on regional and global food production. Rosenzweig and Parry (1994) found that climate change could lead to a reduction in global cereal production by 11–20% without farm-level adaptations. The study also revealed a regional disparity in response to climate change. Cereal production was predicted to decline by 7% in developing countries, while in developed countries, cereal production was expected to increase by 10%. Haile et al. (2017) predict that global production of wheat, maize and rice would decrease by 9% in the 2030s and by 23% in the 2050s. Climate change will also increase annual fluctuations in crop yield by 1–3%. Lee (2009) assessed the potential impact of climate change on food supplies in Asia and predicted a marginal reduction in the production of rice. The study predicted that the production of wheat would increase by 1.3%, while the production of other cereals would decline by 1.3%. Import demand for rice in Asia would increase by 5%, while that of wheat would decline by 7%. Bandara and Cai (2014) found that production of wheat, rice and cereal grains in South Asia is predicted to decline by 7%, 4% and 11%. In India, the production of wheat, rice and cereal grains is predicted to decline by 4%, 5% and 2%, respectively, by 2030. Overall, most studies indicate that climate change will have dangerous consequences for food production in the absence of farm-level adaptations, and the effects are not homogenous across regions.

16.2.1.2 Climate Change Impact on Food Processing, Storage, Transport, and Distribution

Several studies have highlighted various pathways in which the climate change affects various postproduction activities like primary processing, storage and transportation of agricultural commodities leading to possible disruption of supply chains (Vermeulen et al. 2012). In the projected scenarios, the mean precipitation is expected to increase, particularly in tropics and subtropics, which in turn can lead to higher humidity and consequently hastens the chemical and biological deterioration of food commodities, thus resulting in higher postharvest losses (Food and Agricultural Organization of the United Nations 2008b; Stathers et al. 2013). Studies indicate that climate change could lead to postharvest losses of up to 80% in rice and 55% in vegetables (Parfitt et al. 2010). An increase in temperature and moist conditions increases the risk of mycotoxin contamination and bacterial infection making food hazardous for consumption. Rising mean temperatures may also lead to shrivelling of grains due to overdrying Stathers et al. 2013. Temperate regions, on the other hand, are likely to benefit from an increase in humidity, making harvested grains ideal for storage (Food and Agricultural Organization of the United Nations 2008c). However, due to change in seasonal temperatures, there is an increased risk of pest infestations, as the insect's life cycle is shortened leading to faster rates of reproduction (Moses et al. 2015). Thus, climate change would exacerbate the infestation of storage pests both in tropical and temperate regions. This would subsequently lead to an increase in storage cost and postharvest losses.

An increase in the frequency of floods is likely to disrupt food distribution systems in rural areas in South Asia (Ingram 2011). High frequency and intensity of precipitation in regions with inadequate transport infrastructure will curtail the efficacy of food distribution systems (Hendrix and Salehyan 2012). Highly advanced, low-inventory food chains operating a just-in-time delivery mode are particularly susceptible to weather disruptions. Temperate countries are likely to experience a reduction in transportation costs due to the reduction in winter maintenance costs and the opening of river and sea routes for a longer duration (Waters 2011). Therefore, in the face of climate change, developing countries will have to focus on improving transport logistics and the storage so that the losses can be minimised (Godfray et al. 2010).

16.2.2 Climate Change Implications on Food Accessibility

Accessibility to food pertains to an individual's capacity to safeguard his entitlements to food. Cultivation of required quantities of food is not sufficient to ensure food accessibility. Food accessibility is dependent on the rights and capacity of individuals that determine food allocations, food preferences and food affordability (Ericksen 2008). These rights are, in turn, influenced by geographical and socio-economic factors and the proper functioning of food markets. In this section, we discuss the implications of climate change on three dimensions of food accessibility, namely, food affordability, food allocation and food preferences.

Food affordability refers to the capacity of individuals to access healthy and nutritious food at affordable prices. Hence, the impact of climate change on food affordability is closely related to its impact on livelihoods, farm incomes and prices. The impact of climate change on farm incomes and food prices has been extensively studied. Herein we summarise the finding of these studies. Global food prices are expected to increase moderately till 2050, and climate change will be a major contributor to it. The effect of price change due to the direct effects of global warming is likely to be lesser than the price changes due to socio-economic pathways in the medium term (Schmidhuber and Tubiello 2007). Recent studies have found that climate change could lead to an increase in wheat, rice and maize prices by 32%, 55% and 94%, respectively (Nelson et al. 2009). Few studies have also assessed the impact of climate change on food prices specific to South Asia; in India, the prices of wheat, rice and cereal grains are predicted to increase by 2%, 8% and 5%, respectively (Bandara and Cai 2014). Among South Asian countries, Nepal is likely to experience the highest increase in food prices due to climate change. Kumar et al. (2014b) analysed the potential impact of droughts on food prices in India and found that a 10% deficit in rainfall would increase the price of rice by 23%.

Further, a 10% deficit in rainfall would increase the price of maize, sorghum and pearl millet by 15%, 14% and 13%, respectively. Climate change is also expected to affect farm incomes. Burgess et al. (2014) found that one standard deviation increase in high-temperature days reduced real wages by 9.3%. Kumar and Parikh (2001) assessed the impact of global warming on-farm revenues in India and found that global warming could reduce farm revenues by 8%. Studies also indicate a heterogeneous impact on farm incomes, with income reduction by 15% in irrigated regions and by 25% in unirrigated areas (Economic survey 2017). Thus, higher food prices and a fall in farm incomes are expected to increase the threat of hunger and food insecurity in India in the future.

Implications of climate change to food allocations and food preferences have not been extensively documented. From the available sparse studies, we can infer that extreme weather events lead to erosion of farm assets leading to suboptimal food allocation adversely and potentially lead to chronic food insecurity. When climate change-induced shocks affect farm incomes, one of the common coping mechanisms is to reduce the quantum and diversity of food consumption. Farmers in Kenya opine that reduced food intake is a coping mechanism to deal with extreme weather events. Reduction in frequency and variety of meals consumed is frequently adopted coping mechanisms (Thorlakson and Neufeldt 2012). Similarly, in Madagascar, farmers cope with droughts by reducing food intake, changing food composition and substitution with wild foods (Harvey et al. 2014).

On the other hand, the higher vulnerability of the traditional food systems to climate change limits the capacity of indigenous communities to overcome the food crisis induced by extreme weather events (Jernigan et al. 2012). Reduction in food availability also reduces the purchasing power of low-income holders leading to changes in food preferences and a shift to low calorific value food/low diversity products (Ziervogel and Ericksen 2010).

16.2.3 Food Utilisation and Climate Change

Food utilisation is one of the most significant dimensions of the food security that is affected by the climate change, yet the least explored component by researchers (Wheeler and von Braun 2013; Zewdie 2014; Cvitanovic et al. 2016; Schmidhuber and Tubiello 2007). The fourth dimension of food security, i.e. food utilisation, is conceptualised in a number of ways. A few definitions of food use/utilisation include household food treatment and biological processes related to food utilisation, while other researchers restrict the concept of the biological capacity of the person to utilise the food for a prosperous life (Swindale and Paula 2006). Food and Agricultural Organization of the United Nations (2008c) defined food utilisation as 'the nutrient value of the meals, which depends on the ingredients and preparation methods; the value systems of foods which determine what types of foods should be consumed and consumed at different periods of the year and on various occasions; and the standard and safety of the food supply, that may cause loss of nutrients in foods and the transmission of foodborne diseases if not of appropriate standards'. The World Food Programme (WFP 2009) takes a much broader view of use, i.e. 'households' utilisation of the food as well as individual's capability to absorb and metabolise the nutrients'.

Climate change leads to a decline in soil fertility and crop yields and change in the type of crops cultivated (due to changes in suitability) and lowers the nutrient content of the crops (Taub 2010; Clair and Lynch 2010; Lobell and Burke 2010). For instance, Taub (2010) reveals that elevated CO_2 leads to a reduction in protein concentration of rice, wheat, barley and potato tubers by 5–14%. In addition, the percentage of nutritionally essential minerals such as calcium, magnesium and phosphorous may also reduce under-elevated CO_2. Although the degree of decrease is low, combined with decreased yields, the impacts may be significant (Lobell and Burke 2010). Several studies also suggested that new pests and predators that emerged due to changing climatic conditions also affect the quality and availability of food (Rosenzweig et al. 2001; Anderson et al. 2004; Nelson et al. 2009).

The most vulnerable group to climate change impacts are small farmers, tribal communities and those who belong to lower social strata; the ones least capable of coping efficiently are at the highest risk of food and nutritional insecurity (Easterling et al. 2007; Tirado and Meerman 2012). The decline in crop productivity may force vulnerable households to reduce consumption and substitute nutrient-rich foods like fruits and vegetables with cheaper starchy substitutes leading to a decline in dietary diversity and nutritional intake (Thompson and Cohen 2012). Reduced nutritional intake will lower the productive capacity of current as well as future generations leading to a potentially vicious cycle of vulnerability and food insecurity (Hoddinott et al. 2008; Schaible and Kaufmann 2007). This is a vicious cycle of income shocks leading to poor nutrition and poor health, causing further loss in incomes. Recent studies have found that the linkages between food utilisation and economic access to food are much stronger than those between physical access to food and food utilisation (Abbade 2017).

The link between climate change and health is evident from the increased number of patient visits to clinics following excessive heat, rain and cold. The risk of vector-borne and waterborne diseases is enhanced due to climate change (Singh and Dhiman 2012; Kundzewicz et al. 2007; Boko et al. 2007). Increase in exposure to diarrhoeal diseases due to an increase in precipitation and a rise in sea levels, specifically in coastal regions, may reduce the capacity to utilise food efficiently (Aberman and Tirado 2014).

Change in climate will also facilitate an increase in transmission of novel zoonosis, i.e. animal diseases, which can be spread to humans and also shifts the temporal and spatial spread of malaria-like vector-borne diseases. The spread of these diseases will increase due to the alteration in the natural ecosystems, migration pathways and pathogen survival (Mills et al. 2010). The case of Covid-19 pandemic is a case in point.

Climate change is projected to impact food safety as it alters the temperature, pH and salinity, leading to an increased speed of multiplication of microorganisms (Tirado and Meerman 2012). The rise in temperature is also likely to increase the spread of foodborne diseases like salmonellosis by as much as 12% for each unit increase in temperature above ambient levels (Kovats et al. 2004). Increasing temperatures have also increased natural biotoxins seen in several crops that can contaminate the food consumed by humans, leading to foodborne diseases, which further aggravate malnutrition and food insecurity (Tirado and Meerman 2012).

In developing countries, rural women whose livelihood is reliant on subsistence farming are highly vulnerable (Lambrou and Piana 2006). Their capacity to respond to impacts of climate change is constrained by limited access to productive assets. In developing countries, especially in Asia and Africa, women have to travel a long distance to fetch water due to the decline in the water table. As a result, more energy and time is expended to fetch water or compelled to use contaminated water, increasing the possibility of diarrhoeal diseases. This can also lead to a reduction in lesser attention mothers can provide to the children in terms of both time and more and nutrition, which affects the health of the next generation (Levinson et al. 2002; Tirado and Meerman 2012).

Food utilisation also symbolises the religious, sociocultural, moral and nutritional values that society can achieve (Cvitanovic et al. 2016; O'Brien and Wolf 2010). The unavailability of food threatens the consistency and stability of indigenous cultures, as various cultural practices tend to be forgotten (Food and Agricultural Organization of the United Nations 2008c). Cultural elements play a critical role in socialisation and also in formulating adaptation and mitigation strategies of climate change (O'Brien and Wolf 2010).

16.2.4 Climate Change and Stability of Food Systems

Food system stability refers to the capacity of food systems to make adequate quantities of nutritious food available at affordable prices over time. Studies have predicted an increase in the occurrence of floods and droughts in the immediate future (Kundzewicz et al. 2007). An increase in the frequency of these events can increase the price volatility, thereby threatening the stability of food systems. Individuals with low-income levels and lower resource base are more vulnerable to chronic food insecurity due to the uncertainties posed due to climate change (Parker et al. 2019). Further changes in land suitability for cultivation and grazing can threaten livelihoods and reduce household income heightening risks to food security (Jibrillah et al. 2018). The demand and supply shocks triggered by *El Nino* event in 2008 highlight the threat posed by climatic variability to global food system stability (Gilbert and Morgan 2010). The more frequent occurrence of droughts may lead to frequent food emergency (Food and Agricultural Organization of the United Nations 2008c). This will necessitate the maintenance of larger grain reserves leading to an increase in costs of storage and preservation. Developing countries, in particular, are ill-equipped to deal with these shocks. Higher volatility in prices, coupled with the increase in the frequency of food emergencies, will destabilise food systems in these regions. This may lead to a loss of trust in food systems and heightened social conflict and ultimately trigger food riots (Timmer 2017). Several studies have found that a rise in temperature is strongly correlated to an increase in social conflicts (Hsiang and Burke 2014). Another consequence of destabilised food systems is mass migration triggered by food shortages. Several instances of migration triggered by food shortages have been recorded in drought-prone regions of the Sahel and the arid region of India (Sedova and Khalkhul 2020). Similarly, floods and subsequent loss of farm assets have induced migration in parts of Bangladesh, West Bengal and Indonesia.

16.3 Adapting Food Systems to Climate Change and Ensuring Food Security

16.3.1 Adaptation Strategies to Enhance Food Security in Indian Context

Reduction of risk due to climate change is one among India's key challenges, and adaptation strategies to reduce the risk should be given utmost priority. Conversely, India's food insecurity is a complex issue with many dimensions, and developing adaptation strategies to climate change will be challenging. Low-income populations in India are at a higher risk of food insecurity, as they lack resources, which in turn can be traced due to the high population density and small size of holdings (Kumar 2003). Low-income households in a rural area are often trapped in a vicious cycle of poverty characterised by low food intake; infirmity reduced labour productivity and yields. Hence, in addition to food availability, food affordability is a major concern

Table 16.2 Adaptation strategies to enhance food security

Sl. No.	Study	Adaptation strategies
1	Mishra et al. (2012)	(a) Improving the resilience of agriculture in India for changing climatic conditions. (b) Research should focus on developing suitable adaptation and mitigation strategies. (c) Checking for the suitability and field demonstrations of climate-smart technologies for upscaling adoption. (d) Capacity building of farmers, scientists and other stakeholders in climate-smart agriculture (CSA). (e) Formulating policies for out- and upscaling climate-smart technologies
2	ICAR (2018)	(a) Choosing crop varieties and livestock breeds which can withstand climatic variations. (b) Identifying and diffusion of the best practices for CSA in vulnerable areas. (c) Improving the infrastructural facilities at the main institutes to take up research in climate change. (d) Competency building of scientists for climatic change research. (e) Preparing farmers to combat climate change
3	Srinivasa et al. (2016)	(a) Propagation of silviculture to enhance productivity. (b) Improving the standard of living in rural areas. (c) Safeguarding and stabilising ecosystems. (d) Encouraging CSA to reduce the risks
4	GoI (2018)	(a) Focusing on climate-resilient agricultural practises in dryland areas. (b) Climatic risk management. (c) Knowledge diffusion and translation of various climate-smart technologies. (d) More focus on climate-smart research in biotechnology
5	FAO (2015), Shelat (2014)	Adapting climate-smart agricultural (CSA) practices such as: (a) Conservation agriculture. (b) Natural resource management. (c) Climate-smart crops. (d) Farmer prioritisation of CSA technologies. (e) Special focus on conflict management, reducing the risk by ensuring the crop insurance, handling the postharvest losses and food wastages and resolving issues in a public distribution system

for vulnerable populations in the face of climate change-induced income shocks. Thus, the challenge is whether the leap in food production will sustainably cater to the demands of the increasing population. Therefore, a long-term strategy should be devised to reduce the vulnerability of disadvantaged sections and increase agricultural productivity sustainably by conserving natural resources by including all stakeholders (Bryan and Behrman 2013). In the following section, we will discuss adaptation strategies to enhance food security in India. It is important to note that multiple stakeholders have already established several adaptation strategies (Table 16.2).

16.3.2 Promoting Climate-Resilient Agricultural Practices

India should promote government spending in the production and diffusion of crops that can withstand variations in temperature and precipitation and which are also efficient in nutrient and water uptake. It can also be thought of diffusing technologies such as hydroponics and aeroponics in areas with low soil quality and increased soil erosion. Selection of crops and practices that require less water, such as system of rice intensification (SRI) method, aerobic rice, zero-tilled wheat, etc., enables climate resilience and increases yield by minimum use of the scarce natural resource.

16.3.3 Focus on Public Health

The threat of aggravation of human diseases due to climate change could be a serious concern in India. Climate change is likely to aggravate the incidence of diseases like malaria, yellow fever, dengue, chikungunya and cholera, especially among millions of people who are already experiencing pollution, poor hygiene, impoverishment and potable water shortages. Despite the growing threats of climate change, policymakers in the country have often paid scant attention to its direct and indirect causal links to health hazards and disease. Necessary actions should be taken by the government to minimise climate change induced health risks considering the complex interrelationship with the changing climate, zoonotic disease transmission and food absorption. It will be beneficial to educate the local community about essential family health and wellness activities through participatory and planned communication methodologies. Covid-19 pandemic has exposed the lack of preparedness of countries to handle the health emergencies of this scale, and this should be treated as a wake-up call for governments to spend more on primary health care.

16.3.4 Ensuring Rural and Urban Livelihood Security

Ensuring food security in the light of climate change means improving the livelihood opportunities of the deprived and food insecure not just in order to help them avoid poverty but also in order to sustain, survive and respond to the variations in the climate. Providing gainful employment in rural areas can help in achieving this objective. For instance, MGNREGA 2005 has made a remarkable contribution in poverty alleviation and enhancing livelihood security. Apart from the benefits, several gaps are also observed in the implementation of the scheme. Efforts should be made to guarantee employment in the urban areas on the lines of MGNREGA, considering the level of unemployment, poverty and undernourishment.

16.3.5 Ensuring Nutritional Security

Policies in agriculture must be in line with nutrition priorities. They should use device incentive structure for promoting the cultivation of crops rich in nutrients such as millets, oilseeds and pulses and also the production of traditional/local crops for household consumption. Agricultural policies must also foster enhanced productivity, a balanced diet and a sustainable environment, which in turn enhances the food and nutritional security of households.

Some of the strategies that can be employed to enhance nutritional security include aligning the agricultural policy with national nutritional objectives, reaching out to women through extension service to realise nutritional security, boosting private sector engagement in nutrition interventions and organising mass awareness and education campaigns about good nutrition practices (e.g. breastfeeding, hygiene and sanitation).

16.4 Conclusion and Policy Implications

India has made considerable progress in enhancing food production, and it has become not only self-sufficient but also a net exporter of food grains. But climate change has been identified as a significant challenge to Indian agriculture's prosperity. Climate change is considered a major threat to ensuring food security, particularly for the vulnerable sections of society. In this chapter, we have discussed the possible effects of climate change on different dimensions of food security (food availability, food accessibility and food utilisation). Both climate change and its effect on agriculture and food security are location specific and hence difficult to generalise, highlighting the need for more disaggregated level studies. In this chapter, we have also highlighted some of the adaptation strategies to reduce the vulnerability of agriculture to climate change. However, it should be noted that there is no panacea and an integrated approach based on scientific evidence which accommodates the priorities of various stakeholders and concerned institutions are essential in framing policies to ensure food security.

Efforts should be taken to mainstream the climate change adaptation strategies within the processes of national policymaking. For evidence-based policymaking, scientific research needs to be transcribed and communicated transparently. A mixed approach has to be designed to foster the adoption of climate-smart farming technologies and enhance the farming communities' capacity to implement them successfully. The technology for storage pest control and mitigation of storage loss through farm innovations must be identified and validated for their further dissemination. The combination of planned adaptation and the promotion of autonomous approaches by integrating scientific and local knowledge is essential in mitigating risks on food security due to climate change. Institutional aid will also be required to out-scale climate-smart livelihoods.

References

Abbade EB (2017) Availability, access and utilization. W J Sci Tech Sus Dev 44(4):322–335

Aberman NL, Tirado C (2014) Impacts of climate change on food utilization. In: Freedman B (ed) Global Environmental Change. Handbook of Global Environmental Pollution, vol 1. Springer, Dordrecht

Anderson PK, Cunningham AA, Patel NG, Morales FJ, Epstein PR, Daszak P (2004) Emerging infectious diseases of plants: pathogen pollution, climate change and agrotechnology drivers. Trends Eco Evol 19(10):535–544

Bandara JS, Cai Y (2014) The impact of climate change on food crop productivity, food prices and food security in South Asia. J Econ Anal 44(4):451–465

Boko M, Niang I, Nyong A, Vogel C, Githeko A, Medany M, Osman-Elasha B et al (2007) Africa. In: van der Linden PJ, Hanson CE, Parry ML, Canziani OF, Palutikof JP (eds) Climate change 2007: impacts, adaptation and vulnerability, contribution of working group II to the fourth assessment report of the intergovernmental panel on climate change. Cambridge University Press, Cambridge, UK, pp 433–467

Bryan E, Behrman J (2013) Community-based adaptation to climate change: a theoretical framework, overview of key issues and discussion of gender differentiated priorities and participation. International Food Policy Research Institute, Washington, DC

Burgess R, Deschenes O, Donaldson D, Greenstone M (2014) The unequal effects of weather and climate change: evidence from mortality in India. Massachusetts Institute of Technology, Cambridge, MA

Byjesh K, Kumar SN, Aggarwal PK (2010) Simulating impacts, potential adaptation and vulnerability of maize to climate change in India. Mitig Adapt Strateg Glob Chang 15(5):413–431

Chakrabarty M (2016) Climate change and food security in India. Policy Brief No. 15. Observer Research Foundation, New Delhi

Childers K (2015) Changes in yields and their variability at different levels of global Warming EGU general assembly 2015, held April 12–17, 2015 in Vienna, Austria

Clair SBS, Lynch JP (2010) The opening of Pandora's box: climate change impacts on soil fertility and crop nutrition in developing countries. Plant Soil 335(1):101–115

Cvitanovic C, Crimp S, Fleming A, Bell J, Howden M, Hobday AJ et al (2016) Linking adaptation science to action to build food secure Pacific Island communities. Clim Risk Mgmnt 11:53–62

Easterling W, Aggarwal P, Batima P, Brander KM, Erda L, Howden SM, Kirilenko A et al (2007) Food, fibre, and forest products. In: Parry ML, Canziani OF, Palutikof JP, van der Linden PJ, Hanson CE (eds) Climate change 2007: impacts, adaptation, and vulnerability, contribution of working group II to the fourth assessment report of the intergovernmental panel on climate change. Cambridge University Press, Cambridge, UK, pp 273–313

Economic Survey (2017) Economic survey 2017–18. Ministry of Finance of Government of India, New Delhi

Ericksen PJ (2008) Conceptualizing food systems for global environmental change research. Glob Environ Change 18(1):234–245

FAO (2015). Climate-smart agriculture: a call for action.Synthesis of the Asia-Pacific regional workshop, Bangkok, Thailand, June 18–20, 2015

Fischer G, Shah MN, Tubiello F, van Velhuizen H (2005). Socio-economic and climate change impacts on agriculture: an integrated assessment. Philos Trans R Soc B 360(1463):2067–2083

Food and Agricultural Organization of the United Nations (2008a) Food safety and climate change. FAO high level conference on food security and the challenges of climate change and bioenergy. Food and Agricultural Organization of the United Nations, Rome

Food and Agricultural Organization of the United Nations (2008b) Climate change and food security a framework document. Food and Agricultural Organization of the United Nations, Rome

Food and Agricultural Organization of the United Nations (2008c) Climate change and food security: a framework document. Food and Agriculture Organization of the United Nations, Rome Italy
Gilbert CL, Morgan CW (2010) Food price volatility. Philos Trans R Soc B 365(1554):3023–3034
Godfray HCJ, Beddington JR, Crute IR, Haddad L, Lawrence D, Muir JF, Toulmin C (2010) Food security: the challenge of feeding 9 billion people. Science 327(5967):812–818
GoI (2018) National action plan on climate change. Prime Minister's Council on Climate Change, Ministry of Environment, Forest, and Climate Change, Government of India, New Delhi
Gregory PJ, Ingram J, Michael B (2005) Climate change and food security. Philos Trans R Soc B 360(1463):2139–2148
Haile MG, Wossen T, Tesfaye K, von Braun J (2017) Impact of climate change, weather extremes, and price risk on global food supply. Econ Dis Clim Change 1(1):55–75
Harvey CA, Rakotobe ZL, Rao NS, Dave R, Razafimahatratra H, Rabarijohn RH, Mac Kinnon JL (2014) Extreme vulnerability of smallholder farmers to agricultural risks and climate change in Madagascar. Philos Trans R Soc B 369(1639):1–12
Hendrix CS, Salehyan I (2012) Climate change, rainfall, and social conflict in Africa. J Peace Res 49(1):35–50
Hoddinott J, Maluccio JA, Behrman JR, Flores R, Martorell R (2008) Effect of a nutrition intervention during early childhood on economic productivity in Guatemalan adults. Lancet 371(9610):411–416
Hsiang SM, Burke M (2014) Climate, conflict, and social stability: what does the evidence say? Clim Chang 123(1):39–55
ICAR (2018) National innovations in climate resilient agriculture. Indian Council for Agricultural Research, New Delhi
Ingram J (2011) A food systems approach to researching food security and its interactions with global environmental change. J Food Secur 3(4):417–431
IPCC (2001) Climate change: impacts, adaptation and vulnerability, contribution of working group II to the third assessment report of the intergovernmental panel on climate change. Cambridge University Press, Cambridge, UK
IPCC (2007) Climate change: impacts, adaptation and vulnerability, contribution of working group II to the fourth assessment report of the intergovernmental panel on climate change. Cambridge University press, Cambridge, UK
Jernigan V, Salvatore AL, Styne DM, Winkleby M (2012) Addressing food insecurity in a native American reservation using community-based participatory research. Health Educ Res 27 (4):645–655
Jibrillah AM, Choy LK, Jaafar M (2018) Climate change manifestations and impacts in the Sokoto close-settled zone, northwestern Nigeria. Akademika 88(2):21–34
Kovats RS, Edwards SJ, Hajat S, Armstrong BG, Ebi KL, Menne B (2004) The effect of temperature on food poisoning: a time-series analysis of salmonellosis in ten European countries. Epidemiol Infect 132(3):443–453
Krishnan R, Sanjay J, Gnaaseelan C, Majumdar M, Kulkarni A, Chakraborthy S (2020) Assessment of climate change over the Indian region: a report of the Ministry of Earth Sciences. Ministry of Earth Science of Government of India, New Delhi
Kumar MD (2003) Food security and sustainable agriculture in India. The Water Management Challenge, Working Paper No. 60, International Water Management Institute, Colombo, Sri Lanka
Kumar KK, Parikh J (2001) Indian agriculture and climate sensitivity. Glob Environ Change 11 (2):147–154
Kumar P, Joshi PK, Birthal PS (2009) Demand projections for Foodgrains in India. Agric Econ Res Rev 22(2):237–243
Kumar SN, Aggarwal PK, Rani DS, Saxena R, Chauhan N, Jain S (2014a) Vulnerability of wheat production to climate change in India. Clim Res 59(3):173–187

Kumar P, Joshi PK, Aggarwal P (2014b) Projected effect of droughts on supply, demand, and prices of crops in India. Econ Political Wkly 49(52):54–63

Kundzewicz ZW, Mata LJ, Arnell NW, Doll P, Kabat P, Jime'nez B, Miller KA et al (2007) Freshwater resources and their management. In: van der Linden PJ, Hanson CE, Parry ML, Canziani OF, Palutikof JP (eds) Climate change 2007: impacts, adaptation and vulnerability, contribution of working group II to the fourth assessment report of the intergovernmental panel on climate change. Cambridge University Press, Cambridge, UK, pp 173–210

Lambrou Y, Piana G (2006) Energy and gender issues in rural sustainable development. Food and Agriculture Organization of the United Nations, Rome

Lee HL (2009) The impact of climate change on global food supply and demand, food prices, and land use. Paddy Water Environ 7(4):321–331

Levinson M, Halpern O, Mahmud Z, Chowdhury S, Levinson F (2002) Nutrition-related caring practices and women's time constraints: a study in rural Bangladesh. The Gerald J. and Dorothy R. Friedman School of Nutrition Science and Policy Food Policy and Applied Nutrition Program

Lobell D, Burke M (2010) Climate change and food security: adapting agriculture to a warmer world. Springer, New York

Long SP, Zhu X, Naidu SL (2005) Limits to efficiencies of primary production: constraints and opportunities. In: Sylvester-Bradley R, Wiseman J (eds) Yields of farmed species. Nottingham University Press, Nottingham, UK, pp 167–191

Mills JN, Gage KL, Khan AS (2010) Potential influence of climate change on vector-borne and zoonotic diseases: a review and proposed research plan. Environ Health Perspect 118 (11):1507–1514

Mishra A, Singh SRK, Gautam US (2012) Annual report (NICRA) 2011–12. Indian Council of Agricultural Research, New Delhi

Mittal S (2008) Demand supply trends and projections of food in India. Working paper No. 209. Indian Council of Research on International Economic Research, New Delhi

Moses JA, Jayas DS, Alagusundaram K (2015) Climate change and its implications on stored food grains. Agric Res 4(1):21–30

Nelson GC, Rosegrant MW, Koo J, Robertson R, Sulser T, Zhu T, Magalhaes M (2009) Climate change: impact on agriculture and costs of adaptation. International Food Policy Research Institute, Washington, DC

O'Brien KL, Wolf J (2010) A values-based approach to vulnerability and adaptation to climate change. Wiley Interdsp Rev Clim Change 1(2):232–242

Parfitt J, Barthel M, Macnaughton S (2010) Food waste within food supply chains: quantification and potential for change to 2050. Philos Trans R Soc B 365(1554):3065–3081

Parker L, Bourgoin C, Martinez-Valle A, Läderach P (2019) Vulnerability of the agricultural sector to climate change: the development of a pan-tropical climate risk vulnerability assessment to inform sub-national decision making. PLoS One 14(3):1–25

Rosenzweig C, Parry ML (1994) Potential impact of climate change on world food supply. Nature 367(6459):133–138

Rosenzweig C, Iglesias A, Yang XB, Epstein PR, Chivian E (2001) Climate change and extreme weather events - implications for food production, plant diseases, and pests. Glob Chang Hum Health 2(2):90–104

Schaible UE, Kaufmann SHE (2007) Malnutrition and infection: complex mechanisms and global impacts. PLoS Med 4(5):115–126

Schmidhuber J, Tubiello FN (2007) Global food security under climate change. Proc Natl Acad Sci U S A 104(50):19703–19708

Sedova B, Kalkuhl M (2020) Who are the climate migrants and where do they go? Evidence Rural India World Dev 129:1–19

Shelat KN (2014) Climate-smart agriculture, the way forward—The Indian perspectives. National Council for Climate Change, Sustainable Development and Public Leadership (NCCSD), Ahmedabad and Central Research Institute for Dryland Agriculture (CRIDA), Hyderabad

Singh PK, Dhiman RC (2012) Climate change and human health: Indian context. J Vector Borne Dis 49(2):55–60

Soora NK, Aggarwal PK, Saxena R, Rani S, Jain S, Chauhan N (2013) An assessment of regional vulnerability of rice to climate change in India. Clim Chang 118(3):683–699

Srinivasa R, Gopinath KA, Prasad JVNS, Prasannakumar AKS (2016) Sustainable food security in tropical India: concept, process, technologies, institutions, and impacts. In: Sparks DL (ed) Advances in agronomy. Elsevier, Amsterdam, pp 201–217

Srivastava A, Kumar SN, Aggarwal PK (2010) Assessment on vulnerability of sorghum to climate change in India. Agric Ecosyst Environ 138(3):160–169

Stathers T, Lamboll R, Mvumi BM (2013) Post-harvest agriculture in changing climates: its importance to African smallholder farmers. J Food Secur 5(3):361–392

Swindale A, Paula B (2006) Household dietary diversity score (HDDS) for measurement of household food access: indicator guide.Food and Nutrition Technical Assistance Project, Academy for Educational Development, Washington, DC

Taub DR (2010) Effects of rising atmospheric concentrations of carbon dioxide on plants. Nat Educ Knowl 3(10):21–34

Thompson B, Cohen MJ (eds) (2012) The impact of climate change and bioenergy on nutrition. Springer/FAO, Dordrecht/Rome

Thorlakson T, Neufeldt H (2012) Reducing subsistence farmers' vulnerability to climate change: evaluating the potential contributions of agroforestry in western Kenya. Agric Food Security 1 (1):1–15

Timmer CP (2017) Food security, structural transformation, markets and government policy. Asia Pacific Policy Stud 4(1):4–19

Tirado MC, Meerman J (2012) The impact of climate change and bioenergy on nutrition. In: Thompson B, Cohen MJ (eds) The impact of climate change and bioenergy on nutrition. Springer, Dordrecht, pp 43–60

Vermeulen SJ, Campbell BM, Ingram JS (2012) Climate change and food systems. Annu Rev Environ Resour 37:195–222

Waters D (2011) Supply chain risk management: vulnerability and resilience in logistics. Kogan Page, London, UK

WFP (2009) Emergency food security assessment handbook. World Food Programme, Rome, pp 68–70

Wheeler T, von Braun J (2013) Climate change impacts on global food security. Science 341 (6145):508–513

Yadav SS, Hegde VS, Habibi BA, Dia M, Verma S (2018) Climate change agriculture and food security. In: Yadav SS, Redden RJ, Hatfield JL, Ebert AW, Hunter D (eds) Food security and climate change. Wiley, New York, pp 1–24

Zewdie A (2014) Impacts of climate change on food security: a literature review in sub Saharan Africa. J Earth Sci Clim Change 5(8):225

Ziervogel G, Ericksen PJ (2010) Adapting to climate change to sustain food security. Wiley Interdisc Rev Clim Change 1(4):525–540

The manufacturer's authorised representative in the EU is Springer Nature Customer Service Centre GmbH, Europaplatz 3, 69115 Heidelberg, Germany. If you have any concerns regarding our products, please contact ProductSafety@springernature.com

Printed and bound by CPI Group (UK) Ltd, Croydon, CR0 4YY
30/09/2026
02225331-0005